Recent Advances in Biocatalysis and Metabolic Engineering for Biomanufacturing

Recent Advances in Biocatalysis and Metabolic Engineering for Biomanufacturing

Special Issue Editor

Eun Yeol Lee

MDPI • Basel • Beijing • Wuhan • Barcelona • Belgrade

Special Issue Editor
Eun Yeol Lee
Kyung Hee University
Korea

Editorial Office
MDPI
St. Alban-Anlage 66
4052 Basel, Switzerland

This is a reprint of articles from the Special Issue published online in the open access journal *Catalysts* (ISSN 2073-4344) from 2018 to 2019 (available at: https://www.mdpi.com/journal/catalysts/special_issues/Biocatalysis_Metabolic_Engineering)

For citation purposes, cite each article independently as indicated on the article page online and as indicated below:

LastName, A.A.; LastName, B.B.; LastName, C.C. Article Title. *Journal Name* **Year**, *Article Number*, Page Range.

ISBN 978-3-03921-574-4 (Pbk)
ISBN 978-3-03921-575-1 (PDF)

Contents

About the Special Issue Editor . . . . **vii**

Eun Yeol Lee
Recent Advances on Biocatalysis and Metabolic Engineering for Biomanufacturing
Reprinted from: *Catalysts* **2019**, *9*, 707, doi:10.3390/catal9090707 . . . . **1**

Prajakatta Mulay, Gayatri Shrikhande and Judit E. Puskas
Synthesis of Mono- and Dithiols of Tetraethylene Glycol and Poly(ethylene glycol)s via Enzyme Catalysis
Reprinted from: *Catalysts* **2019**, *9*, 228, doi:10.3390/catal9030228 . . . . **4**

Sung-Yeon Joo, Hee-Wang Yoo, Sharad Sarak, Byung-Gee Kim and Hyungdon Yun
Enzymatic Synthesis of ω-Hydroxydodecanoic Acid By Employing a Cytochrome P450 from *Limnobacter* sp. 105 MED
Reprinted from: *Catalysts* **2019**, *9*, 54, doi:10.3390/catal9010054 . . . . **16**

Md Murshidul Ahsan, Mahesh D. Patil, Hyunwoo Jeon, Sihyong Sung, Taeowan Chung and Hyungdon Yun
Biosynthesis of Nylon 12 Monomer, ω-Aminododecanoic Acid Using Artificial Self-Sufficient P450, AlkJ and ω-TA
Reprinted from: *Catalysts* **2018**, *8*, 400, doi:10.3390/catal8090400 . . . . **26**

Moritz Senger, Konstantin Laun, Basem Soboh and Sven T. Stripp
Infrared Characterization of the Bidirectional Oxygen-Sensitive [NiFe]-Hydrogenase from *E. coli*
Reprinted from: *Catalysts* **2018**, *8*, 530, doi:10.3390/catal8110530 . . . . **39**

Marco Antonio Seiki Kadowaki, Mariana Ortiz de Godoy, Patricia Suemy Kumagai, Antonio José da Costa-Filho, Andrew Mort, Rolf Alexander Prade and Igor Polikarpov
Characterization of a New Glyoxal Oxidase from the Thermophilic Fungus *Myceliophthora thermophila* M77: Hydrogen Peroxide Production Retained in 5-Hydroxymethylfurfural Oxidation
Reprinted from: *Catalysts* **2018**, *8*, 476, doi:10.3390/catal8100476 . . . . **53**

Yeo Reum Park, Hee Seon Yoo, Min Young Song, Dong-Heon Lee and Seung Jae Lee
Biocatalytic Oxidations of Substrates through Soluble Methane Monooxygenase from *Methylosinus sporium* 5
Reprinted from: *Catalysts* **2018**, *8*, 582, doi:10.3390/catal8120582 . . . . **68**

Fei-Long Li, Meng-Yao Zhuang, Jia-Jia Shen, Xiao-Man Fan, Hyunsoo Choi, Jung-Kul Lee and Ye-Wang Zhang
Specific Immobilization of *Escherichia coli* Expressing Recombinant Glycerol Dehydrogenase on Mannose-Functionalized Magnetic Nanoparticles
Reprinted from: *Catalysts* **2019**, *9*, 7, doi:10.3390/catal9010007 . . . . **84**

Anna Dzionek, Jolanta Dzik, Danuta Wojcieszyńska and Urszula Guzik
Fluorescein Diacetate Hydrolysis Using the Whole Biofilm as a Sensitive Tool to Evaluate the Physiological State of Immobilized Bacterial Cells
Reprinted from: *Catalysts* **2018**, *8*, 434, doi:10.3390/catal8100434 . . . . **96**

Sara Arana-Peña, Yuliya Lokha and Roberto Fernández-Lafuente
Immobilization of Eversa Lipase on Octyl Agarose Beads and Preliminary Characterization of Stability and Activity Features
Reprinted from: *Catalysts* **2018**, *8*, 511, doi:10.3390/catal8110511 . . . . **111**

Murilo Amaral-Fonseca, Willian Kopp, Raquel de Lima Camargo Giordano, Roberto Fernández-Lafuente and Paulo Waldir Tardioli
Preparation of Magnetic Cross-Linked Amyloglucosidase Aggregates: Solving Some Activity Problems
Reprinted from: *Catalysts* **2018**, *8*, 496, doi:10.3390/catal8110496 . . . . **126**

Meng-Qiu Xu, Shuang-Shuang Wang, Li-Na Li, Jian Gao and Ye-Wang Zhang
Combined Cross-Linked Enzyme Aggregates as Biocatalysts
Reprinted from: *Catalysts* **2018**, *8*, 460, doi:10.3390/catal8100460 . . . . **147**

Soo-Jung Kim, Seong Keun Kim, Wonjae Seong, Seung-Gyun Woo, Hyewon Lee, Soo-Jin Yeom, Haseong Kim, Dae-Hee Lee and Seung-Goo Lee
Enhanced (−)-α-Bisabolol Productivity by Efficient Conversion of Mevalonate in *Escherichia coli*
Reprinted from: *Catalysts* **2019**, *9*, 432, doi:10.3390/catal9050432 . . . . **167**

Ji Hoon Lee, Sanghak Cha, Chae Won Kang, Geon Min Lee, Hyun Gyu Lim and Gyoo Yeol Jung
Efficient Conversion of Acetate to 3-Hydroxypropionic Acid by Engineered *Escherichia coli*
Reprinted from: *Catalysts* **2018**, .*8*, 525, doi:10.3390/catal8110525 . . . . **182**

Kei-Anne Baritugo, Hee Taek Kim, Mi Na Rhie, Seo Young Jo, Tae Uk Khang, Kyoung Hee Kang, Bong Keun Song, Binna Lee, Jae Jun Song, Jong Hyun Choi, Dae-Hee Lee, Jeong Chan Joo and Si Jae Park
Construction of a *Vitreoscilla* Hemoglobin Promoter-Based Tunable Expression System for *Corynebacterium glutamicum*
Reprinted from: *Catalysts* **2018**, *8*, 561, doi:10.3390/catal8110561 . . . . **192**

Si-si Xie, Lingyun Zhu, Xin-yuan Qiu, Chu-shu Zhu and Lv-yun Zhu
Advances in the Metabolic Engineering of *Escherichia coli* for the Manufacture of Monoterpenes
Reprinted from: *Catalysts* **2019**, *9*, 433, doi:10.3390/catal9050433 . . . . **204**

Hyang-Mi Lee, Phuong N. L. Vo and Dokyun Na
Advancement of Metabolic Engineering Assisted by Synthetic Biology
Reprinted from: *Catalysts* **2018**, *8*, 619, doi:10.3390/catal8120619 . . . . **218**

Yong Sun, Jun He, Gang Yang, Guangzhi Sun and Valérie Sage
A Review of the Enhancement of Bio-Hydrogen Generation by Chemicals Addition
Reprinted from: *Catalysts* **2019**, *9*, 353, doi:10.3390/catal9040353 . . . . **237**

Rina Mariyana, Min-Sik Kim, Chae Il Lim, Tae Wan Kim, Si Jae Park, Byung-Keun Oh, Jinwon Lee and Jeong-Geol Na
Mass Transfer Performance of a String Film Reactor: A Bioreactor Design for Aerobic Methane Bioconversion
Reprinted from: *Catalysts* **2018**, *8*, 490, doi:10.3390/catal8110490 . . . . **258**

About the Special Issue Editor

Eun Yeol Lee (Ph.D.) is a Kyung Hee Fellow of Kyung Hee University in South Korea. He obtained his bachelor in 1989 from Seoul National University and his Ph.D. in 1995 from Seoul National University. He was a visiting scholar at Microbiology Institute at Goettingen University (Germany), Industrial Microbiology and Food Biotechnology Institute at Wageningen University (The Netherlands), and Biomedical Engineering Department at Cornell University (USA). He was a senior researcher at Samsung Advanced Institute of Science and Technology, and then an associate professor of Department of Food Science and Biotechnology at Kyungsung University. Currently, he is a professor of the Chemical Engineering Department at Kyung Hee University. His research interests are biocatalysis, metabolic engineering, and biomanufacturing of biofuels, chemicals, bioplastics, and functional biomaterials from lignocellulosic biomass, macro-/microalgae, and waste gas.

catalysts

MDPI

Editorial

Recent Advances on Biocatalysis and Metabolic Engineering for Biomanufacturing

Eun Yeol Lee

Department of Chemical Engineering, Kyung Hee University, Yongin-si, Gyeonggi-do 17104, Korea; eunylee@khu.ac.kr; Tel.: +82-31-201-3839

Received: 19 August 2019; Accepted: 21 August 2019; Published: 23 August 2019

The use of biocatalysts, including enzymes and metabolically engineered cells, has attracted a great deal of attention in chemical and bio-industry, because biocatalytic reactions can be conducted under environmentally-benign conditions and in more sustainable ways. The catalytic efficiency and chemo-, regio-, and stereo-selectivity of enzymes can be enhanced and modulated using protein engineering. Metabolic engineering seeks to enhance cellular biosynthetic productivity of target metabolites via controlling and redesigning metabolic pathways using multi-omics analysis, genome-scale modeling, metabolic flux control, and reconstruction of novel pathways.

The aim of this Special Issue was to deal with the recent advances in biocatalysis and metabolic engineering for biomanufacturing of biofuels, chemicals, biomaterials, and pharmaceuticals. Reviews and original research articles on the development of new strategies to improve the catalytic efficiency of enzyme, biosynthetic capability of cell factory, and their applications in production of various bioproducts and chemicals have been published.

This special issue on "Recent Advances on Biocatalysis and Metabolic Engineering for Biomanufacturing" includes 18 published articles including review and original research papers. Among the research articles presented in this issue, there is a set of studies on enzyme catalysis, which was a powerful tool to effectively synthesize various target products. In more detail, Mulay et al. investigated *Candida antarctica* Lipase B-catalyzed transesterification of methyl 3-mercaptopropionate with tetraethylene glycol (TEG) and poly(ethylene glycol)s (PEG)s to synthesize thiol-functionalized TEGs and TEGs without use of solvent [1]. Joo et al. reported the biosynthesis of ω-hydroxydodecanoic acid via whole-cell biotransformations using a novel monooxygenase CYP153AL.m from *Limnobacter* sp. 105 MED [2]. ω-Aminododecanoic acid can be used as Nylon 12 monomers. The biotransformation of dodecanoic acid to ω-aminododecanoic acid has been achieved by using an artificial self-sufficient P450, ω-transaminase, and alcohol dehydrogenase, as reported by Ahsan et al. [3].

This issue also covers several studies concerning the characterization of novel enzymes that become more attractive biocatalysts to serve as an alternative platform for chemical synthesis. Senger et al. successfully analyzed the infrared characterization of [NiFe]-hydrogenase from *Escherichia coli* HYD-2 by in situ attenuated total reflection Fourier-transform infrared spectroscopy which proved as an efficient and powerful technique for the analysis of biological macromolecules and enzymatic small molecule catalysis [4]. Glyoxal oxidase, an extracellular oxidoreductase that oxidizes aldehydes and α-hydroxy carbonyl substrates coupled to the reduction of O_2 to H_2O_2, from *Myceliophthora thermophyla*, has been characterized by Kadowaki et al. [5]. In addition, hydroxylation mechanism of soluble methane monooxygenase from *Methylosinus sporium* strain 5, a type II methanotrophs, was reported by Park et al., which revealed that two molar equivalents of methane monooxygenase regulatory protein B (MMOB) are necessary to achieve catalytic activities toward a broad range of substrates including alkanes, alkenes, halogens, and aromatics [6].

Several investigations in this issue focused on the development of immobilization methods for better biocatalytic performance. By using mannose-functionalized magnetic nanoparticles, Li et al. successfully immobilized *E. coli* cells harboring recombinant glycerol dehydrogenase gene, which showed two-fold higher production of 1,3-dihydroxyacetone from glycerol, compared to the free cells [7]. An optimized procedure of fluorescein diacetate hydrolysis for quantifying total enzymatic activity in the whole biofilm on the carrier without disturbing immobilization was reported by Dzionek et al., which can serve as a promising method to evaluate the physiological state of immobilized bacterial cells [8]. Additionally, Arana-Peña et al. reported the immobilization of Eversa lipase on octyl and aminated agarose beads for the first time, which greatly enhanced the stability of the enzyme [9]. The immobilized enzymes prepared by the cross-linked enzyme aggregates (CLEA) have become more attractive due to their simple preparation and high catalytic efficiency. In this issue, the magnetic cross-linked aggregates of amyloglucosidase was successfully achieved by Amaral-Fonseca et al. [10]. Especially the conditions or factors for the preparation of combi-CLEAs, such as the proportion of enzymes, the type of cross-linker, and coupling temperature, were intensively reviewed by Xu et al. [11].

The last part of this special issue focuses on metabolic engineering of various microorganism for the production of value-added products. Kim et al. reported the enhancement of (-)-α-bisabolol productivity by creating a more efficient heterologous mevalonate pathway [12]. An engineered *E. coli* strain for the conversion of acetate to 3-hydroxypropionic acid by heterologous expression of malonyl-CoA reductase from *Chloroflexus aurantiacus* and the activation of acetate assimilating pathway and glyoxylate shunt pathway was developed by Lee et al. [13]. Baritugo et al. developed a novel tunable promoter system based on repeats of the *Vitreoscilla* hemoglobin promoter and subsequently used for 5-aminovaleric acid and gamma-aminobutyric acid production in several *C. glutamicum* strains [14]. Three intensive reviews on various aspects on metabolic engineering have been published in this issue. Xie et al. highlighted insights into the current advances of monoterpene bioproduction and future outlook to promote the industrial production of valuable monoterpenes [15]. Recent advances in synthetic biology are greatly useful for achieving metabolic engineering purposes, which have been intensively reviewed by Lee et al. [16]. The technological gaps and effective approaches for process intensification of bio-hydrogen production were reviewed by Sun et al., particularly on the latest methods of chemicals/metal addition for improving hydrogen generation during dark fermentation processes [17]. Furthermore, this special issue includes the investigation on mass transfer performance of a novel string film reactor for the aerobic conversion of methane gas, investigated by Mariyana et al., to address process intensification issue on biomanufacturing [18].

Funding: This research was funded by the C1 Gas Refinery Program through the National Research Foundation of Korea (NRF), funded by the Ministry of Science and ICT (2015M3D3A1A01064882).

Acknowledgments: The Guest Editor thanks all the authors contributing in this Special Issue, the Editorial staff of *Catalysts*, and A.D. Nguyen for their kind support.

Conflicts of Interest: The authors declare no conflict of interest.

References

1. Mulay, P.; Shrikhande, G.; Puskas, J.E. Synthesis of Mono- and Dithiols of Tetraethylene Glycol and Poly(ethylene glycol)s via Enzyme Catalysis. *Catalysts* **2019**, *9*, 228. [CrossRef]
2. Joo, S.-Y.; Yoo, H.-W.; Sarak, S.; Kim, B.-G.; Yun, H. Enzymatic Synthesis of ω-Hydroxydodecanoic Acid By Employing a Cytochrome P450 from *Limnobacter* sp. 105 MED. *Catalysts* **2019**, *9*, 54. [CrossRef]
3. Ahsan, M.M.; Patil, M.D.; Jeon, H.; Sung, S.; Chung, T.; Yun, H. Biosynthesis of Nylon 12 Monomer, ω-Aminododecanoic Acid Using Artificial Self-Sufficient P450, AlkJ and ω-TA. *Catalysts* **2018**, *8*, 400. [CrossRef]
4. Senger, M.; Laun, K.; Soboh, B.; Stripp, S.T. Infrared Characterization of the Bidirectional Oxygen-Sensitive [NiFe]-Hydrogenase from *E. coli*. *Catalysts* **2018**, *8*, 530. [CrossRef]

5. Kadowaki, M.A.S.; Godoy, M.; Kumagai, P.S.; Costa-Filho, A.; Mort, A.; Prade, R.A.; Polikarpov, I. Characterization of a New Glyoxal Oxidase from the Thermophilic Fungus *Myceliophthora thermophila* M77: Hydrogen Peroxide Production Retained in 5-Hydroxymethylfurfural Oxidation. *Catalysts* **2018**, *8*, 476. [CrossRef]
6. Park, Y.R.; Yoo, H.S.; Song, M.Y.; Lee, D.-H.; Lee, S.J. Biocatalytic Oxidations of Substrates through Soluble Methane Monooxygenase from *Methylosinus sporium* 5. *Catalysts* **2018**, *8*, 582. [CrossRef]
7. Li, F.; Zhuang, M.; Shen, J.; Fan, X.; Choi, H.; Lee, J.; Zhang, Y. Specific immobilization of *Escherichia coli* expressing recombinant glycerol dehydrogenase on mannose-functionalized magnetic nanoparticles. *Catalysts* **2019**, *9*, 7. [CrossRef]
8. Dzionek, A.; Dzik, J.; Wojcieszyńska, D.; Guzik, U. Fluorescein Diacetate Hydrolysis Using the Whole Biofilm as a Sensitive Tool to Evaluate the Physiological State of Immobilized Bacterial Cells. *Catalysts* **2018**, *8*, 434. [CrossRef]
9. Arana-Peña, S.; Lokha, Y.; Fernández-Lafuente, R. Immobilization of Eversa Lipase on Octyl Agarose Beads and Preliminary Characterization of Stability and Activity Features. *Catalysts* **2018**, *8*, 511. [CrossRef]
10. Amaral-Fonseca, M.; Kopp, W.; Giordano, R.D.L.C.; Fernández-Lafuente, R.; Tardioli, P.W. Preparation of Magnetic Cross-Linked Amyloglucosidase Aggregates: Solving Some Activity Problems. *Catalysts* **2018**, *8*, 496. [CrossRef]
11. Xu, M.-Q.; Wang, S.-S.; Li, L.-N.; Gao, J.; Zhang, Y.-W. Combined Cross-Linked Enzyme Aggregates as Biocatalysts. *Catalysts* **2018**, *8*, 460. [CrossRef]
12. Kim, S.-J.; Kim, S.K.; Seong, W.; Woo, S.-G.; Lee, H.; Yeom, S.-J.; Kim, H.; Lee, D.-H.; Lee, S.-G. Enhanced (−)-α-Bisabolol Productivity by Efficient Conversion of Mevalonate in *Escherichia coli*. *Catalysts* **2019**, *9*, 432. [CrossRef]
13. Lee, J.H.; Cha, S.; Kang, C.W.; Lee, G.M.; Lim, H.G.; Jung, G.Y. Efficient Conversion of Acetate to 3-Hydroxypropionic Acid by Engineered *Escherichia coli*. *Catalysts* **2018**, *8*, 525. [CrossRef]
14. Baritugo, K.-A.; Kim, H.T.; Na Rhie, M.; Jo, S.Y.; Khang, T.U.; Kang, K.H.; Song, B.K.; Lee, B.; Song, J.J.; Choi, J.H.; et al. Construction of a Vitreoscilla Hemoglobin Promoter-Based Tunable Expression System for *Corynebacterium glutamicum*. *Catalysts* **2018**, *8*, 561. [CrossRef]
15. Xie, S.-S.; Zhu, L.; Qiu, X.-Y.; Zhu, C.-S.; Zhu, L.-Y. Advances in the Metabolic Engineering of *Escherichia coli* for the Manufacture of Monoterpenes. *Catalysts* **2019**, *9*, 433. [CrossRef]
16. Lee, H.-M.; Vo, P.N.L.; Na, D. Advancement of Metabolic Engineering Assisted by Synthetic Biology. *Catalysts* **2018**, *8*, 619. [CrossRef]
17. Sun, Y.; He, J.; Yang, G.; Sun, G.; Sage, V. A Review of the Enhancement of Bio-Hydrogen Generation by Chemicals Addition. *Catalysts* **2019**, *9*, 353. [CrossRef]
18. Mariyana, R.; Kim, M.-S.; Lim, C.I.; Kim, T.W.; Park, S.J.; Oh, B.-K.; Lee, J.; Na, J.-G. Mass Transfer Performance of a String Film Reactor: A Bioreactor Design for Aerobic Methane Bioconversion. *Catalysts* **2018**, *8*, 490. [CrossRef]

catalysts

MDPI

Article

Synthesis of Mono- and Dithiols of Tetraethylene Glycol and Poly(ethylene glycol)s via Enzyme Catalysis

Prajakatta Mulay [1], Gayatri Shrikhande [1] and Judit E. Puskas [1,2,*]

1 Department of Chemical and Biomolecular Engineering, The University of Akron, Akron, OH 44325, USA; pm62@zips.uakron.edu (P.M.); gss26@zips.uakron.edu (G.S.)

2 Department of Food, Agricultural and Biological Engineering, The Ohio State University, Wooster, OH 44691, USA

* Correspondence: puskas.19@osu.edu; Tel.: +1-330-263-3861

Received: 11 February 2019; Accepted: 26 February 2019; Published: 2 March 2019

Abstract: This paper investigates the transesterification of methyl 3-mercaptopropionate (MP-SH) with tetraethylene glycol (TEG) and poly(ethylene glycol)s (PEG)s catalyzed by *Candida antarctica* Lipase B (CALB) without the use of solvent (in bulk). The progress of the reactions was monitored by ^{1}H-NMR spectroscopy. We found that the reactions proceeded in a step-wise manner, first producing monothiols. TEG-monothiol was obtained in 15 min, while conversion to dithiol took 8 h. Monothiols from PEGs with M_n = 1000 and 2050 g/mol were obtained in 8 and 16 h, respectively. MALDI-ToF mass spectrometry verified the absence of dithiols. The synthesis of dithiols required additional fresh CALB and MP-SH. The structure of the products was confirmed by ^{1}H-NMR and ^{13}C-NMR spectroscopy. Enzyme catalysis was found to be a powerful tool to effectively synthesize thiol-functionalized TEGs and PEGs.

Keywords: *Candida antarctica* Lipase B; transesterification; polymer functionalization; tetraethylene glycol; poly(ethylene glycol)

1. Introduction

Poly(ethylene glycol) (PEG) is the most frequently used polymer for biomedical research and applications because it is soluble in organic as well as aqueous media [1], is not cytotoxic and immunogenic [2], and is easily excreted from living organisms [3]. Click chemistries and Michael addition reactions are often used for PEGylation of drugs to make them more water soluble [4]. Thiol-functionalized PEGs have an important role in these reactions [5–10] and can be used as a 'Michael donor' or in thiol-ene click reactions to synthesize conjugates for targeted drug delivery [7]. Other uses include the following: An anti-fouling biosensor coating [8], to stabilize gold nanorods used to test water for chemical pollutants [9], and to stabilize gold nanoparticles used as drug delivery vehicles [10]. Thiol-functionalized PEG is also a favorite to make self-assembling monolayers on gold surfaces [8]. Mahou et al. [11] reported the single synthetic strategy to obtain PEG-monothiol. They tosylated one hydroxyl end-group of the PEG-diol using *p*-toluenesulfonyl chloride in the presence of silver oxide and potassium iodide catalyst and toluene solvent. The tosylated PEG was then reduced with sodium hydrosulfide at 60 °C to yield PEG-monothiol with 84% yield. PEG-dithiols have been synthesized by various methods. In one method, the hydroxyl end groups were reacted with allyl bromide at 120 °C, followed by a radical-mediated addition of thioacetic acid and subsequent reduction to thiol using sodium hydroxide/sodium thiomethoxide, with a 56% yield [12–14]. Another route reported tosylation of the hydroxyl end groups, followed by a reaction with a xanthate and de-protection with an alkyl amine that gave 98% yield [15,16]. The simplest method used esterification of mercapto-acids in toluene at 120 °C using p-toluenesulfonic acid or sulfuric acid as catalysts: An example is shown in Figure 1 [17–25].

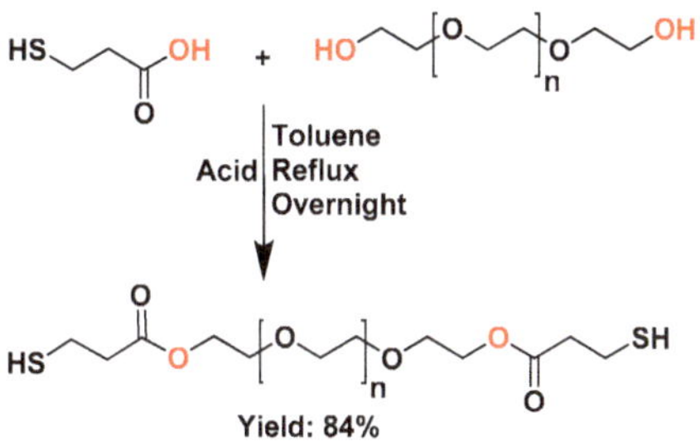

Figure 1. Synthesis of poly(ethylene glycol) (PEG)-dithiol [22].

These methods employ acid catalysts; and hence are not "green". Against this background, we investigated the synthesis of thiol-functionalized tetraethylene glycol (TEG) and PEGs by transesterification of methyl 3-mercaptopropionate (MP-SH) under solvent-less conditions using a heterogeneous catalyst, namely, *Candida antarctica* Lipase B (CALB). CALB-catalyzed functionalization of low molecular weight polymers was first reported by our research group yielding pure products with high efficiency [26–32]. For example, halogen-functionalized PEGs were made by the transesterification of halo-esters with PEG monomethyl ether under solvent-less conditions at 65 °C for 4 h under vacuum (70 milliTorr) [31]. Methacrylate, acrylate, and crotonate functionalization of PEGs was also achieved under solvent-less conditions within 4 h at 50 °C by reacting PEG with the corresponding vinyl esters (vinyl methacrylate, vinyl acrylate, and vinyl crotonate) in the presence of immobilized CALB [32].

Precise thiol-functionalization of TEG and PEGs by enzyme catalysis has not been reported previously in the literature. This study presents the first examples of precision synthesis of TEG and PEG monothiols and dithiols. In this study, two types of PEGs (M_n = 1000 g/mol and M_n = 2050 g/mol) were used to evaluate the effect of PEG chain length on the kinetics of the CALB-catalyzed transesterification reaction of methyl 3-mercaptopropionate (MP-SH) with PEGs at 50 °C, an optimum temperature for CALB-catalyzed reactions [33]. MP-SH was selected because of its low cost and the convenient removal of the methanol side product by vacuum. CALB supported on various carriers were reported to be more effective than the native enzyme [34–36] and depending on the specific conditions were shown to be recyclable four [37] or twenty times [36]. We have been using the only commercially available CALB (20 wt% immobilized on a macroporous acrylic resin, Novozyme® 435).

2. Results and Discussion

2.1. CALB-Catalyzed Transesterification of MP-SH with TEG

First, MP-SH was transesterified with TEG under solvent-less conditions using CALB as the catalyst. The catalytic triad for transesterification of CALB was shown to consist of serine (Ser105), histidine (His224), and aspartate (Asp187) [38]. Figure 2 illustrates our rendition of the mechanism of transesterification of MP-SH by TEG [33]. The top (dark shaded) portion of the enzyme is the so-called "carbonyl pocket" while the bottom (lighter shaded) is the "hydroxyl pocket". First, the nucleophilic serine (Ser105) in the free enzyme interacts with the carbonyl group of the thioester, forming the first tetrahedral intermediate (THI-1) that is stabilized by the so-called oxyanion hole (three hydrogen bonds: One from glutamine (Gln106) and two from threonine (Thr40)) [38]. In the second step, the ester bond in THI-1 is cleaved to form an acyl-enzyme complex (AEC) that releases the first product, methanol in this case, which is removed due to the applied vacuum (420 Torr), making the reaction irreversible. In the third step, the HO- group of the diol positioned in the hydroxyl pocket interacts with the carbonyl group of the AEC, forming the second tetrahedral intermediate (THI-2), which is also stabilized by the oxyanion hole. In the last step, the enzyme is deacylated to form a TEG-monothiol that is released from the THI-2 and the enzyme is regenerated.

The second -OH group of the TEG-monothiol will then be converted to thiol in a second cycle in a similar manner as the first cycle as shown in Figure 3. However, the first and second cycle may proceed simultaneously in a competitive reaction between the hydroxyl groups of unreacted TEG and TEG-monothiol. Thus, we first studied the kinetics of CALB-catalyzed transesterification of MP-SH with TEG.

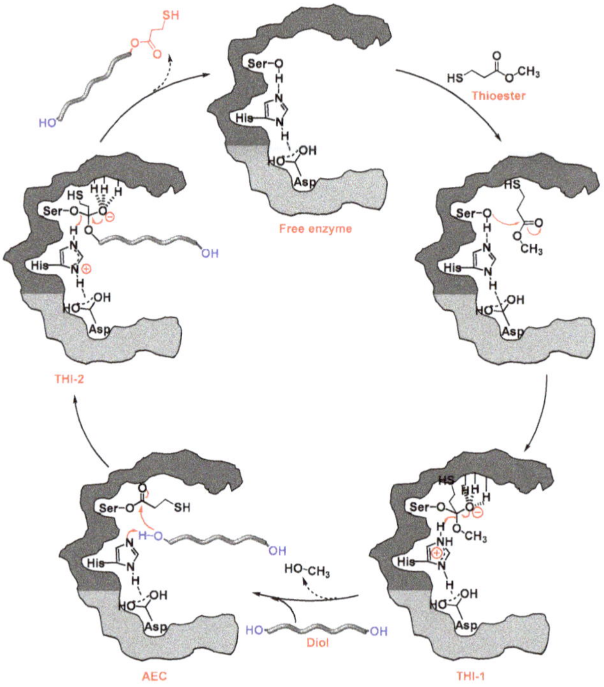

Figure 2. Reaction mechanism of *Candida antarctica* Lipase B (CALB)-catalyzed transesterification of methyl 3-mercaptopropionate (MP-SH) with tetraethylene glycol (TEG)—first cycle.

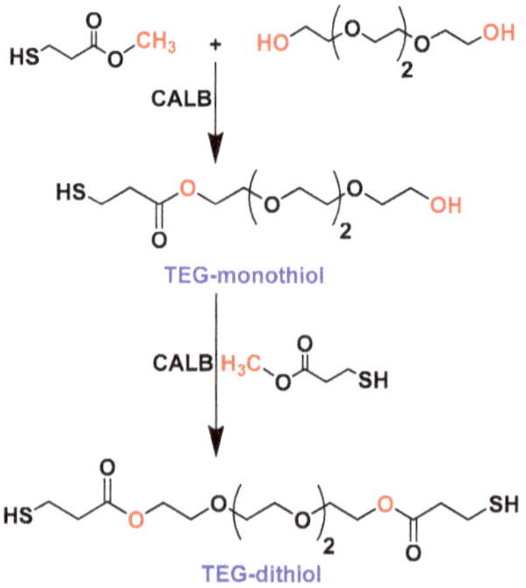

Figure 3. CALB-catalyzed transesterification of MP-SH with TEG.

2.1.1. Kinetics of CALB-Catalyzed Transesterification of MP-SH with TEG

The progress of the reaction was monitored by ^{1}H-NMR spectroscopy. At time 0, the protons from MP-SH (thiol proton triplet at 1.60 ppm (a), methylene protons—quartet at δ = 2.73 ppm (b) and triplet at δ = 2.61 ppm (c)) can be seen together with the proton signals of TEG (CH_2 protons next to the -OH end group at δ = 3.57 ppm (e) and at δ = 3.64 ppm (f) and the internal CH_2 protons of TEG at δ = 3.63 ppm (g)). The methyl protons of MP-SH (h) also appear in this region at δ = 3.66 ppm, overlapping with the methylene proton signals (f) of TEG. It can be observed from Figure 4 that the intensity of the signal at δ = 3.57 ppm (e) gradually decreases as the reaction time increases. The formation of the ester bond is demonstrated by the appearance of a new signal at δ = 4.23 ppm, corresponding to the methylene protons next to the carbonyl group in the product (e′, Figure 4). The proton signals (b) and (c) slightly shift to 2.75 ppm (b′) and δ = 2.64 ppm (c′). After 15 min of reaction time, the ratio of the internal CH_2 protons of TEG at δ = 3.61 ppm (g) to (e′) in the product was 8:1.98, indicating the formation of TEG-monothiol. After the formation of TEG-monothiol, the reaction slowed down considerably. Complete conversion to dithiol took 450 min, and the relative integrals of (g): (e′) at 8:3.88 indicated the formation of TEG-dithiol (Figure 4).

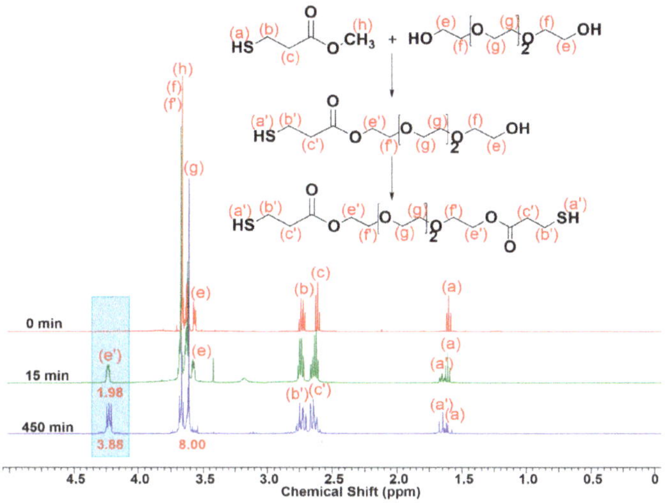

Figure 4. ^{1}H-NMR monitoring of the kinetics of the transesterification of MP-SH with TEG [15 min: ^{1}H-NMR (500 MHz, $CDCl_3$): δ 4.23 (2H) 3.61(8H); 450 min: ^{1}H-NMR (500 MHz, $CDCl_3$): δ 4.23 (4H) 3.61 (8H)].

We theorize that in the second cycle the carbonyl group of the free MP-SH competes with the carbonyl group of the TEG-monothiol for complexation in the carbonyl pocket of CALB, thereby slowing down the second cycle of the reaction. Another reason might be the deactivation of CALB by the methanol released in the reactions that is not completely removed in the vacuum. Thus, the reaction proceeds sequentially in a consecutive manner.

2.1.2. Synthesis of TEG-monothiol and TEG-dithiol

Figure S1 shows the ^{1}H- and ^{13}C-NMR spectra of TEG-monothiol synthesized with a reaction time of 15 min after filtering the enzyme and removing the excess thioester but without further purification (93% reaction yield because some material is lost with the enzyme). In the ^{1}H-NMR spectrum of the monothiol (Figure S1A), the ratio of the integral of the methylene protons next to the SH group in the product at 2.75 ppm (b′) and δ = 2.64 ppm (c′) to the integral of the methylene protons next to the carbonyl group at δ = 4.23 ppm (e′) is 4.00:2.00, indicating the formation of the TEG-monothiol with 100% conversion. In the ^{13}C-NMR spectrum of the monothiol (Figure S1B), signals corresponding to the carbons in the thiol end group (B, C, D, E′ and F′) and the carbons next to the -OH end group (E and F) appears distinctly, that demonstrates the formation of the TEG-monothiol.

Figure S2 shows the ^{1}H- and ^{13}C-NMR spectra of the TEG-dithiol that was synthesized with a reaction time of 7.5 h (88% reaction yield). The ratio of the integral values of signals (b′) + (c′) to (e′) are 8.00:3.88, indicating the formation of the TEG-dithiol. The ^{13}C-NMR spectrum in Figure S2B shows only the signals corresponding to the thiol end groups, with only traces of signals corresponding to the carbons next to the -OH (E and F) at δ = 72.38 ppm and δ = 61.16 ppm, possibly from traces of residual TEG-monothiol, indicating 100% conversion of the -OH groups to the thiols.

In summary, CALB-catalyzed transesterification of MP-SH with TEG in bulk yielded TEG-monothiol in 15 min with 93% reaction yield, and TEG-dithiol in 7.5 h with 88% reaction yield and 100% conversion without purification, such as column chromatography.

2.2. CALB-Catalyzed Transesterification of MP-SH with PEGs

2.2.1. Kinetics of CALB-catalyzed transesterification of MP-SH with PEG

MP-SH was reacted with PEG_{1000} using enzyme catalysis, and the reaction was monitored over 24 h with ^{1}H-NMR spectroscopy (see Figure 5). Because low molecular weight PEGs (<3000 g/mol) are liquid at the reaction temperature and are miscible with MP-SH, no solvent was necessary as a medium for the reaction. The main chain protons (g) and the methylene protons next to the –OH (e and f) and the thioester (f′) appear at δ = 3.61 ppm. For PEG_{1000}-monothiol, the new methylene protons next to the carbonyl group (e′) appear at δ = 4.23 ppm, which makes the integral value of internal protons of PEG_{1000} (g, e, f, and f′): 88 − 2 = 86. Therefore, the integral value of the internal protons was set to 86 for calculating the extent of the reaction. Based on the integral ratio of (g, e, f, and f′): (e′), about 60% of the PEG_{1000} was converted to PEG_{1000}-monothiol in 60 min (Figure 5). Then the reaction slowed down, and it took 8 h to convert all PEG_{1000} into monothiol. Dithiol was not detected even after 24 h. The mechanism presented in Figure 2 for TEG also applies for PEG. Thus, we theorize that in the second cycle the carbonyl group of the free MP-SH competes with the carbonyl group of the PEG-monothiol for complexation in the carbonyl pocket of CALB, thereby slowing down the second cycle of the reaction. In addition, the CALB may be deactivated by the methanol released in the reactions that is not completely removed by the vacuum.

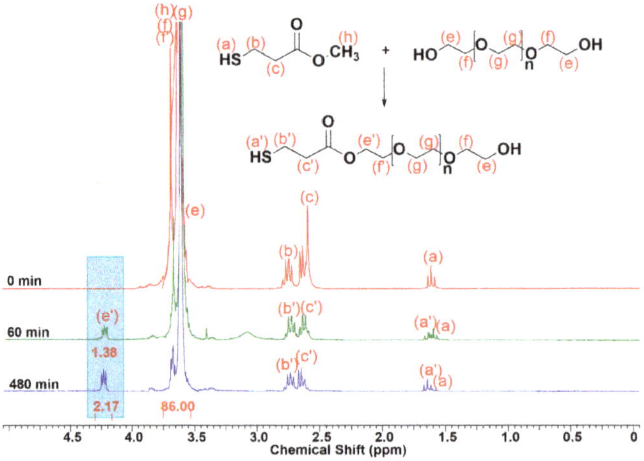

Figure 5. ^{1}H-NMR monitoring of the kinetics of transesterification of MP-SH with PEG_{1000} [480 min: ^{1}H-NMR (500 MHz, $CDCl_3$): δ 4.23 (2H) 3.61(86H)].

MP-SH was also transesterified with PEG_{2050} and the reaction was monitored over 24 h by ^{1}H-NMR spectroscopy (not shown). Similarly to the PEG_{1000}, only monothiol was obtained. In addition, complete conversion to monothiol was achieved in 16 h, which suggests higher molecular weight required longer reaction time.

2.2.2. Synthesis of PEG_{1000}-monothiol

The ^{1}H-NMR of the PEG_{1000}-monothiol is shown in Figure 6A. The integral ratio of (b′) + (c′) to the methylene protons in the new ester bond at δ = 4.23 ppm is 4:00:1.86, indicating the formation of PEG_{1000}-monothiol. Figure 6B shows the ^{13}C-NMR spectrum of PEG_{1000}-monothiol. Signals corresponding to the carbons next to the thioester (E′ and F′) and –OH end groups (E and F) appear simultaneously, indicating the formation of monothiol.

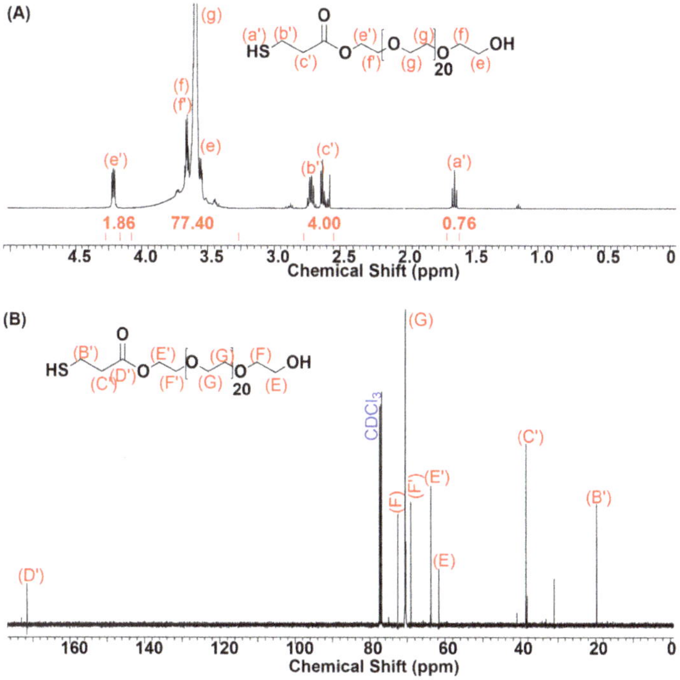

Figure 6. (**A**) ^{1}H-NMR and (**B**) ^{13}C-NMR of PEG_{1000}-monothiol [^{1}H-NMR (500 MHz, $CDCl_3$): δ 4.21 (2H) 3.61(86H) 2.72 (2H) 2.63 (2H) 1.65 (1H); ^{13}CNMR (500 MHz, $CDCl_3$): δ 171.5, 72.66, 70.6, 69.1, 63.8, 61.72, 38.5, 19.7].

The product was further analyzed by MALDI-ToF mass spectrometry and Figure 7 shows the spectrum. There are two major distributions of peaks (Figure 7A), each separated by 44 m/z units (Figure 7B). The peak at m/z 1097.63 corresponds to the Na complex of the 22-mer fraction of PEG_{1000} monothiol [1097.63 = 22 × 44.03 (C_2H_4O repeat unit) + 89.14 (HSC_2H_4CO- end group) + 17 (HO- end group) + 22.99 (Na^+)]. The peak at m/z 560.31 corresponds to the doubly charged Na complex of the 22-mer fraction of PEG_{1000} monothiol [560.31 = [(1097.63 ([M + Na]$^+$) + 22.99 (Na^+)]/2]. The small distribution of peaks appearing under the doubly charged Na complex distribution corresponds to traces of unreacted PEG_{1000} from the reaction mixture (<5%) that could not be detected by NMR. Thus, based on the MALDI mass spectrometry data, over 95% conversion of one of the OH groups to thiol was achieved in 24 h. No traces of PEG-dithiol were found. Therefore, it can be concluded that the product was exclusively PEG_{1000}-monothiol with no traces of dithiol, with 100% yield.

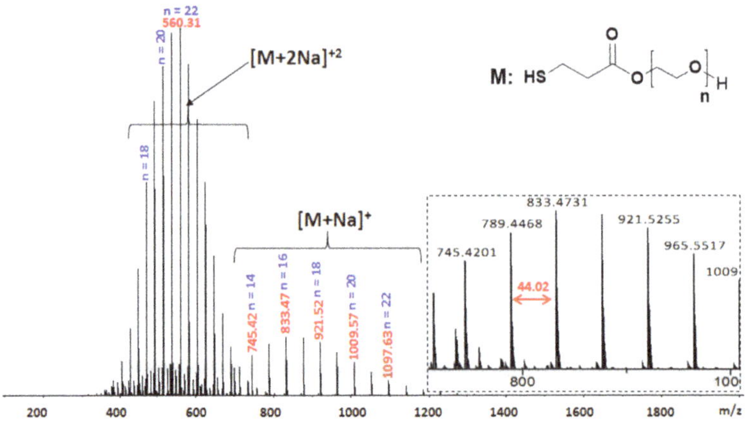

Figure 7. MALDI-ToF mass spectra of PEG_{1000} monothiol. Inset: The zoomed version of the 14- to 20-mer fractions, 44 m/z = PEG repeat unit.

2.2.3. Synthesis of PEG_{1000}-dithiol

PEG_{1000}-dithiol was obtained by reacting PEG_{1000}-monothiol with fresh MP-SH and CALB for 24 h under solvent-less conditions. Figure 8 shows the ^{13}CNMR spectrum of the PEG_{1000}-dithiol. The disappearance of the signals (F and E, Figure 6) at δ = 72.66 ppm and δ = 61.72 ppm, corresponding to the methylene protons next to the hydroxyl end-groups from the PEG_{1000}-monothiol indicates full conversion to PEG_{1000}-dithiol in 24 h with 85% reaction yield.

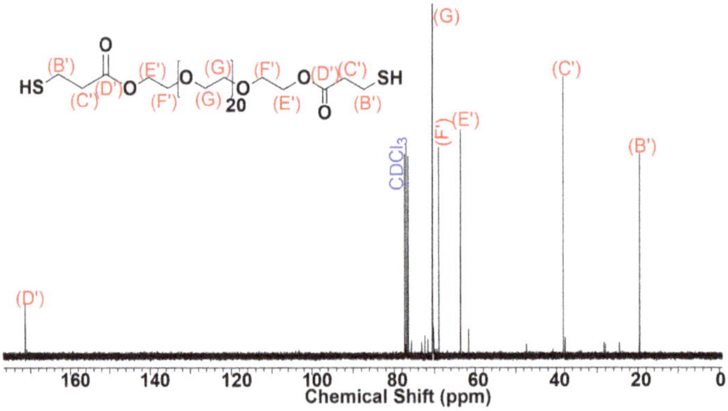

Figure 8. ^{13}C-NMR of PEG_{1000} dithiol [^{13}C-NMR (500 MHz, $CDCl_3$): δ 171.3, 70.4, 68.9, 63.6, 38.3, 19.6].

PEG_{2050} mono- and di-thiols were also synthesized as described in the Experimental section. The ^{1}H-NMR spectra shown in Figure S3 verified the structure of the PEG_{2050} mono- and di-thiols that were obtained with 100% and 94% reaction yield.

3. Materials and Methods

3.1. Materials

Candida antarctica Lipase B (CALB, 33273 Da, 20 wt% immobilized on a macroporous acrylic resin Novozyme® 435) was obtained from Sigma Chemicals (St. Louis, MO, USA). Poly(ethylene glycol)s (PEG_{1000}, $\overline{M_n}$ = 1000 g/mol, Đ = 1.14; and PEG_{2050}, $\overline{M_n}$ = 2050 g/mol, Đ = 1.09), and methyl-3-mercaptopropionate (MP-SH, 98%) were obtained from Aldrich Chemicals (St. Louis, MO, USA). Tetraethylene glycol (TEG) and tetrahydrofuran (THF, ≥99%) were obtained from Sigma-Aldrich (St. Louis, MO, USA). Diethyl ether (95.8%) was obtained from Fisher Chemicals (Hampton, NH, USA). Deuterated chloroform ($CDCl_3$, D 99.8%) was obtained from Cambridge Isotope Laboratories Inc.

3.2. *Methods*

3.2.1. CALB-catalyzed transesterification of MP-SH with TEG

1. Kinetic study

TEG (1.9782 g, 10.2 mmol) was dried under vacuum (Schlenk line) at 65 °C and 0.2 Torr until bubble formation ceased. It was then mixed with MP-SH (3.6204 g, 30.1 mmol) at 50 °C and 420 Torr in the presence of CALB (0.2549 g resin @ 20 wt% enzyme, 0.0015 mmol). After 1 min, the vacuum was removed, N_2 gas was passed through the system and an aliquot was collected. The vacuum was reinstated, and the procedure was repeated to collect aliquots at 3, 5, 10, 15, 30, 60, 120, 240, 300, 390, and 450 min. ^{1}H-NMR spectroscopy was used to check the extent of the reaction.

2. Synthesis of TEG-monothiol

TEG (3.8805 g, 20 mmol) was dried under vacuum (Schlenk line) at 65 °C and 0.2 Torr until bubble formation ceased. It was then mixed with MP-SH (7.4007 g, 61.6 mmol) at 50 °C and 420 Torr in the presence of CALB (0.4912 g resin @ 20 wt% enzyme, 0.0029 mmol). After 15 min, the reaction mixture was diluted with 3 mL of dried THF, filtered over a Q5 filter paper and then dried under vacuum (Schlenk line) at 50 °C for two hours. The product was then dried in a vacuum oven for further analysis (4.1685 g, 93% reaction yield).

3. Synthesis of TEG-dithiol

TEG (1.9782 g, 10.2 mmol) was dried under vacuum (Schlenk line) at 65 °C and 0.2 Torr until bubble formation ceased. It was then mixed with MP-SH (3.6204 g, 30.1 mmol) at 50 °C and 420 Torr in presence of CALB (0.2549 g resin @ 20 wt% enzyme, 0.0015 mmol). After 450 min, the reaction mixture was diluted with 3 mL of dried THF, filtered over a Q5 filter paper and then dried under vacuum (Schlenk line) at 50 °C for two hours. The product was then dried in a vacuum oven for further analysis (3.5327 g, 88% reaction yield).

3.2.2. CALB-catalyzed transesterification of MP-SH with PEG

1. Kinetic study

- PEG_{1000}

PEG_{1000} (3.9893 g, 4.04 mmol) was dried under vacuum (Schlenk line) at 65 °C and 0.2 Torr for 16 h. It was then mixed with MP-SH (1.4563 g, 12.11 mmol) and reacted at 50 °C and 420 Torr in the presence of CALB (0.0977 g resin @ 20 wt% enzyme, 0.00058 mmol). After 1 min, the vacuum was removed and N_2 gas was passed through the system and an aliquot was collected. The vacuum was reinstated, and the procedure was repeated to collect aliquots at 3, 5, 10, 15, 30, 60, 120, 240, 360, 480, 600, 720, 960, and 1440 min. ^{1}H-NMR spectroscopy was used to check the extent of the reaction.

- PEG_{2050}

PEG_{2050} (4.6117 g, 2.25 mmol) was dried under vacuum (Schlenk line) at 65 °C and 0.2 Torr for 16 h. It was then mixed with MP-SH (1.6550 g, 13.7 mmol) and reacted at 50 °C and 420 Torr in the presence of CALB (0.1134 g resin @ 20 wt% enzyme, 0.00068 mmol). After 1 min, the vacuum was removed and N_2 gas was passed through the system and an aliquot was collected. The vacuum was reinstated, and the procedure was repeated to collect aliquots at 3, 5, 10, 15, 30, 60, 120, 240, 360, 480, 600, 720, 960, and 1440 min. ^{1}H-NMR spectroscopy was used to check the extent of the reaction.

2. Synthesis of PEG-monothiols

- PEG_{1000}-monothiol

PEG_{1000} (6.0882 g, 6.16 mmol) was dried under vacuum (Schlenk line) at 65 °C and 0.2 Torr for 16 h. It was then mixed with MP-SH (2.1793 g, 18.13 mmol) and reacted at 50 °C and 420 Torr in the presence of CALB (0.1501 g resin @ 20 wt% enzyme, 0.00090 mmol). After 24 h, the reaction mixture was diluted with 3 mL of dried THF, filtered over a Q5 filter paper and then dried under vacuum (Schlenk line) at 50 °C for 16 h. The product was then dried in a vacuum oven for further analysis (6.0967 g, ~100% reaction yield).

- PEG_{2050}-monothiol

PEG_{2050} (4.6117 g, 2.25 mmol) was dried under vacuum (Schlenk line) at 65 °C and 0.2 Torr for 16 h. It was then mixed with MP-SH (1.6550 g, 13.7 mmol) and reacted at 50 °C and 420 Torr in the presence of CALB (0.1134 g resin @ 20 wt% enzyme, 0.00068 mmol). After 24 h, the reaction mixture was diluted with 3 mL of dried THF, filtered over a Q5 filter paper and precipitated in 100 mL of diethyl ether. The precipitate was then dried in a vacuum oven for further analysis (4.2034 g, ~100% reaction yield).

3. Synthesis of PEG-dithiols

- PEG_{1000}-dithiol

PEG_{1000}-monothiol (2.1005 g, 1.95 mmol) was dried under vacuum (Schlenk line) at 65 °C and 0.2 Torr for 16 h. It was then mixed with MP-SH (3.1031 g, 25.8 mmol) at 50 °C and 420 Torr in presence of CALB (0.0940 g resin @ 20 wt% enzyme, 0.00056 mmol) for 24 h. After 24 h of reaction time, the reaction mixture was diluted with 3 mL of dried THF, filtered over a Q5 filter paper and then dried under vacuum (Schlenk line) at 50 °C for 16 h. The product was then dried in a vacuum oven for further analysis. (2.1461 g, 85% reaction yield).

- PEG_{2050}-dithiol

PEG_{2050}-monothiol (4.000 g, 1.87 mmol) was dried under vacuum (Schlenk line) at 65 °C and 0.2 Torr for 16 h. It was then mixed with MP-SH (2.7125 g, 22.5 mmol) at 50 °C and 420 Torr in the presence of CALB (0.2274 g resin @ 20 wt% enzyme, 0.0013 mmol) for 24 h. After 24 h, the reaction mixture was diluted with 3 mL of dried THF, filtered over a Q5 filter paper and precipitated in 100 mL of diethyl ether. The precipitate was then dried in a vacuum oven for further analysis (3.7791 g, 94% reaction yield).

3.3. *Characterization*

3.3.1. Nuclear Magnetic Resonance (NMR) Spectroscopy

Varian Mercury 300 MHz and 500 MHz spectrometer (Palo Alto, CA, USA) was used to record the ^{1}H-NMR and ^{13}C-NMR spectra in $CDCl_3$ at 20 mg/ml and 60 mg/ml respectively with the following parameters: 10 second relaxation time, 128 scans (5000 scans for ^{13}C-NMR) and 90° angle. The internal reference for chloroform was δ = 7.26 ppm (^{1}H-NMR) and δ = 77 ppm (^{13}C-NMR).

3.3.2. Mass Spectrometry

Matrix-Assisted Laser Desorption Ionization Mass Spectrometry (MALDI-MS) experiments were performed on a Bruker UltraFlex III MALDI tandem time-of-flight (ToF/ToF) mass spectrometer (Bruker Daltonics, Billerica, MA, USA) equipped with a Nd:YAG laser emitting at 355 nm. *Trans*-2-[3-(4-*tert*-Butylphenyl)-2-methyl-2-propenylidene] malononitrile (98%; Sigma-Aldrich, St.

Louis, MO, USA) and sodium trifluoroacetic acid (98%; Sigma-Aldrich, St. Louis, MO, USA) (NaTFA) served as a matrix and cationizing salt, respectively. Solutions of the matrix (20 mg/mL), cationizing salt (10 mg/mL), and the sample (10 mg/mL) were prepared in THF (Fisher, Fair Lawn, NJ, USA). The matrix/sample/cationizing agent solutions were mixed in the ratio 10:2:1 (vol/vol/vol), and 0.5–1.0 μL of the final mixture were applied to the MALDI sample target and allowed to dry at ambient conditions before spectral acquisition. This sample preparation protocol led to the formation of $[M + Na]^+$ ions. Spectral acquisition was carried out at reflectron mode and ion source 1 (IS 1), ion source 2 (IS 2), source lens, reflectron 1, and reflectron 2 potentials were set at 25.03 kV, 21.72 kV, 9.65 kV, 26.32, and 13.73 kV, respectively.

4. Conclusions

In conclusion, we successfully prepared thiol-functionalized TEGs and PEGs via enzyme catalyzed transesterification of methyl 3-mercaptopropionate with TEG, and PEGs having M_n = 1000, and 2050 g/mol. These reactions were performed without using solvents (in bulk) and using *Candida antartica* Lipase B (CALB) as an enzyme catalyst. The progress of the reactions was monitored using ^{1}H-NMR spectroscopy. The transesterification was found to be a step-wise consecutive reaction. TEG-monothiol was exclusively formed in 15 min, followed by a slower second cycle yielding TEG-dithiol in 7.5 h. PEG_{1000}-monothiol was obtained within 8 h; however, dithiol formation was not observed even after 24 h of reaction. PEG_{1000}-dithiol was obtained by reacting PEG_{1000}-monothiol with fresh CALB and MP-SH for 24 h. PEG_{2050}-monothiol was formed in 16 h, and dithiol formation required additional CALB and MP-SH. Based on our data, it can be concluded that enzyme catalyzed transesterification is a convenient and green method to effectively synthesize PEG mono- and di-thiols that are suitable candidates for thiol-ene click reactions and Michael addition type of reactions [4].

Supplementary Materials: The following are available online at http://www.mdpi.com/2073-4344/9/3/228/s1, Figure S1: (A) ^{1}H-NMR and (B) ^{13}C-NMR spectra of TEG-monothiol, Figure S2: (A) ^{1}H-NMR and (B) ^{13}C-NMR of TEG-dithiol, Figure S3: ^{1}H-NMR spectrum of (A) PEG_{2050} monothiol and (B) PEG_{2050} dithiol.

Author Contributions: Conceptualization, J.E.P. and P.M.; methodology, P.M. and J.E.P.; validation, P.M., G.S., and J.E.P.; formal analysis, P.M. and J.E.P.; investigation, P.M.; resources, J.E.P.; data curation, J.E.P.; writing—original draft preparation, P.M.; writing—review and editing, P.M., G.S., and J.E.P.; visualization, P.M.; supervision, J.E.P.; project administration, J.E.P.; funding acquisition, J.E.P.

Funding: This research was funded by the BREAST CANCER INNOVATION FOUNDATION, Akron, OH. The NMR instrument used in this work was purchased from the funds provided by THE OHIO BOARD OF REGENTS, grant number CHE-0341701 and DMR-0414599.

Acknowledgments: The contribution to this work by Sanghamitra Sen is acknowledged and appreciated. Financial support by the Breast Cancer Innovation Foundation is greatly appreciated. The authors would also like to thank Chrys Wesdemiotis for helping with mass spectrometry analysis of the samples.

Conflicts of Interest: Puskas is the Chief Scientific Officer of Enzyme Catalyzed Polymers LLC, a start-up company that has exclusive license from the University of Akron for US Patents 8,710,156 (2014) and 9,885,070 (2018). The other authors declare no conflict of interest. The funders had no role in the design of the study; in the collection, analyses, or interpretation of data; in the writing of the manuscript, or in the decision to publish the results.

References

1. Zalipsky, S.; Harris, J.M. Introduction to Chemistry and Biological Applications of Poly (Ethylene Glycol). In *Poly(Ethylene Glycol) Chemistry and Biological Applications*; Harris, J.M., Zalipsky, S., Eds.; American Chemical Society: Washington, DC, USA, 1997; Volume 680, ISBN 978–0841235373.
2. Dreborg, S.; Akerblom, E.B. Immunotherapy with Monomethoxypolyethylene Glycol Modified Allergens. *Crit. Rev. Ther. Drug Carr. Syst.* **1989**, *6*, 315–365.
3. Yamaoka, T.; Tabata, Y.; Ikada, Y. Distribution and Tissue Uptake of Poly(Ethylene Glycol) with Different Molecular Weights after Intravenous Administration to Mice. *J. Pharm. Sci.* **1994**, *83*, 601–606. [CrossRef] [PubMed]

4. Puskas, J.E.; Castano, M.; Mulay, P.; Dudipala, V.; Wesdemiotis, C. Method for the Synthesis of γ-PEGylated Folic Acid and Its Fluorescein-Labeled Derivative. *Macromolecules* **2018**, *51*, 9069–9077. [CrossRef]
5. Anseth, K.S.; Klok, H.A. Click Chemistry in Biomaterials, Nanomedicine, and Drug Delivery. *Biomacromolecules* **2016**, *17*, 1–3. [CrossRef] [PubMed]
6. Mather, B.D.; Viswanathan, K.; Miller, K.M.; Long, T.E. Michael Addition Reactions in Macromolecular Design for Emerging Technologies. *Prog. Polym. Sci.* **2006**, *31*, 487–531. [CrossRef]
7. Nair, D.P.; Podgorski, M.; Chatani, S.; Gong, T.; Xi, W.; Fenoli, C.R.; Bowman, C.N. The Thiol-Michael Addition Click Reaction: A Powerful and Widely used Tool in Materials Chemistry. *Chem. Mater.* **2014**, *26*, 724–744. [CrossRef]
8. Oliverio, M.; Perotto, S.; Messina, G.C.; Lovato, L.; De Angelis, F. Chemical Functionalization of Plasmonic Surface Biosensors: A Tutorial Review on Issues, Strategies, and Costs. *ACS Appl. Mater. Interfaces* **2017**, *9*, 29394–29411. [CrossRef] [PubMed]
9. Wang, C.; Yu, C. Detection of Chemical Pollutants in Water using Gold Nanoparticles as Sensors: A Review. *Rev. Anal. Chem.* **2013**, *32*, 1–14. [CrossRef]
10. Manson, J.; Kumar, D.; Meenan, B.J.; Dixon, D. Polyethylene glycol Functionalized Gold Nanoparticles: The Influence of Capping Density on Stability in Various Media. *Gold Bull.* **2011**, *44*, 99–105. [CrossRef]
11. Mahou, R.; Wandrey, C. Versatile Route to Synthesize Heterobifunctional Poly(Ethylene Glycol) of Variable Functionality for Subsequent PEGylation. *Polymers* **2012**, *4*, 561–589. [CrossRef]
12. Goessl, A.; Tirelli, N.; Hubbell, J.A. A Hydrogel System for Stimulus-Responsive, Oxygen-Sensitive In Situ Gelation. *J. Biomater. Sci. Polym. Ed.* **2004**, *15*, 895–904. [CrossRef] [PubMed]
13. Buwalda, S.J.; Dijkstra, P.J.; Feijen, J. In Situ Forming Poly(Ethylene Glycol)-Poly(L-Lactide) Hydrogels via Michael Addition: Mechanical Properties, Degradation, and Protein Release. *Macromol. Chem. Phys.* **2012**, *213*, 766–775. [CrossRef]
14. Hiemstra, C.; van der Aa, L.J.; Zhong, Z.; Dijkstra, P.J.; Feijen, J. Novel In Situ Forming, Degradable Dextran Hydrogels by Michael Addition Chemistry: Synthesis, Rheology, and Degradation. *Macromolecules* **2007**, *40*, 1165–1173. [CrossRef]
15. Yoshimoto, K.; Hirase, T.; Nemoto, S.; Hatta, T.; Nagasaki, Y. Facile Construction of Sulfanyl-Terminated Poly(Ethylene Glycol)-Brushed Layer on a Gold Surface for Protein Immobilization by the Combined use of Sulfanyl-Ended Telechelic and Semitelechelic Poly(Ethylene Glycol)s. *Langmuir* **2008**, *24*, 9623–9629. [CrossRef] [PubMed]
16. Hirase, T.; Nagasaki, Y. Construction of Mercapto-Ended Poly(Ethylene Glycol) Tethered Chain Surface for High Performance Bioconjugation. In Proceedings of the AIChE Annual Meeting, San Francisco, CA, USA, 12–17 November 2006.
17. Nie, T.; Baldwin, A.; Yamaguchi, N.; Kiick, K.L. Production of Heparin-Functionalized Hydrogels for the Development of Responsive and Controlled Growth Factor Delivery Systems. *J. Control. Release* **2007**, *122*, 287–296. [CrossRef] [PubMed]
18. Belair, D.G.; Miller, M.J.; Wang, S.; Darjatmoko, S.R.; Binder, B.Y.; Sheibani, N.; Murphy, W.L. Differential Regulation of Angiogenesis using Degradable VEGF-Binding Microspheres. *Biomaterials* **2016**, *93*, 7–37. [CrossRef] [PubMed]
19. Yu, H.; Feng, Z.G.; Zhang, A.Y.; Sun, L.G.; Qian, L. Synthesis and Characterization of Three-Dimensional Crosslinked Networks Based on Self-Assemly of α-Cyclodextrins with Thiolated 4-arm PEG using a Three-Step Oxidation. *Soft Matter* **2006**, *2*, 343–349. [CrossRef]
20. Du, Y.J.; Brash, J.L. Synthesis and Characterization of thiol-terminated Poly(Ethylene Oxide) for Chemisorption to Gold Surface. *J. Appl. Polym. Sci.* **2003**, *90*, 594–607. [CrossRef]
21. Wan, J.K.S.; Depew, M.C. Some Mechanistic Insights in the Behaviour of Thiol Containing Antioxidant Polymers in Lignin Oxidation Processes. *Res. Chem. Intermed.* **1996**, *22*, 241–253. [CrossRef]
22. Yang, T.; Long, H.; Malkoch, M.; Kristofer Gamstedt, E.; Berglund, L.; Hult, A. Characterization of Well-Defined Poly(Ethylene Glycol) Hydrogels Prepared by Thiol-Ene Chemistry. *J. Polym. Sci. Part A Polym. Chem.* **2011**, *49*, 4044–4054. [CrossRef]
23. Zhang, H.J.; Xin, Y.; Yan, Q.; Zhou, L.L.; Peng, L.; Yuan, J.Y. Facile and Efficient Fabrication of Photoresponsive Microgels via Thiol–Michael Addition. *Macromol. Rapid Commun.* **2012**, *33*, 1952–1957. [CrossRef] [PubMed]
24. Zustiak, S.P.; Leach, J.B. Hydrolytically Degradable Poly(Ethylene Glycol) Hydrogel Scaffolds with Tunable Degradation and Mechanical Properties. *Biomacromolecules* **2010**, *11*, 1348–1357. [CrossRef] [PubMed]

25. Zustiak, S.P. Hydrolytically Degradable Polyethylene Glycol (PEG) Hydrogel: Synthesis, Gel Formation, and Characterization. In *Extracellular Matrix*; Leach, J., Powell, E., Eds.; Humana Press: New York, NY, USA, 2015; Volume 93, ISBN 978–1–4939–2082–2.
26. Puskas, J.E.; Sen, M. Process of Preparing Functionalized Polymers via Enzymatic Catalysis. U.S. Patents 8,710,156, 29 April 2014.
27. Puskas, J.E.; Sen, M.Y.; Seo, K.S. Green Polymer Chemistry using Nature's Catalysts, Enzymes. *J. Polym. Sci. Part A Polym. Chem.* **2009**, *47*, 2959–2976. [CrossRef]
28. Castano, M.; Seo, K.S.; Guo, K.; Becker, M.L.; Wesdemiotis, C.; Puskas, J.E. Green Polymer Chemistry: Synthesis of Symmetric and Asymmetric Telechelic Ethylene Glycol Oligomers. *Polym. Chem.* **2015**, *6*, 1137–1142. [CrossRef]
29. Puskas, J.E.; Sen, M.Y.; Kasper, J.R. Green Polymer Chemistry: Telechelic Poly(Ethylene Glycol)s via Enzymatic Catalysis. *J. Polym. Sci. Part A Polym. Chem.* **2008**, *46*, 3024–3028. [CrossRef]
30. Puskas, J.E.; Seo, K.S.; Sen, M.Y. Green Polymer Chemistry: Precision Synthesis of Novel Multifunctional Poly(Ethylene Glycol)s using Enzymatic Catalysis. *Eur. Polym. J.* **2011**, *47*, 524–534. [CrossRef]
31. Castano, M.; Seo, K.S.; Kim, E.H.; Becker, M.L.; Puskas, J.E. Green Polymer Chemistry VIII: Synthesis of Halo-Ester-Functionalized Poly(Ethylene Glycol)s via Enzymatic Catalysis. *Macromol. Rapid Commun.* **2013**, *34*, 1375–1380. [CrossRef] [PubMed]
32. Puskas, J.E.; Seo, K.S.; Castano, M.; Casiano, M.; Wesdemiotis, C. Green Polymer Chemistry: Enzymatic Functionalization of Poly(ethylene glycol)s under solventless conditions. In *Green Polymer Chemistry: Biocatalysis and Materials II*; Cheng, H.N., Gross, R.A., Smith, P.B., Eds.; ACS Symposium Series; American Chemical Society: Washington, DC, USA, 2013; Volume 1144, ISBN 978–0–8412–2895–5.
33. Sen, S.; Puskas, J.E. Green Polymer Chemistry: Enzyme Catalysis for Polymer Functionalization. *Molecules* **2015**, *20*, 9358–9379. [CrossRef] [PubMed]
34. Zdarta, J.; Wysokowski, M.; Norman, M.; Kołodziejczak-Radzimska, A.; Moszyński, D.; Maciejewski, H.; Ehrlich, H.; Jesionowski, T. *Candida antarctica* Lipase B Immobilized onto Chitin Conjugated with POSS® Compounds: Useful Tool for Rapeseed Oil Conversion. *Int. J. Mol. Sci.* **2016**, *17*, 1581. [CrossRef] [PubMed]
35. Jesionowski, T.; Zdarta, J.; Krajewska, B. Enzyme Immobilization by Adsorption: A Review. *Adsorption* **2014**, *20*, 801–821. [CrossRef]
36. Zdarta, J.; Klapiszewski, L.; Jedrzak, A.; Nowicki, M.; Moszynski, D.; Jesionowski, T. Lipase B from *Candida antarctica* Immobilized on a Silica-Lignin Matrix as a Stable and Reusable Biocatalytic System. *Catalyst* **2017**, *7*, 14. [CrossRef]
37. Wolfson, A.; Atyya, A.; Dlugy, C.; Tavor, D. Glycerol Triacetate as Solvent and Acyl Donor in the Production of Isoamyl Acetate with *Candida antarctica* Lipase B. *Bioprocess Biosyst. Eng.* **2010**, *33*, 363–366. [CrossRef] [PubMed]
38. Uppenberg, J.; Oehrner, N.; Norin, M.; Hult, K.; Kleywegt, G.J.; Patkar, S.; Waagen, V.; Anthonsen, T.; Jones, T.A. Crystallographic and Molecular-Modeling Studies of Lipase B from Candida antarctica reveal a Stereospecificity Pocket for Secondary Alcohols. *Biochemistry* **1995**, *34*, 16838–16851. [CrossRef] [PubMed]

catalysts

MDPI

Article

Enzymatic Synthesis of ω-Hydroxydodecanoic Acid By Employing a Cytochrome P450 from *Limnobacter* sp. 105 MED

Sung-Yeon Joo [1,†], Hee-Wang Yoo [2,†], Sharad Sarak [3], Byung-Gee Kim [1,2,*] and Hyungdon Yun [3,*]

1 School of Chemical and Biological Engineering, Seoul National University, Seoul 08826, Korea; celberose@gmail.com
2 Bioengineering Institute, Seoul National University, Seoul 08826, Korea; wsiy@naver.com
3 Department of Systems Biotechnology, Konkuk University, Seoul 05029, Korea; sharad.niper2014@gmail.com
* Correspondence: byungkim@snu.ac.kr (B.-G.K.); hyungdon@konkuk.ac.kr (H.Y.)
† These authors contributed equally to this work.

Received: 12 December 2018; Accepted: 3 January 2019; Published: 8 January 2019

Abstract: ω-Hydroxylated fatty acids are valuable and versatile building blocks for the production of various adhesives, lubricants, cosmetic intermediates, etc. The biosynthesis of ω-hydroxydodecanoic acid from vegetable oils is one of the important green pathways for their chemical-based synthesis. In the present study, the novel monooxygenase CYP153AL.m from *Limnobacter* sp. 105 MED was used for the whole-cell biotransformations. We constructed three-component system that was comprised of CYP153AL.m, putidaredoxin and putidaredoxin reductase from *Pseudomonas putida*. This in vivo study demonstrated that CYP153AL.m is a powerful catalyst for the biosynthesis of ω-hydroxydodecanoic acid. Under optimized conditions, the application of a solid-state powdered substrate rather than a substrate dissolved in DMSO significantly enhanced the overall reaction titer of the process. By employing this efficient system, 2 g/L of 12-hydroxydodecanoic acid (12-OHDDA) was produced from 4 g/L of its corresponding fatty acid, which was namely dodecanoic acid. Furthermore, the system was extended to produce 3.28 g/L of 12-OHDDA using 4 g/L of substrate by introducing native redox partners. These results demonstrate the utility of CYP153AL.m-catalyzed biotransformations in the industrial production of 12-OHDDA and other valuable building blocks.

Keywords: 12-hydroxydodecanoic acid; dodecanoic acid; CYP153A; whole-cell biotransformation

1. Introduction

ω-Hydroxylated fatty acids (ω-OHFAs) obtained from medium- and long-chain length fatty acids are versatile building blocks that are used as precursors for bioplastics [1] and high-end polymers in the chemical industry [2,3]. In addition, oxygenated fatty acids can be used in the cosmetics industry to produce perfumes and for pharmaceutical applications as anticancer agents and polyketide antibiotics [4,5]. To synthesize ω-OHFAs, various chemical routes have been reported, including cross-metathesis of unsaturated fatty acid esters, followed by the hydroformylation and hydrogenation of the carbonyl group [6] or by the reduction of α, ω-dicarboxylic acids [7]. However, the chemical-based processes for the oxidation of the unreactive carbon atom require very harsh conditions and multiple steps; depend on nonrenewable feedstocks and have poor selectivity [8].

For these reasons, attention has been focused on biological approaches. Wenhua Lu and co-workers have reported biotransformation of 200 g/L methyl tetradecanoate, which resulted into 174 g/L and 6 g/L of its corresponding OHFAs and α, ω-dicarboxylic acids (ω-DCAs), respectively, using an engineered *Candida tropicalis* [9]. Although remarkable progress has been made, these production platforms have not been exploited to a larger extent yet. Factors affecting the applicability of these processes for the large-scale production include low productivity, instability of biocatalyst and the requirement of economically feasible production facilities [10].

To overcome these limitations, alternative bacterial-based processes have been investigated. In these processes, the main key enzyme is CYP153As. The CYP153s are bacterial class I P450 enzymes that operate as three-component systems, containing a heme-dependent monooxygenase core (CYP) and two additional redox partners and/or domains, which are namely an iron–sulfur electron carrier (ferredoxin, Fdx) and a FAD-containing reductase (ferredoxin reductase, FdR). These transfer electrons from NAD(P)H to the monooxygenase active site [11]. This subfamily displays excellent activity towards the ω-hydroxylation of alkanes, primary alcohols and fatty acids [12].

Recently, Bernhard Hauer and co-workers have demonstrated the construction of a chimeric protein, where the heme domain of CYP153AM.aq. was fused to the reductase domain of CYP102A1 isolated from *Bacillus megaterium* (*B. megaterium*) (CPR_{BM3}), the most catalytically active P450 reported to date [13]. Additionally, they introduced the G307A mutant based on the GGNDT motif in the I-helix, conserved in CYP153A subfamily, which showed 2- to 20-fold increase in the activity toward medium chain fatty acids [14].

The CYP153A.M. aq.-CPR_{BM3} fusion construct has been used for in vivo hydroxylation to produce ω-OHDDA. This reaction produced 1.2 g/L of ω-hydroxydodecanoic acid (ω-OHDDA) with high regioselectivity (>95% ω-regioselectivity) for the terminal position by using 10 g/L of its corresponding free fatty acid as a substrate [10]. However, feasible productivity and yield at an industrial scale have not been reached yet. In the current study, an enzyme mining approach revealed that an excellent CYP153A can be a powerful catalyst in the three-component systems. During this study, the application of a solid-state powdered substrate rather than a substrate dissolved in DMSO and introducing the native redox partner were shown to be effective strategies (Figure 1).

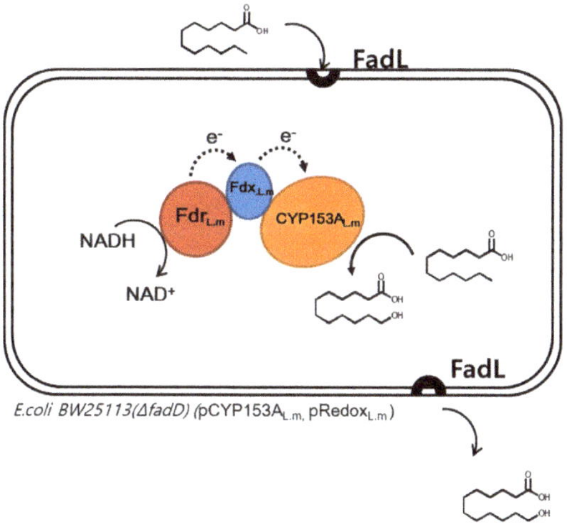

Figure 1. Synthesis of ω-hydroxydodecanoic acid (ω-OHDDA) from dodecanoic acid (DDA) using CYP153A three-component system.

2. Results and Discussion

2.1. Construction of a CYP153A Three-Component System

Bioinformatic tools have been routinely used to search for new enzymes and for conducting analysis of the similarity among already available enzymes [15–17]. Previously, the CYP153A from *Marinobacter aquaeolei* (CYP153AM.aq.) was reported to possess high activity and selectivity toward ω-hydroxylation of C12-FA (63% conversion, 95% ω-regioselectivity) [14]. For the identification of more suitable CYP153As, Basic Local Alignment Search Tool (BLAST) search was performed using CYP153AM.aq as a query sequence and 100 candidates showing resemblance to the query sequence were selected. Among them, seven CYP153As were randomly selected from various groups based on a phylogenetic tree (Figure S1). The genes encoding CYP153As were cloned into pCDF_duet vector and expressed in *E. coli BW25113* ΔfadD (DE3). As CYP153As from *Oceanococcus atlanticus*, *Alcanivorax jadensis* T9 and *Nocardioides luteus* (CYP53AO.a, CYP153AA.j and CYP153AN.l, respectively) were expressed as inclusion bodies, they were neglected in the further studies. SDS-PAGE analysis and CO-binding assays confirmed the production of the other three CYP153As in soluble and active forms (Figures S2 and S3).

To compare the performance of the active P450s, whole-cell (0.11 g_{CDW}/mL) reactions were carried out at 30 °C and 200 rpm using cells with the active P450s, which co-expressed CamA (putidaredoxin reductase) and CamB (putidaredoxin) from *Pseudomonas putida* in potassium phosphate buffer (100 mM, pH 7.5) in the presence of 1% (*w/v*) glucose. The Cells with CYP153AA.d from *Alcanivorax dieselolei* produced the lowest ω-OHDDA (0.38 g/L) in 24 h, while those of CYP153AS.f from *Solimonas flava* and CYP153AM.aq produced 0.72 g/L and 1.13 g/L from 4 g/L of DDA (Figure 2a).

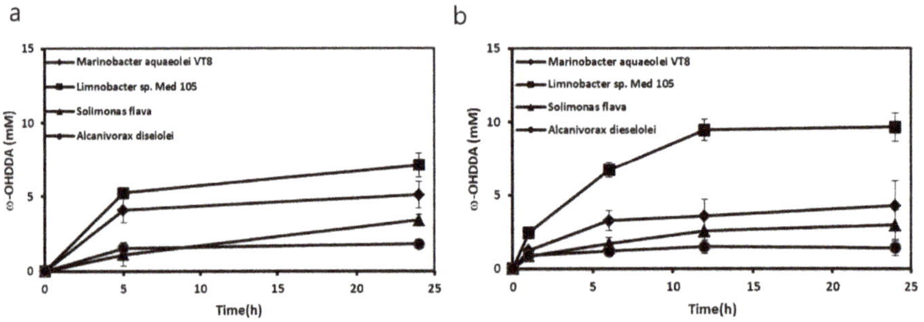

Figure 2. ω-Hydroxylation of 20 mM DDA by CYP153A and Cam AB containing cells, Reaction conditions: Volume, 10 mL in 100 mL flask; Temp, 30 °C; Cell type, BW25113 (ΔfadD, DE3) with CYP153AM.aq, CYP153AL.m, CYP153AS.f and CYP153AA.d, respectively, with CamA/B; Cell OD600, 30; Phosphate buffer, 100 mM; pH, 7.5; glucose 1% (*w/v*). (**a**) 20 mM DDA (DMSO 1%), (**b**) 20 mM of solid state powdered substrate.

The performance of CYP153AM.aq expressing cells was similar to that of the CYP153A.M. aq.–CPR_{BM3} fusion construct [10]. Furthermore, those with CYP153AL.m showed the largest ω-OHDDA production (1.5 g/L) from 4 g/L of DDA (Figure 2a). Interestingly, the yields achieved herein are the highest via batch reaction so far to best of our knowledge. We also attempted to make a more detailed comparison and determined the concentration of the active P450s by CO-binding assay (Figure S4). After normalization by the amount of active P450s, CYP153AL.m was shown to be more active than the other P450s (Figure S5). Therefore, through the enzyme mining approach, the CYP153AL.m containing three-component system was identified as an efficient catalyst for the bioconversion of DDA to ω-OHDDA.

2.2. Use of Solid State Powdered Substrate in Bioconversion

In many cases, substrates for reactions catalyzed by P450 have limited water solubility. For example, medium chain fatty acids (C_{10}–C_{16}) are water soluble from ~7 mg/L to 61 mg/L. Moreover, to increase the industrial feasibility of the process, two-liquid phase systems have been previously utilized by Maurer et al. [18] and von Buhler et al. [19]. Furthermore, the addition of co-solvents or cyclodextrins was reported by Donova et al. [20] and Kuehnel et al. [21]. However, the excessive use of DMSO as a co-solvent could damage the catalyst and host cell [22]. Moreover,

Lundemo and co-workers demonstrated that a DMSO free strategy was advantageous in whole-cell P450 catalyzed reactions [23].

Based on these previous reports, cells were subjected to resting cell reaction in the absence of DMSO. The CYP153AL.m expressing strain was shown to give the highest yields in the conversion of DDA (4 g/L) to ω-OHDDA (2 g/L), whereas CYP153AM.aq, CYP153AS.f and CYP153AA.d gave 0.9 g/L, 0.63 g/L and 0.3 g/L, respectively (Figure 2b). There is no significant difference in the amount of product obtained using a solid-state powdered substrate or a substrate dissolved in DMSO. The use of substrate in the solid state appeared to have a negligible effect on CYP153AM.aq, CYP153A.d and CYP153AS.f, having showed a slightly declined amount of conversion. Nevertheless, only the CYP153A.Lm-catalyzed reaction titer was shown to be improved using powdered substrate. Any efficient biocatalytic conversion requisites the effective uptake of the substrate followed by its conversion to the desired product [24,25]. The better uptake of a powdered substrate in the absence of DMSO could be the plausible reason for the better titer achieved by the cells expressing CYP153A.Lm.

2.3. *Effect of Homogeneous Redox Partners*

Generally, it has been accepted that the optimal redox partners for a P450 enzyme should be homogeneous ones [26–28]. Therefore, we tried to introduce native redox partners of *Limnobacter* sp. *105* MED. Using the functional protein association network database STRING v10.5 [29], a network comprising CYP153AL.m was obtained and we found that there are two Fdxs and one FdR from *Limnobacter* sp. 105 MED (Figure S6). The FdR (LimA) and a Fdx (LimB) with high values were codon-optimized and synthesized into pETduet_vector. After this, they were expressed and purified together with CYP153AL.m, CYP153AM.aq and CamA/B (Figure S7).

To evaluate the native redox partner chains, in vitro biotransformation was carried out with a final volume of 0.5 mL of 100 mM potassium phosphate buffer (pH 7.5), containing 2 μM CYP153A, ferredoxin reductase and ferredoxin (1:10:5 ratio). DDA was added at a final concentration of 0.5 mM (25 mM stock in DMSO). The reaction was started by the addition of 0.2 mM NADH and the coupling efficiencies were obtained by determining the ratio of the initial product forming rate and the NADH consumption rate as previously described [30].

Although the coupling efficiencies of CYPs are similar regardless of redox partners, the CYP153AL.m has a higher initial product forming rate (8.91 ± 0.89 μM/min) than CYP153AM.aq (7.41 ± 0.70 μM/min) and a higher NADH consumption than CYP153AM.aq (17.04 ± 0.32 μM/min and 13.83 ± 1.16 μM/min, respectively). Moreover, the native redox partners of CYP153A.L.m improved the performance with the highest initial product forming rate (9.81 ± 1.60 μM/min) and the highest NADH consumption (20.26 ± 0.57 μM/min) (Table 1).

Table 1. In vitro evaluation of native redox partners of CYP153AL.m.

Entry	CYP153AM.aq CamA + CamB	CYP153AL.m CamA + CamB	CYP153AL.m LimA + LimB
Coupling efficiency (%)	53.6 ± 5.5	52.3 ± 3.6	48.4 ± 5.4
Initial product forming rate (μM/min)	7.41 ± 0.70	8.91 ± 0.89	9.81 ± 1.60
NADH consumption rate (μM/min)	13.83 ± 1.16	17.04 ± 0.32	20.26 ± 0.57

After confirming the performance of the native redox partner chain, the proteins of the redox partner chain were further investigated in vivo. Whole-cell (0.11 g_{CDW}/mL) reactions were carried out at 30 °C and 200 rpm using cells with CYP153A.L.m, co-expressing LimA and LimB in potassium phosphate buffer (100 mM, pH 7.5) in the presence of 1% (*w/v*) glucose. Ideally, the in vivo endogenous three-component system produced 3.28 g/L of 12-OHDDA from 4 g/L DDA (Figure 3). Therefore, our data demonstrated that the application of naturally occurring redox partners can enhance the reactivity in vivo and in vitro.

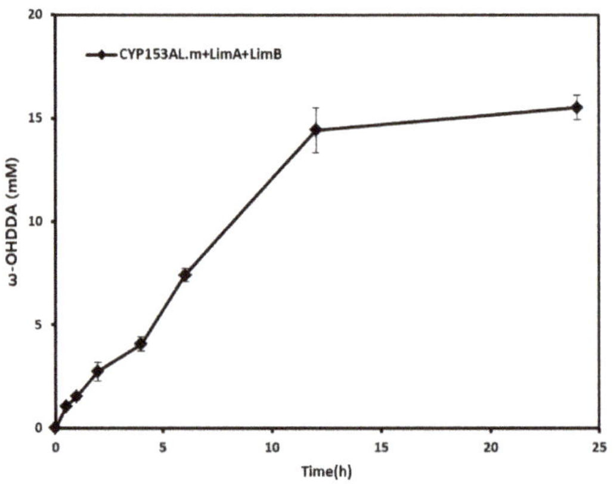

Figure 3. ω-Hydroxylation of 20 mM DDA by CYP153AL.m and LimAB containing cells. Reaction conditions: Substrate concentration, 20 mM; Volume, 10 mL in 100 mL flask; Temp, 30 °C; Cell type, BW25113 (ΔfadD, DE3) expressing CYP153AL.m and Lim A/B; Cell OD600, 30; Phosphate buffer, 100 mM; pH, 7.5; glucose 1% (*w/v*).

2.4. Limitations of the CYP153AL Three-Component System

The stability issues of CYPs have been frequently mentioned, which are caused by uncoupling of the NADH oxidation during product formation [23,31]. To evaluate the stability of our system, a series of four rounds was carried out with whole cells expressing CYP153AL.m and Cam A/B. In each round, the cells were purified and reused every 6 h in the next round as equilibrium is reached in a relatively short time period of 6 h (Figure 2b). In the second round, the amount of product obtained from the whole cells was considerably reduced (53%) compared to the first round and further decreased to 35% in the third round (Figure 4a). This data implies that our system suffered from instability and thus, a relatively short process time is desirable.

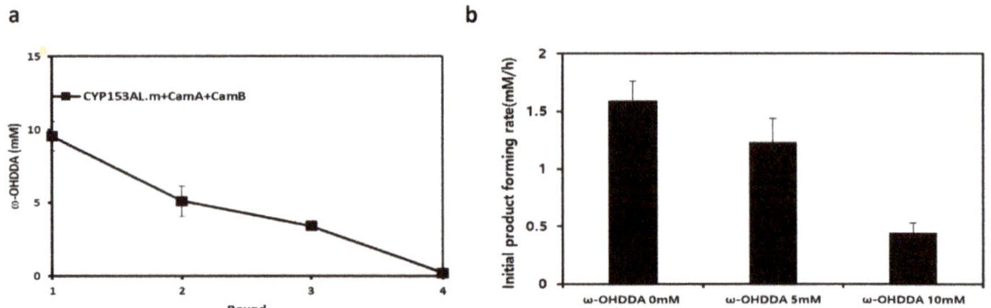

Figure 4. (**a**) Residual stability of whole cells expressing CYP153AL.m. Reaction conditions: Substrate concentration, 20 mM; Volume, 10 mL in 100 mL flask; Temp, 30 °C; Cell type, BW25113 (ΔfadD, DE3) having CYP153AL.m and CamA/B; Cell OD600, 30; Phosphate buffer, 100 mM; pH, 7.5; glucose 1% (*w/v*), The biocatalysts in the each round (every 6 h) were re-used in next round, (**b**) Product inhibited ω-hydroxylation of 20 mM DDA. Reaction condition is the same with 3 ω-hydroxylation of 20 mM fatty acids by CYP153A and Cam AB containing cells, Note that: 0, 5 and 10 mM of ω-OHDDA was initially added.

Additionally, the residual activity in the third round did not compensate for the product conversion after 12 h (Figure 2b). We speculate that the phenomenon is caused by product inhibition, which was previously mentioned by Lundemo [23] and the resting cell reaction was carried out in the additional presence of 5 mM and 10 mM ω-OHDDA. The product forming rate significantly decreased from 1.59 ± 0.09 mM/h to 0.44 ± 0.17 mM/h as the initial product concentration increased from 0 to 10 mM (Figure 4b), indicating significant product inhibition. The results of these studies demonstrated that CYP153AL.m three-component system suffered from severe product inhibition, which needs to be overcome to produce high concentrations of the products.

3. Materials and Methods

3.1. Chemicals and Media

All chemicals, such as DDA, ω-OHDDA, dimethyl sulfoxide (DMSO), isopropyl-thio-β-D -galactopyranoside (IPTG), 5-aminolevulinic acid (5-ALA), N, O-Bis (trimethylsilyl)-trifluoroacetamide (BSTFA, were purchased from Sigma-Aldrich (St. Louis, MO, USA). Chloroform was obtained from Junsei (Tokyo, Japan). Bacteriological agar, Luria Bertani (LB) broth and terrific broth (TB) media were purchased from BD Difco (Franklin Lakes, NJ, USA). All chemicals used in this study were of analytical grade.

3.2. Plasmid Construction

For efficient substrate uptake, IPTG-inducible pCDF_duet1 vector with outer membrane long chain fatty acid (LCFA) transporter, *fadL* [30] was used. The genes encoding various CYPs were codon-optimized and synthesized from Bionics (Seoul, Korea), before each gene was inserted into the IPTG-inducible pCDF_duet1 vectors using *Nde*I and *Xho*I as a restriction site. For redox partners, CamA (ferredoxin reductase) and CamB (ferredoxin) from *Pseudomonas putida* and LimA and LimB from *Limnobacter* sp. MED 105 were cloned into pET_duet1 vector using BamHI & SacI and NdeI & EcorV, respectively [30].

3.3. Protein Expression and Purification

E. coli BW25113 (DE3) ΔfadD strain [32] was utilized for the biotransformation studies, wherein the fatty acid degrading β-oxidation pathway was blocked. Plasmid DNA were transformed into host strains using the standard heat shock method. For the expression and purification of enzymes, fresh colonies from agar plates of transformants were cultured in 2 mL of LB medium containing appropriate antibiotics at 37 °C overnight.

The seed-cultured cells were inoculated into 50 mL of Terrific-Broth (TB) in a 250-mL baffle flask and cultured at 37 °C until the cell concentration reached an optical density at 600 nm (OD_{600}) of 0.6–0.8 for IPTG induction. The induction was performed by adding 0.01 mM IPTG, 0.25 mM 5-ALA as heme precursor and 0.5 mM $FeSO_4$ at 30 °C for 16 h. MaqCYP153A, CYP153ALim, CamA/CamB, and LimA/LimB expression were carried out in 1 L of Terrific-Broth (TB) in a 3-L flask and induction was performed by adding 0.01 mM of IPTG, 0.25 mM 5-ALA as heme precursor and 0.5 mM $FeSO_4$ or only 0.01 mM for the redox-partners at 30 °C for 16 h. For the purification of enzymes, cells were harvested by centrifugation at 4 °C and 3000 rpm for 15 min. After this, they were washed and re-suspended in 20 mL of potassium phosphate buffer (100 mM, pH 7.5) before finally being disrupted by sonication.

The soluble fraction of each lysate was collected after centrifugation at 16,000 rpm for 30 min and the enzymes were purified using a His-Trap-™ HP column (GE Healthcare Bio-Sciences AB, Uppsala, Sweden). The Ni-NTA-bound enzymes were washed twice with 50 mM sodium phosphate buffer (pH 8.0) containing 300 mM NaCl and 20 mM imidazole. Next, the enzymes were eluted out with the same buffer containing 250 mM imidazole. Finally, the purified protein was concentrated by ultra-filtration, before the imidazole and sodium chloride was removed via sequential dialysis. Purified proteins were subjected to SDS-PAGE analysis.

3.4. CO-Binding Assay

The UV absorption spectra of CO-bound recombinant CYP proteins after sodium dithionite reduction were measured by Multi-scan UV–Vis spectrometry (Thermo Fisher Scientific, MA, USA) by scanning wavelengths of 400–500 nm at intervals of 5 nm. The concentration of P450 was measured using an extinction coefficient of 91.9 mM^{-1} cm^{-1} at 450 nm.

3.5. In-Vitro Oxidation Assay of CYP153As

The coupling efficiency of each CYPs with a redox partner was evaluated using C12:0 saturated fatty acids as a substrate. The biotransformations were performed using a final volume of 0.5 mL of 100 mM potassium phosphate buffer (pH 7.5), containing 2 μM CYP153A, ferredoxin reductase and ferredoxin i.e., P450, CamA and CamB (1:5:10 ratio) or P450, LimA and LimB (1:5:10 ratio) at 25 °C [28]. Fatty acids were added at a final concentration of 0.5 mM (25 mM stock in DMSO). The reaction was started by the addition of the 0.2 mM NADH. Furthermore, the consumption of NADH was monitored by measuring absorbance at 340 nm (e340 = 6.22 mM^{-1} cm^{-1}). After completion, the reaction was stopped by adding HCl and samples were analyzed by GC/FID analysis.

3.6. Resting Cell Reaction

Cells were harvested by centrifugation and washed with 100 mM potassium phosphate buffer (pH 7.5), followed by resuspension in the same buffer. The cell buffer resuspension was placed into a 100-mL shake flask, with the cell density adjusted to OD_{30} in the final 10 mL volume. The resting cell reaction was initiated by adding 20 mM dodecanoic acid (1 M stock in DMSO or powder form). The reactants were incubated at 30 °C and 200 rpm, with 0.5 mL sample aliquots collected every 2 or 6 h. The sample preparation for product analysis was done as follows: 500 μL of the whole cell culture was acidified with 6 M HCl to a pH of 2 and extracted with chloroform by vigorous vortexing for 1 min. The organic phases were then collected and derivatized by using trimethylsilyl (TMS) and incubation at 50 °C for 20 min, with an excess of BSTFA.

3.7. Product Identification and Quantification

Quantitative analysis was performed by gas chromatography, HP 6890 Series (Agilent Technologies, Santa Clara, CA, USA) with flame ionization detector (GC/FID). One μL of the sample was injected by the split less mode (a split less time of 0.8 min) and analyzed using a nonpolar capillary column (5% phenyl methyl siloxane capillary 30 m × 320 μm i.d. 0.25 μm film thickness, HP-5 ms).

The oven temperature was maintained at 50 °C for 1 min, before being increased by 15 °C/min to 250 °C, with this temperature held for 10 min. The temperature of the inlet was kept at 250 °C and the temperature of the detector was 280 °C. The flow rate of the carrier gas was 1.0 mL/min, while flow rates of hydrogen, air and helium in the FID were 45, 400 and 20 mL/min, respectively. Each peak was identified by comparison of the GC chromatogram with that of an authentic sample. Errors in the analysis were corrected by using heptadecanoic acid as an internal standard.

4. Concluding Remarks

This study reports on a ω-hydroxy fatty acid production platform employing efficient CYP153A from *Limnobacter* sp. MED 105. Furthermore, the application of a powder substrate and introducing a native redox partner was successfully examined in the study. However, product inhibition is a hurdle for the industrial feasibility of this biocatalytic process. In conclusion, with the help of the heterologous expression of novel non-native enzymes in *E. coli*, we successfully produced 3.28 g/L of ω-OHDDA from 4 g/L of DDA.

Supplementary Materials: The following are available online at http://www.mdpi.com/2073-4344/9/1/54/s1, Figure S1: Phylogenetic tree used in this study, Figure S2: A SDS-PAGE analysis of protein expression of CYP153As, Figure S3:A CO-binding analysis of CYP153A expressing strains used in this study, Figure S4: An active P450 concentration used in this study, Figure S5: ω-OHDDA production normalized by amount of active P450s, Figure S6: Protein–protein network of CYP153AL.m(LMED105_04587), Figure S7: SDS-PAGE gel picture of purified protein of CamB (12.75 kDa), CamA (47 kDa), CYP153AM.aq (52.28 kDa), LimB (11.87 kDa), LimA (45.61 kDa), and CYP153AL.m (52.28 kDa).

Author Contributions: H.Y. and B.-G.K. designed the experiments of the project. S.-Y.J. carried out the research works as part of his master's project. H.-W.Y. supervised the whole studies reported in the manuscript. S.-Y.J. and S.S. wrote and revised the manuscript. H.-W.Y. assisted in experimental tools.

Funding: This work was supported by the Ministry of Trade, Industry and Energy of South Korea (MOTIE, Korea) under the industrial Technology Innovation Program (No. 10062550 and 10076343).

Conflicts of Interest: The authors declare no conflict of interest.

Abbreviations

BSTFA	N, O-bis(trimethylsilyl)trifluoroacetamide
CDW	cell dry weight
CYP	cytochrome P450 enzyme
CYP153AL.m	*Limnobacter* sp. MED 105 CYP153A
CYP153AM.aq	*Marinobacter aquaeolei* CYP153A
CYP153AA.d	*Alcanivorax dieselolei* CYP153A
CYP153AS.f	*Solimonas flava* CYP153A
DCA	dicarboxylic acid
DDA	dodecanoic acid
ω-OHDDA	omega-hydroxydodecanoic acid
OHFAs	omega-hydroxy fatty acids
TMS	trimethylsilyl

References

1. Soliday, C.L.; Kolattukudy, P.E. Biosynthesis of Cutin ω-Hydroxylation of Fatty Acids by a Microsomal Preparation from Germinating *Vicia faba*. *Plant Physiol.* **1977**, *59*, 1116–1121. [CrossRef] [PubMed]
2. Liu, C.; Liu, F.; Cai, J.; Xie, W.; Long, T.E.; Turner, S.R.; Lyons, A.; Gross, R.A. Polymers from Fatty Acids: Poly(ω-hydroxyl tetradecanoic acid) Synthesis and Physico-Mechanical Studies. *Biomacromolecules* **2011**, *12*, 3291–3298. [CrossRef] [PubMed]
3. Ebata, H.; Toshima, K.; Matsumura, S. Lipase-Catalyzed Synthesis and Properties of Poly[(12-hydroxydodecanoate)-*co*-(12-hydroxystearate)] Directed Towards Novel Green and Sustainable Elastomers. *Macromol. Biosci.* **2008**, *8*, 38–45. [CrossRef] [PubMed]
4. Abe, A.; Sugiyama, K. Growth inhibition and apoptosis induction of human melanoma cells by omega-hydroxy fatty acids. *Anti-Cancer Drugs* **2005**, *16*, 543–549. [CrossRef] [PubMed]
5. Bordeaux, M.; Galarneau, A.; Fajula, F.; Drone, J. A Regioselective Biocatalyst for Alkane Activation Under Mild Conditions. *Angew. Chem. Int. Ed. Engl.* **2011**, *50*, 2075–2079. [CrossRef] [PubMed]
6. Metzger, J.O.; Bornscheuer, U. Lipids as renewable resources: Current state of chemical and biotechnological conversion and diversification. *Appl. Microbiol. Biotechnol.* **2006**, *71*, 13–22. [CrossRef] [PubMed]
7. Yokota, T.; Watanabe, A. Process for Producing Omega-Hydroxy Fatty Acids. U.S. Patent 5191096, 2 March 1993.
8. Labinger, J.A. Selective alkane oxidation: Hot and cold approaches to a hot problem. *J. Mol. Catal. A Chem.* **2004**, *220*, 27–35. [CrossRef]
9. Lu, W.; Ness, J.E.; Xie, W.; Zhang, X.; Minshull, J.; Gross, R.A. Biosynthesis of Monomers for Plastics from Renewable Oils. *J. Am. Chem. Soc.* **2010**, *132*, 15451–15455. [CrossRef] [PubMed]
10. Scheps, D.; Honda Malca, S.; Richter, S.M.; Marisch, K.; Nestl, B.M.; Hauer, B. Synthesis of ω-hydroxy dodecanoic acid based on an engineered CYP153A fusion construct. *Microb. Boithechnol.* **2013**, *6*, 694–707. [CrossRef] [PubMed]
11. Funhoff, E.G.; Salzmann, J.; Bauer, U.; Witholt, B.; van Beilen, J.B. Hydroxylation and epoxidation reactions catalyzed by CYP153 enzymes. *Enzyme Microb. Technol.* **2007**, *40*, 806–812. [CrossRef]
12. Scheps, D.; Malca, S.H.; Hoffmann, H.; Nestl, B.M.; Hauer, B. Regioselective ω-hydroxylation of medium-chain n-alkanes and primary alcohols by CYP153 enzymes from *Mycobacterium marinum* and *Polaromonas* sp. strain JS666. *Org. Biomol. Chem.* **2011**, *9*, 6727–6733. [CrossRef] [PubMed]

13. Narhi, L.O.; Fulco, A.J. Identification and Characterization of Two Functional Domains in Cytochrome P-450BM-3, a Catalytically Self-sufficient Monooxygenase Induced by Barbiturates in Bacillus megaterium. *J. Biol. Chem* **1987**, *262*, 6683–6690. [PubMed]
14. Malca, S.H.; Scheps, D.; Kühnel, L.; Venegas-Venegas, E.; Seifert, A.; Nestl, B.M.; Hauer, B. Bacterial CYP153A monooxygenases for the synthesis of omega-hydroxylated fatty acids. *Chem. Commun.* **2012**, *48*, 5115–5117. [CrossRef] [PubMed]
15. Zhang, T.; Wei, D.-Q. Recent Progress on Structural Bioinformatics Research of Cytochrome P450 and Its Impact on Drug Discovery. *Adv. Struct. Eng.* **2015**, *827*, 327–339.
16. Darabi, M.; Seddigh, S.; Abarshahr, M. Structural, functional, and phylogenetic studies of cytochrome P450 (CYP) enzyme in seed plants by bioinformatics tools. *Caryologia* **2017**, *70*, 62–76. [CrossRef]
17. Fischer, M.; Knoll, M.; Sirim, D.; Wagner, F.; Funke, S.; Pleiss, J. The Cytochrome P450 Engineering Database: A navigation and prediction tool for the cytochrome P450 protein family. *Bioinformatics* **2007**, *23*, 2015–2017. [CrossRef]
18. Ma, Q.H.; Tian, B. Biochemical characterization of a cinnamoyl-CoA reductase from wheat. *Biol. Chem.* **2005**, *386*, 553–560. [CrossRef]
19. von Bühler, C.; Le-Huu, P.; Urlacher, V.B. Cluster Screening: An Effective Approach for Probing the Substrate Space of Uncharacterized Cytochrome P450s. *ChemBioChem* **2013**, *14*, 2189–2198. [CrossRef]
20. Donova, M.V.; Nikolayeva, V.M.; Dovbnya, D.V.; Gulevskaya, S.A.; Suzina, N.E. Methyl-beta-cyclodextrin alters growth, activity and cell envelope features of sterol-transforming mycobacteria. *Microbiology* **2007**, *153*, 1981–1992. [CrossRef]
21. Kühnel, K.; Maurer, S.C.; Galeyeva, Y.; Frey, W.; Laschat, S.; Urlacher, V.B. Hydroxylation of Dodecanoic Acid and (2*R*,4*R*,6*R*,8*R*)-Tetramethyldecanol on a Preparative Scale using an NADH-Dependent CYP102A1 Mutant. *Adv. Synth. Catal.* **2007**, *349*, 1451–1461. [CrossRef]
22. Rammler, D.H.; Zaffaroni, A. Biological implications of DMSO based on a review of its chemical properties. *Ann. N. Y. Acad. Sci.* **1967**, *141*, 13–23. [CrossRef] [PubMed]
23. Lundemo, M.T.; Notonier, S.; Striedner, G.; Hauer, B.; Woodley, J.M. Process limitations of a whole-cell P450 catalyzed reaction using a CYP153A-CPR fusion construct expressed in *Escherichia coli*. *Appl. Microbiol. Biotechnol.* **2016**, *100*, 1197–1208. [CrossRef] [PubMed]
24. Rimal, H.; Lee, S.W.; Lee, J.H.; Oh, T.J. Understanding of real alternative redox partner of *Streptomyces peucetius* DoxA: Prediction and validation using in silico and in vitro analyses. *Arch. Biochem. Biophys.* **2015**, *585*, 64–74. [CrossRef] [PubMed]
25. Lin, B.; Tao, Y. Whole-cell biocatalysts by design. *Microb. Cell Fact.* **2017**, *16*, 106. [CrossRef] [PubMed]
26. Wachtmeister, J.; Rother, D. Recent advances in whole cell biocatalysis techniques bridging from investigative to industrial scale. *Curr. Opin. Biotechnol.* **2016**, *42*, 169–177. [CrossRef] [PubMed]
27. Zhang, W.; Du, L.; Li, F.; Zhang, X.; Qu, Z.; Han, L.; Li, Z.; Sun, J.; Qi, F.; Yao, Q.; et al. Mechanistic Insights into Interactions between Bacterial Class I P450 Enzymes and Redox Partners. *ACS Catal.* **2018**, *8*, 9992–10003. [CrossRef]
28. Pandey, B.P.; Lee, N.; Choi, K.Y.; Kim, J.N.; Kim, E.J.; Kim, B.G. Identification of the specific electron transfer proteins, ferredoxin, and ferredoxin reductase, for CYP105D7 in *Streptomyces avermitilis* MA4680. *Appl. Microbiol. Biotechnol.* **2014**, *98*, 5009–5017. [CrossRef] [PubMed]
29. Szklarczyk, D.; Morris, J.H.; Cook, H.; Kuhn, M.; Wyder, S.; Simonovic, M.; Santos, A.; Doncheva, N.T.; Roth, A.; Bork, P.; et al. The STRING database in 2017: Quality-controlled protein-protein association networks, made broadly accessible. *Nucleic. Acids Res.* **2017**, *45*, D362–D368. [CrossRef] [PubMed]
30. Jung, E.; Park, B.G.; Ahsan, M.M.; Kim, J.; Yun, H.; Choi, K.Y.; Kim, B.G. Production of ω-hydroxy palmitic acid using CYP153A35 and comparison of cytochrome P450 electron transfer system in vivo. *Appl. Microbiol. Biotechnol.* **2016**, *100*, 10375–10384. [CrossRef] [PubMed]

31. Bernhardt, R.; Urlacher, V.B. Cytochromes P450 as promising catalysts for biotechnological application: Chances and limitations. *Appl. Microbiol. Biotechnol.* **2014**, *98*, 6185–6203. [CrossRef] [PubMed]
32. Bae, J.H.; Park, B.G.; Jung, E.; Lee, P.-G.; Kim, B.-G. *fadD* deletion and *fadL* overexpression in *Escherichia coli* increase hydroxy long-chain fatty acid productivity. *Appl. Microbiol. Biotechnol.* **2014**, *98*, 8917–8925. [CrossRef] [PubMed]

catalysts

MDPI

Article

Biosynthesis of Nylon 12 Monomer, ω-Aminododecanoic Acid Using Artificial Self-Sufficient P450, AlkJ and ω-TA

Md Murshidul Ahsan [1], Mahesh D. Patil [2], Hyunwoo Jeon [2], Sihyong Sung [2], Taeowan Chung [1] and Hyungdon Yun [2,*]

1 School of Biotechnology, Yeungnam University, Gyeongsan 38541, Korea; murshidvet@gmail.com (M.M.A.); twchung001@gmail.com (T.C.)

2 Department of Systems Biotechnology, Konkuk University, Seoul 05029, Korea; mahi1709@gmail.com (M.D.P.); hw5827@naver.com (H.J.); sihyong21@naver.com (S.S.)

* Correspondence: hyungdon@konkuk.ac.kr; Tel.: +82-2-450-0496; Fax: +82-2-450-0686

Received: 25 August 2018; Accepted: 15 September 2018; Published: 18 September 2018

Abstract: ω-Aminododecanoic acid is considered as one of the potential monomers of Nylon 12, a high-performance member of the bioplastic family. The biosynthesis of ω-aminododecanoic acid from renewable sources is an attractive process in the polymer industry. Here, we constructed three artificial self-sufficient P450s (ArtssP450s) using CYP153A13 from *Alcanivorax borkumensis* and cytochrome P450 reductase (CPR) domains of natural self-sufficient P450s (CYP102A1, CYP102A5, and 102D1). Among them, artificial self-sufficient P450 ($CYP_{153A13}BM3_{CPR}$) with CYP102A1 CPR showed the highest catalytically activity for dodecanoic acid (DDA) substrate. This form of ArtssP450 was further co-expressed with ω-TA from *Silicobacter pomeroyi* and AlkJ from *Pseudomonas putida* GPo1. This single-cell system was used for the biotransformation of dodecanoic acid (DDA) to ω-aminododecanoic acid (ω-AmDDA), wherein we could successfully biosynthesize 1.48 mM ω-AmDDA from 10 mM DDA substrate in a one-pot reaction. The productivity achieved in the present study was five times higher than that achieved in our previously reported multistep biosynthesis method (0.3 mM).

Keywords: artificial self-sufficient P450; bioplastics; dodecanoic acid; Nylon 12; ω-aminododecanoic acid

1. Introduction

Production of biochemicals from vegetable oil derivatives (e.g., free fatty acids (FFAs)) have drawn great attention as an alternative means to develop sustainable and green production processes, known as biorefinery [1–4]. Fatty acids and fatty acid derivatives are employed for the production of various polymer intermediates and precursors with broad commercial and pharmaceutical implications, including cosmetics, adhesives, lubricants, surfactants, coatings, biofuels, and anticancer agents [5–7]. In particular, long chain ω-hydroxy fatty acids (ω-OHFAs) contain two functional groups—hydroxy and carboxyl groups—at their ends, and they can therefore be further oxidized to fatty aldehydes, dicarboxylic acids, and oxo-fatty acids. Recently, it was reported that dodecanoic acid (DDA) can be efficiently transformed to ω-amino dodecanoic acid (ω-AmDDA) by utilizing different enzymes like Cytochrome P450 monooxygenases, alcohol dehydrogenases (AlkJ), alkane hydroxylase AlkBGT, Baeyer–Villiger monooxygenases (BVMOs), esterases, and ω-transaminases (ω-TAs) [8–10]. ω-AmDDA is a very important monomer for the synthesis of Nylon 12, along with other aliphatic polyamides. Owing to its extraordinary heat-, abrasion-, chemical-, UV-, and scratch-resistance capabilities, Nylon 12 is frequently used as a coating agent on fuel and braking systems in most passenger cars [9,10]. Usually, Nylon 12 is industrially produced from the monomer ω-laurolactam through ring opening polymerization, the chemical synthesis of which is initiated by the trimerization of 1,3-butadiene originating from steam cracking (200–300 °C) of crude oil [9,10]. However, this method has raised many environmental concerns. Considering the "green" alternatives, synthesis of ω-amino dodecanoic acid from DDA is a potential approach to address this issue. Moreover, DDA shows relatively high solubility in water compared to other medium- to long-chain fatty acid substrates. Earlier, our group reported that CYP153A13 from *Alcanivorax borkumensis* SK2 efficiently transformed DDA to ω-hydroxy dodecanoic acid (ω-OHDDA) in an in vivo reaction [10–12].

Multistep biocatalytic synthesis often negatively impacts practical industrial applications [13,14]. With increasing number of steps in biocatalytic reactions, the number of generated by-products also increases, many of which may be toxic to cells and make it difficult to separate desirable products. Moreover, the expected expression of multiple recombinant proteins and efficient bioconversion from multistep enzymatic reaction itself is a challenging task [10,15]. Hence, the use of lesser number of recombinant proteins is of immense importance for the biosynthesis of industrially important monomers in a one-pot reaction [16]. Recently, we reported the biosynthesis of ω-AmDDA using the cascade of novel CYP153A, AlkJ, and ω-TA enzymes [10]. In that study, we could achieve more than 87% conversion of ω-OHDDA to ω-AmDDA in a sequential biocatalytic reaction. However, the conversion was negligible in a one-pot reaction (Figure 1a). We therefore envisioned to biosynthesize ω-AmDDA in a single-cell, one-pot cascade reaction. In the present study, we changed the reaction pattern of CYP153A13 from a three-protein system to a single-protein system for ω-hydroxylation of DDA. As CYP153A belongs to the class I CYPs, three component systems are required for the activation of the enzyme (Figure S1). Two additional proteins—putidaredoxin reductase (CamA) and putidaredoxin (CamB)—are related to electron transfer system [10–12]. Self-sufficient P450s, belonging to Class VIII CYPs, are catalytically self-sufficient monooxygenase that contain a heme domain (P450 domain) and a flavin reductase domain (cytochrome P450 reductase (CPR) domain) on a single polypeptide chain ([11,12,17–19]; Figure S1). Recently published findings have shown that the fusion construct of CYP153A with the CPR domain of natural self-sufficient P450s results in the generation of the active form of the enzyme and successfully catalyzes a biocatalytic reaction [11,12,17]. In the present study, we also co-expressed the artificial self-sufficient P450s (ArtssP450s) with an alcohol dehydrogenase (AlkJ) from *Pseudomonas putida* and an omega transaminase (ω-TA) from *Silicobacter pomeroyi* for the synthesis of ω-AmDDA from DDA in a one-pot single-cell reaction (Figure 1b).

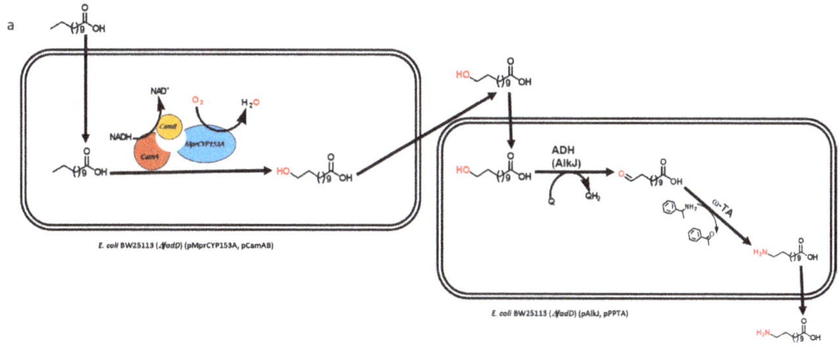

Figure 1. *Cont.*

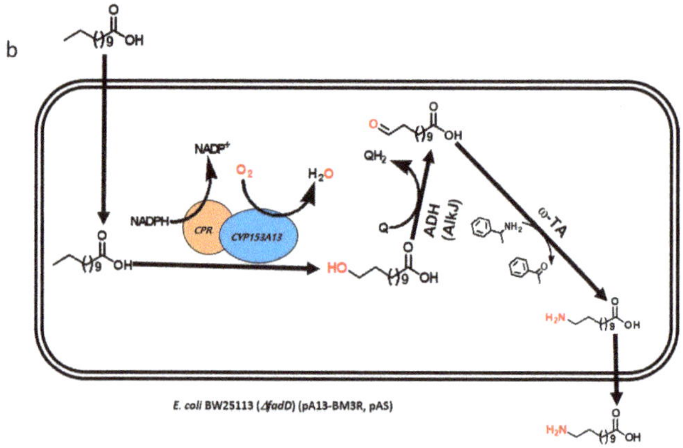

Figure 1. Schematic diagram showing the biosynthesis of ω-amino dodecanoic acid (ω-AmDDA) from dodecanoic acid (DDA) using (**a**) three-protein system (previous study [10]) and (**b**) a single-protein system (this study) for ω-hydroxylation of DDA.

2. Results and Discussion

2.1. Construction and Expression of Artificial Self-Sufficient P450s (ArtssP450s)

CYP153A13 of *Alcanivorax borkumensis* (Abk) was cloned into *Nde*I and *Hind*III restriction sites of pET24ma plasmid vector. The *Hind*III restriction site was chosen to make it a common restriction site to fuse CYP153A13 with CPR domains of CYP102A1, CYP102A5, and CYP102D1 (Figures S2 and S3). After successful cloning of CYP153A13, CYP153A13-containing plasmid was used again to make artificial self-sufficient fusion construct of CYP153A13 and CPRs in the same vector. All three CPR domains were amplified; RE digested and ligated into *Hind*III and *Not*I sites for CYP102A5 and CYP102D1 and into *Hind*III and *Xho*I sites for CYP102A1. To remove *Hind*III sites within the CPR domain of CYP102A1, the 243th alanine residue DNA codon was changed from GCT to GCA by utilizing overlap extension PCR method. Next, all the ArtssP450s were confirmed by sequencing and nomenclatured newly as $Abk_{CYP153A13}BM3_{CPR}$ (for CYP102A1 CPR), $Abk_{CYP153A13}Bc21_{CPR}$ (for CYP102A5 CPR), and $Abk_{CYP153A13}Sav_{CPR}$ (for CYP102D1 CPR) on the basis of their gene source (See Materials and Method). All three different CPRs were chosen from three different natural self-sufficient P450s based on previous reports. Among them, BM3 is a well reported self-sufficient P450, and its CPR domain is frequently used for making artificial constructs [11,12,17]. The self-sufficient P450 of Bc21 (CYP102A5) has been reported to possess the highest turnover rates among P450s [19]. In addition, a codon-optimized Sav (CYP102D1) has been reported to show similar catalytic activity for saturated and unsaturated fatty acids as that of other members of the CYP102A family [20].

After successful construction of ArtssP450s, the expression of all recombinant proteins was tested at three different temperature (20, 25, and 30 °C) and three different isopropyl-thio-β-D-galactopyranoside (IPTG) concentrations—0.01, 0.1, and 0.5 mM—in terrific broth (TB) bacterial expression media. All three proteins showed better expression at 20 °C with 0.1 mM IPTG concentration. The protein expression was confirmed through SDS-PAGE gel analysis (Figure S4). SDS-PAGE analysis demonstrated clear protein expression band at expected size (~120 kDa). Additionally, the conventional carbon monoxide (CO)-differential spectral assay with sodium dithionite was used in cell-free extracts as a standard method to determine the concentration of active P450. The characteristic absorbance at 450 nm for CO complexed to the reduced ferrous state of P450 was observed (Figure 2), corroborating their active expression. The amount of active P450 was 1.51, 2.25, and 0.75 nmol mL^{-1} for $Abk_{CYP153A13}BM3_{CPR}$, $Abk_{CYP153A13}Bc21_{CPR}$, and $Abk_{CYP153A13}Sav_{CPR}$, respectively. These results were well correlated with the reported titer range of the active P450 expression (0.6–2.33 nmol mL^{-1}) in *E. coli* system [18–21].

It has also been reported that, in many cases, the co-expression of CPRs lead to formation of high titers of active P450 [21]. In the CO-binding analysis, $Abk_{CYP153A13}Bc21_{CPR}$ construct seemed to be the most active, being nearly three times higher than $Abk_{CYP153A13}Sav_{CPR}$ and two times higher than $Abk_{CYP153A13}BM3_{CPR}$.

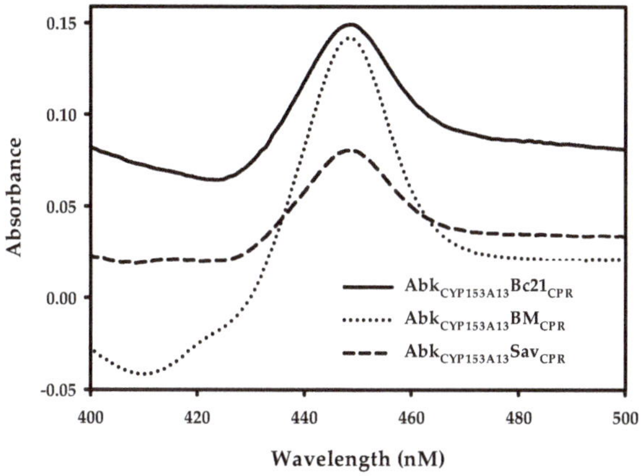

Figure 2. Carbon monoxide (CO)-binding analysis of ArtssP450s. $Abk_{CYP153A13}$, *Alcanivorax borkumensis* CYP153A13; $Abk_{CYP153A13}BM3_{CPR}$, $Abk_{CYP153A13}$ fused with BM3 (CYP102A1) cytochrome P450 reductase (CPR); $Abk_{CYP153A13}Bc21_{CPR}$, $Abk_{CYP153A13}$ fused with *Bacillus cereus* (CYP102A5) CPR; and $Abk_{CYP153A13}Sav_{CPR}$, $Abk_{CYP153A13}$ fused with *Streptomyces avermitilis* (CYP102D1) CPR.

2.2. Production of ω-OHDDA Using ArtssP450s

It has been reported that $Abk_{CYP153A13}$ exhibits better ω-OHDDA productivity among CYP153A13, CYP153A33, and CYP153A35 in an in vivo biocatalytic reaction employing three protein system [11]. In order to know the ω-OHDDA productivity of newly constructed ArtssP450s in a single protein system, all three enzymes were tested with 10 mM DDA substrate in a whole-cell reaction. Additionally, a control whole-cell reaction was performed using $Abk_{CYP153A13}$ in CamAB system. The initial specific rates of ω-hydroxylation were 0.10, 0.09, and 0.037 mmol/g_{CDW}/h for $Abk_{CYP153A13}BM3_{CPR}$, $Abk_{CYP153A13}Bc21_{CPR}$, and $Abk_{CYP153A13}Sav_{CPR}$, respectively. Accordingly, 24-h reactions resulted in 3.95, 2.64, and 1.40 mM ω-OHDDA, respectively (Figure 3). Although the initial specific rates of ω-hydroxylation and production of ω-OHDDA after 24 h were better for $Abk_{CYP153A13}$ in CamAB system (0.16 mmol/g_{CDW}/h and 6.21 mM, respectively) (Data not shown), we used single-protein system due to its obvious advantages. Among ArtssP450s, the initial specific rate of $Abk_{CYP153A13}BM3_{CPR}$ and $Abk_{CYP153A13}Bc21_{CPR}$ were very close. Unexpectedly, the highest productivity (3.95 mM) was observed in $Abk_{CYP153A13}BM3_{CPR}$. As $Abk_{CYP153A13}Bc21_{CPR}$ showed the highest amount of active P450 in CO-binding spectrum analysis, it was estimated that the productivity would also be the highest for $Abk_{CYP153A13}Bc21_{CPR}$ (Figure 2). However, the published findings have demonstrated that BM3 natural self-sufficient P450 (CYP102A1) shows the highest catalytic activity among all P450s [20]. Moreover, CamAB system showed better ω-OHDDA productivity than ArtssP450 systems developed in the present study. Despite these findings, the use of ArtssP450s for ω-OHDDA biosynthesis was of utmost importance in consideration of the one-pot biocatalytic cascade reaction in order to get the final product of this study, i.e., ω-AmDDA (Figure 1b). As $Abk_{CYP153A13}BM3_{CPR}$ system showed the highest ω-OHDDA productivity among all the ArtssP450 constructs, it was further used to produce ω-AmDDA from DDA (Figure 1).

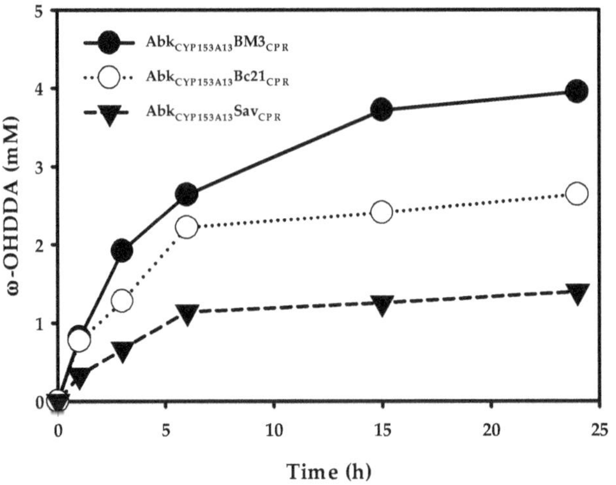

Figure 3. ω-hydroxy dodecanoic acid (ω-OHDDA) production by ArtssP450s. Reaction conditions: substrate concentration, 10 mM; volume, 10 mL in 100 mL flask; temperature, 30 °C; cell type, BW25113 (ΔfadD, DE3) containing pA13-BM3R, pA13-Bc21R, pA13-SavR; cell OD_{600}: 30 and buffer, 100 mM potassium phosphate buffer (pH 7.5) with 1% (w/v) glucose.

2.3. *Biosysnthesis of ω-AmDDA in One-Pot Reaction*

2.3.1. Co-expression of $Abk_{CYP153A13}BM3_{CPR}$ with AlkJ and Sp ω-TA

The production of ω-AmDDA from DDA in a one-pot reaction requires co-expression of $Abk_{CYP153A13}BM3_{CPR}$ with AlkJ and ω-TA. In this reaction, the DDA substrate is initially converted to ω-OHDDA by $Abk_{CYP153A13}BM3_{CPR}$. AlkJ then catalyzes the transformation of ω-OHDDA to ω-OxoDDA, which is further transaminated by ω-TA to yield ω-AmDDA. Therefore, $Abk_{CYP153A13}BM3_{CPR}$ was co-expressed with AlkJ and ω-TA at 20 °C using 0.1 mM IPTG induction in TB media. The expression of all three proteins in the soluble form was confirmed by SDS-PAGE gel analysis (Figure 4). Next, the co-expressed cells were used in a whole-cell reaction to produce ω-AmDDA from DDA in a one-pot reaction. The reaction was initiated by adding 5 mM DDA (stock in dimethyl sulfoxide (DMSO), 5% (v/v) final concentration), 0.1 mM pyridoxal-5′-phosphate (PLP), and 40 mM benzylamine using co-expressed cells (0.3 g_{DCW}/mL) in 100 mM Tris-HCl buffer. The reaction mixture was incubated at 35 °C and 200 rpm. After 6 h, 0.8 and 0.3 mM ω-OHDDA and ω-AmDDA, respectively, were detected (data not shown). This can be explained by the fact that $Abk_{CYP153A13}BM3_{CPR}$ transformed DDA to ω-OHDDA. AlkJ converted this ω-OHDDA to ω-OxoDDA, which was utilized by Sp ω-TA to form ω-AmDDA. Albeit with low conversion percentage, the successful biotransformation of DDA to ω-AmDDA encouraged us to further improve the productivity by optimizing different parameters.

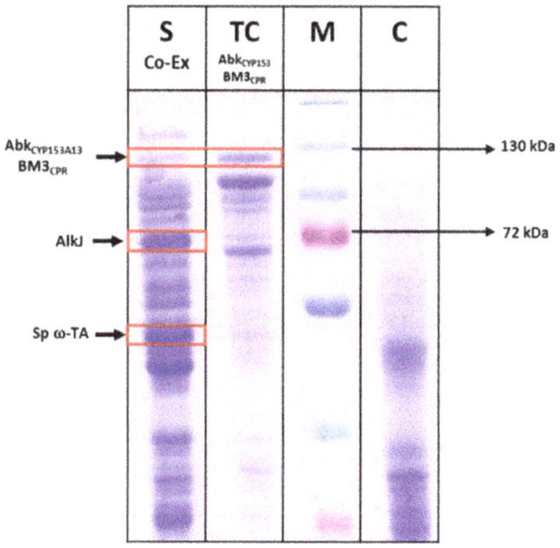

Figure 4. SDS-PAGE analysis of recombinant *E. coli* expressing $Abk_{CYP153A13}BM3_{CPR}$ (~120 kDa), AlkJ (~61.2 kDa), and Sp ω-TA (~52 kDa) co-expression. C: control (*E. coli* BW25113(DE3) ΔfadD cells without any plasmid were used as control); M: Marker; TC: total cells; SP: soluble proteins. Protein expression was carried out using 0.1 mM isopropyl-thio-β-D-galactopyranoside (IPTG) at 20 °C and 200 rpm for 16 h.

2.3.2. Effect of Benzylamine (BA) on ω-OHDDA Productivity

The selection of a suitable amino donor is of immense importance to enhance the yield of ω-TA reaction. Organic syntheses by ω-TAs may suffer from unfavorable reaction equilibrium due to insufficient amount of amino donor [16]. To maximize the productivity of ω-TAs, shifting the equilibrium to the product using higher concentration of amino donor is extremely important [22]. Previously, while synthesizing ω-AmDDA from ω-OHDDA using AlkJ and Sp ω-TA, we found benzylamine (BA) as the best amino donor [10]. It has been reported that the performance of P450 employed in the amination cascade is influenced by the amino donor used [23]. Therefore, it was presumed that the productivity of ω-OHDDA might be hindered due to excess amount of BA as amino donor for ω-TA in the one-pot reaction demonstrated herein. Before optimizing ω-AmDDA productivity, the biosynthesis of ω-OHDDA by $Abk_{CYP153A13}BM3_{CPR}$ was tested in 100 mM (pH 7.5) potassium phosphate buffer and Tris-HCl buffer using five different concentrations of BA. It was demonstrated that 0–5 mM BA had negligible effect on the production of ω-OHDDA in both types of buffers, but the productivity was better in the phosphate buffer. Moreover, 50 mM BA had remarkable effect on ω-OHDDA in both the cases (Figure 5). This could be explained by the interception of the hydrogen-borrowing step ($NADP^+$/NADPH) of the cofactors in endogenous *E. coli* enzyme systems, which may lead to imbalanced regeneration of the cofactors [23].

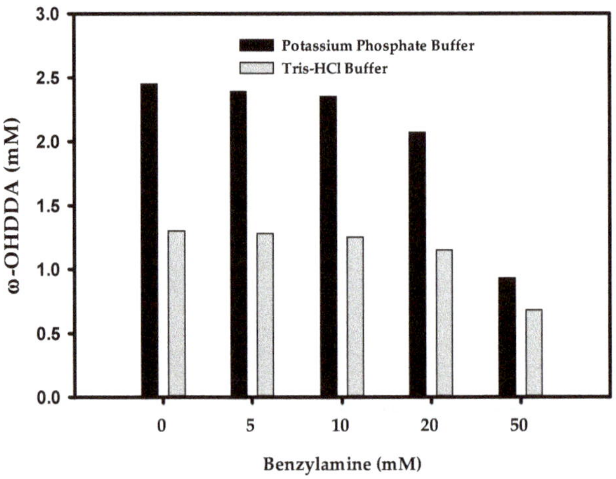

Figure 5. Effect of BA on ω-OHDDA productivity. Reaction conditions: substrate concentration: 10 mM; volume: 10 mL in 100 mL flask; temperature: 35 °C; cell type: BW25113 (ΔfadD, DE3) containing pA13-BM3R; cell OD_{600}: 30 and buffer, 100 mM (pH 7.5) with 1% (w/v) glucose; reaction time: 5 h.

We have previously reported that ω-TA enzyme is more active in Tris-HCl buffer with alkaline pH [10]. In the present study, it was revealed that ArtssP450s did not show any activity for ω-hydroxylation of DDA in an alkali pH (>8) (data not shown). Therefore, further optimization of ω-AmDDA production in a one-pot reaction was continued in 100 mM Tris-HCl buffer (pH 7.5), which seemed to be a convenient reaction condition for all the enzymes.

2.3.3. Optimization of ω-AmDDA Biosynthesis

Finally, in order to check the optimum productivity of ω-AmDDA in a one-pot reaction, cells (0.3 g_{CDW}/mL) co-expressing $Abk_{CYP153A13}BM3_{CPR}$, AlkJ, and Sp ω-TA were used in various concentration of BA (0–50 mM) in a whole-cell reaction. The reaction was initiated by adding 10 mM DDA (stock in DMSO, 5% (v/v) final concentration), 0.1 mM PLP, and 20 mM benzylamine. The reactants were incubated at 35 °C and 200 rpm. After 3 h of whole-cell reaction, the amount ω-AmDDA was 0.03, 0.07, 0.15, and 0.55 mM when the BA was used in the reaction was 0, 5, 10, and 20 mM, respectively. In particular, it was observed that the amount of ω-OHDDA after 3 h was 1.21, 1.10, 0.95, and 0.82 mM, respectively, with the subsequent concentration of used BA. The production of both ω-AmDDA and ω-OHDDA increased up to 5 h, and the amount of ω-AmDDA was 0.10, 0.20, 0.36, and 0.84 mM, respectively, whereas the amount of ω-OHDDA was 2.35, 2.22, 1.98, and 1.64 mM, respectively, for the corresponding amount of BA used. Interestingly, the biosynthesis of ω-AmDDA was increasing over the 15 h reaction period (Figure 6), while the production of ω-OHDDA dropped after 5 h. The whole-cell reaction yielded 0.17, 0.33, 1.03, and 1.17 mM of ω-AmDDA and 1.87, 1.61, 1.12, and 0.56 mM of ω-OHDDA after 15 h when 0, 5, 10, and 20 mM BA was used, respectively (Figure 6). The present study demonstrated that the use of 20 mM BA as an amino donor in a one-pot whole-cell reaction gave the best productivity for the synthesis of ω-AmDDA from DDA substrate. Moreover, the best whole-cell reaction condition was continued up to 24 h. It was found that 1.48 mM ω-AmDDA was biosynthesized from 10 mM DDA, which is five times more than our last report [10]. Whole-cell reaction of *E. coli* cells expressing MprCYP153A/CamAB for ω-hydroxylation of DDA and *E. coli* cells expressing AlkJ/mll1207 ω-TA resulted in the generation of ~0.3 mM ω-AmDDA from DDA in a 24 h reaction [10].

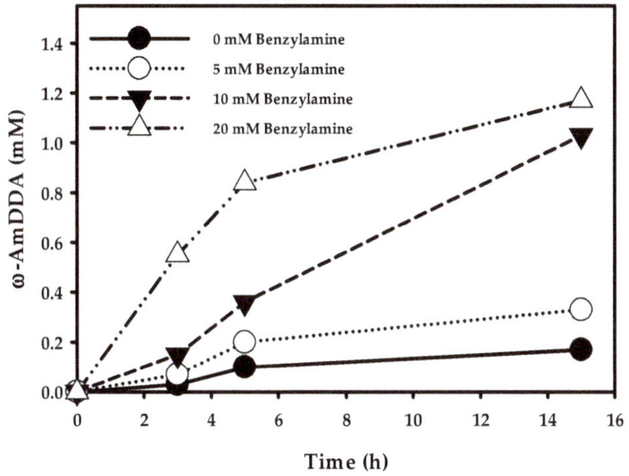

Figure 6. Biosysnthesis of ω-AmDDA from DDA using co-expressed $Abk_{CYP153A13}BM3_{CPR}$, AlkJ, and Sp ω-TA enzymes. Reaction conditions: substrate concentration: 10 mM; volume: 10 mL in 100 mL flask; temperature: 35 °C; cell type, BW25113 (ΔfadD, DE3) containing pA13-BM3R and pAS; cell OD_{600}: 30; cofactor: 0.1 mM pyridoxal-5′-phosphate (PLP); amino donor: 20 mM benzylamine and buffer, 100 mM Tris-HCl buffer (pH 7.5) with 1% (w/v) glucose.

It has been reported that AlkJ from *Pseudomonas putida* GPo1 shows thermal denaturation at moderate temperatures, and T_{stab}—the temperature where the enzyme shows 50% residual activity—was found to be 34 °C [24]. Thus, biotransformations carried out at lower temperatures could have resulted in improved overall yield in the present study. However, the major challenge in one-pot biotransformations employing more than one protein—as demonstrated in the present study—is the satisfactory performance of all proteins involved. In these multiprotein biotransformations, not every protein can be optimally functional in the given set of reaction parameters. Hence, there is always a compromise involved in yield of final product and the optimal performance of the proteins employed in the multiprotein biotransformations. In the addition, the final outcome of multiprotein biotransformations is severely compromised when proteins exhibit differential performance at specific reaction conditions, such as temperature and/or reaction pH. It should also be noted that whole-cell biocatalysts are more stable than their purified counterparts. Nevertheless, in our previously reported study [10], it was observed that all of the initially added ω-OHDDA (5 mM) was consumed within 1 h when whole-cell reaction was performed at 35 °C, implying that AlkJ activity was enough to carry out transformation of ω-OHDDA.

It has been well reported that the *fadL* is required for the transport of long chain fatty acids in *E. coli* [25,26]. Although overexpression of *fadL* was one of the possible approaches to improve the mass transfer of DDA substrate across cell membrane in the present study, it could have adversely affected the expression of other proteins. As expression of a recombinant protein may impart a metabolic burden on the host microorganism [27], the primary object of multiprotein, one-pot biotransformations should be the use of lesser number of recombinant proteins to achieve the synthesis of a desired product. Recently, Janßen et al. [28] reported the improved transfer of long-chain fatty acids in *fadL*-overexpressed *E. coli* cells. However, a recently reported study by our group [10] demonstrated that *E. coli* cells can satisfactorily uptake DDA without overexpression of *fadL*.

3. Materials and Methods

3.1. Chemicals and Media

All chemicals such as DDA, ω-OHDDA, ω-AmDDA, dimethyl sulfoxide (DMSO), IPTG, 5-aminolevulinic acid (5-ALA), *N,O*-Bis(trimethylsilyl)-trifluoroacetamide (BSTFA), *N*-Methyl-*N*-(trimethylsily) trifluoroacetamide (MSTFA), PLP, benzylamine, and pyridine were purchased from Sigma-Aldrich (St. Louis, MO, USA). Chloroform was obtained from Junsei (Tokyo, Japan). Bacteriological agar, Luria-Bertani (LB) broth, and terrific broth (TB) media were bought from BD Difco (Franklin Lakes, NJ, USA). All chemicals used in this study were of analytical grade.

3.2. Artificial Self-Sufficient P450 Construction and Gene Manipulation

All the bacterial strains and plasmid vectors used in this study are listed in Table 1. *Alcanivorax borkumensis* SK2 (KACC no. 12864) was procured from the Korean Agricultural Culture Collection (KACC, Jeonju, Korea). Genomic DNA was extracted from lyophilized commercial cell stock using a kit (G-Spin™ Genomic DNA Extraction Kit for bacteria (iNtRON Biotechnology, Suwon, Korea)). The gene encoding CYP153A13 (GI: CAL15649.1) from *A. borkumensis* SK2 was amplified by PCR with oligonucleotides (Table S1). After restriction digestion and ligation with T4 DNA ligase, the plasmid was utilized to transform competent *E. coli* DH5α cells. Successful cloning was confirmed by DNA sequencing. Construction of artificial self-sufficient fusion proteins (pA13-BM3R, pA13-Bc21R, and pA13-SavR) with CYP153A13 and the reductase domains (CPRs) of self-sufficient P450s (CYP102A1 from *Bacillus megaterium*, CYP102A5 from *Bacillus cereus*, and CYP102D1 from *Streptomyces avermitilis*) were performed following previously described procedures [18]. CPR domains of self-sufficient P450s were analyzed using online bioinformatics tools Pfam 31.0 site (http://pfam.xfam.org/). Gene synthesis and codon optimization for *E. coli* codon preferences was performed by Cosmo Genetech (Cosmo Genetech, Seoul, Korea).

Table 1. Plasmids and strains used in this study.

Plasmids/Strains	Description	Reference
Plasmids		
pET24ma	P15A ori lacI T7 promoter, KmR	[10]
pETDuet-1	pBR322 ori lacI T7 promoter, AmpR	Novagen
pA13	pET24ma encoding CYP153A13 ($Abk_{CYP153A13}$)	[11]
pA13-BM3R	pET24ma encoding $Abk_{CYP153A13}BM3_{CPR}$	This study
pA13-Bc21R	pET24ma encoding $Abk_{CYP153A13}Bc21_{CPR}$	This study
pA13-SavR	pET24ma encoding $Abk_{CYP153A13}Sav_{CPR}$	This study
pAlkJ	pET24ma encoding AlkJ	[10]
pPPTA	pETDuet-1 encoding Sp ω-TA	[10]
pAS	pETDuet-1 encoding Sp ω-TA and AlkJ	This study
***E. coli* strains**		
DH5α	F^-, *endA1*, *gln*V44, *thi*-1, *rec*A1, *rel*A1, *gyrA*96 *deo*R, *supE*44, Φ80d*lacZ*ΔM15 Δ(*lacZYA*$^-$-*arg*F)-U169, *hsd*R17(r_K^- m_K^+), λ^-	[29]
BW25113(DE3)	rrnB3 ΔlacZ4787 hsdR514 Δ(araBAD)567 Δ(rhaBAD)568 rph-1 λ(DE3)	[11]
DL	BW25113(DE3) ΔfadD	[10]
A13	DL carrying pA13	This study
A13-BM3R	DL carrying pA13-BM3R	This study
A13-Bc21R	DL carrying pA13-Bc21R	This study
A13-BSavR	DL carrying pA13-BSavR	This study
AlkJ-PPTA	DL carrying pAlkJ and pPPTA	This study
BM-AS	DL carrying pA13-BM3R and pAS	This study

3.3. Expression of Enzymes

E. coli BW25113 (DE3) ΔfadD strain [10] was utilized for the biotransformation studies, wherein fatty acid degrading β-oxidation pathway was blocked. Plasmid DNA were transformed into host strains using standard heat shock method. Transformants were selected based on their antibiotic resistance [30]. Transformants were grown overnight at 37 °C in 10 mL LB medium containing 50 μg/mL of kanamycin (for pA13-BM3R, for pA13-Bc21R, and for pA13-SavR) and/or 100 μg/mL ampicillin (for pAS, pAlkJ, and pPPTA). The seed cultures were added to expression flask into 1/3 ratio. All artificial self-sufficient P450 protein expression were carried out in 200 mL of Terrific Broth in a 1 L flask. They were cultured at 37 °C until cell concentration reached an OD_{600} of 0.6–0.8. The induction was performed by adding 0.1 mM IPTG, 0.5 mM 5-ALA as heme precursor, and 0.1 mM $FeSO_4$, and cells were grown at 20 °C for 16 h. The expression of Sp ω-TA/AlkJ with or without artificial self-sufficient P450 was carried out in 200 mL of LB medium in 1 L flasks. Induction was performed by adding 0.1 mM IPTG, and cells were allowed to grow for 16 h at 20 °C.

After 16 h, cells were harvested by centrifugation (4000 rpm, 20 min, 4 °C), washed with phosphate-buffered saline (PBS), and resuspended in 100 mM potassium phosphate buffer (pH 7.5) containing 1% (w/v) glucose.

3.4. CO-Binding Assay and Gel Electrophoresis

The expressed proteins were subjected to SDS-PAGE and spectrophotometric analysis in order to measure the CO-binding activity. UV absorption spectra of CO-bound artificial self-sufficient P450 proteins after sodium dithionite reduction were measured by Cary 100 UV-Vis spectrometry (Agilent Technologies, CA, USA) by wavelength scan from 400 to 500 nm. The concentration of P450 was measured by CO-binding affinity using an extinction coefficient of 91.9 mM^{-1} cm^{-1} at 450 nm. SDS-PAGE (Bio-Rad Laboratories, Inc, Hercules, CA, USA.) was carried out with 12% polyacrylamide gel as described elsewhere [31]. Proteins were visualized by Coomassie® brilliant blue R-250 staining.

3.5. Biotransformation

Biotransformation of FFAs to ω-OHFAs was performed according to the previously reported method [10]. For the biotransformation of ω-OHDDA acid to ω-AmDDA, *E. coli* BM-AS (Table 1) cotransformed strain was grown and harvested as described above. The cell pellets were resuspended in 100 mM Tris-HCl buffer (pH 7.5 and 8.0), and biotransformation (300 mg_{CDW}/mL) was performed at 35 °C at 150 rpm in a shaking incubator following our previous published protocols [10]. The whole-cell reaction was started by the addition of ω-OHDDA (stock in DMSO, 5% (v/v) final concentration).

To assess the biotransformation of DDA to ω-OHDDA by ArtssP450s, whole cell reaction was performed according to the previously reported method [10]. Briefly, induced cells expressing ArtssP450s in a single protein system were grown for 16 h at 20 °C. Cells were harvested by centrifugation (4000 rpm, 20 min, 4 °C) and washed with PBS. To initiate the biotransformation, resting cells (300 mg_{CDW}/mL), DDA (10 mM; stock in DMSO, 5% (v/v) final concentration) and buffer (Potassium phosphate; 100 mM, pH 7.5) with 1% (w/v) glucose were added to a 100 mL flask to a final volume of 10 mL. This reaction mixture was incubated at 30 °C and 200 rpm for 24 h. For one-pot biotransformation of DDA to ω-AmDDA, *E. coli* BM-AS (Table 1) cotransformed strain was grown and harvested as described above. Biotransformation was initiated by adding resting cells (300 mg_{CDW}/mL), DDA (10 mM; stock in DMSO, 5% (v/v) final concentration) and buffer (Tris-HCl; 100 mM, pH 7.5 and 8.0) to a 100 mL flask to a final volume of 10 mL. This reaction mixture was incubated at 35 °C and 150 rpm for 24 h.

3.6. Product Identification and Quantification

Whole-cell reactions of both the biotransformations, i.e., DDA to ω-OHDDA and DDA to ω-AmDDA, were stopped and acidified with 6 M HCl to pH 2.0. The substrates and products of DDA biotransformation were extracted with an equal volume of chloroform (200 μL) after vigorous vortexing for 1 min (Table S2). After centrifugation, the extracted sample in chloroform (bottom layer) was transferred to a new microcentrifuge tube for derivatization. These samples were transformed to their trimethylsilyl (TMS) derivatives by incubation with an excess of BSTFA at 50 °C for 20 min. In the case of biotransformation generating ω-AmDDA, the acidified reaction samples (with 6 M HCl) were centrifuged, and the supernatant was dried in a vacuum concentrator. After complete drying, the supernatant was dissolved again to the original volume with pyridine. The sample was then mixed with an equal volume of MSTFA by vigorous vortexing for 1 min and converted to the TMS derivatives by incubation at 50 °C for 20 min.

Analytical conditions for free fatty acids and their derivatives using gas chromatography are well established and reported in our previously published reports [4,7,10–12,20]. Quantitative analysis was performed using a gas chromatography instrument with a flame ionization detector (GC/FID) fitted with an AOC-20i series auto sampler injector (GC 2010 plus Series, Shimadzu Scientific Instruments, Kyoto 604-8511, Japan). Two-microliter samples were inserted by split mode (split ratio 20:1) and examined by means of a nonpolar capillary column (5% phenyl methyl siloxane capillary 30 m × 320 μm i.d., 0.25-μm film thickness, HP-5). The oven temperature program for fatty acid analysis was 50 °C for 1 min, an increase by 10 °C/min to 250 °C, and hold for 10 min. The inlet temperature was 250 °C, and the detector temperature was 280 °C. The flow rate of the carrier gas (He) was 1 mL/min, and the flow rates of H_2, air, and He in FID were 45 mL/min, 400 mL/min, and 20 mL/min, respectively. For the analysis of DDA and ω-OHDDA, commercial decanoic acid was used as internal standard. Internal standard was added after stopping the reaction and before the centrifugation step of chloroform extraction. For ω-AmDDA analysis, the oven temperature program was modified. The initial oven temperature was 90 °C, which was then increased by 15 °C/min to 250 °C, holding at this temperature for 5 min. Internal standard was not used. As chloroform extraction was not carried out in this analysis, there was presumably no loss of product and substrate as in the chloroform extraction. Pyridine was used for dilution of the product after vacuum evaporation of buffer–water, and dilution factor was considered while quantifying the product yield. Products of the biotransformations were confirmed by comparing the GC chromatograms with authentic references (Figures S5–S8).

4. Conclusions

In summary, we constructed artificial self-sufficient P40s by fusing CYP153A13 with various CPR domains—CYP102A1, CYP102A5, and CYP102D1—and tested them for the hydroxylation of DDA to ω-OHDDA. Furthermore, a single-cell system was generated for the biosynthesis of Nylon 12 monomer ω-AmDDA in a one-pot reaction by co-expressing the best artificial self-sufficient P40 generated (i.e., $Abk_{CYP153A13}BM3_{CPR}$), AlkJ, and Sp ω-TA enzymes. This highly efficient biocatalytic cascade produced ω-AmDDA in an easy and cost-effective way. It is worth emphasizing that the productivity of the biocatalytic cascade reported herein was five times higher than that of our previously reported sequential cascade method [10].

Supplementary Materials: The following are available online at http://www.mdpi.com/2073-4344/8/9/400/s1, Figure S1: Electron transportation system bacterial CYP. A. class I, and B. class VIII; Figure S2: Co-expression plasmid diagram for the synthesis of ω-AmDDA; Figure S3: Schematic diagram of ArtssP450s construction strategy using restriction sites into pET24ma vector; Figure S4: SDS-PAGE analysis of recombinant *E. coli* expressing ArtssP450s (~120 kDa). C: control; M: marker; TC: total cells; SP: soluble proteins. Protein expression was carried out using 0.1 mM IPTG, 0.1 mM $FeSO_4$, and 0.5 mM 5-ALA at 20 °C and 170 rpm; Figure S5: GC chromatogram of the chemical standard of DDA. The peak at 13.5 min is the internal standard decanoic acid; Figure S6: GC chromatogram of the chemical standard of ω-AmDDA. The peak at 13.5 min is the internal standard decanoic acid; Figure S7: GC chromatogram of authentic ω-AmDDA and reaction mixture of DDA to ω-AmDDA; Figure S8: GC/MS analysis of authentic ω-AmDDA and reaction mixture of DDA to ω-AmDDA; Table S1: Primers used for the construction of ArtssP450s; Table S2: Retention time of the substrates and products by gas chromatography.

Author Contributions: H.Y., and T.C. designed the experiments of the project. M.M.A. carried out the research works as part of his PhD project. H.Y. supervised the whole studies reported in the manuscript. M.M.A. wrote the manuscript. M.D.P. revised this manuscript. H.J., and S.S. assisted in experimental tools.

Funding: This research was funded by the Ministry of Trade, Industry and Energy of South Korea (MOTIE, Korea) under the Industrial Technology Innovation Program, Grant Nos. 10062550 and 10076343.

Conflicts of Interest: The authors declare no financial or commercial conflict of interest.

References

1. Steen, E.J.; Kang, Y.; Bokinsky, G.; Hu, Z.; Schirmer, A.; McClure, A.; del Cardayre, S.B.; Keasling, J.D. Microbial production of fatty-acid-derived fuels and chemicals from plant biomass. *Nature* **2010**, *463*, 559–562. [CrossRef] [PubMed]
2. Chung, H.; Yang, J.E.; Ha, J.Y.; Chae, T.U.; Shin, J.H.; Gustavsson, M.; Lee, S.Y. Bio-based production of monomers and polymers by metabolically engineered microorganisms. *Curr. Opin. Biotechnol.* **2015**, *36*, 73. [CrossRef] [PubMed]
3. Zorn, K.; Oroz-Guinea, I.; Brundiek, H.; Bornscheuer, U.T. Engineering and application of enzymes for lipid modification, an update. *Prog. Lipid Res.* **2016**, *63*, 153–164. [CrossRef] [PubMed]
4. Ahsan, M.; Sung, S.; Jeon, H.; Patil, M.D.; Chung, T.; Yun, H. Biosynthesis of medium- to long-chain α,ω-diols from free fatty acids using CYP153A monooxygenase, carboxylic acid reductase, and *E. coli* endogenous aldehyde reductases. *Catalysts* **2018**, *8*, 4. [CrossRef]
5. Metzger, J.O.; Bornscheuer, U. Lipids as renewable resources: Current state of chemical and biotechnological conversion and diversification. *Appl. Microbiol. Biotechnol.* **2006**, *71*, 13–22. [CrossRef] [PubMed]
6. Lu, W.; Ness, J.E.; Xie, W.; Zhang, X.; Minshull, J.; Gross, R.A. Biosynthesis of monomers for plastics from renewable oils. *J. Am. Chem. Soc.* **2010**, *132*, 15451–15455. [CrossRef] [PubMed]
7. Durairaj, P.; Malla, S.; Nadarajan, S.P.; Lee, P.-G.; Jung, E.; Park, H.H.; Kim, B.-G.; Yun, H. Fungal cytochrome P450 monooxygenases of *Fusarium oxysporum* for the synthesis of ω-hydroxy fatty acids in engineered *Saccharomyces cerevisiae*. *Microb. Cell Fact.* **2015**, *14*, 1–16. [CrossRef] [PubMed]
8. Song, J.-W.; Lee, J.-H.; Bornscheuer, U.T.; Park, J.-B. Microbial Synthesis of Medium-Chain α,ω-Dicarboxylic Acids and ω-Aminocarboxylic Acids from Renewable Long-Chain Fatty Acids. *Adv. Synth. Catal.* **2014**, *356*, 1782–1788. [CrossRef]
9. Ladkau, N.; Assmann, M.; Schrewe, M.; Julsing, M.K.; Schmid, A.; Bühler, B. Efficient production of the Nylon 12 monomer ω-aminododecanoic acid methyl ester from renewable dodecanoic acid methyl ester with engineered *Escherichia coli*. *Metab. Eng.* **2016**, *36*, 1–9. [CrossRef] [PubMed]
10. Ahsan, M.M.; Jeon, H.; Nadarajan, S.P.; Chung, T.; Yoo, H.-W.; Kim, B.-G.; Patil, M.D.; Yun, H. Biosynthesis of the Nylon 12 monomer, ω-aminododecanoic acid with novel CYP153A, AlkJ, and ω-TA enzymes. *Biotechnol. J.* **2018**, *13*, 1–11. [CrossRef] [PubMed]
11. Jung, E.; Park, B.G.; Ahsan, M.M.; Kim, J.; Yun, H.; Choi, K.Y.; Kim, B.G. Production of ω-hydroxy palmitic acid using CYP153A35 and comparison of cytochrome P450 electron transfer system in vivo. *Appl. Microbiol. Biotechnol.* **2016**, *100*, 10375–10384. [CrossRef] [PubMed]
12. Jung, E.; Park, B.G.; Yoo, H.-W.; Kim, J.; Choi, K.-Y.; Kim, B.-G. Semi-rational engineering of CYP153A35 to enhance ω-hydroxylation activity toward palmitic acid. *Appl. Microbiol. Biotechnol.* **2018**, *102*, 269–277. [CrossRef] [PubMed]
13. Patil, M.D.; Grogan, G.; Yun, H. Biocatalyzed C-C bond formation for the production of alkaloids. *ChemCatChem* **2018**. [CrossRef]
14. Patil, M.D.; Dev, M.J.; Shinde, A.S.; Bhilare, K.D.; Patel, G.; Chisti, Y.; Banerjee, U.C. Surfactant-mediated permeabilization of *Pseudomonas putida* KT2440 and use of the immobilized permeabilized cells in biotransformation. *Process Biochem.* **2017**, *63*, 113–121. [CrossRef]
15. Lin, B.; Tao, Y. Whole-cell biocatalysts by design. *Microb. Cell Fact.* **2017**, *16*, 1–16. [CrossRef] [PubMed]
16. Patil, M.D.; Grogan, G.; Bommarius, A.; Yun, H. Recent advances in ω-transaminase-mediated biocatalysis for the enantioselective synthesis of chiral amines. *Catalysts* **2018**, *8*, 254. [CrossRef]
17. Honda Malca, S.; Scheps, D.; Kuhnel, L.; Venegas-Venegas, E.; Seifert, A.; Nestl, B.M.; Hauer, B. Bacterial CYP153A monooxygenases for the synthesis of omega-hydroxylated fatty acids. *Chem. Commun.* **2012**, *48*, 5115–5117. [CrossRef] [PubMed]

18. Scheps, D.; Honda Malca, S.; Richter, S.M.; Marisch, K.; Nestl, B.M.; Hauer, B. Synthesis of ω-hydroxydodecanoic acid based on an engineered CYP153A fusion construct. *Microb. Biotechnol.* **2013**, *6*, 694–707. [PubMed]
19. Chowdhary, P.K.; Alemseghed, M.; Haines, D.C. Cloning, expression and characterization of a fast self-sufficient P450: CYP102A5 from *Bacillus cereus*. *Arch. Biochem. Biophys.* **2007**, *468*, 32–43. [CrossRef] [PubMed]
20. Choi, K.Y.; Jung, E.; Yun, H.; Yang, Y.H.; Kim, B.G. Engineering class I cytochrome P450 by gene fusion with NADPH-dependent reductase and *S. avermitilis* host development for daidzein biotransformation. *Appl. Microbiol. Biotechnol.* **2014**, *98*, 8191. [CrossRef] [PubMed]
21. Hausjell, J.; Halbwirth, H.; Spadiut, O. Recombinant production of eukaryotic cytochrome P450s in microbial cell factories. *Biosci. Rep.* **2018**, *38*, 1–13. [CrossRef] [PubMed]
22. Mathew, S.; Jeong, S.-S.; Chung, T.; Lee, S.-H.; Yun, H. Asymmetric synthesis of aromatic β-amino acids using ω-transaminase: Optimizing the lipase concentration to obtain thermodynamically unstable β-keto acids. *Biotechnol. J.* **2016**, *11*, 185–190. [CrossRef] [PubMed]
23. Tavanti, M.; Mangas-Sanchez, J.; Montgomery, S.L.; Thompson, M.P.; Turner, N.J. A biocatalytic cascade for the amination of unfunctionalised cycloalkanes. *Org. Biomol. Chem.* **2017**, *15*, 9790–9793. [CrossRef] [PubMed]
24. Kirmair, L.; Skerra, A. Biochemical analysis of recombinant AlkJ from *Pseudomonas putida* reveals a membrane-associated, FAD-dependent dehydrogenase suitable for the biosynthetic production of aliphatic aldehydes. *Appl. Environ. Microbiol.* **2014**, *80*, 2468–2477. [CrossRef] [PubMed]
25. Tan, Z.; Black, W.; Yoon, J.M.; Shanks, J.V.; Jarboe, L.R. Improving *Escherichia coli* membrane integrity and fatty acid production by expression tuning of FadL and OmpF. *Microb. Cell Fact.* **2017**, *16*, 38. [CrossRef] [PubMed]
26. Jeon, E.Y.; Song, J.W.; Cha, H.J.; Lee, S.M.; Lee, J.; Park, J.B. Intracellular transformation rates of fatty acids are influenced by expression of the fatty acid transporter FadL in *Escherichia coli* cell membrane. *J. Biotechnol.* **2018**, *281*, 161–167. [CrossRef] [PubMed]
27. Rosano, G.L.; Ceccarelli, E.A. Recombinant protein expression in *Escherichia coli*: Advances and challenges. *Front. Microbiol.* **2014**, *5*, 172. [CrossRef] [PubMed]
28. Janßen, H.; Steinbüchel, A. Fatty acid synthesis in escherichia coli and its applications towards the production of fatty acid based biofuels. *Biotechnol. Biofuels* **2014**, *7*, 7. [CrossRef] [PubMed]
29. Hanahan, D. Studies on transformation of *Escherichia coli* with plasmids. *J. Mol. Biol.* **1983**, *166*, 557–580. [CrossRef]
30. Sambrook, J. *Molecular Cloning: A Laboratory Manual/joseph Sambrook, David w. Russell*; Cold Spring Harbor Laboratory: Cold Spring Harbor, NY, USA, 2001.
31. Bollag, D.M.; Rozycki, M.D.; Edelstein, S.J. *Protein Methods*, 2nd ed.; Wiley-Liss: Hoboken, NJ, USA, 1996; ISBN 978-0-471-11837-4.

catalysts

MDPI

Article

Infrared Characterization of the Bidirectional Oxygen-Sensitive [NiFe]-Hydrogenase from *E. coli*

Moritz Senger [1,†], Konstantin Laun [1,†], Basem Soboh [2] and Sven T. Stripp [1,*]

1 Department of Physics, Experimental Molecular Biophysics, Freie Universität Berlin, 14195 Berlin, Germany; senger@zedat.fu-berlin.de (M.S.); konstantin.laun@fu-berlin.de (K.L.)

2 Department of Physics, Genetic Biophysics, Freie Universität Berlin, 14195 Berlin, Germany; basem.soboh@fu-berlin.de

* Correspondence: sven.stripp@fu-berlin.de; Tel.: +49-030-838-55069

† These authors contributed equally.

Received: 16 October 2018; Accepted: 6 November 2018; Published: 8 November 2018

Abstract: [NiFe]-hydrogenases are gas-processing metalloenzymes that catalyze the conversion of dihydrogen (H_2) to protons and electrons in a broad range of microorganisms. Within the framework of green chemistry, the molecular proceedings of biological hydrogen turnover inspired the design of novel catalytic compounds for H_2 generation. The bidirectional "O_2-sensitive" [NiFe]-hydrogenase from *Escherichia coli* HYD-2 has recently been crystallized; however, a systematic infrared characterization in the presence of natural reactants is not available yet. In this study, we analyze HYD-2 from *E. coli* by in situ attenuated total reflection Fourier-transform infrared spectroscopy (ATR FTIR) under quantitative gas control. We provide an experimental assignment of all catalytically relevant redox intermediates alongside the O_2- and CO-inhibited cofactor species. Furthermore, the reactivity and mutual competition between H_2, O_2, and CO was probed in real time, which lays the foundation for a comparison with other enzymes, e.g., "O_2-tolerant" [NiFe]-hydrogenases. Surprisingly, only Ni-B was observed in the presence of O_2 with no indications for the "unready" Ni-A state. The presented work proves the capabilities of in situ ATR FTIR spectroscopy as an efficient and powerful technique for the analysis of biological macromolecules and enzymatic small molecule catalysis.

Keywords: redox enzymes; FTIR spectroscopy; small molecules

---

1. Introduction

Hydrogenases are gas-processing metalloenzymes that catalyze "hydrogen turnover" ($H_2 \rightleftharpoons 2H^+ + 2e^-$) in various organisms [1]. Oxidation of molecular hydrogen is typically referred to as "H_2 uptake" while proton reduction leads to "H_2 release". Under physiological conditions, [FeFe]-hydrogenases show pronounced H_2 release activity while [Fe]- and [NiFe]-hydrogenases are typically employed in H_2 uptake [2–4]. These three classes are unrelated and differ significantly in protein fold and composition of the catalytic transition metal cofactor [5]. An understanding of the molecular proceedings of hydrogen turnover may inspire novel catalytic compounds for an industrial generation of H_2 as a fuel [6–8].

The gram-negative, facultative anaerobic bacterium *Escherichia coli* synthesizes at least three membrane-associated [NiFe]-hydrogenases [9]. Facing the cytoplasm, HYD-3 catalyzes the oxidation of formic acid into H_2 and CO_2 as part of the formate hydrogenlyase complex. HYD-1 and HYD-2 are respiratory hydrogenases of the periplasm whose hydrogen turnover activity has been suggested to be linked to the energy metabolism of the cell (HYD-2) or the reduction of trace amounts of O_2 (HYD-1) [10]. Accordingly, HYD-1 is an "O_2-tolerant" [NiFe]-hydrogenase that maintains significant turnover activity in the presence of O_2 while HYD-2 is inhibited by O_2 and has been classified

"O_2-sensitive" [11–13]. In O_2-sensitive [NiFe]-hydrogenases the structural placidity of an iron–sulfur cluster proximal to the active site cofactor was shown to play a key role in the immediate recovery from O_2-inhibited species and reduction of O_2 under hydrogen turnover conditions [14–16].

Figure 1A depicts the recently crystallized HYD-2 heterodimer [13]. Subunit HybO (~40 kDa) carries three iron–sulfur clusters that facilitate electron exchange between redox partners and catalytic center. In difference to HYD-1 [17,18] the moiety most proximal to the [NiFe] cofactor is a conventional [4Fe-4S]-cluster. Subunit HybC (~63 kDa) binds the catalytic center in tunneling distance to the proximal iron–sulfur cluster (Figure 1B). Four conserved cysteine residues (C61/C64, C546/C549) coordinate the bimetallic [NiFe] active site. The iron ion binds two cyanide (CN^-) and one carbon monoxide ligand (CO), which facilitates infrared (IR) spectroscopic studies on [NiFe]-hydrogenases [19]. The CO/CN^- stretching frequencies are sensitive "reporters" for changes in electron density distribution across the cofactor and can be addressed to characterize different redox- and protonation states [20–22]. In the second coordination sphere, several homologous residues have been proposed to be involved in proton transfer between bulk water and active site cofactor. One possible trajectory includes C546 and E14 [23–25], an alternative route comprises R479, D103, and D544 [26].

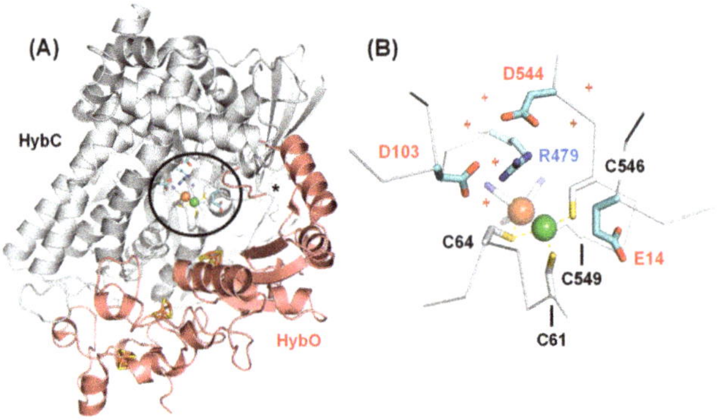

Figure 1. Crystal structure of the [NiFe]-hydrogenase HYD-2 from *E. coli* (pdb coordinates 6EHQ) [13]. (**A**) Subunit HybC (light grey) binds the catalytic cofactor whereas subunit HybO (red) carries three iron–sulfur clusters. In this representation HybO is truncated by 48 amino acids (*) to reveal the view onto the active site niche (black circle); (**B**) Catalytic cofactor including amino acids of the first and second coordination sphere (cyan) that are potentially involved in proton transfer [23–26]. Non-bonded molecules (crosses, H_2O) may be involved in proton transfer as well.

Protein crystallography, electrochemistry, EPR- and IR spectroscopy on [NiFe]-hydrogenases significantly contributed to the understanding of biological hydrogen turnover [27–30]. The H_2 uptake reaction was suggested to include 3–4 redox species (Figure 2) [20,21]. **Ni-SI** represents the active-ready, oxidized state (Ni^{2+}/Fe^{2+}) that stabilizes an open coordination site between nickel and iron ion (dotted circle in Figure 2). Dihydrogen reacts at the nickel ion and forms a Ni-Fe bridging hydride species in the "super-reduced" **Ni-R** state [31,32]. One proton is released in the process; for **Ni-R1**, high resolution crystallography suggested a protonated cysteine (C546 in HYD-2) [25] but protonation of R479 has also been proposed [26]. The very location of the proton in **Ni-R2** and **Ni-R3** is unclear; however, the shift to lower IR frequencies from R1 to R3 may indicate an increase in distance relative to the $Fe(CN)_2CO$ reporter group.

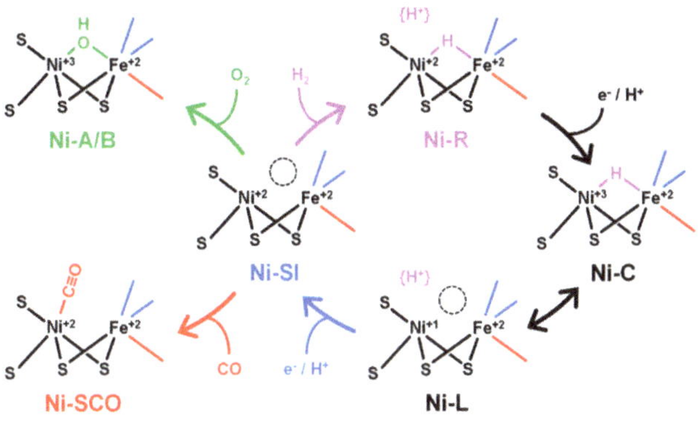

Figure 2. Intermediates of the [NiFe] cofactor and catalytic cycle [20,21]. Based on crystallographic data the cartoon shows the bimetallic cofactor as coordinated by four cysteines. The iron ion is equipped with two CN^- (blue) and one CO ligand (red). Catalytic intermediates Ni-SI, Ni-R, and Ni-C/Ni-L constitute for the H_2 uptake reaction. In the active-ready oxidized state Ni-SI the cofactor reacts with CO or O_2 to form the inhibited species Ni-SCO or Ni-A/B, respectively.

Oxidization and deprotonation of **Ni-R** by one electron forms the **Ni-C** state [33]. The assignment of Ni^{2+}/Fe^{2+} in **Ni-R** and Ni^{3+}/Fe^{2+} in **Ni-C** is reflected by a CO frequency up-shift of up to ~40 cm^{-1} for **Ni-R3**, compensated by the presence of a proton in **Ni-R1** (~Δ15 cm^{-1}) and **Ni-R2** (~Δ30 cm^{-1}) [20]. Both **Ni-R** and **Ni-C** carry a bridging hydride. The release of this hydride is accompanied by a two-fold reduction of the cofactor in the **Ni-L** states (Ni^{1+}/Fe^{2+}) and protonation of the protein fold [34,35]. Infrared data on **Ni-L2** and **Ni-L3** is rare; however, previous studies suggest a down-shift relative to **Ni-C** by ~Δ70 cm^{-1} and ~Δ85 cm^{-1}, respectively [36–39]. This is compatible with a formal difference of two electrons. Oxidation and deprotonation restores **Ni-SI** and completes the cycle as shown in Figure 2. Whether **Ni-L** and/or **Ni-C** are included is subject to ongoing discussions [20,21].

Figure 2 additionally depicts the reaction of [NiFe]-hydrogenases with CO and O_2. Carbon monoxide binds the terminal coordination site of Ni^{2+} in the **Ni-SI** state to form the oxidized, CO-inhibited **Ni-SCO** state [40]. This species differs from the parent **Ni-SI** state by a weakly coordinated Ni-CO ligand at high IR frequencies [41]. In the presence of O_2, two different O_2-inhibited, "super-oxidized" states may be enriched: **Ni-A** and **Ni-B** (Ni^{3+}/Fe^{2+}). **Ni-A** represents an "unready" species that converts slowly into **Ni-SI** while **Ni-B** is readily activated under reducing conditions [42,43]. [NiFe]-hydrogenases that do not form **Ni-A** maintain catalytic activity under aerobic conditions and have been classified O_2-tolerant [14]. Standard [NiFe]-hydrogenases, e.g., as isolated from strict anaerobes are inhibited by O_2 and have been referred to as O_2-sensitive [11]. Both **Ni-A** and **Ni-B** carry a Ni-Fe bridging ligand (most likely a hydroxo species) that reacts to H_2O upon reductive activation [44–46]. On structural level the kinetic activation differences between **Ni-A** and **Ni-B** remain elusive.

In this work we present the first conclusive IR characterization of the periplasmatic [NiFe]-hydrogenase HYD-2 from *E. coli*. In difference to HYD-1 [37], only tentative IR band assignments have been reported for HYD-2 [47,48]. We suggest an experimentally verified assignment of catalytic (**Ni-SI**, **Ni-C**, **Ni-R**) and inhibited states (**Ni-SCO**, **Ni-B**) making use of steady-state and real-time attenuated total reflection Fourier-transform infrared spectroscopy (ATR FTIR) under gas control. The reaction and mutual competition with H_2, CO, and O_2 is probed in kinetic experiments and forms the basis for a comparative analysis with other [NiFe]-hydrogenases. Our methodology has mainly been applied to [FeFe]-hydrogenases in the past [49–51]. Here, we show that all advantages (low sample demand, simple experimental design, measurements under biological conditions, etc.) hold true for the analysis of [NiFe]-hydrogenases as well.

2. Results and Discussion

2.1. FTIR Steady-State and In Situ Difference Spectra

Synthesized and isolated in the absence of O_2, HYD-2 from *E. coli* adopted a steady-state mixture of at least two species (Figure 3a, first spectrum "as-isolated"). The carbonyl stretching frequencies are distinct enough (υCO = 1966 and 1945 cm^{-1}) whereas the cyanide regime from 2100–2040 cm^{-1} does not immediately suggest four individual υCN^- contributions. In the presence of 1% CO ambient partial pressure, the as-isolated spectrum completely converted into the oxidized, CO-inhibited state **Ni-SCO** (Figure 3a, second spectrum) [41]. This species is characterized by a high-frequency Ni-CO band (υCO = 2054 cm^{-1}) and a low-frequency Fe-CO band (υCO = 1944 cm^{-1}). The CN^- stretching frequencies can be found at 2084 and 2073 cm^{-1}. Isotope editing with ^{13}CO confirmed the existence of two individual, vibrationally uncoupled CO ligands at the active site cofactor (Figure S1). When as-isolated HYD-2 was brought in contact with 1% O_2 (Figure 3a, third spectrum), an oxygen-inhibited state was populated (υCO = 1957 cm^{-1} with υCN^- = 2092 and 2082 cm^{-1}) that has been suggested to be **Ni-B** by Hexter and co-workers [47]. Incubation under 100% O_2 for up to four hours did not induce any further changes in the spectrum. In the presence of 1% H_2, ambient partial pressure the species observed in the as-isolated sample were found to be significantly diminished (Figure 3a, fourth spectrum). Even under 100% H_2, these states did not vanish completely (Figure S2). Three novel CO bands were detected at 1950, 1936, and 1927 cm^{-1} alongside a larger number of cyanide bands.

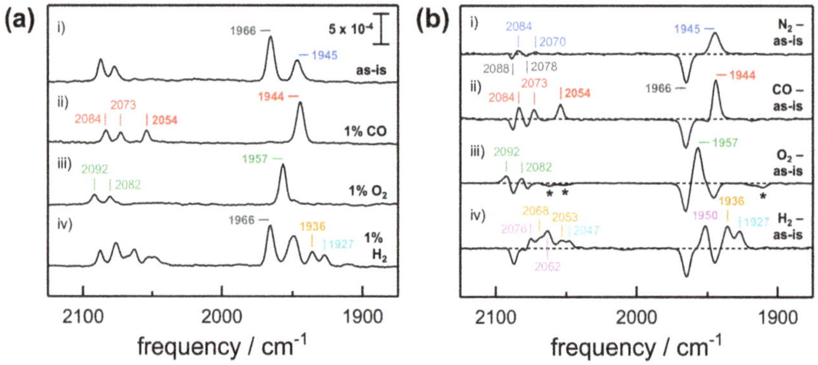

Figure 3. HYD-2 from *E. coli* in the presence of different gases. (**a**) Absorbance spectra under (i) N_2, (ii) CO, (iii) O_2, and (iv) H_2. All spectra are corrected for the broad contribution of liquid H_2O (see Appendix A). (**b**) Difference spectra. (i) Upon extensive treatment with N_2, the 1945 cm^{-1} species (CO marker band for Ni-SI) was enriched over the 1966 cm^{-1} species (Ni-C). (ii) The 1944 cm^{-1} species was populated to 100% under CO (Ni-SCO). (iii) The reaction with O_2 lead to a pure state as well (1957 cm^{-1}, CO marker band for Ni-B). (*) Note the decrease of a species at 1911, 2062, and 2052 cm^{-1}. (iv) The reduced species Ni-R1–R3 (1950, 1936, 1927 cm^{-1}) have been found to be enriched over Ni-C and Ni-SI in the presence of H_2.

In order to refine the band assignments, we recorded in situ difference spectra as a function of gas. For this an as-isolated absorbance spectrum was subtracted from the absorbance spectrum after the respective gas treatment (e.g., "H_2 − as-is"). The first spectrum in Figure 3b indicates a very slow enrichment of the 1945 cm^{-1} band over the 1966 cm^{-1} band upon elongated treatment with N_2 gas (the spectrum depicts a treatment with N_2 of more than four hours). This allows correlating CO and CN^- frequencies as suggested in the Figure 3b and Table 1. The enrichment under auto-oxidizing conditions (an inert gas is exploited to remove traces of evolved H_2 that otherwise would back-react with the enzyme [50]) indicates that the 1945 cm^{-1} band represents the active-ready, oxidized state **Ni-SI**. This assignment is supported by the spectral similarity to **Ni-SCO** (Figure 3a) [41]. The concomitant decrease of the 1966 cm^{-1} band under oxidizing conditions suggests a reduced cofactor intermediate. According to the analyses of Ash et al. [20], the difference in CO stretching frequency of 21 cm^{-1} to **Ni-SI** strongly supports an assignment to the one-electron reduced hydride state **Ni-C**. Interestingly, the second spectrum in Figure 3b shows the loss of **Ni-SI** and **Ni-C** in the presence of CO although it is known that only **Ni-SI** is sensitive to CO inhibition [41]. Most likely, this reflects changes in the steady-state equilibrium between **Ni-C** and **Ni-SI** upon CO reacting with the later: once CO binds to the oxidized cofactor, reduced enzyme converts into **Ni-SI** and becomes reactive to CO. This process is related to intramolecular electron transfer and auto-oxidation of **Ni-C** into **Ni-SI** as discussed above, however much more effective because the CO affinity outrivals the H_2 release activity of HYD-2 [52], in particular under non-reducing conditions. Under O_2, HYD-2 is oxidized into either **Ni-A** or **Ni-B** (Figure 3b, third spectrum). The difference spectrum shows a pronounced loss of **Ni-C** and **Ni-SI**, as well as a species at lower frequencies (υCO = 1911 cm^{-1} with υCN^- = 2062 and 2052 cm^{-1}) that may represent an **Ni-L** state [34,35]. As HYD-2 is an O_2-sensitive [NiFe]-hydrogenases, it is not immediately possible to conclude whether **Ni-A** or **Ni-B** was formed. In the presence of H_2, the difference spectrum represents the enrichment of at least three novel species (Figure 3b, fourth spectrum). The corresponding three CO bands are clearly visible in the difference spectrum, and the downshift relative to **Ni-C** suggests an assignment of **Ni-R1** (1950 cm^{-1}), **Ni-R2** (1936 cm^{-1}), and **Ni-R3** (1927 cm^{-1}) [32]. In Figure S2, the correlation of CO and CN^- bands for the R-states is presented.

Table 1. Experimentally identified redox species. All frequencies given in cm^{-1}.

Species	υCO	υCN^-	υCN^-
Ni-B	1957	2082	2092
Ni-SI	1945	2070	2084
Ni-SCO [1]	2054/1944	2073	2084
Ni-C	1966	2078	2088
Ni-R1	1950	2062	2076
Ni-R2	1936	2053	2068
Ni-R3	1929	2047	2063
Ni-L	1911	2052	2062

[1] In the presence of ^{13}CO the Ni-CO band shifts to 2009 cm^{-1}.

2.2. Kinetic Traces for the Reaction with H_2, CO, and O_2

Figure 4 illustrates the reaction of HYD-2 to changes of the N_2/H_2 gas composition in the head phase above the protein film. The sum of peak area of one CO and two CN^- bands for each species is followed over time. In as-isolated sample, **Ni-C** and **Ni-SI** were populated in a ratio of approximately 2:1. When the inert N_2 atmosphere was enriched with 1% H_2 ambient partial pressure a steep decrease of both these species was observed; however, while **Ni-SI** vanished from the spectrum completely **Ni-C** lost only ~40% sum of peak area and remained to be the most prominent species. **Ni-R1** rose in the presence of H_2 over **Ni-SI** and **Ni-C** as dominant "super-reduced" state while **Ni-R2** and **Ni-R3** are omitted for clarity (see Figure S2 for the complete data set). Removal of H_2 resulted in a slow conversion of the R-states back into **Ni-C** and **Ni-SI**, compared to the fast decrease under H_2. **Ni-SI** only reaches ~50% sum of peak area in comparison to as-isolated sample. No traces of **Ni-A**, **Ni-B**, or **Ni-SCO** were observed.

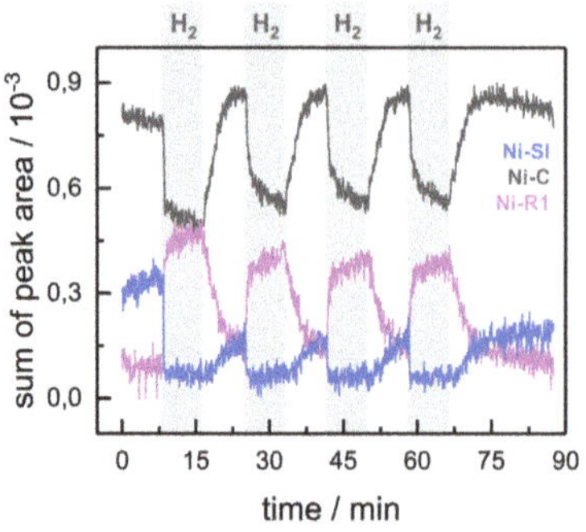

Figure 4. Repeated N_2/H_2 cycles on HYD-2 from *E. coli*. The sum of peak area of CO/CN^- bands assigned to each state is plotted against time. In the presence of 1% H_2 ambient partial pressure, Ni-C and Ni-SI are diminished in favor of the "super-reduced" R-states. The one-electron reduced hydride state Ni-C (black) is the dominating species throughout the experiment. The experiment demonstrates the robust nature of the Ni-C/Ni-SI ↔ Ni-R1 conversion.

Figure 5 depicts on the inhibition and reactivation kinetics upon contact with CO in either N_2 or H_2 carrier gas. Panel (a) shows how **Ni-C** and **Ni-SI** were populated under N_2 and converted into **Ni-SCO** at 1% CO. After 500 s, CO was removed from the gas phase, which prompted the film to swing into an equilibrium between **Ni-SCO** and **Ni-SI**. The CO-inhibited species remained to be the dominant, though. After 500 s, 1% H_2 was introduced to the gas stream and gave rise to **Ni-C/Ni-R** over **Ni-SCO/Ni-SI**. The reaction of CO with the reduced active site cofactor is addressed in Figure 5b. Here, the percentage of CO was stepped up systematically in H_2 carrier gas. Half-max intensity for **Ni-SCO** is achieved at ~30% CO (Figure S3), whereas only 1% CO resulted in a full conversion under N_2 (Figure 5a). The observed decline of reduced states cannot be explained by the relative decrease of H_2 ambient partial pressure between 100 and 50%: Figure 4 and Figure S2 showed that 1% H_2 is sufficient for a reduction of HYD-2, therefore no significant changes in the effective H_2 redox potential can be assumed. Similar to the equilibrium between **Ni-C** and **Ni-SI** (see above) the R-states are in equilibrium with the "unprotected" **Ni-SI** state under reducing conditions (Figure 5b). In the presence of CO **Ni-SI** reacts to **Ni-SCO**, which induces a continuous auto-oxidation of the reduced species into **Ni-SI** (note the relative stable population of **Ni-SI** up to 10% CO). The enrichment of **Ni-C** and **Ni-R** under H_2 affects this equilibrium and delays the process of auto-oxidation. This results in an indirect protection against CO inhibition although H_2 and CO do not compete for the same binding site (compare Figure 2) [40]. No reaction with CO is observed in the absence of **Ni-SI**, i.e., under O_2 (Figure S3).

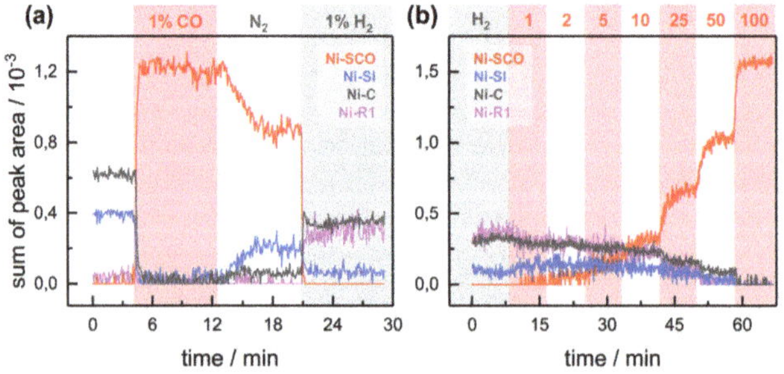

Figure 5. Inhibition of HYD-2 from *E. coli* with CO in either N_2 or H_2 carrier gas. (**a**) With 1% CO in N_2 carrier gas HYD-2 converts into Ni-SCO immediately (red trace). The species was semi-stable in the absence of CO (100% N_2) adopting an equilibrium with Ni-SI (blue trace). Under 1% H_2 an immediate conversion into the catalytic species Ni-C and Ni-R was observed. (**b**) In contrast, using H_2 as carrier gas it takes significantly higher concentrations of CO to convert the enzyme into Ni-SCO.

In a similar set of experiments, Figure 6 depicts on the inhibition and reactivation kinetics of HYD-2 upon contact with O_2 in the presence of either N_2 or H_2 carrier gas. Under 1% O_2, **Ni-C** and **Ni-SI** were converted into a single species, i.e., **Ni-A** or **Ni-B** (Figure 6a, N_2 carrier gas). In difference to the reaction with CO, no changes in equilibrium were observed upon removal of O_2 from the gas stream and the oxygen-inhibited species remained stable even in the absence of O_2. After 500 s, 1% H_2 was introduced to the gas stream, resulting in a conversion of the oxygen-inhibited species into the typical mixture of **Ni-C** and **Ni-R** as seen before (Figure 4). Based on the apparent reactivity of the oxygen-inhibited species, we propose to assign the identified IR signature under O_2 to **Ni-B**, in agreement with earlier suggestions [47]. HYD-2 was stable under 100% O_2 for at least four hours with no notable decrease or conversion into any other species, e.g., **Ni-A**.

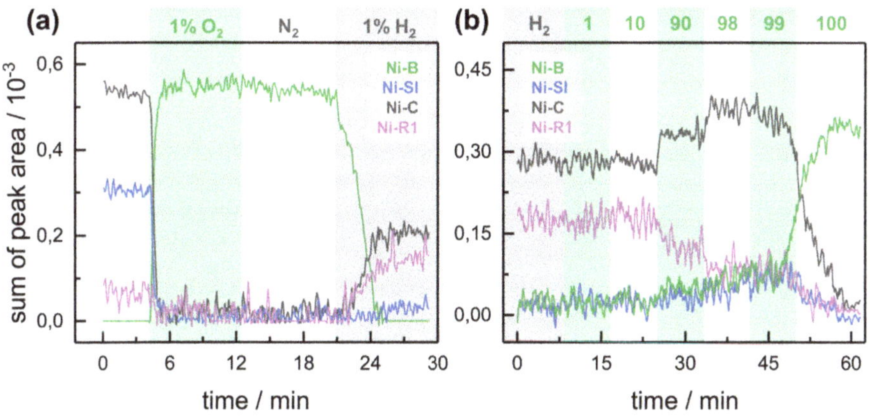

Figure 6. Inhibition of HYD-2 from *E. coli* with O_2 in either N_2 or H_2 carrier gas. (**a**) With 1% CO in N_2 carrier gas HYD-2 converts into Ni-B immediately (green trace). Ni-B is stable under 100% N_2 and converted in the presence of 1% H_2 into reduced species Ni-C/Ni-R within minutes. (**b**) The mutual robustness of the hydride-binding species Ni-C and Ni-R shows in the right panel as well: Ni-B is enriched in the film only at 99–100% O_2. Potentially explosive mixtures of 15–85% O_2 in H_2 were avoided.

In the next step, HYD-2 was subjected to increasing concentration of O_2 in the presence of H_2 carrier gas (Figure 6b). The system did not convert into **Ni-B** until 99–100% O_2 was reached. In difference to the reaction with CO, the lack of O_2 inhibition under reducing conditions can be explained by competition of O_2 and H_2 for the same binding site at the [NiFe] cofactor [40,53]. Interestingly, **Ni-C** is populated over **Ni-R** from 90–98% O_2, most likely reflecting the instability of the "super-reduced" R-states under increasingly oxidizing conditions. This hints at an equilibrium between hydrogenase- and oxygenase-activity that has been suggested to explain the hydrogen turnover activity of "O_2-tolerant" [NiFe]-hydrogenases [54]. Accordingly, the diverging reaction kinetics from CO- and O_2 inhibition as highlighted in Figure 7 may indicate the difference between CO release (fast) and O_2 reduction (slow).

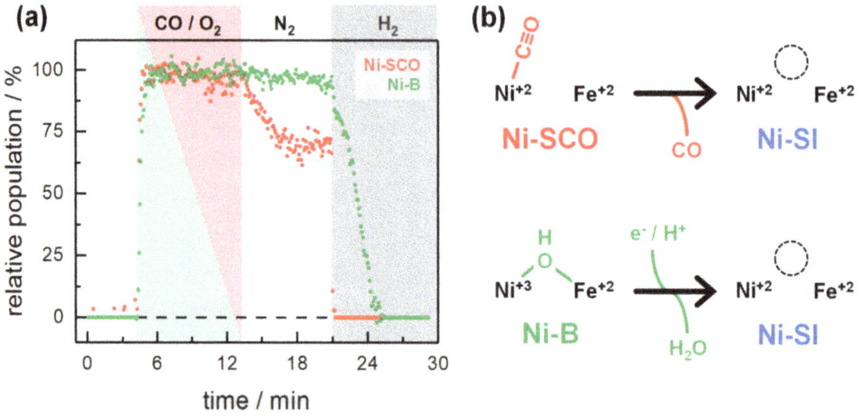

Figure 7. (**a**) Direct comparison of the reaction with CO (Ni-SCO, red trace) and O_2 (Ni-B, green trace). Introduction of either 1% CO or O_2 to the N_2 gas stream resulted in a fast and complete conversion into Ni-SCO or Ni-B, respectively. Under 100% N_2, the CO-inhibited state is semi-stable whereas no decrease of Ni-B was observed. Ni-SCO was immediately lost in the presence of 1% H_2 whereas the reductive activation of Ni-B took significantly longer. (**b**) While CO binds to Ni-SI with no changes in the electronic structure of the [NiFe] cofactor (top) [41], O_2 oxidizes the active site and forms a Ni-Fe bridging hydroxo ligand (bottom) [44–46]. Reactivation from O_2 inhibition may include proton-coupled electron transport and the release of water [16]. This is reflected in the pronounced difference in activation time under 1% H_2 (left panel). Note that reductive reactivation of Ni-B comprises additional species most likely. Furthermore, Ni-SI is a transient intermediate in the reaction with H_2.

3. Materials and Methods

3.1. *Synthesis and Isolation of HYD-2 from E. coli*

To synthesize and isolate preparative amounts of active [NiFe]-hydrogenase, 5 mL "reaction mix" were used. As reported earlier [48], this mixture contained ^{Strep}HybC (HYD-2 catalytic subunit) and the purified maturases HybG-HypDE (GDE complex) and HypEF. An "activation mix" comprising ATP, carbamoylphosphate, $NiCl_2$, $FeSO_4$, and sodium dithionite was added. After incubation at room temperature in the absence of air, cofactor synthesis was initiated adding endopeptidase HybD. After 30 min, the mixture was supplemented with HYD-2 subunit ^{His}HybO. Active ^{Strep}HybC-^{His}HybO heterocomplex (HYD-2 holoenzyme) was isolated from the reaction mix by strep-tactin affinity chromatography and histidine affinity chromatography. The sample was concentrated to ~200 μM protein concentration in 100 mM Tris/HCl buffer (pH 8) including 1 mM dodecyl maltoside as detergent. See ref. [48,55] for further details.

3.2. *Infrared Spectroscopy*

All spectroscopy was performed at room temperature and in the dark on a Bruker Tensor27 FTIR spectrometer (Ettlingen, Germany) housed in a CoyLab anaerobic chamber. A DuraSamplIR 2 optical cell with a three-reflections silicon mircocrystal was used for ATR spectroscopy. Beam path, spectrometer, and anaerobic chamber were purged with dry N_2 gas as provided by an Inmatec (Herrsching, Germany) nitrogen generator (gas purity 5.0). For each experiment, 1 μL of HYD-2 protein sample (~200 μM) was pipetted onto the silicon crystal, dried under N_2, and rehydrated in the presence of an aerosol by running the gas mixture through a wash bottle with a buffer solution of 10 mM Tris/HCl (pH 8). Digital mass flow controllers (Sierra Instruments, Monterey, CA, USA) were used to adjust the absolute amount of gas and ratio between reactants. All gas treatments were performed at ambient partial pressure (1.013 bar). Following this procedure, concentrated and stable protein films were formed (Figure A1). See ref. [50] for details of the experimental setup.

Data were recorded with a spectral resolution of 2 cm^{-1} (80 kHz scanner velocity) and 1.000 interferometer scans (steady-state spectra) or 25 scans in time-resolved experiments. Difference spectra were calculated from single channel spectra via OPUS software. In order to analyze spectral changes in the cofactor regime from 2150–1850 cm^{-1}, absorbance spectra were corrected for the background contribution of liquid water by a low frequency spline function using a home-written routine. The CO/CN^- signature of all redox species was trained on pure spectra to evaluate the individual frequencies, peak areas, and peak ratios. The "sum of peak area" as obtained by Gaussian fits was plotted to follow the conversion of species over time. In Figure A2, this procedure is demonstrated on a representative data set.

4. Conclusions

The membrane-associated, bidirectional [NiFe]-hydrogenase HYD-2 from *E. coli* has been analyzed by in situ ATR FTIR spectroscopy. Based on the reactivity with H_2, CO, and O_2, an experimental band assignment is suggested that agrees well with other [NiFe]-hydrogenases (Table 1). In contrast to HYD-1 from *E. coli*, the one-electron reduced **Ni-L** states were barely observed and the formally isoelectronic hydride state **Ni-C** represents the most stable redox species. This suggests superior stabilization of the Ni-Fe bridging hydride in HYD-2 and may be related to the diverging catalytic properties of HYD-1 (H_2 uptake) and HYD-2 (bidirectional) [52]. In the future, we will explore the differences between Hyd-1 and Hyd-2, making use of the unique possibilities of in vitro maturation (e.g., site-specific isotope editing or Hyd-1/Hyd-2 hybrid constructs) [48].

We have no reason to conclude that anything other than **Ni-B** was enriched in the presence of O_2. **Ni-A** and **Ni-B** are difficult to distinguish by IR spectroscopy. However, in comparison to the O_2-sensitive hydrogenases of strict anaerobes that recover from O_2 inhibition over the time course of hours, the spectral assignment to **Ni-B** is easily compatible with the fast reactivation kinetics of HYD-2 observed in our experiments. The crystal structure of HYD-2 clearly indicates a standard [4Fe-4S] cluster proximal to the [NiFe] cofactor, well in agreement with protein film electrochemistry on HYD-2 that prompted a classification as O_2-sensitive. Thus, HYD-2 was expected to show low rates of reductive reactivation and to form **Ni-A** upon reacting with O_2. This is not the case. With respect to O_2 tolerance, the role of the proximal iron–sulfur cluster in the membrane-bound [NiFe] hydrogenase of *Ralstonia eutropha* has been questioned recently [56]. The example of HYD-2 from *E. coli* shows that the reaction with O_2 may involve additional check screws like proton- and electron transfer that remain to be evaluated.

Supplementary Materials: The following are available online at http://www.mdpi.com/2073-4344/8/11/530/s1, Figure S1: 13CO isotope editing, Figure S2: Further H2 titrations and Ni-R band assignment, Figure S3: Further CO titrations.

Author Contributions: Conceptualization, S.T.S. and M.S.; methodology, S.T.S. and M.S.; formal analysis, S.T.S.; investigation, M.S. and K.L.; resources, B.S.; supervision, S.T.S.

Funding: Basem Soboh acknowledges funding by the DFG priority program "FeS for life" (SPP-1927).

Conflicts of Interest: The authors declare no conflict of interest.

Appendix A

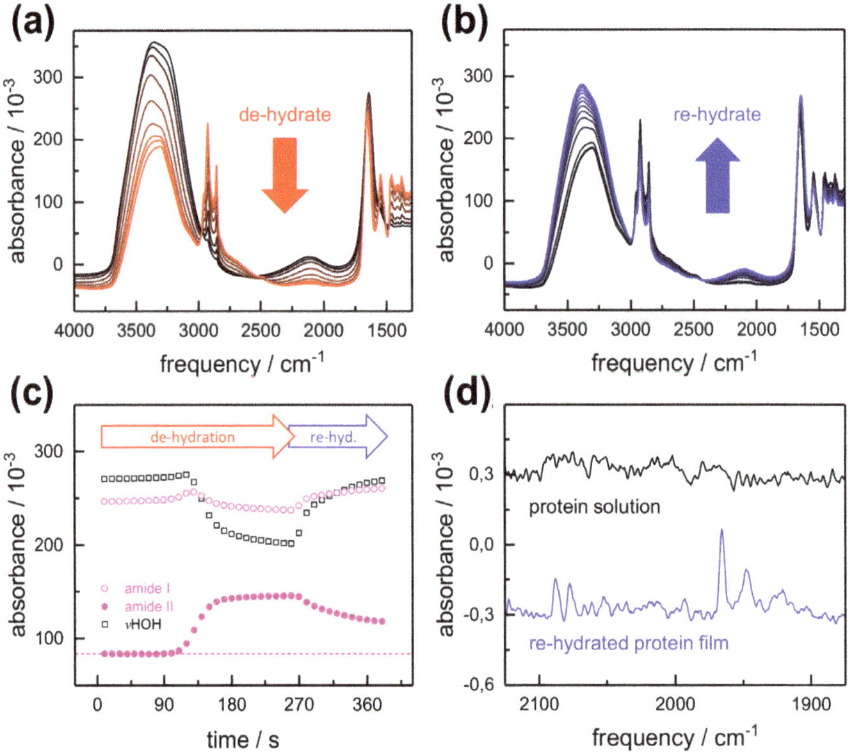

Figure A1. Preparation of the protein film. (**a**) protein solution (1 µL) of ~200 µM HYD-2 was pipetted to the ATR crystal and carefully "de-hydrated" under dry N_2. The spectra in the graph run from black (protein solution) to red (protein film). (**b**) Humidification of carrier N_2 induces a "re-hydration" of the protein film. This is an important prerequisite to follow the specific reactivity of HYD-2 to varying gases. The spectra in the graph run from black (dry protein film) to blue (re-hydrated film). (**c**) Due to the overlap of water and protein bands, the changes in the film are not trivial to analyze. In particular, following the amide I band (~1655 cm^{-1}, open circles) and HOH bending mode intensity (~1635 cm^{-1}, open squares) is not helpful. The amide II band (1540 cm^{-1}, full circles) shares less overlap with liquid water and can be addressed to follow the changes in protein concentration more reliably. Upon de-hydration the band intensity increases by ~60 × 10^{-3} absorbance units while the decrease upon re-hydration accounts to only ~20 × 10^{-3} absorbance units. For a comparable hydration level, we observe a ~40 × 10^{-1} net-increase of amide II band intensity. (**d**) In the cofactor regime, the CO/CN^- become visible in the re-hydrated film (blue spectrum) where no such bands are observed with protein solution (black spectrum). Both spectra are baseline-corrected (see Figure A2). The inferior signal-to-noise level in comparison to Figure 3a stems from the number of averages (here: 50 interferometer scans).

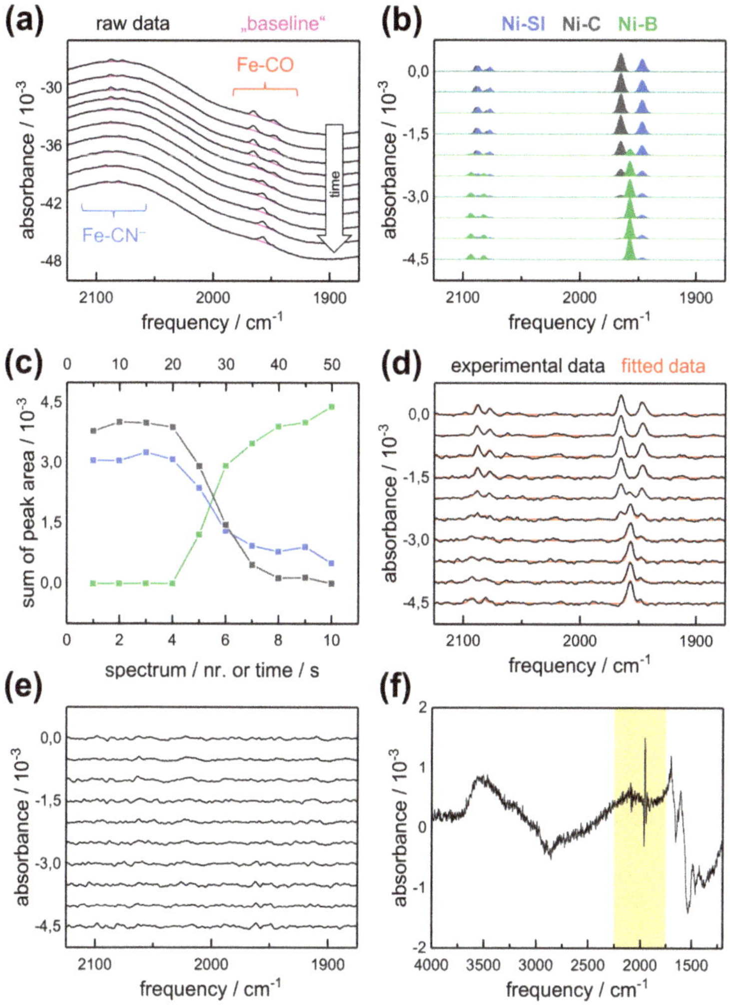

Figure A2. Baseline correction, data fitting, kinetic description, and quality assurance at the example of the formation of **Ni-B** in the presence of O_2. All spectra are plotted from "as-isolated" (top) to "O_2-inhibited" (bottom). (**a**) Absorbance spectra (raw data, black) were corrected for the broad contribution of liquid water by spline functions (magenta). The CO- and CN^- regime is indicated. (**b**) The $Fe(CN)_2CO$ signature of the cofactor is described by three Gaussian fits including frequency, peak area, and peak ratio for each redox species. This data was obtained from pure spectra for **Ni-SI** (blue), **Ni-C** (black), and **Ni-B** (green) beforehand. (**c**) The "sum of peak area" (CO + CN + CN) for each species is then plotted over the number of spectra. Depending on scanner velocity (80 kHz), spectral resolution (2 cm^{-1}), and number of averages (e.g., 25 interferometer scans), this value is converted into time. (**d**) Overlay of baseline-subtracted data (black) and fitted trace (envelope, red). (**e**) The residuals ("data – fit") do not suggest any additional species and allow estimating the signal-to-noise ratio. (**f**) A difference spectrum between the last and first spectrum indicates a small increase of water (OH stretching, ~3500 cm^{-1}) that is accompanied by a decrease of protein signals (amide I, ~1655 cm^{-1} and amide II, ~1540 cm^{-1}). These unspecific changes do not significantly affect the cofactor regime from 2150–1850 cm^{-1} (yellow mark-up).

References

1. Lubitz, W.; Ogata, H.; Ru, O.; Reijerse, E. Hydrogenases. *Chem. Rev.* **2014**, *114*, 4081–4148. [CrossRef] [PubMed]
2. Shima, S.; Thauer, R.K. A third type of hydrogenase catalyzing H_2 activation. *Chem. Rec.* **2007**, *7*, 37–46. [CrossRef] [PubMed]
3. Ogata, H.; Lubitz, W.; Higuchi, Y. Structure and function of [NiFe] hydrogenases. *J. Biochem.* **2016**, *160*, 251–258. [CrossRef] [PubMed]
4. Stripp, S.T.; Happe, T. How Algae Produce Hydrogen—News from the Photosynthetic Hydrogenase. *Dalton Trans.* **2009**, *45*, 9960–9969. [CrossRef] [PubMed]
5. Vignais, P.M.; Billoud, B. Occurrence, classification, and biological function of hydrogenases: An overview. *Chem. Rev.* **2007**, *107*, 4206–4272. [CrossRef] [PubMed]
6. Tard, C.; Pickett, C.J. Structural and functional analogues of the active sites of the [Fe]-, [NiFe]-, and [FeFe]-hydrogenases. *Chem. Rev.* **2009**, *109*, 2245–2274. [CrossRef] [PubMed]
7. Simmons, T.R.; Berggren, G.; Bacchi, M.; Fontecave, M.; Artero, V. Mimicking hydrogenases: From biomimetics to artificial enzymes. *Coord. Chem. Rev.* **2014**, *270–271*, 127–150. [CrossRef]
8. Schilter, D.; Camara, J.M.; Huynh, M.T.; Hammes-Schiffer, S.; Rauchfuss, T.B. Hydrogenase Enzymes and Their Synthetic Models: The Role of Metal Hydrides. *Chem. Rev.* **2016**, *116*, 8693–8749. [CrossRef] [PubMed]
9. Sargent, F. *The Model [NiFe]-Hydrogenases of Escherichia coli*, 1st ed.; Elsevier Ltd.: Amsterdam, The Netherlands, 2016; Volume 68, ISBN 9780128048238.
10. Laurinavichene, T.V.; Tsygankov, A.A. H_2 consumption by *Escherichia coli* coupled via hydrogenase 1 or hydrogenase 2 to different terminal electron acceptors. *FEMS Microbiol. Lett.* **2001**, *202*, 121–124. [CrossRef] [PubMed]
11. Vincent, K.A.; Parkin, A.; Lenz, O.; Albracht, S.P.J.; Fontecilla-Camps, J.C.; Cammack, R.; Friedrich, B.; Armstrong, F.A. Electrochemical definitions of O_2 sensitivity and oxidative inactivation in hydrogenases. *J. Am. Chem. Soc.* **2005**, *127*, 18179–18189. [CrossRef] [PubMed]
12. Lukey, M.J.; Roessler, M.M.; Parkin, A.; Evans, R.M.; Davies, R.A.; Lenz, O.; Friedrich, B.; Sargent, F.; Armstrong, F.A. Oxygen-Tolerant [NiFe]-Hydrogenases: The Individual and Collective Importance of Supernumerary Cysteines at the Proximal Fe-S Cluster. *J. Am. Chem. Soc.* **2011**, *133*, 16881–16892. [CrossRef] [PubMed]
13. Beaton, S.E.; Evans, R.M.; Finney, A.J.; Lamont, C.M.; Armstrong, F.A.; Sargent, F.; Carr, S.B. The Structure of Hydrogenase-2 from *Escherichia coli*: Implications for H_2-Driven Proton Pumping. *Biochem. J.* **2018**, *2*, BCJ20180053. [CrossRef] [PubMed]
14. Fritsch, J.; Lenz, O.; Friedrich, B. Structure, function and biosynthesis of O_2-tolerant hydrogenases. *Nat. Rev.* **2013**, *11*, 106–114. [CrossRef] [PubMed]
15. Fritsch, J.; Scheerer, P.; Frielingsdorf, S.; Kroschinsky, S.; Friedrich, B.; Lenz, O.; Spahn, C.M.T. The crystal structure of an oxygen-tolerant hydrogenase uncovers a novel iron-sulphur centre. *Nature* **2011**, *479*, 249–252. [CrossRef] [PubMed]
16. Qiu, S.; Olsen, S.; MacFarlane, D.R.; Sun, C. The oxygen reduction reaction on [NiFe] hydrogenases. *Phys. Chem. Chem. Phys.* **2018**, *20*, 23528–23534. [CrossRef] [PubMed]
17. Volbeda, A.; Amara, P.; Darnault, C.; Mouesca, J.-M.; Parkin, A.; Roessler, M.M.; Armstrong, F.A.; Fontecilla-Camps, J.C. X-ray crystallographic and computational studies of the O_2-tolerant [NiFe]-hydrogenase 1 from *Escherichia coli*. *Proc. Natl. Acad. Sci. USA* **2012**, *109*, 5305–5310. [CrossRef] [PubMed]
18. Volbeda, A.; Darnault, C.; Parkin, A.; Sargent, F.; Armstrong, F.A.; Fontecilla-Camps, J.C. Crystal structure of the O_2-tolerant membrane-bound hydrogenase 1 from *Escherichia coli* in complex with its cognate cytochrome *b*. *Structure* **2013**, *21*, 184–190. [CrossRef] [PubMed]
19. Pierik, A.J.; Roseboom, W.; Happe, R.P.; Bagley, K.A.; Albracht, S.P.J. Carbon monoxide and cyanide as intrinsic ligands to iron in the active site of [NiFe]-hydrogenases. *J. Biol. Chem.* **1999**, *274*, 3331–3337. [CrossRef] [PubMed]
20. Ash, P.A.; Hidalgo, R.; Vincent, K.A. Proton Transfer in the Catalytic Cycle of [NiFe] Hydrogenases: Insight from Vibrational Spectroscopy. *ACS Catal.* **2017**, *7*, 2471–2485. [CrossRef] [PubMed]
21. Tai, H.; Higuchi, Y.; Hirota, S. Comprehensive reaction mechanisms at and near the Ni-Fe active sites of [NiFe] hydrogenases. *Dalton Trans.* **2018**, *47*, 4408–4423. [CrossRef] [PubMed]

22. Pandelia, M.-E.; Ogata, H.; Lubitz, W. Intermediates in the catalytic cycle of [NiFe] hydrogenase: Functional spectroscopy of the active site. *ChemPhysChem* **2010**, *11*, 1127–1140. [CrossRef] [PubMed]
23. De Lacey, A.L.; Pardo, A.; Fernández, V.M.; Dementin, S.; Adryanczyk-Perrier, G.; Hatchikian, E.; Rousset, M. FTIR spectroelectrochemical study of the activation and inactivation processes of [NiFe] hydrogenases: Effects of solvent isotope replacement and site-directed mutagenesis. *J. Biol. Chem.* **2004**, *9*, 636–642. [CrossRef] [PubMed]
24. Greene, B.L.; Wu, C.; Vansuch, G.E.; Adams, M.W.W.; Dyer, R.B. Proton Inventory and Dynamics in the Ni_a-S to Ni_a-C Transition of a [NiFe] Hydrogenase. *Biochemistry* **2016**, *55*, 1813–1825. [CrossRef] [PubMed]
25. Ogata, H.; Nishikawa, K.; Lubitz, W. Hydrogens detected by subatomic resolution protein crystallography in a [NiFe] hydrogenase. *Nature* **2015**, *520*, 571–574. [CrossRef] [PubMed]
26. Evans, R.M.; Brooke, E.J.; Wehlin, S.A.M.; Nomerotskaia, E.; Sargent, F.; Carr, S.B.; Phillips, S.E.V.; Armstrong, F.A. Mechanism of hydrogen activation by [NiFe] hydrogenases. *Nat. Chem. Biol.* **2016**, *12*, 46–50. [CrossRef] [PubMed]
27. Lubitz, W.; Reijerse, E.; van Gastel, M. [NiFe] and [FeFe] hydrogenases studied by advanced magnetic resonance techniques. *Chem. Rev.* **2007**, *107*, 4331–4365. [CrossRef] [PubMed]
28. Vincent, K.A.; Parkin, A.; Armstrong, F.A. Investigating and exploiting the electrocatalytic properties of hydrogenases. *Chem. Rev.* **2007**, *107*, 4366–4413. [CrossRef] [PubMed]
29. Fontecilla-Camps, J.C.; Volbeda, A.; Cavazza, C.; Nicolet, Y. Structure/function relationships of [NiFe]- and [FeFe]-hydrogenases. *Chem. Rev.* **2007**, *107*, 4273–4303. [CrossRef] [PubMed]
30. Van der Spek, T.M.; Arendsen, A.F.; Happe, R.P.; Yun, S.; Bagley, K.A.; Stufkens, D.J.; Hagen, W.R.; Albracht, S.P. Similarities in the architecture of the active sites of Ni-hydrogenases and Fe-hydrogenases detected by means of infrared spectroscopy. *Eur. J. Biochem.* **1996**, *237*, 629–634. [CrossRef] [PubMed]
31. Brecht, M.; Van Gastel, M.; Buhrke, T.; Friedrich, B.; Lubitz, W. Direct Detection of a Hydrogen Ligand in the [NiFe] Center of the Regulatory H_2-Sensing Hydrogenase from Ralstonia eutropha in Its Reduced State by HYSCORE and ENDOR Spectroscopy. *J. Am. Chem. Soc.* **2003**, *125*, 13075–13083. [CrossRef] [PubMed]
32. De Lacey, A.L.; Fernandez, V.M.; Rousset, M.; Cammack, R. Activation and inactivation of hydrogenase function and the catalytic cycle: Spectroelectrochemical studies. *Chem. Rev.* **2007**, *107*, 4304–4330. [CrossRef] [PubMed]
33. Albracht, S.P.J. Nickel hydrogenases: In search of the active site. *BBA Bioenerg.* **1994**, *1188*, 167–204. [CrossRef]
34. Fichtner, C.; van Gastel, M.; Lubitz, W. Wavelength dependence of the photo-induced conversion of the Ni-C to the Ni-L redox state in the [NiFe] hydrogenase of Desulfovibrio vulgaris Miyazaki F. *Phys. Chem. Chem. Phys.* **2003**, *5*, 5507–5513. [CrossRef]
35. Schröder, O.; Bleijlevens, B.; De Jongh, T.E.; Chen, Z.; Li, T.; Fischer, J.; Förster, J.; Friedrich, C.G.; Bagley, K.A.; Albracht, S.P.J.; et al. Characterization of a cyanobacterial-like uptake [NiFe] hydrogenase: EPR and FTIR spectroscopic studies of the enzyme from Acidithiobacillus ferrooxidans. *J. Biol. Inorg. Chem.* **2007**, *12*, 212–233. [CrossRef] [PubMed]
36. Pandelia, M.E.; Infossi, P.; Stein, M.; Giudici-Orticoni, M.T.; Lubitz, W. Spectroscopic characterization of the key catalytic intermediate Ni-C in the O_2-tolerant [NiFe] hydrogenase i from Aquifex aeolicus: Evidence of a weakly bound hydride. *Chem. Commun.* **2012**, *48*, 823–825. [CrossRef] [PubMed]
37. Hidalgo, R.; Ash, P.A.; Healy, A.J.; Vincent, K.A. Infrared spectroscopy during electrocatalytic turnover reveals the Ni-L active site state during H_2 oxidation by a NiFe hydrogenase. *Angew. Chem. Int. Ed.* **2015**, *54*, 7110–7113. [CrossRef] [PubMed]
38. Murphy, B.J.; Hidalgo, R.; Roessler, M.M.; Evans, R.M.; Ash, P.A.; Myers, W.K.; Vincent, K.A.; Armstrong, F.A. Discovery of Dark pH-Dependent H+ Migration in a [NiFe]-Hydrogenase and Its Mechanistic Relevance: Mobilizing the Hydrido Ligand of the Ni-C Intermediate. *J. Am. Chem. Soc.* **2015**, *137*, 8484–8489. [CrossRef] [PubMed]
39. Tai, H.; Nishikawa, K.; Inoue, S.; Higuchi, Y.; Hirota, S. FT-IR Characterization of the Light-Induced Ni-L2 and Ni-L3 States of [NiFe] Hydrogenase from *Desulfovibrio vulgaris* Miyazaki F. *J. Phys. Chem. B* **2015**, 150430101225004. [CrossRef] [PubMed]

40. Ogata, H.; Mizoguchi, Y.; Mizuno, N.; Miki, K.; Adachi, S.; Yasuoka, N.; Yagi, T.; Yamauchi, O.; Hirota, S.; Higuchi, Y. Structural studies of the carbon monoxide complex of [NiFe]hydrogenase from *Desulfovibrio vulgaris* Miyazaki F: Suggestion for the initial activation site for dihydrogen. *J. Am. Chem. Soc.* **2002**, *124*, 11628–11635. [CrossRef] [PubMed]
41. Pandelia, M.E.; Ogata, H.; Currell, L.J.; Flores, M.; Lubitz, W. Inhibition of the [NiFe] hydrogenase from Desulfovibrio vulgaris Miyazaki F by carbon monoxide: An FTIR and EPR spectroscopic study. *Biochim. Biophys. Acta Bioenerg.* **2010**, *1797*, 304–313. [CrossRef] [PubMed]
42. Lamle, S.E.; Albracht, S.P.J.; Armstrong, F.A. Electrochemical potential-step investigations of the aerobic interconversions of [NiFe]-hydrogenase from allochromatium vinosum: Insights into the puzzling difference between unready and ready oxidized inactive states. *J. Am. Chem. Soc.* **2004**, *126*, 14899–14909. [CrossRef] [PubMed]
43. Bleijlevens, B.; van Broekhuizen, F.A.; De Lacey, A.L.; Roseboom, W.; Fernandez, V.M.; Albracht, S.P.J. The activation of the [NiFe]-hydrogenase from Allochromatium vinosum. An infrared spectro-electrochemical study. *J. Biol. Inorg. Chem.* **2004**, *9*, 743–752. [CrossRef] [PubMed]
44. Volbeda, A.; Martin, L.; Cavazza, C.; Matho, M.; Faber, B.W.; Roseboom, W.; Albracht, S.P.J.; Garcin, E.; Rousset, M.; Fontecilla-Camps, J.C. Structural differences between the ready and unready oxidized states of [NiFe] hydrogenases. *J. Biol. Inorg. Chem.* **2005**, *10*, 239–249. [CrossRef] [PubMed]
45. Ogata, H.; Hirota, S.; Nakahara, A.; Komori, H.; Shibata, N.; Kato, T.; Kano, K.; Higuchi, Y. Activation process of [NiFe] hydrogenase elucidated by high-resolution X-ray analyses: Conversion of the ready to the unready state. *Structure* **2005**, *13*, 1635–1642. [CrossRef] [PubMed]
46. Barilone, J.L.; Ogata, H.; Lubitz, W.; Van Gastel, M. Structural differences between the active sites of the Ni-A and Ni-B states of the [NiFe] hydrogenase: An approach by quantum chemistry and single crystal ENDOR spectroscopy. *Phys. Chem. Chem. Phys.* **2015**, *17*, 16204–16212. [CrossRef] [PubMed]
47. Hexter, S.V.; Chung, M.-W.; Vincent, K.A.; Armstrong, F.A. Unusual Reaction of [NiFe]-Hydrogenases with Cyanide. *J. Am. Chem. Soc.* **2014**, *136*, 10470–10477. [CrossRef] [PubMed]
48. Senger, M.; Stripp, S.T.; Soboh, B. Proteolytic cleavage orchestrates cofactor insertion and protein assembly in [NiFe]-hydrogenase biosynthesis. *J. Biol. Chem.* **2017**, *292*, 11670–11681. [CrossRef] [PubMed]
49. Senger, M.; Mebs, S.; Duan, J.; Wittkamp, F.; Apfel, U.-P.; Heberle, J.; Haumann, M.; Stripp, S.T. Stepwise isotope editing of [FeFe]-hydrogenases exposes cofactor dynamics. *Proc. Natl. Acad. Sci. USA* **2016**, *113*, 8454–8459. [CrossRef] [PubMed]
50. Senger, M.; Mebs, S.; Duan, J.; Shulenina, O.; Laun, K.; Kertess, L.; Wittkamp, F.; Apfel, U.-P.; Happe, T.; Winkler, M.; et al. Protonation/reduction dynamics at the [4Fe–4S] cluster of the hydrogen-forming cofactor in [FeFe]-hydrogenases. *Phys. Chem. Chem. Phys.* **2018**, *20*, 3128–3140. [CrossRef] [PubMed]
51. Haumann, M.; Stripp, S.T. The Molecular Proceedings of Biological Hydrogen Turnover. *Acc. Chem. Res.* **2018**, *51*, 1755–1763. [CrossRef] [PubMed]
52. Lukey, M.J.; Parkin, A.; Roessler, M.M.; Murphy, B.J.; Harmer, J.; Palmer, T.; Sargent, F.; Armstrong, F.A. How *Escherichia coli* is equipped to oxidize hydrogen under different redox conditions. *J. Biol. Chem.* **2010**, *285*, 3928–3938. [CrossRef] [PubMed]
53. Higuchi, Y.; Ogata, H.; Miki, K.; Yasuoka, N.; Yagi, T. Removal of the bridging ligand atom at the Ni-Fe active site of [NiFe] hydrogenase upon reduction with H_2, as revealed by X-ray structure analysis at 1.4 Å resolution. *Structure* **1999**, *7*, 549–556. [CrossRef]
54. Shafaat, H.S.; Rüdiger, O.; Ogata, H.; Lubitz, W. [NiFe] hydrogenases: A common active site for hydrogen metabolism under diverse conditions. *Biochim. Biophys. Acta* **2013**, *1827*, 986–1002. [CrossRef] [PubMed]
55. Soboh, B.; Lindenstrauss, U.; Granich, C.; Javed, M.; Herzberg, M.; Thomas, C.; Stripp, S.T. [NiFe]-hydrogenase maturation in vitro: Analysis of the roles of the HybG and HypD accessory proteins. *Biochem. J.* **2014**, *464*, 169–177. [CrossRef] [PubMed]
56. Hartmann, S.; Frielingsdorf, S.; Ciaccafava, A.; Lorent, C.; Fritsch, J.; Siebert, E.; Priebe, J.; Haumann, M.; Zebger, I.; Lenz, O. O_2-tolerant H_2 activation by an isolated large subunit of a [NiFe]-hydrogenase. *Biochemistry* **2018**, *57*, 5339–5349. [CrossRef] [PubMed]

catalysts

MDPI

Article

Characterization of a New Glyoxal Oxidase from the Thermophilic Fungus *Myceliophthora thermophila* M77: Hydrogen Peroxide Production Retained in 5-Hydroxymethylfurfural Oxidation

Marco Antonio Seiki Kadowaki [1,2], Mariana Ortiz de Godoy [1,2], Patricia Suemy Kumagai [1], Antonio José da Costa-Filho [3], Andrew Mort [2], Rolf Alexander Prade [2,*] and Igor Polikarpov [1,*]

1 São Carlos Institute of Physics, University of São Paulo, Av. Trabalhador São-Carlense, 400, São Carlos 13566-590, Brazil; marcokadowaki@gmail.com (M.A.S.K.); marianaort@gmail.com (M.O.d.G.); patysuemy@gmail.com (P.S.K.)

2 Departments of Biochemistry & Molecular Biology and Microbiology & Molecular Genetics, Oklahoma State University, Stillwater, OK 74078, USA; andrew.mort@okstate.edu

3 Department of Physics, Ribeirão Preto School of Philosophy, Sciences and Literature, University of São Paulo, Ribeirão Preto BR-14040901, Brazil; ajcosta@usp.br

* Correspondence: rolf.prade@okstate.edu (R.A.P.); ipolikarpov@ifsc.usp.br (I.P.); Tel.: +1-405-744-7522 (R.A.P.); +55-16-3373-8088 (I.P.)

Received: 2 October 2018; Accepted: 16 October 2018; Published: 19 October 2018

Abstract: *Myceliophthora thermophyla* is a thermophilic industrially relevant fungus that secretes an assortment of hydrolytic and oxidative enzymes for lignocellulose degradation. Among them is glyoxal oxidase (*Mt*GLOx), an extracellular oxidoreductase that oxidizes several aldehydes and α-hydroxy carbonyl substrates coupled to the reduction of O_2 to H_2O_2. This copper metalloprotein belongs to a class of enzymes called radical copper oxidases (CRO) and to the "auxiliary activities" subfamily AA5_1 that is based on the Carbohydrate-Active enZYmes (CAZy) database. Only a few members of this family have been characterized to date. Here, we report the recombinant production, characterization, and structure-function analysis of *Mt*GLOx. Electron Paramagnetic Resonance (EPR) spectroscopy confirmed *Mt*GLOx to be a radical-coupled copper complex and small angle X-ray scattering (SAXS) revealed an extended spatial arrangement of the catalytic and four N-terminal WSC domains. Furthermore, we demonstrate that methylglyoxal and 5-hydroxymethylfurfural (HMF), a fermentation inhibitor, are substrates for the enzyme.

Keywords: *Myceliophthora*; glyoxal oxidase; 5-hydroxymethylfurfural

1. Introduction

Waste plant biomass could be a massive resource for biofuels and commodity chemicals. Two-thirds of a typical biomass is composed of polysaccharides. The predominant polysaccharide is cellulose, a β(1,4) glucan followed by hemicelluloses a group of β(1,4) linked polysaccharides that interact with cellulose but can be solubilized in strong alkali [1,2]. Another abundant biomass components is lignin, a polyphenolic complex, which forms an insoluble network that confers rigidity to plant cell walls [3].

Natural recycling of biomass entails decomposition of the polysaccharides into simple sugars by microorganisms that use them as a carbon source for growth. Many microorganisms produce a wide variety of hydrolytic enzymes to degrade the polymers. These enzymes can be used in some industrial settings to extract sugars from biomass, which can then be converted into useful commodities, such as fuels (typically ethanol and biodiesel), organic acids (citric and succinic acid), and other bio commodities with high aggregate value. The barriers for biomass conversion to valuable products are the high cost of the enzymatic cocktails needed to break down the polysaccharides and the generation of inhibitory compounds, such as furfural and 5-hydroxymethylfurfural (HMF), which can inhibit enzymatic action or microbial growth in fermentation processes [4,5]. Furfural and 5-hydroxymethylfurfural (HMF) are derived from pentoses and hexoses, respectively, via dehydration during, for example, dilute acid pretreatment [6] or catalytic pyrolysis of lignocellulosic biomass [7]. These chemicals can follow different oxidation/reduction pathways for the production of renewable building blocks for the polymer industry [8]. 2,5-diformylfuran (DFF), a direct product from HMF oxidation, is applied for polymer resin synthesis [9]. 2,5-furandicarboxylic acid (FDCA), originated from two oxidation steps from DFF to 2,5-formylfurancarboxylic acid (FFCA) and FDCA, co-polymerize and produce a polyester with plastic properties [10] and 2-methylfuran, a product of furfural hydrogenolysis, can react with ketones and produce branched-chain liquid hydrocarbons with fuel properties [7,11].

A variety of microorganisms produce an arsenal of enzymes to obtain nutrients from biomass. Not all degrade biomass in the same way. Some use predominantly hydrolytic mechanisms for breaking down glycosidic bonds, while others utilize a combination of hydrolytic and oxidative mechanisms. Based on genome and secretome analysis, the thermophilic fungus, *Myceliophthora thermophila*, isolated from soil and natural composts with high temperatures and humidity [12,13], is an oxidative enzyme producer. Its enzymes are thermo tolerant, which is a desirable characteristic in the biotechnological field.

The classical hydrolytic mechanism employs enzymes that hydrolyze the glycosidic bonds through a general acid/base mechanism [14]. On the other hand, as predicted in 1974, oxidoreductive enzymes that directly oxidize glycans and lignin have been discovered, confirming the role of oxidation reactions in the breakdown of biomass components [15]. The so-called oxizymes play a fundamental role as auxiliary enzymes enhancing cellulase action and cellulose accessibility [16,17]. Oxizymes can also target lignin providing greater access of cellulases to cellulose. Some of these enzymes generate hydrogen peroxide, as a by-product that can feed lignin peroxidases for depolymerization of lignin, or, can generate hydroxyl radicals by Fenton chemistry that can directly attack the biomass structure [18,19].

Glyoxal oxidase (GLOx) (E.C. 1.2.3.15) is an extracellular copper metalloenzyme that belongs to a class of enzymes called radical copper oxidases (CRO) [20,21]. Based on the Carbohydrate-Active enZYmes database (CAZy), these enzymes fall into the "auxiliary activities" subfamily AA5_1 in the CAZy database [22,23]. Family AA5 also includes a second subfamily, AA5_2, containing galactose and alcohol oxidases. GLOx couples the two-electron oxidation of aldehydes to carboxylic acids to the reduction of O_2 to H_2O_2 [21,24,25], α-Hydroxy carbonyl and α-diol substrates can also be oxidized in two steps to carboxylic acids. Previous reports on the enzymatic activity of GLOxes from *Phanerochaete chrysosporium* [26] and *Pycnoporus cinnabarinus* [27] indicate that glyoxal and methylglyoxal are the best substrates for the enzymes. Both aldehydes are recognized as products of glucose degradation through a combination of retroaldol condensation and auto-oxidation, and are found in ligninolytic cultures [28,29]. Another reported substrate that is oxidized by GLOx is glycerol, formed in bulk during biodiesel production [25]. Experimental evidence suggests that GLOxes need to be activated by other enzymes such as lignin peroxidases and can enhance lignolytic activity of white-rot fungi in biomass degradation [26,29,30]. Glycoaldehydes, products from lignin degradation and inhibitors of bioethanol fermentation, have been suggested to be GLOx substrates [31]. Thus, GLOx could play a detoxifying role in the inhibitory aldehydes [21]. Spectroscopic and sequence analysis show that GLOx, *Cgr*AlcOx (Alcohol oxidase from *Colletotrichum graminicola*), and galactose-6-oxidase share similar active sites containing a phenoxyl radical-copper motif [20,32,33]. This copper ion is coordinated by two histidine residues and a tyrosine, which is cross-linked to a cysteine residue by an unusual thioester bond. These residues are conserved in GLOxes, *Cgr*AlcOx, and galactose-6-oxidases [27,31,32].

The central physiological function of GLOx remains unknown, but experimental reports reveal that the enzyme is important for hyphal tip development and pathogenicity [34,35]. Moreover, some GLOxes are linked to the cell Wall Stress-responsive Component (WSC) domain that is predicted to be a cell-wall and membrane protein in yeast and fungi associated with cell wall integrity and stress responses [36,37]. Only three catalytic domains of GLOx enzymes from *Phanaerochaete chrysosporium*, *Ustilago maydis*, and *Pycnoporus cinnabarinus* fungi have been studied biochemically to date [26,27,34,38], but none were associated with such WSC domains. In the present study, we report the structural and functional characterization of a multidomain glyoxal oxidase from the thermo tolerant *Myceliophthora thermophila* M77 and show that 5-hydroxymethylfurfural as a new described substrate for this enzyme, demonstrating its potential in green enzymatic synthesis.

2. Results and Discussion

2.1. Comparative Analysis of *Mt*GLOx with Other Copper Radical Oxidases

Complete AA5_1 glyoxal oxidase genes are present in the genomes of various fungi (mainly Basidiomycete and also some Ascomycetes) and several plant genomes. The three-dimensional (3D) structure of the *Mt*GLOx catalytic domain has not yet been solved by experimental methods but the sequence shares 25% sequence identity with the galactose oxidases from *Streptomyces lividans* (PDBid: 4UNM) [39] and *Fusarium graminearum* (PDBid: 1GOF) and alcohol oxidase from *Colletotrichum graminicola* (PDBid: 5C92) [33] indicating a similar fold, albeit with possible different substrate specificities.

Phylogenetic analysis of 47 predicted AA5_1 domains of GLOx enzymes shows that *Mt*GLOx is grouped within a distinct cluster that does not include the characterized GLOx from *P. chrysosporium*, *P. cinnabarinus*, and *U. maydis* (Figure 1). Interestingly, all WSC associated enzymes were clustered together, based on the alignment of just AA5_1 domains (Figure 1). Fungal glyoxal oxidase AA5_1 genes analyzed lack a WSC domain or have one to six of them, followed by a linker that connects with the classical C-terminal copper-radical oxidase catalytic domain (Figure 1). The *M. thermophila* glyoxal oxidase AA5 has four WSC domains. On the other hand, GLOxes from higher plants, *Arabidopsis thaliana*, *Oryza sativa*, *Theobroma cacao*, and *Zea mays* all have just the catalytic domain as do the characterized enzymes from *P. chrysosporium*, *U. maydis*, and *P. cinnabarinus*. Four GLOxes from the Ascomycete group, closely related to representatives from the plant group, have chitin-binding domains (ChtBD) instead of WSC domains. The WSC domain is considered as a putative carbohydrate binding domain that contains up to eight conserved cysteines involved into disulfides bridges formation and is crucial for fungal adaptation [40]. Two WSC domains which putatively mediate interactions with glucan chains have been previously identified in *Trichodema harzianum* β-1,3-exoglucanase [41]. WSC proteins from *A. nidulans* involved with hypo-osmotic and acidic pH stress tolerance presents extra regions: a potentially glycosylated serine/threonine-rich, transmembrane, and highly charged C-terminal cytoplasmic region besides WSC domain [37]. Moreover, the WSC domains are required for Wsc1 protein clustering that signalize stress conditions [42]. These occurrences and extra associations are suggested to be correlated with the metabolism type of the diverse fungi [43]. The relationship between fungal peroxidases and GLOXes are also been described [21] and it suggests a coupled reaction that is associated with lignin degradation. However, *M. thermophyla* do not codify or secrete this class of peroxidases [12]. Thus, *Mt*GLOx could work as a H_2O_2 provider to other oxidative enzymes, such as lytic polysaccharide monooxigenases (LPMO) [44] or used to attack putative pathogens [45].

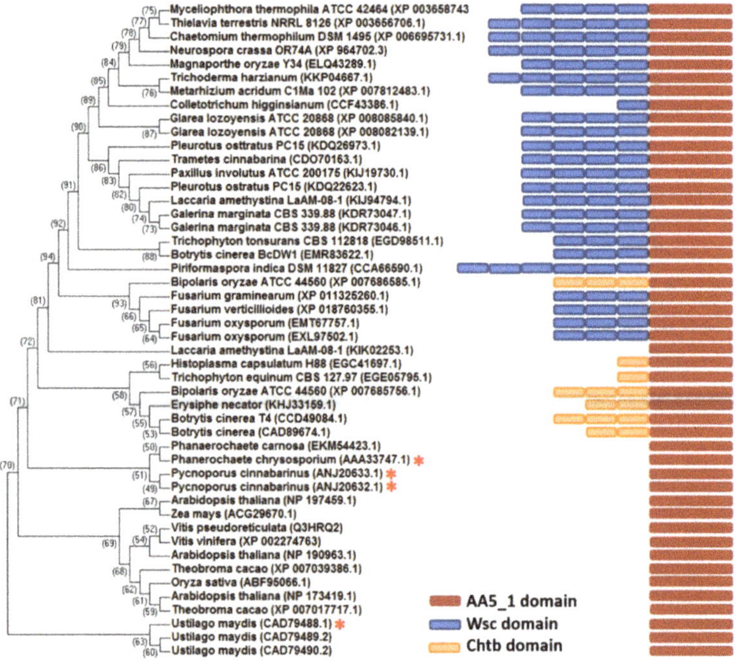

Figure 1. Domain organization of glyoxal oxidases. Phylogenetic tree of predicted AA5 domains of glyoxal oxidase (GLOx) enzymes. The N-terminal Wall Stress-responsive Component (WSC)/Chtb domains were removed to avoid alignment bias. Muscle alignment and tree constructed using MEGA are shown. The GLOx domains were annotated using the CDD (Conserved Domain Database) tool in NCBI. The bootstrap values are indicated on the nodes based on 1000 trials. The asterisk indicates the characterized GLOxes.

2.2. Catalytic Properties of MtGLOx

The Amplex red assay was used to quantify H_2O_2 production to provide more sensitive measurement than the ABTS assay used in other studies [27,46]. Moreover, no addition of H_2O_2 was necessary to activate the purified *Mt*GLOx (Figure S1) or eliminate the lag period, as described for other characterized glyoxal oxidases [26,27]. From our HPLC analysis and unlike other described glyoxal oxidases, the native enzyme from *M. thermophyla* is fully active even in the absence of auxiliary peroxidases or oxidizing agents. Activity screening of *Mt*GLOx against common substrates for AA5_1 family members (Table 1) demonstrated higher activity of the enzyme against small aldehydes, such as methylglyoxal (~3.5 U mg^{-1}) than against glycerol. Moreover, HMF was found to be a previously unidentified substrate for GLOxes. The kinetic constants on the tested substrates are given in Table 2. *Mt*GLOx oxidizes methylglyoxal and 5-HMF with almost the same catalytic efficiency and 10-fold higher when compared with glycerol as substrate. The oxidation product of the new substrate 5-HMF was 2,5-diformylfuran (DFF) and not 5-hydroxymethyl-2-furancarboxylic acid (HMFCA) (Figure 2). DFF is also a valuable compound with several applications for pharmaceuticals production [47], polymer resin synthesis [9], and material science [48]. Interestingly, this result shows that the oxidation pathway of 5-HMF leads to preferential conversion of the primary alcohol arm to an aldehyde instead of aldehyde to carboxylic acid (Figure 2A,B). The time course reaction shows that *Mt*GLOx oxidized 56% of the HMF in 24 h (Figure 2C) but it was unable to further oxidize DFF. Low conversion levels were also observed for other characterized GLOxes [25,27] and for a new AA5_2 from *Colletotrichum graminicola* able to oxidize 1-butanol to butanal [33]. A possible autooxidation of 5-HMF in solution was discarded, as can be seen in the reaction control (Figure S2A). The low yields have been attributed to oxidative damage that is caused by H_2O_2, acidification, in the case of carboxylic acid products, or end-product inhibition [25]. In the case of *Mt*GLOx, the hypothesis of H_2O_2 oxidative damage was not supported by our experimental results, since catalase addition was unable to promote an increased *Mt*GLOx activity (Figure S2B). The same phenomenon was observed for GLOx from *P. chrysosporium* [25].

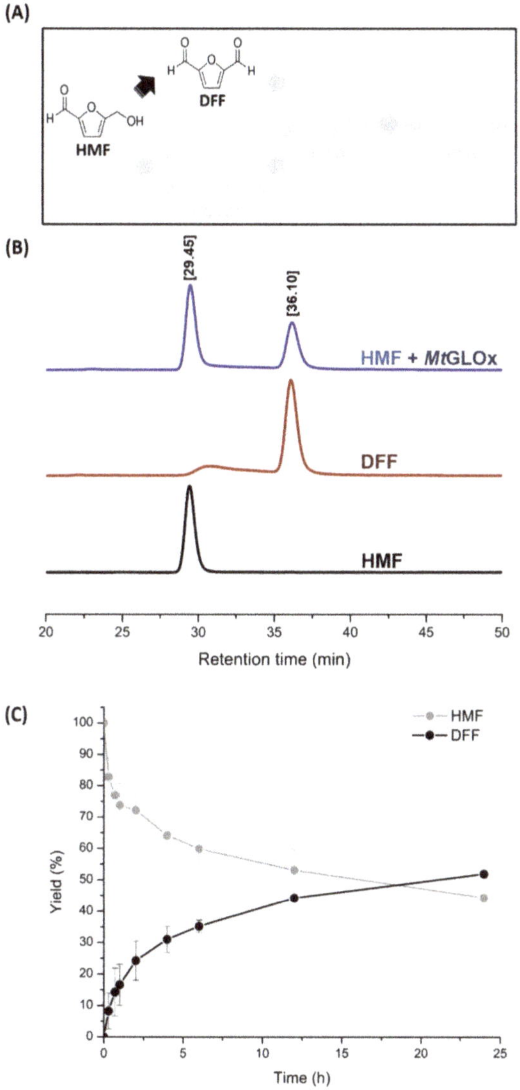

Figure 2. Pathway followed by *Mt*GLOx for hydroxymethylfurfural (HMF) oxidation. (**A**) Representation of possible pathways of hydroxymethylfurfural (HMF) oxidation and products. *Mt*GLOx oxidizes HMF only to 2,5-dimormylfuran (DFF) (black path). (**B**) High Performance Liquid Chromatography (HPLC) chromatograms of the products generated from HMF oxidation. Blue line: product generated from HMF oxidation. Red and black lines: DFF and HMF standards, respectively. Reaction mixture were incubated for 24 h (blue). (**C**) Time course reactions were monitored for oxidation of 1 mM HMF to DFF by 0.5 μM *Mt*GLOx. Standard deviations are shown by error bars (n = 3).

Table 1. Substrate specificity of *Mt*GLOx.

Substrate	μmol $H_2O_2 \cdot min^{-1} \cdot mg^{-1}$ enz	Relative Activity (%)
5-HMF	3.58 ± 0.07	100
Methylglyoxal	3.45 ± 0.11	96
Glycerol	0.57 ± 0.06	16
Formaldehyde	0.16 ± 0.01	4
Furfural	ND	ND
DFF	ND	ND
Glutaraldehyde	ND	ND

The specificity test was done with 25 mM of each substrate and 0.5 μM enzyme at 50 °C for 5 min. The values are the mean of three replicates. ND: not detected activity.

Table 2. Kinetic parameters of *Mt*GLOx with different substrates.

Substrate	Vmax (nkat·mg^{-1})	K_M (mM)	*kcat* (s^{-1})	*kcat*/K_M ($mM^{-1} \cdot s^{-1}$)
Methylglyoxal	123.3 ± 17.0	12.8 ± 2.6	12.6	0.99
5-HMF	156.1 ± 15.8	20.2 ± 9.0	15.9	0.86
Glycerol	333.7 ± 16.4	471.3 ± 88.7	34.1	0.07

A selective oxidation mechanism of primary alcohols to aldehydes has been previously described for a flavoenzyme, aryl-alcohol oxidase [49]. The preferential oxidation of HMF to DFF has also been shown for galactose oxidases [49], and in combination with aldehyde oxidases, can lead to FDCA (2,5-furandicarboxylic acid), a bioplastic precursor, production [10] (Figure 2A).

The maximum activity of *Mt*GLOx was observed at pH 6.0 and 50 °C (Figure 3A,B). These optimum conditions for *Mt*GLOx were practically the same as described for the two glyoxal oxidases from *P. cinnabarinus* [27], but distinct from the *P. chrysosporium* enzyme (30 °C and pH 5.0) [38]. The residual activity after incubation at various temperatures was assayed at pH 6.0 using methylglyoxal as a substrate. *Mt*GLOx maintained its activity at 50 °C and 60 °C for 4 h, but lost 50% of its initial activity after 15 min at 70 °C (Figure 3B). *Mt*GLOx reveals higher thermostability than *P. cinnabarinus* glyoxal oxidases.

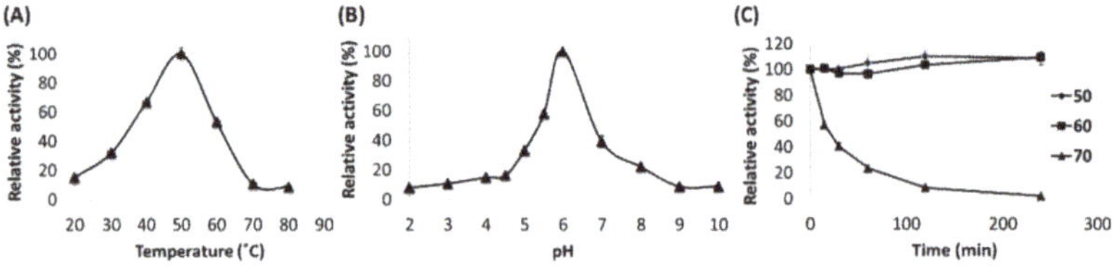

Figure 3. Effect of pH and temperature on enzymatic activity. Effect of temperature (**A**) and pH (**B**) on enzymatic activity of *Mt*GLOx. Values calculated as a percentage of the activity at the maximum. (**C**) The enzyme residual activity after incubation at different temperature is represented as a percentage with respect to the enzyme initial activity at different incubation times.

2.3. Catalytic Domain and Copper Site of MtGLOx

The three-dimensional structure of the *Mt*GLOx catalytic domain was homology modeled using the crystal structure of the cuproenzyme GlxA from *Streptomyces lividans* (PDBid: 4UNM) [39] as a template. The model suggested a similar fold to that of galactose and alcohol oxidases from family AA5 that is composed mostly of β-sheets [33,50]. The model highlights an N-terminal β-propeller structure containing the catalytic copper center linked to a C-terminal immunoglobulin-like domain, with both domains being involved in forming the active site of *Mt*GLOx (Figure 4A). The copper region is exposed to the solvent environment. Based on sequence alignment, and the three-dimensional structure model, residues His804, His889, and Tyr803 that compose the first-shell coordination of the copper ion and Cys522-Tyr581 that stabilize the free-radical species [50] are all conserved between glyoxal, galactose, alcohol oxidases, and GlxA copper-radical oxidase from the AA5 family (Figure 4B). In the second-shell coordination, a tryptophan residue that lies over the Tyr581-Cys522 is associated with substrate recognition, catalysis, and radical stabilization [51–53]. In AA5_1 members, the Trp580 residue is conserved and is adjacent to Tyr581 unlike AA5_2 members (Figure 4B). The galactose oxidase structure from *F. graminearum* [50] shows this residue conserved in the same position but 17 residues away from Tyr. On the other hand, the alcohol oxidase from *C. graminicola* [33] shows the Tyr residue replaced by Phe138.

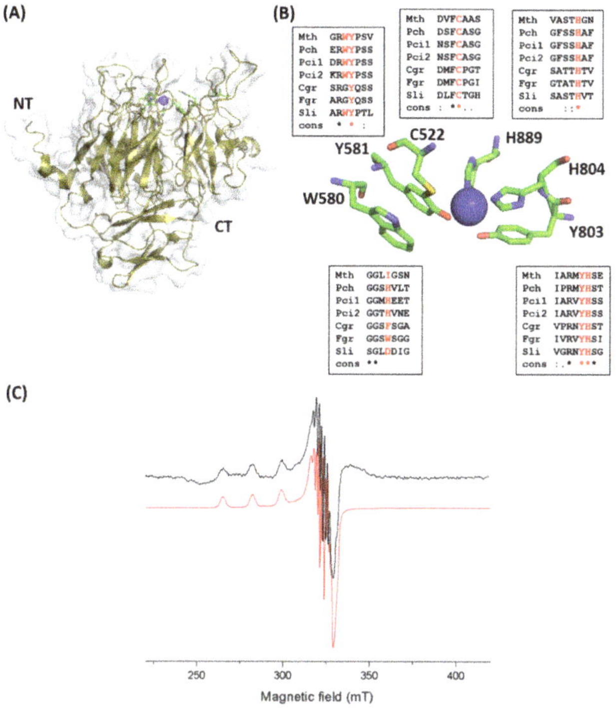

Figure 4. Structural model and spectroscopy of the *Mt*GLOx Cu center. (**A**) Cartoon representation of the model of the catalytic domain of *Mt*GLOx. (**B**) Stick model of the substrate pocket showing the conserved residues coordinating the copper ion (blue sphere). Sections of the sequence from the characterized glyoxal oxidases from *P. chrysosporium* (Pch) and *Pycnoporus cinnabarinus* (Pci1 and 2), alcohol oxidase from *Colletotrichum graminicola* (Cgr), galactose oxidase from *Fusarium graminearum* (Fgr) and cuproenzyme from *Streptomyces lividans* (Sli) showing conserved amino acids. (**C**) Cu(II)-*Mt*GLOx EPR spectrum (black) with simulation (red).

The EPR spectrum of *Mt*GLOx (Figure 4C) is characteristic of a mononuclear Cu(II) center with an axial coordination, which is in good agreement with previous EPR data from related enzymes [33]. Although the Cu(I) state cannot be directly detected by EPR, our experimental EPR spectra clearly demonstrate a strong signal from a Cu(II) ion that is bound to the active site of the enzyme. The superhyperfine lines that were observed in the perpendicular direction of the spectrum indicate the interaction of the copper ion with other nuclei in its vicinity. To further characterize the paramagnetic center, we performed spectral simulation using the Easy Spin package. A spin Hamiltonian containing terms that took into account the Zeeman, hyperfine, and superhyperfine interactions was used. A good agreement with the experimental spectrum was obtained by including two nitrogen nuclei and the following magnetic parameters: $g_x = g_y = 2.05$ and $g_z = 2.28$; $|A_x| = |A_y| = 60$ MHz and ∣Az∣ = 540 MHz. The superhyperfine couplings to the nitrogen nuclei were 40 MHz. These parameters indicate a main d(x2-y2) character for the copper molecular orbital, as observed for other copper-radical oxidases [32,33], and suggest that the *Mt*GLOx active site contains the main structural features that are characteristic for the AA5_1 enzyme family.

2.4. Structural Insights of the Multi-Domain MtGLOx by SAXS

SAXS were used to infer a low-resolution envelope of the enzyme, determine the relative arrangements of the four WSC domains to the catalytic domain in solution. This method was used following unsuccessful crystallization attempts. The SAXS curve profile and linear radius of gyration (R_g) of 45.3 Å indicates an aggregation-free state and a maximal dimension (D_{max}) of 155 Å suggesting an elongated shape (Figure 5A,B). The pair distribution function (P(r)) also shows a maximum at ~30 Å from pair electron distances within the catalytic domain (CD) and a secondary shoulder at ~90 Å that is attributed to distances between the CD and the WSC domains (Figure 5B). The molecular weight that was predicted by SAXS_MoW for *Mt*GLOx was 99 kDa, in agreement with the theoretical 102 kDa of the monomer. Ab initio molecular envelope reconstruction of *Mt*GLOx allowed modeling two regions that could accommodate the WSC domains and the catalytic domain (Figure 5D). To better address this issue, the structure of each domain was modeled based on its amino acid sequence, linked by polyalanine linkers, and allowed to move as rigid bodies using molecular dynamic simulations. The best model generated shows an excellent fit (χ = 1.7) with the experimental curve and fits closely to the low-resolution SAXS molecular envelope, discarding significant large-amplitude inter-domain dynamic or conformational changes. The WSC1-WSC2 and WSC4-WSC3-AA5_1 three-dimensional structural models are connected by a longer and flexible linker, as supported by the Kratky plot (Figure 5C), but not perfectly aligned and appear instead in a bent conformation. The SAXS data also show that the WSC domains are probably connected to the catalytic domain by their upper region (Figure 5B), where the buried copper co-factor resides. Moreover, the low-resolution model allows us to propose that the WSC domains could act as an anchor that orient the catalytic site at the substrate's surface.

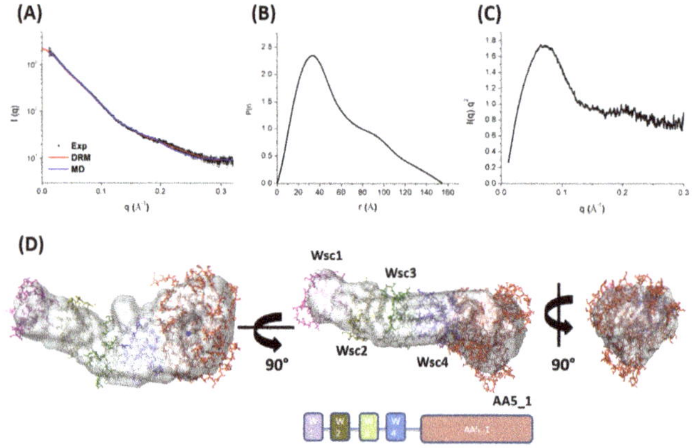

Figure 5. Solution structure of *Mt*GLOx. (**A**) Small Angle X-ray Scattering (SAXS) data. Raw data: plot of scattered intensity vs. scattering angle q. Experimental SAXS curve is shown in black filled circles. The fit of the molecular envelope (red line) and molecular dynamic model (blue line). Exp: experimental raw data. DRM: dammy residue modelling fit. MD: molecular dynamic model fit. (**B**) Pair distribution function P(r). (**C**) Kratky plot. (**D**) Ab initio envelope models based on SAXS data. Molecular envelope superimposed on the three-dimensional model of each domain. The Ab initio envelope is represented in gray. Each domain is represented in stick form. The blue sphere highlights the position of the copper co-factor center.

3. Material and Methods

3.1. Chemicals and Reagents

Chemicals, biochemicals and enzymes were obtained from Fisher Scientific (Pittsburgh, PA, USA), Invitrogen, Merck (Darmstadt, Germany) or Sigma-Aldrich (St. Louis, MO, USA).

3.2. Expression and Purification

The *Mt*GLOx gene (MYCTH_2294895) from *Myceliophthora thermophila* M77 was PCR amplified from genomic DNA without the original signal peptide. The PCR product was amplified using the oligonucleotide primers forward (5′-GGGTTGGCACAGCTCTCAATCCCTACGGACCTTCCGGA-3′) and reverse (5-GTCCCGTGCCGGTTATCAGACGCCGGGGACAGAAAGTCGGGCGC-3) and was cloned into the pEXPYR vector [54] using the Ligation-Independent Cloning protocol (LIC) [55]. The clone was transformed in *A. nidulans* A773, as described [54].

Approximately 10^7 spores/mL were inoculated in liquid minimal medium supplemented with 5% maltose and maintained in static culture at 37 °C for 40 h. The culture medium was filtered using Miracloth membrane (Calbiochem, San Diego, CA, USA). The secreted proteins were concentrated by tangential flow filtration (GE Healthcare, Uppsala, Sweden) and were immediately applied to a DEAE-Sephadex column (GE Healthcare). The enzyme was eluted with a stepwise gradient (200, 300, 400, and 500 mM) NaCl in 50 mM Tris-HCl pH 8.0. The purified samples were concentrated by ultrafiltration (50 kDa cutoff Centricon-Millipore, Billerica, MA, USA) and further purified using size exclusion chromatography on a HiLoad 16/60 Sephadex75 column (GE Healthcare) with a running buffer consisting of 150 mM NaCl and 20 mM Tris–HCl pH 8.0. The total protein was quantified spectrophotometrically at 280 nm using a molar extinction coefficient of 168550. The protein purity was analyzed by SDS-PAGE [56], stained with Coomassie blue G-250, and the protein identity was confirmed by mass spectrometry.

3.3. Mass Spectrometry

The peptide mass fingerprint was performed by in-gel digestion. The Coomassie stained protein band was removed from the SDS-PAGE 15% gel and was submitted to in-gel trypsin (20 ng/μL) digestion after its reduction and alkylation. An aliquot of digested product (1 μL) was desalted by a ZipTip C18 column and mixed with 1 μL of alpha-cyano-4-hydroxy cinnamic acid (HCCA) matrix at 10 mg/mL and was allowed to dry over the sample plate. The measurement was done in the linear positive-ion mode at room temperature within a range of 400–3300 m/z. Average masses were assigned and processed using flexAnalysisTM software (Bruker Daltonics, Bremen, Germany). The mass fingerprint search was done using BioToolsTM (Bruker Daltonics) and the peptide m/z list generated in silico by computational tryptic digestion of protein amino acid sequence. The peptide fingerprinting match was also performed using the MASCOT (Matrix Science Inc.). The analysis was performed using a Microflex LT MALDI-TOF (Bruker Daltonics).

3.4. Enzyme Activity Assay and Steady-State Kinetics

*Mt*GLOx activity was determined measuring hydrogen peroxide (H_2O_2) production using a subsequent reaction with horseradish peroxidase (HRP) and Amplex® Red reagent (Thermo Fisher Scientific, Bremen Germany). The enzyme-coupled reaction protocol was modified due to the differences in optimum reaction conditions between *Mt*GLOx and HRP. The reaction mixture (100 μL) containing 0.5 μM *Mt*GLOx, 10 mM Bis-Tris (pH 6.0), and 25 mM substrate was incubated for 5 min at optimum temperature. The second reaction was started by addition of 10 μL of the previous reaction to 90 μL of Amplex Red working solution according to the manufacturer. The H_2O_2 production was measured using oxidized Amplex Red absorbance at 560 nm on a Tecan Infinite M200 microtiter plate reader. All of the measurements were performed in triplicate. The optimum pH and temperature were determined using 25 mM methylglyoxal as substrate in citrate-glycine-phosphate buffer. The temperature stability of the enzyme was measured by determining residual activity after incubating the enzyme at 50, 60, and 70 °C, followed by cooling in ice and activity measurement.

The kinetic constants were determined at optimum temperature and pH using the protocol described above. The kinetic parameters were determined using a substrate range of 1–40 mM for methylglyoxal/5-HMF and 10–500 mM for glycerol. Data analysis was performed using a non-linear regression of the Michaelis-Menten equation on GraphPad Prism v5.1 software (GraphPad Software, La Jolla, CA, USA).

3.5. Analysis of Oxidized Products

The oxidized products of *Mt*GLOx catalytic activity were analyzed by High Performance Liquid Chromatography (HPLC), using a Bio-Rad Aminex HPX-87H column (300 mm × 7.8 mm) (Bio-Rad, Hercules, CA, USA). Chromatography was carried out at 65 °C with 5 mM sulfuric acid as mobile phase at a flow rate of 0.6 mL/min. Eluted peaks of 5-HMF and DFF were detected by UV absorption at 276 nm. The time course of oxidation of HMF (1 mM) by *Mt*GLOx (0.5 μM) was performed in 100 mM Bis-Tris pH 6.0. The same reaction was monitored using 670 U/mL of catalase from *Aspergillus niger* (Sigma-Aldrich). Aliquots (60 μL) of the reaction were diluted with 300 μL water and 60 μL 1 M HCl. The solution containing denatured protein was centrifuged and the supernatant was used for injection onto the column. Purified HMF and DFF (Sigma-Aldrich) were used as standards.

3.6. Small Angle X-ray Scattering (SAXS) Experiments

*Mt*GLOx was prepared at concentrations of 1, 2, and 4 mg/mL in 50 mM Tris-HCl (pH 8.0) and 150 mM NaCl. SAXS data were collected through mail-in-SAXS on the 12.3.1 SIBYLS beamline at the Advanced Light Source, Lawrence Berkeley National Laboratory [57]. Scattering from the buffer was subtracted from sample scattering and was checked for agreement. The raw data were processed using PRIMUS [58] and GNOM [59]. Twenty low-resolution envelope models were generated using GASBOR [60] and were averaged with DAMAVER [61]. The three-dimensional model of each *Mt*GLOx domain was generated using the iTasser server [62] and available structures of the cuproenzyme GalxA from *Streptomyces lividans* (PDBid: 4UNM) [39] and WSC domain from the human Wnt modulator (PDBid: 5FWW) [63] as templates. The relative position of each domain based on SAXS data was determined using a combination of conformational sampling and molecular dynamics performed on the BILBOMD server [64]. The final model and low resolution envelope superposition was performed using SUPERCOMB [65]. The molecular weight based on the SAXS data was calculated using SAXS MoW [66].

3.7. Electron Paramagnetic Resonance (EPR) Spectroscopy

EPR experiments were performed on a Varian E109 spectrometer equipped with a cryogenic system, which allowed for low-temperature data collection. The spectrometer was operated at 9.26 GHz, with a modulation amplitude of 4 G and microwave power of 10 mW, at 70 K. Samples were drawn into quartz tubes and were then frozen in liquid nitrogen. The EPR parameters were optimized to avoid line saturation and distortion. The spectrum of the buffer only was used as a baseline and was subtracted from all other spectra. Spectral simulations of the EPR data were performed using the EasySpin package [67]. The spin Hamiltonian included terms to account for the Zeeman and hyperfine interactions, which yielded the calculated g- and A-values.

3.8. Sequence and Domain Analysis

The AA5_1 domain sequence of 47 glyoxal oxidases from 39 different organisms were aligned using MUSCLE software [68]. For comparison purposes, an identity tree was generated using the Neighbor-joining (NJ) method that was implemented using the MEGA software [69]. The domain composition of each GLOX was annotated using the CDD (Conserved Domain Database) tool in NCBI.

4. Conclusions

The core cellulase secretome of *M. thermophila* comprises the classical set of hydrolytic cellobiohydrolases (GH6/GH7), beta-glucosidase (GH3), and the oxidative enzymes cellobiose dehydrogenase (AA3), lytic polysaccharide monooxygenases (AA9), aryl-alcohol oxidase (AA3), glyoxal oxidase (AA5_1), and an unknown GMC oxidoreductase. Oxizymes are able to produce hydrogen peroxide and have been suggested to be coupled to ligninolytic peroxidases. However, fungi such as *M. thermophyla* do not secrete this class of peroxidases. Suggesting another unknown role for *Mt*GLOx. Here, we characterize a new multi-domain glyoxal oxidase that produces hydrogen peroxide as a second reaction and oxidizes the fermentation inhibitor HMF. This activity suggests possible green applications of *Mt*GLOx as an enzyme for raw biomass byproduct conversion into sustainable chemical product. Moreover, *Mt*GLOx is the first glyoxal oxidase reported to be connected to four unknown domains, called WSC. However, based on sequence analysis, a substantial number of fungal multidomain AA5_1 members are linked to WSC domains at the N-terminus. This is the same location where either a transmembrane helix [39] or a globular domain 1 related to binding to extracellular carbohydrate [70] can be found in the closely related galactose oxidases. A WSC domain is considered to be a functionally variable putative carbohydrate binding domain [40] able to mediate sensor clustering in stress conditions [42]. However, the specific target of such a domain remains to be elucidated. The overall architecture and spatial arrangement of this multi-domain enzyme was revealed by SAXS data and it describes *Mt*GLOx as a monomeric enzyme in an extended conformation in solution.

Supplementary Materials: The following are available online at http://www.mdpi.com/2073-4344/8/10/476/s1, Figure S1. Identification of the purified *Mt*GLOx heterologously expressed and secreted by *Aspergillus nidulans*. (A) SDS-PAGE showing the purified *Mt*GLOx. MW: molecular weight. (B) MALDI-TOF/MS peptide mass fingerprint analysis of *Mt*GLOx. Mass spectrum profile in the m/z range 800–3300 Da. The peptide mass fingerprint was made from fragments of *Mt*GLOx after tryptic digestion. Green dots mark the expected tryptic masses that matches the theoretical m/z with a maximum 2 Da tolerance. (C) Protein identification and sequence coverage after m/z list analysis. The sequence coverage of the tryptic fragments is shown in bold red (31% coverage). Figure S2. Reaction controls of the time course oxidation of HMF by *Mt*GLOx. (A) Effect of HMF incubation in 100 mM Bis-Tris pH 6.0. (B) Effect of catalase on time course conversion of HMF to DFF. The time course oxidation of HMF (1 mM) by *Mt*GLOx (0.5 μM) in 100 mM Bis-Tris pH 6.0. The reaction was monitored by DFF quantification by High Performance Liquid Chromatography (HPLC), using a Bio-Rad Aminex HPX-87H column with and without 670 U/mL of catalase from *Aspergillus niger* (Sigma-Aldrich).

Author Contributions: M.A.S.K., I.P. and R.A.P. designed the experiments; M.A.S.K. and M.O.d.G. performed the gene cloning, expression and purification; M.A.S.K. and M.O.d.G. characterized the enzymatic activities; M.A.S.K. performed HPLC analysis; M.A.S.K. treated SAXS data; J.A.C.F. and P.S.K. performed EPR experiments and data analysis; M.A.S.K, I.P. and R.A.P. wrote the manuscript with the input from all the other authors; A.M., I.P. and R.A.P. supervised the project.

Funding: This research was financially supported by Fundação de Amparo à Pesquisa do Estado de São Paulo (FAPESP) via grants 2011/20505-4 and 2015/13684-0 and by Conselho Nacional de Desenvolvimento Científico e Tecnológico (CNPq) via grants 405191/2015-4, 303988/2016-9, 440977/2016-9 and 151963/2018-5 and by Coordenação de Aperfeiçoamento de Pessoal de Nível Superior (CAPESP).

Conflicts of Interest: The authors declare no conflicts of interest.

References

1. Sanchez, C. Lignocellulosic residues: Biodegradation and bioconversion by fungi. *Biotechnol. Adv.* **2009**, *27*, 185–194. [CrossRef] [PubMed]
2. Scheller, H.V.; Ulvskov, P. Hemicelluloses. *Annu. Rev. Plant Biol.* **2010**, *61*, 263–289. [CrossRef] [PubMed]
3. Higuchi, T. Lignin biochemistry: Biosynthesis and biodegradation. *Wood Sci. Technol.* **1990**, *24*, 3–63. [CrossRef]
4. Jonsson, L.J.; Alriksson, B.; Nilvebrant, N.O. Bioconversion of lignocellulose: Inhibitors and detoxification. *Biotechnol. Biofuels* **2013**, *6*, 16. [CrossRef] [PubMed]
5. Mohanram, S.; Amat, D.; Choudhary, J.; Arora, A.; Nain, L. Novel perspectives for evolving enzyme cocktails for lignocellulose hydrolysis in biorefineries. *Sustain. Chem. Process.* **2013**, *1*, 15. [CrossRef]

6. Klinke, H.B.; Thomsen, A.B.; Ahring, B.K. Inhibition of ethanol-producing yeast and bacteria by degradation products produced during pre-treatment of biomass. *Appl. Microbiol. Biotechnol.* **2004**, *66*, 10–26. [CrossRef] [PubMed]
7. Agblevor, F.A.; Jahromi, H. Aqueous-Phase Synthesis of Hydrocarbons from Furfural Reactions with Low-Molecular-Weight Biomass Oxygenates. *Energy Fuels* **2018**, *32*, 8552–8562. [CrossRef]
8. Rosatella, A.A.; Simeonov, S.P.; Frade, R.F.M.; Afonso, C.A.M. 5-Hydroxymethylfurfural (HMF) as a building block platform: Biological properties, synthesis and synthetic applications. *Green Chem.* **2011**, *13*, 754–793. [CrossRef]
9. Amarasekara, A.S.; Green, D.; Williams, L.D. Renewable resources based polymers: Synthesis and characterization of 2,5-diformylfuran–urea resin. *Eur. Polym. J.* **2009**, *45*, 595–598. [CrossRef]
10. McKenna, S.M.L.S.; Herter, S.; Turner, N.J.; Carnell, A.J. Enzyme cascade reactions: Synthesis of furandicarboxylic acid (FDCA) and carboxylic acids using oxidases in tandem. *Green Chem.* **2015**, *17*, 3271–3275. [CrossRef]
11. Jahromi, H.; Agblevor, F.A. Hydrodeoxygenation of Aqueous-Phase Catalytic Pyrolysis Oil to Liquid Hydrocarbons Using Multifunctional Nickel Catalyst. *Ind. Eng. Chem. Res.* **2018**, *57*, 13257–13268. [CrossRef]
12. Berka, R.M.; Grigoriev, I.V.; Otillar, R.; Salamov, A.; Grimwood, J.; Reid, I.; Ishmael, N.; John, T.; Darmond, C.; Moisan, M.C.; et al. Comparative genomic analysis of the thermophilic biomass-degrading fungi *Myceliophthora thermophila* and *Thielavia terrestris*. *Nat. Biotechnol.* **2011**, *29*, 922–927. [CrossRef] [PubMed]
13. Dos Santos, H.B.; Bezerra, T.M.; Pradella, J.G.; Delabona, P.; Lima, D.; Gomes, E.; Hartson, S.D.; Rogers, J.; Couger, B.; Prade, R. *Myceliophthora thermophila* M77 utilizes hydrolytic and oxidative mechanisms to deconstruct biomass. *AMB Express* **2016**, *6*, 1–12. [CrossRef] [PubMed]
14. Divne, C.; Stahlberg, J.; Reinikainen, T.; Ruohonen, L.; Pettersson, G.; Knowles, J.K.; Teeri, T.T.; Jones, T.A. The three-dimensional crystal structure of the catalytic core of cellobiohydrolase I from *Trichoderma reesei*. *Science* **1994**, *265*, 524–528. [CrossRef] [PubMed]
15. Eriksson, K.E.; Pettersson, B.; Westermark, U. Oxidation: An important enzyme reaction in fungal degradation of cellulose. *FEBS Lett.* **1974**, *49*, 282–285. [CrossRef]
16. Villares, A.; Moreau, C.; Bennati-Granier, C.; Garajova, S.; Foucat, L.; Falourd, X.; Saake, B.; Berrin, J.G.; Cathala, B. Lytic polysaccharide monooxygenases disrupt the cellulose fibers structure. *Sci. Rep.* **2017**, *7*, 1–9. [CrossRef] [PubMed]
17. Vaaje-Kolstad, G.; Westereng, B.; Horn, S.J.; Liu, Z.L.; Zhai, H.; Sorlie, M.; Eijsink, V.G.H. An Oxidative Enzyme Boosting the Enzymatic Conversion of Recalcitrant Polysaccharides. *Science* **2010**, *330*, 219–222. [CrossRef] [PubMed]
18. Eastwood, D.C.; Floudas, D.; Binder, M.; Majcherczyk, A.; Schneider, P.; Aerts, A.; Asiegbu, F.O.; Baker, S.E.; Barry, K.; Bendiksby, M.; et al. The plant cell wall-decomposing machinery underlies the functional diversity of forest fungi. *Science* **2011**, *333*, 762–765. [CrossRef] [PubMed]
19. Gudrun, G.; Willem, J.H.V.B. Oxizymes for Biotechnology. *Curr. Biotechnol.* **2015**, *4*, 100–110. [CrossRef]
20. Whittaker, M.M.; Kersten, P.J.; Cullen, D.; Whittaker, J.W. Identification of catalytic residues in glyoxal oxidase by targeted mutagenesis. *J. Biol. Chem.* **1999**, *274*, 36226–36232. [CrossRef] [PubMed]
21. Daou, M.; Faulds, C.B. Glyoxal oxidases: Their nature and properties. *World J. Microbiol. Biotechnol.* **2017**, *33*, 87. [CrossRef] [PubMed]
22. Cantarel, B.L.; Coutinho, P.M.; Rancurel, C.; Bernard, T.; Lombard, V.; Henrissat, B. The Carbohydrate-Active EnZymes database (CAZy): An expert resource for Glycogenomics. *Nucleic Acids Res.* **2009**, *37*, D233–D238. [CrossRef] [PubMed]
23. Levasseur, A.; Drula, E.; Lombard, V.; Coutinho, P.M.; Henrissat, B. Expansion of the enzymatic repertoire of the CAZy database to integrate auxiliary redox enzymes. *Biotechnol. Biofuels* **2013**, *6*, 41. [CrossRef] [PubMed]
24. Whittaker, J.W. The radical chemistry of galactose oxidase. *Arch. Biochem. Biophys.* **2005**, *433*, 227–239. [CrossRef] [PubMed]
25. Roncal, T.; Munoz, C.; Lorenzo, L.; Maestro, B.; Diaz de Guerenu Mdel, M. Two-step oxidation of glycerol to glyceric acid catalyzed by the *Phanerochaete chrysosporium* glyoxal oxidase. *Enzym. Microb. Technol.* **2012**, *50*, 143–150. [CrossRef] [PubMed]

26. Kersten, P.J. Glyoxal oxidase of *Phanerochaete chrysosporium*: Its characterization and activation by lignin peroxidase. *Proc. Natl. Acad. Sci. USA* **1990**, *87*, 2936–2940. [CrossRef] [PubMed]
27. Daou, M.; Piumi, F.; Cullen, D.; Record, E.; Faulds, C.B. Heterologous production and characterization of two glyoxal oxidases from *Pycnoporus cinnabarinus*. *Appl. Environ. Microbiol.* **2016**. [CrossRef] [PubMed]
28. Thornalley, P.J.; Langborg, A.; Minhas, H.S. Formation of glyoxal, methylglyoxal and 3-deoxyglucosone in the glycation of proteins by glucose. *Biochem. J.* **1999**, *344 Pt 1*, 109–116. [CrossRef]
29. Kersten, P.J.; Kirk, T.K. Involvement of a new enzyme, glyoxal oxidase, in extracellular H_2O_2 production by *Phanerochaete chrysosporium*. *J. Bacteriol.* **1987**, *169*, 2195–2201. [CrossRef] [PubMed]
30. Yamada, Y.; Wang, J.; Kawagishi, H.; Hirai, H. Improvement of ligninolytic properties by recombinant expression of glyoxal oxidase gene in hyper lignin-degrading fungus *Phanerochaete sordida* YK-624. *Biosci. Biotechnol. Biochem.* **2014**, *78*, 2128–2133. [CrossRef] [PubMed]
31. Kersten, P.; Cullen, D. Copper radical oxidases and related extracellular oxidoreductases of wood-decay Agaricomycetes. *Fungal Genet. Biol. FG B* **2014**, *72*, 124–130. [CrossRef] [PubMed]
32. Whittaker, M.M.; Kersten, P.J.; Nakamura, N.; Sanders-Loehr, J.; Schweizer, E.S.; Whittaker, J.W. Glyoxal oxidase from *Phanerochaete chrysosporium* is a new radical-copper oxidase. *J. Biol. Chem.* **1996**, *271*, 681–687. [CrossRef] [PubMed]
33. Yin, D.T.; Urresti, S.; Lafond, M.; Johnston, E.M.; Derikvand, F.; Ciano, L.; Berrin, J.G.; Henrissat, B.; Walton, P.H.; Davies, G.J.; et al. Structure-function characterization reveals new catalytic diversity in the galactose oxidase and glyoxal oxidase family. *Nat. Commun.* **2015**, *6*, 10197. [CrossRef] [PubMed]
34. Leuthner, B.; Aichinger, C.; Oehmen, E.; Koopmann, E.; Muller, O.; Muller, P.; Kahmann, R.; Bolker, M.; Schreier, P.H. A H_2O_2-producing glyoxal oxidase is required for filamentous growth and pathogenicity in *Ustilago maydis*. *Mol. Genet. Genom. MGG* **2005**, *272*, 639–650. [CrossRef] [PubMed]
35. Song, X.S.; Xing, S.; Li, H.P.; Zhang, J.B.; Qu, B.; Jiang, J.H.; Fan, C.; Yang, P.; Liu, J.L.; Hu, Z.Q.; et al. An antibody that confers plant disease resistance targets a membrane-bound glyoxal oxidase in Fusarium. *New Phytol.* **2016**, *210*, 997–1010. [CrossRef] [PubMed]
36. Verna, J.; Lodder, A.; Lee, K.; Vagts, A.; Ballester, R. A family of genes required for maintenance of cell wall integrity and for the stress response in *Saccharomyces cerevisiae*. *Proc. Natl. Acad. Sci. USA* **1997**, *94*, 13804–13809. [CrossRef] [PubMed]
37. Futagami, T.; Nakao, S.; Kido, Y.; Oka, T.; Kajiwara, Y.; Takashita, H.; Omori, T.; Furukawa, K.; Goto, M. Putative stress sensors WscA and WscB are involved in hypo-osmotic and acidic pH stress tolerance in *Aspergillus nidulans*. *Eukaryot. Cell* **2011**, *10*, 1504–1515. [CrossRef] [PubMed]
38. Son, Y.L.; Kim, H.Y.; Thiyagarajan, S.; Xu, J.J.; Park, S.M. Heterologous Expression of *Phanerochaete chrysoporium* Glyoxal Oxidase and its Application for the Coupled Reaction with Manganese Peroxidase to Decolorize Malachite Green. *Mycobiology* **2012**, *40*, 258–262. [CrossRef] [PubMed]
39. Chaplin, A.K.; Petrus, M.L.; Mangiameli, G.; Hough, M.A.; Svistunenko, D.A.; Nicholls, P.; Claessen, D.; Vijgenboom, E.; Worrall, J.A. GlxA is a new structural member of the radical copper oxidase family and is required for glycan deposition at hyphal tips and morphogenesis of *Streptomyces lividans*. *Biochem. J.* **2015**, *469*, 433–444. [CrossRef] [PubMed]
40. Tong, S.M.; Chen, Y.; Zhu, J.; Ying, S.H.; Feng, M.G. Subcellular localization of five singular WSC domain-containing proteins and their roles in *Beauveria bassiana* responses to stress cues and metal ions. *Environ. Microbiol. Rep.* **2016**, *8*, 295–304. [CrossRef] [PubMed]
41. Cohen-Kupiec, R.; Broglie, K.E.; Friesem, D.; Broglie, R.M.; Chet, I. Molecular characterization of a novel beta-1,3-exoglucanase related to mycoparasitism of *Trichoderma harzianum*. *Gene* **1999**, *226*, 147–154. [CrossRef]
42. Heinisch, J.J.; Dupres, V.; Wilk, S.; Jendretzki, A.; Dufrene, Y.F. Single-molecule atomic force microscopy reveals clustering of the yeast plasma-membrane sensor Wsc1. *PLoS ONE* **2010**, *5*, e11104. [CrossRef] [PubMed]
43. Floudas, D.; Binder, M.; Riley, R.; Barry, K.; Blanchette, R.A.; Henrissat, B.; Martinez, A.T.; Otillar, R.; Spatafora, J.W.; Yadav, J.S.; et al. The Paleozoic origin of enzymatic lignin decomposition reconstructed from 31 fungal genomes. *Science* **2012**, *336*, 1715–1719. [CrossRef] [PubMed]
44. Bissaro, B.; Rohr, A.K.; Muller, G.; Chylenski, P.; Skaugen, M.; Forsberg, Z.; Horn, S.J.; Vaaje-Kolstad, G.; Eijsink, V.G.H. Oxidative cleavage of polysaccharides by monocopper enzymes depends on H_2O_2. *Nat. Chem. Biol.* **2017**, *13*, 1123–1128. [CrossRef] [PubMed]

45. Lamb, C.; Dixon, R.A. The Oxidative Burst in Plant Disease Resistance. *Annu. Rev. Plant Physiol. Plant Mol. Biol.* **1997**, *48*, 251–275. [CrossRef] [PubMed]
46. Holland, J.T.; Harper, J.C.; Dolan, P.L.; Manginell, M.M.; Arango, D.C.; Rawlings, J.A.; Apblett, C.A.; Brozik, S.M. Rational redesign of glucose oxidase for improved catalytic function and stability. *PLoS ONE* **2012**, *7*, e37924. [CrossRef] [PubMed]
47. Hopkins, K.T.; Wilson, W.D.; Bender, B.C.; McCurdy, D.R.; Hall, J.E.; Tidwell, R.R.; Kumar, A.; Bajic, M.; Boykin, D.W. Extended aromatic furan amidino derivatives as anti-*Pneumocystis carinii* agents. *J. Med. Chem.* **1998**, *41*, 3872–3878. [CrossRef] [PubMed]
48. Ma, J.; Du, Z.; Xu, J.; Chu, Q.; Pang, Y. Efficient aerobic oxidation of 5-hydroxymethylfurfural to 2,5-diformylfuran, and synthesis of a fluorescent material. *ChemSusChem* **2011**, *4*, 51–54. [CrossRef] [PubMed]
49. Kalum, L.; Morant, M.D.; Lund, H.; Jensen, J.; Lapainaite, I.; Soerensen, N.H.; Pedersen, S.; Østergaard, L.H.; Xu, F. Enzymatic Oxidation of 5-Hydroxymethylfurfural and Derivatives Thereof. Google Patents WO2014015256A3, 23 January 2014.
50. Ito, N.; Phillips, S.E.; Stevens, C.; Ogel, Z.B.; McPherson, M.J.; Keen, J.N.; Yadav, K.D.; Knowles, P.F. Novel thioether bond revealed by a 1.7 A crystal structure of galactose oxidase. *Nature* **1991**, *350*, 87–90. [CrossRef] [PubMed]
51. Chaplin, A.K.; Svistunenko, D.A.; Hough, M.A.; Wilson, M.T.; Vijgenboom, E.; Worrall, J.A. Active-site maturation and activity of the copper-radical oxidase GlxA are governed by a tryptophan residue. *Biochem. J.* **2017**, *474*, 809–825. [CrossRef] [PubMed]
52. Rogers, M.S.; Tyler, E.M.; Akyumani, N.; Kurtis, C.R.; Spooner, R.K.; Deacon, S.E.; Tamber, S.; Firbank, S.J.; Mahmoud, K.; Knowles, P.F.; et al. The stacking tryptophan of galactose oxidase: A second-coordination sphere residue that has profound effects on tyrosyl radical behavior and enzyme catalysis. *Biochemistry* **2007**, *46*, 4606–4618. [CrossRef] [PubMed]
53. Baron, A.J.; Stevens, C.; Wilmot, C.; Seneviratne, K.D.; Blakeley, V.; Dooley, D.M.; Phillips, S.E.; Knowles, P.F.; McPherson, M.J. Structure and mechanism of galactose oxidase. The free radical site. *J. Biol. Chem.* **1994**, *269*, 25095–25105. [PubMed]
54. Segato, F.; Damasio, A.R.; Goncalves, T.A.; De Lucas, R.C.; Squina, F.M.; Decker, S.R.; Prade, R.A. High-yield secretion of multiple client proteins in Aspergillus. *Enzym. Microb. Technol.* **2012**, *51*, 100–106. [CrossRef] [PubMed]
55. Aslanidis, C.; De Jong, P.J. Ligation-independent cloning of PCR products (LIC-PCR). *Nucleic Acids Res.* **1990**, *18*, 6069–6074. [CrossRef] [PubMed]
56. Laemmli, U.K. Cleavage of structural proteins during the assembly of the head of bacteriophage T4. *Nature* **1970**, *227*, 680–685. [CrossRef] [PubMed]
57. Perry, J.J.; Tainer, J.A. Developing advanced X-ray scattering methods combined with crystallography and computation. *Methods* **2013**, *59*, 363–371. [CrossRef] [PubMed]
58. Konarev, P.V.; Volkov, V.V.; Sokolova, A.V.; Koch, M.H.J.; Svergun, D.I. PRIMUS: A Windows PC-based system for small-angle scattering data analysis. *J. Appl. Crystallogr.* **2003**, *36*, 1277–1282. [CrossRef]
59. Svergun, D.I. Determination of the regularization parameter in indirect-transform methods using perceptual criteria. *J. Appl. Crystallogr.* **1992**, *25*, 495–503. [CrossRef]
60. Svergun, D.I. Restoring low resolution structure of biological macromolecules from solution. *Biophys. J.* **1999**, *76*, 2879–2886. [CrossRef]
61. Volkov, V.V.; Svergun, D.I. Uniqueness of ab initio shape determination in small-angle scattering. *J. Appl. Crystallogr.* **2003**, *36*, 860–864. [CrossRef]
62. Yang, J.; Yan, R.; Roy, A.; Xu, D.; Poisson, J.; Zhang, Y. The I-TASSER Suite: Protein structure and function prediction. *Nat. Methods* **2015**, *12*, 7–8. [CrossRef] [PubMed]
63. Zebisch, M.; Jackson, V.A.; Zhao, Y.; Jones, E.Y. Structure of the Dual-Mode Wnt Regulator Kremen1 and Insight into Ternary Complex Formation with LRP6 and Dickkopf. *Structure* **2016**. [CrossRef] [PubMed]
64. Pelikan, M.; Hura, G.L.; Hammel, M. Structure and flexibility within proteins as identified through small angle X-ray scattering. *Gen. Physiol. Biophys.* **2009**, *28*, 174–189. [CrossRef] [PubMed]
65. Kozin, M.B.; Svergun, D.I. Automated matching of high- and low-resolution structural models. *J. Appl. Crystallogr.* **2001**, *34*, 33–41. [CrossRef]

66. Fischer, H.; Oliveira Neto, M.D.; Napolitano, H.B.; Polikarpov, I.; Craievich, A.F. Determination of the molecular weight of proteins in solution from a single small-angle X-ray scattering measurement on a relative scale. *J. Appl. Crystallogr.* **2009**, *43*, 101–109. [CrossRef]
67. Stoll, S.; Schweiger, A. EasySpin, a comprehensive software package for spectral simulation and analysis in EPR. *J. Magn. Reson.* **2006**, *178*, 42–55. [CrossRef] [PubMed]
68. Edgar, R.C. MUSCLE: A multiple sequence alignment method with reduced time and space complexity. *BMC Bioinform.* **2004**, *5*, 113. [CrossRef] [PubMed]
69. Tamura, K.; Stecher, G.; Peterson, D.; Filipski, A.; Kumar, S. MEGA6: Molecular Evolutionary Genetics Analysis version 6.0. *Mol. Biol. Evol.* **2013**, *30*, 2725–2729. [CrossRef] [PubMed]
70. Whittaker, J.W. Galactose oxidase. *Adv. Protein Chem.* **2002**, *60*, 1–49. [PubMed]

catalysts

MDPI

Article

Biocatalytic Oxidations of Substrates through Soluble Methane Monooxygenase from *Methylosinus sporium* 5

Yeo Reum Park [1], Hee Seon Yoo [1], Min Young Song [1], Dong-Heon Lee [1] and Seung Jae Lee [1,2,*]

1 Department of Chemistry and Institute for Molecular Biology and Genetics, Chonbuk National University, Jeonju 54896, Korea; parkyr@jbnu.ac.kr (Y.R.P.); heeseon0202@jbnu.ac.kr (H.S.Y.); mysong0823@gmail.com (M.Y.S.); dhl@jbnu.ac.kr (D.-H.L.);

2 Department of Chemistry and Research Center for Bioactive Materials, Chonbuk National University, Jeonju 54896, Korea

* Correspondence: slee026@jbnu.ac.kr; Tel.: +82-63-270-3412; Fax: +82-63-270-3407

Received: 27 October 2018; Accepted: 21 November 2018; Published: 26 November 2018

Abstract: Methane, an important greenhouse gas, has a 20-fold higher heat capacity than carbon dioxide. Earlier, through advanced spectroscopy and structural studies, the mechanisms underlying the extremely stable C–H activation of soluble methane monooxygenase (sMMO) have been elucidated in *Methylosinus trichosporium* OB3b and *Methylococcus capsulatus* Bath. Here, sMMO components—including hydroxylase (MMOH), regulatory (MMOB), and reductase (MMOR)—were expressed and purified from a type II methanotroph, *Methylosinus sporium* strain 5 (*M. sporium* 5), to characterize its hydroxylation mechanism. Two molar equivalents of MMOB are necessary to achieve catalytic activities and oxidized a broad range of substrates including alkanes, alkenes, halogens, and aromatics. Optimal activities were observed at pH 7.5 for most substrates possibly because of the electron transfer environment in MMOR. Substitution of MMOB or MMOR from another type II methanotroph, *Methylocystis species* M, retained specific enzyme activities, demonstrating the successful cross-reactivity of *M. sporium* 5. These results will provide fundamental information for further enzymatic studies to elucidate sMMO mechanisms.

Keywords: biocatalysts; biocatalytic reaction; *Methylosinus sporium* strain 5; soluble methane monooxygenase; C–H activation; O_2 activation

1. Introduction

Methanotrophic bacteria (methanotrophs) require methane gas (CH_4) as the sole carbon and energy source [1–3]. Hydroxylation of methane by methanotrophs is a crucial process that regulates the carbon cycle in ecological systems and can be applied in the fields of bioremediation and bioenergy [4–6]. Although carbon dioxide (CO_2) is the most abundant greenhouse gas, with 20-fold higher heat capacity, CH_4 accelerates global warming [7,8]. The level of CH_4 in ecological systems is regulated by methane monooxygenases (MMOs). The membrane-bound methane monooxygenase (pMMO) is expressed in most methanotrophs, including type I, II, and X, and recent studies have demonstrated detailed structures of this enzyme and its active sites [9–11]. The other type of MMO, soluble methane monooxygenase (sMMO), has a non-heme diiron active site and is expressed in copper-limited conditions in limited type II and type X methanotrophs [1,4,6,12]. To achieve full biological conversion from methane to methanol, sMMO requires hydroxylase (MMOH), regulatory (MMOB), and reductase (MMOR) components, similar to other bacterial multi-component monooxygenase (BMM) superfamily members that utilize molecular oxygen (O_2), electrons (e^-), and protons (H^+), for the oxygenation of various substrates [13–17].

Type II methanotrophs, including *Methylosinus* and *Methylocystis*, and a type X methanotroph, such as *Methylococcus capsulatus* Bath, can express sMMO for hydroxylating methane [1,18]. The structures and functions of sMMO have been intensively investigated to elucidate the mechanisms using two methanotrophs, *M. trichosporium* OB3b and *M. capsulatus* Bath [16,19–21]. Hydroxylase from natural methanotrophs has been applied for enzymology studies in most reports because of solubility issues in heterologous systems. MMOH consists of α-, β-, and γ-subunits as a homodimer ($\alpha_2\beta_2\gamma_2$), and the non-heme diiron active sites are positioned in the α-subunit for the hydroxylation of methane [20,21]. Preliminary studies have shown that the absence of regulatory and/or reductase proteins decelerates the catalytic activities and that these component interactions are elaborately concerted for methane oxidation [13,22]. The oxidized MMOH (Fe^{III}-Fe^{III}) were reduced through electron transfer from MMOR to reduce the diiron active site. O_2 activates $MMOH_{red}$ (Fe^{II}-Fe^{II}) to get transient intermediates including H_{peroxo} and Q intermediates. Methane can be converted to methanol only through Q intermediate [17,19]. A structural study between MMOH and MMOB through X-ray crystallography proposed a pathway of substrates to the diiron active sites [23]. MMOB regulates gatekeeper residues, which control the opening of cavities for substrate access to the diiron active sites and pore regions.

M. sporium 5, a type II methanotroph, oxidizes methane to methanol through pMMO and sMMO [24–27]. The presence of a brown–black water-soluble pigment in nitrate mineral salt (NMS) medium is a distinct feature of *M. sporium* 5, although details of this mechanism require further studies [24,27]. Essential components of sMMO, including MMOH (248 kDa), MMOR (37.4 kDa), and MMOB (14.9 kDa), exhibit high sequence identity to those from other type II methanotrophs, including *M. species* M and *M. trichosporium* OB3b. When compared with *M. sporium* 5, type X methanotrophs show relatively low sequence identity for MMOB (67%), [2Fe–2S] cluster containing ferredoxin domain (MMOR-Fd, 58%), and the FAD-binding domain (MMOR-FAD, 46%) (Figure 1 and Figures S1–S3). The three subunits of MMOH are encoded by *mmoX*, *mmoY*, and *mmoZ* and express α-, β-, and γ-subunits, respectively. For the expression of MMOB, *mmoB* is positioned between *mmoY* and *mmoZ* in all sMMO-expressing methanotrophs. Alignment analyses show that the α-subunit has high sequence identity with β- and γ-subunits in both type II and type X methanotrophs. MMOR consists of a MMOR-Fd domain at the *N*-terminus and a MMOR-FAD domain at the *C*-terminus, both of which are required for diiron reduction within the four-helix bundle at the α-subunit of MMOH. The reduced diferrous state (Fe^{II}–Fe^{II}) can in turn activate O_2 and initiate a catalytic cycle [28–31]. MMOB does not have any metal ions and coenzymes; however, this regulatory enzyme acts on the pore region near to the diiron active sites, containing the residues Thr213, Asn214, and Glu240, which are considered key regulators for the electron and/or proton transfer [23,32]. Catalytic activities are improved in the presence of MMOB and MMOH–MMOB complex, indicating that two MMOB molecules can bind to the hydroxylase component at the diiron active site, regulating methane hydroxylation [23]. The structural information between MMOH and MMOR is required to discover the electron transfer pathway that is crucial for turnover number. Recent studies have proposed that the ferredoxin domain of MMOR shares the binding site with MMOB located in the canyon region of MMOH for electron transfer [32].

In this study, *M. sporium* 5 was cultured in a tightly regulated NMS media by supplying methane and air to understand its growth and the expression levels of multi-component enzymes [24,26,27]. MMOH was found to be highly expressed in *M. sporium* 5, and it was purified to evaluate its catalytic activities using diverse substrates. MMOB and MMOR were also expressed in *E. coli* via constructed plasmids to obtain highly purified enzymes. The successfully expressed and purified enzymes were utilized to measure specific enzyme activities (SEA), and these results showed that *M. sporium* 5 exhibits optimal activity at pH 7.5. The electron transfer environment of MMOR is crucial for the activity of sMMO, and different acidities may change the electron transfer environment. In vitro activity measurements demonstrated that alkanes, halogens, benzene, and toluene are oxidized through sMMO, and 2 mol equivalents of MMOB showed optimal activity. The pathways of methane oxidation through sMMO have been studied intensively in limited methanotrophic bacteria to date, but the present study provides fundamental information for elucidating these complicated and elaborate mechanisms of C–H activation through structural and functional approaches from *M. sporium* 5.

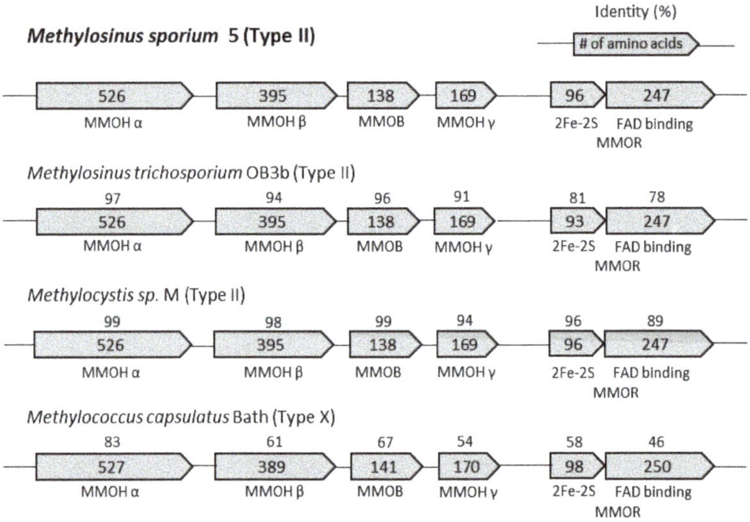

Figure 1. Multiple sequence alignments of amino acid sequences of soluble methane monooxygenase (sMMO) from type II and type X methanotrophs. Numbers above boxes represent the sequence identity to sMMO from *M. sporium* 5, and numbers inside the box represent the number of amino acids in each enzyme and domain.

2. Results and Discussion

2.1. *Cultures of* M. sporium *5 and Iron Concentration in NMS Medium*

For the expression of sMMO, *M. sporium* 5 was cultured in copper-limited NMS media, and results showed that using a feeding gas with ratio of CH_4:air = 1:5 yielded optimal cell growth. *M. sporium* 5 can proliferate to an optical density at 600 nm (OD_{600}) of more than 10.0 under methane/air. Early studies demonstrated that *M. sporium* 5 produces a brown–black pigment during growth, and these phenotypes were also detected under certain growth conditions (Figure 2A and Figure S4) [24]. Cultures of *M. sporium* 5 showed white media at OD_{600} of approximately 1.0, which changed to a faint orange color beyond OD_{600} of 3.0. Additionally, the turbidity increased gradually by feeding with methane and air. The culture showed a brown–black color over OD_{600} of 8.0, and the color turned darker with increased cell growth (Figure 2A). As a control experiment, *M. sporium* 5 was cultured with different amounts of methane feeding, with once and twice per day in a rubber cap-sealed flask, and the growth was directly affected by methane amount (Figure S4). The one-time feeding culture proliferated to an OD_{600} of approximately 3.0, but two-times methane feeding resulted in an OD_{600} of 8.2 with brown–black pigmentation. This change in color is a characteristic feature of *M. sporium* 5, and previous studies have proposed that this phenomenon is associated with the bioavailability of iron and the expression of sMMO [24,27].

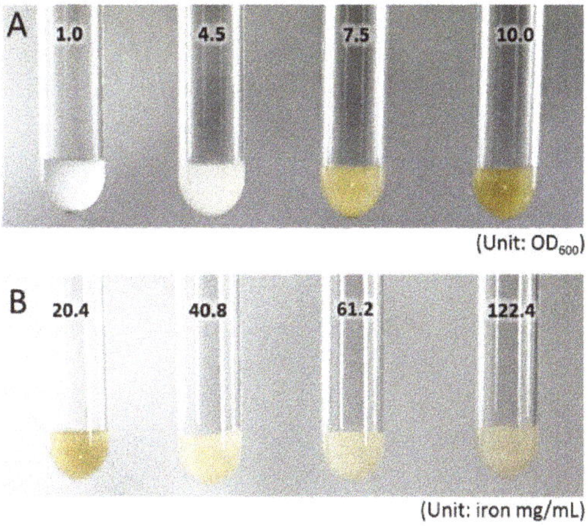

Figure 2. Pigment changes in *M. sporium* 5 cultured in NMS media during sMMO expression. (**A**) Normal media show color changes by increasing optical density at 600 nm (OD_{600}, presented as numbers). The white pigment around OD_{600} of 1.0 changed to pink–orange around 3.0, and media changed to brown–black at OD_{600} values over 8.0. These changes were only observed in *M. sporium* 5 among other *Methylosinus*. (**B**) All cells were cultured to an OD_{600} of 8.5 in media with different iron concentrations. Normal NMS media for growing *M. sporium* 5 contained 20.4 mg/mL (1×) of iron and showed a dark brown color. Iron-supplemented NMS media (40.8, 61.2, and 122.4 mg/mL of iron in media) did not show color changes upon increasing optical density.

Ali et al. showed that mutation of *mmoX* results in different phenotypes based on the concentration of the iron supplemented in the agar plate, and they proposed that siderophores can be secreted from *M. sporium* 5 under iron-limited conditions to acquire iron from the extracellular medium [24]. To test this hypothesis, *M. sporium* 5 was cultured in iron-rich NMS media, and results suggested that iron is a key regulator of this pigment generation. *M. sporium* 5 was cultured to OD_{600} of 8.5 under normal iron concentrations (20.4 mg) in NMS media and iron-rich conditions (2-, 4-, and 6-fold higher iron concentration) to observe pigment changes (Figure 2B). Growth in normal NMS media resulted in brown–black color changes at OD_{600} of approximately 8.0, but *M. sporium* 5 cultured in iron-rich media did not exhibit brown–black color changes. These results provide an indirect evidence that the presence of iron at the diiron active site is required for the catalytic activity of the sMMO. Secretion of siderophores could be a defense mechanism of *M. sporium* 5 in growth condition of high bacterial concentrations and limited iron concentration. A possible explanation is that *M. sporium* 5 responds more sensitively than other type II methanotrophs because this methanotroph generates a brown–black pigment in response to a high cell:iron ratio (Figure 2 and Figure S4). Although transcriptomic studies of pMMO have proposed that iron requirements are achieved through the FecR iron sensor protein and FecR-like protein (fecR), biosynthetic pathways and secretion of siderophores from sMMO, including *M. sporium* 5, still need to be determined [33]. Another study proposed that this change in color may be caused by methanobactin. Following cell growth, the iron level decreases, whereas the amount of methanobactin increases; methanobactin is then secreted to the extracellular media to recruit copper and work as a 'copper-switch'. It has been demonstrated that in the absence of copper, methanobactin is able to bind to other transition metals, and this may be linked to the color change phenomenon [34].

2.2. Expression and Purification of Hydroxylase from M. sporium 5

Cultured *M. sporium* 5 was harvested to monitor expression levels of sMMO enzymes, and subunits of MMOH, including the α- (59.9 kDa), β- (45.2 kDa), and γ-subunits (19.3 kDa), showed significant expression levels (Figure 3A). Sodium dodecyl sulfate-polyacrylamide gel electrophoresis

(SDS-PAGE) demonstrated positive expression of MMOH, MMOB, and MMOR. The three subunits of MMOH were expressed highly compared with that of other essential components, including MMOB and MMOR. MMOH was purified from the extracts of native bacteria owing to its limited solubility in heterologous expression systems, such as *E. coli*, for further characterization. Preliminary studies showed that highly purified MMOH is essential to understand optimal activities and substrate hydroxylation by sMMO [35]. A study of *Methylocystis species* WI 14, a type II methanotroph, showed that SEA (mU/mg) of the cell extract improved by more than 10-fold by a three-step purification process using two anion-exchange columns and one size-exclusion column [35]. In enzyme activity assays using cell lysates, unpurified sMMO enzymes demonstrated that cell lysate exhibits low activity because of the presence of inhibitors and other metal components. Metal ions—such as Zn^{2+}, Cu^{2+}, and Ni^{2+}—are considered as strong inhibitors of sMMO activity, although these metal ions were strongly controlled for the growth of *M. sporium* 5 through chelex resin in this study. Other possible inhibitory enzymes, such as MMOD (*orfY*), could inhibit its activity in the soluble portion of cell lysates [36]. For optimal activity of proteins from *M. sporium* 5, essential enzymes were purified through a multi-step purification process. *M. sporium* 5 was cultured to an OD_{600} of approximately 8–10 and purified through DEAE sepharose fast flow, Superdex 200, and Q sepharose resins to obtain highly purified MMOH (Figure 3B). The iron concentration of MMOH was determined through a ferrozine assay that indicated four iron atoms per MMOH (3.8–4.1 Fe/MMOH; Figure S5). Structural information has shown that this diiron active site is located 12 Å below the surface with helices E and F, which consists of the pore region of MMOH, and this region is considered as the electron transfer pathway and egress route of the product [23,32,37]. These results are in concurrence with the observation that purified MMOH contains proper iron-coordinated active sites and could be further applied for measuring catalytic activities.

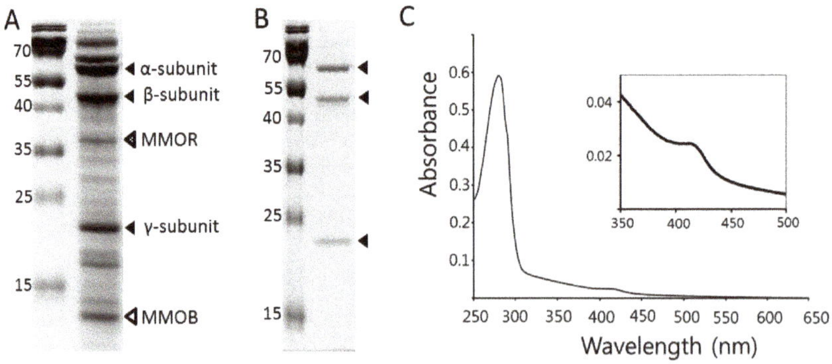

Figure 3. Expressed sMMO enzymes from *M. sporium* 5 and purified hydroxylase, regulatory, and reductase components. (**A**) Left lane represents the protein size-marker (unit: kDa), and right lane represents the cell lysates from *M. sporium* 5. MMOH including α- (59.9 kDa), β- (45.2 kDa), and γ- (19.3 kDa) subunits (black arrowheads) showed high expression in cell lysates. Two essential components, including reductase (MMOR; 37.4 kDa; a dotted arrowhead) and regulatory (MMOB; 14.9 kDa; an empty arrowhead) components, were also detected after expression. (**B**) Purified MMOH subunits after a three-step purification process are represented as arrows, and these highly purified enzymes were further applied to measure enzyme activities. (**C**) UV–visible absorption spectrum of purified MMOH from *M. sporium* 5 in 25 mM sodium phosphate buffer at pH 7.0. Inset shows the enlarged spectrum from 350 to 500 nm.

The UV–visible spectrum of purified MMOH is shown in Figure 3C, which exhibited optimal absorption at 280 nm with a calculated extinction coefficient of 561,220 cm^{-1} M^{-1}, and a weak charge transfer band was observed at approximately 390–430 nm. Characterization of MMOH from *M. trichosporium* IMV3011, a type II methanotroph, showed that oxidized MMOH exhibits optimal intensity at 281 nm and weak absorption is detected at approximately 395–420 nm due to the oxo-bridged iron clusters [38]. The yellow color of MMOH arises from the charge transfer band around the 300–350 nm region, and spectral studies have evinced that this phenomenon is usually observed for the (μ-oxo)diiron(III) centers of the synthetic analogues. Preliminary reports have shown that the weak absorption spectrum from 400 to 500 nm by hydroxylase represents tyrosine radical formation and ligation, and *M. sporium* 5 showed these weak spectra, but *M. trichosporium* OB3b did not [17,38–40].

2.3. Expression and Purification of MMOB and MMOR from E. coli

MMOB does not have cofactors, such as metal ions and prosthetic groups, as shown by solution structures, although it affects O_2 activation, product distribution, and reaction rates [14,17,22,41,42]. The complex structure of MMOH–MMOB confirmed that this regulatory component interacts near diiron active sites to control the positions of the side chains from certain amino acids in MMOH that are crucial for catalytic activity [23]. The structure of MMOB was initially analyzed from *M. capsulatus* Bath [43], which showed an unstructured *N*-terminal tail (Met1-Ser35), well-folded core region (Asp36-Leu129) with seven β-strands and three α-helices, and short *C*-terminal sequences (Met130-Ala141). The MMOH–MMOB complex demonstrated that the *N*-terminal long sequence binds to helices H and 4 in MMOH through specific hydrogen bonds and hydrophobic interactions. Thus, the *N*-terminal tail is essential for catalytic activity and product hydroxylation. Because the expression level of MMOB is low from the native strain, a synthetic MMOB-encoding nucleotide sequence was inserted into pET30a plasmid (*pET30a-mmoB*, Figure S6A) to achieve high MMOB expression in *E. coli*. The purified MMOB showed a typical non-cofactor UV–visible absorption spectrum with maximal intensity at 280 nm with a calculated extinction coefficient of ε_{280} = 15,220 $cm^{-1}M^{-1}$ (Figure 4A and Figure S6B). MMOB could be obtained at a concentration of >10 mM as it is highly soluble in 25 mM phosphate and 100 mM NaCl buffer. These results imply that MMOH has reasonable solubility in bacterial cells, but MMOB has higher solubility compared to MMOH.

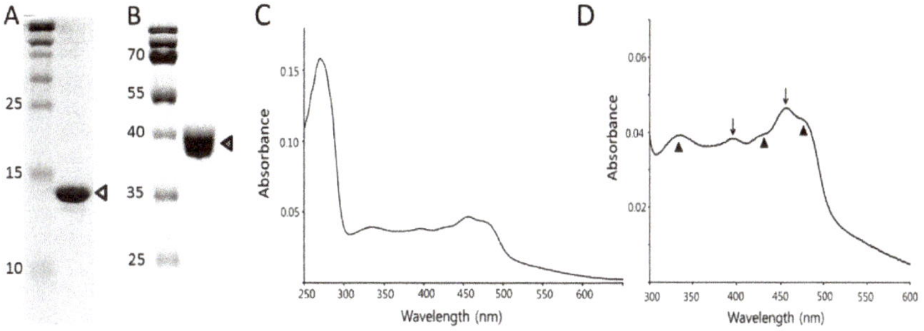

Figure 4. Purification of regulatory (MMOB) and reductase (MMOR) components. (**A**) Purified MMOB (empty triangle) through use of a recombinant plasmid (*pET30a-mmoB*) expression in *E. coli* (unit: kDa). (**B**) Purified MMOR (dotted triangle) through recombinant plasmid (*pET30a-mmoC*). (**C**) UV–visible spectrum of purified MMOR from *M. sporium* 5 showing the oxidized MMOR. (**D**) Enlarged UV–visible spectrum of MMOR from 300 to 600 nm. Arrows and triangles represent the oxidized FAD-containing domain and [2Fe–2S] cluster domain, respectively.

MMOR is an essential component for the catalytic cycle owing to its electron transfer abilities, which are accomplished by FAD-containing and [2Fe–2S] cluster ferredoxin domains to reduce diiron active sites in MMOH [13,17,44,45]. Kinetic studies have proposed that NADH binds to the MMOR-FAD in MMOH to transfer hydride, and the conformational change of NADH–FAD generates charge transfer bands [28,29,46]. The overall shape and volume of the MMOR-Fd are similar to those of the MMOB core region, suggesting that MMOB and MMOR share the binding sites of MMOH [11]. A synthetic nucleotide sequence encoding MMOR was inserted into pET30a and expressed in *E. coli* (*pET30a-mmoC*, Figure S7A) to achieve highly concentrated MMOR (ε_{458} = 21,200 $cm^{-1}M^{-1}$) for enzymology assays and was purified using two-step anionic exchange columns and a single-step size exclusion column (Figure 4B). Ferrozine assay confirmed the presence of two irons (2.03–2.14 Fe/MMOR; Figure S7B) in the [2Fe–2S] cluster of MMOR-Fd. Two domains of MMOR, including MMOR-Fd and MMOR-FAD, show specific absorbance at approximately 300–600 nm (Figure 4C,D). Three major peaks were generated from MMOR at 334, 395, and 458 nm (indicated by arrows in Figure 4D), and these values were similar to those of MMOR from *M. capsulatus* Bath (332, 394, and 458 nm) and *M. trichosporium* OB3b [17,29]. Reported data from *M. capsulatus* Bath have shown that an oxidized MMOR-FAD generates peaks at 394 and 458 nm and the [2Fe–2S] cluster yields peaks at 332, 418, and 467 nm. These results confirmed the existence of cofactors of MMOR.

2.4. Activities of Essential Enzymes from M. sporium 5

Purified enzymes were used to measure SEA of MMOH from *M. sporium* 5, and optimal activity (494 ± 19 mU/mg) was found with the substrate propylene in 25 mM sodium phosphate buffer with 2 mol equiv. of MMOB to MMOH at pH 7.5 (Figure 5A). MMOB affects the redox potential of MMOH and controls the formation of intermediates that finally alters product distribution significantly in *M. trichosporium* OB3b [22,47,48]. Hydroxylation of NO_2–benzene produces *m*-NO_2–phenol as the major product (>90%) in the absence of MMOB, but 2 mol equiv. of MMOB with MMOH produce 89% *p*-NO_2–phenol [22]. Structural studies of the MMOH–MMOB complex demonstrated that 2 mol equiv. of MMOB bind to MMOH symmetrically to each $\alpha\beta\gamma$ promoter to regulate the substrate pathways through hydrophobic cavities. This study also confirmed that less than 2 mol equiv. of MMOB are not sufficient for optimal activity of MMOH ($\alpha_2\beta_2\gamma_2$) and 3 mol equiv. of MMOB may block the interaction of MMOR to MMOH for efficient electron transfer. The activities of 1 and 3 mol equiv. of MMOB/MMOH showed reduced SEA by more than 35% and 20%, respectively. The ratio of MMOR/MMOH in the presence of 2 mol equiv. of MMOB influenced SEA of MMOH, but more than 1 mol equiv. of MMOR did not significantly increase MMOH activity (Figure S8). SEA of 2 mol equiv. of MMOR (515 ± 15 mU/mg) was similar to that of 1 mol equiv. of MMOR (494 ± 19 mU/mg). The proposed mechanisms of sMMO demonstrated that MMOB binds to MMOH for intermediate generation, but MMOR/MMOH interaction is required for a relatively short time compared with that of MMOB in the resting states of MMOH (Fe^{III}-Fe^{III}), and 1 mol equiv. of MMOR was applied to measure activity in this study.

A report demonstrated that highly purified MMOH exhibits SEA of 185 mU/mg because all multi-components are not successfully purified from *M. sporium* 5 [26]. Our results demonstrated SEA of 28 mU/mg in the absence of MMOB (Figure 5B); reports have also proposed that MMOB is essential for the activity because it can improve activity by more than 20-fold. Another type II methanotroph, *M. trichosporium* IMV 3011, exhibits SEA of 603.6 mU/mg, and the optimal activity is obtained at pH 7.2 in 25 mM Tris-HCl buffer [38]. Highly purified enzymes are required for sMMO studies, possibly because of other components that are co-expressed in methanotrophs. Although the specific function of MMOD (*orfY*) has yet to be uncovered, it is considered as a catalytic inhibitor of MMOH. In addition, MMOH stability is a crucial factor for achieving high SEA in sMMO, and $Fe(NH_4)_2(SO_4)_2$, cysteine, dithiothreitol (DTT), and sodium thioglycolate are added to maintain the stability of cell extracts during purification [35,49].

The FAD-containing domain (MMOR-FAD; MMOR residues 97–343) provides crucial clues to understand these activity changes at different pH conditions (Figure 6). MMOR-FAD was expressed using the constructed *pET30a-mmoC-FAD* plasmid (Figure S9) and was purified to measure the UV–visible absorption spectra in solutions with different acidities. MMOR-FAD yielded distinct peaks at 396 nm and 456 nm from pH 6 to 8 (Figure 6). The absorption ratio (A_{270}/A_{458}) of MMOR-FAD was approximately 6.8, implying that flavin co-factors are fully complemented in the purified MMOR [29]. Oxidized MMOR-FAD spectra in an earlier study proposed that buffer acidity changes the absorption at 394 nm because of the subtle perturbation of the flavin-binding site in MMOR, and alterations of hydrogen bonding or solvent accessibility occur due to the conformational changes of isoalloxazine in FAD [29]. The intensity at 396 nm increased up to pH 7.0, and absorption reduced at pH 8.0. The absorption ratio between 458 and 396 nm (A_{458}/A_{396}) changed from 1.29 at pH 4.0 to 1.4 at pH 8.0, and these results agree with the optical spectrum of oxidized flavin, which shows absorption changes based on acidity. Lowest activity was measured at around pH 5.0 (Figure 5; 72 ± 5.8 mU/mg), and MMOR-FAD showed the lowest absorption at 396 nm and absorption at 456 nm at pH 4–5 (Figure 6). The spectra showed distinct peaks at approximately 456 nm at pH 6–8, and a small peak was observed at approximately 476 nm, similar to the peak observed at the same pH range usually detected for oxidized FAD, but these absorptions were not observed at low pH values. These results indicated that electron transfer thorough the flavin cofactor is an important aspect for diiron activation. The reported optical spectra of MMOR-FAD from *M. capsulatus* Bath demonstrated that absorption decreases at 500 nm and absorption increases gradually at 450 nm by changing the pH from 5 to 9.

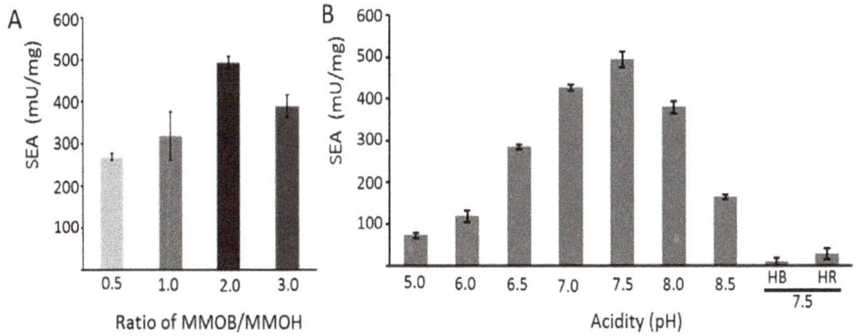

Figure 5. SEA of purified sMMO, including MMOH, MMOB, and MMOR, from *M. sporium* 5. sMMO converts propylene to propylene oxide in the presence of NADH. (**A**) Purified MMOH (1.0 mol equiv.), MMOR (1.0 mol equiv.), and different mole ratios of MMOB were incubated with propylene in 25 mM sodium phosphate buffet at 30 °C at pH 7.5. (**B**) The maximal epoxidation activity (mol ratio of MMOH:MMOB:MMOR = 1:2:1) of propylene was monitored at different acidities, and control experiments confirmed that MMOB and MMOR are essential enzymes for catalytic activity. HB and HR represent the absence of MMOR (HB) and MMOB (HR), respectively. All experiments were performed at least three times, and error bars represent standard deviations.

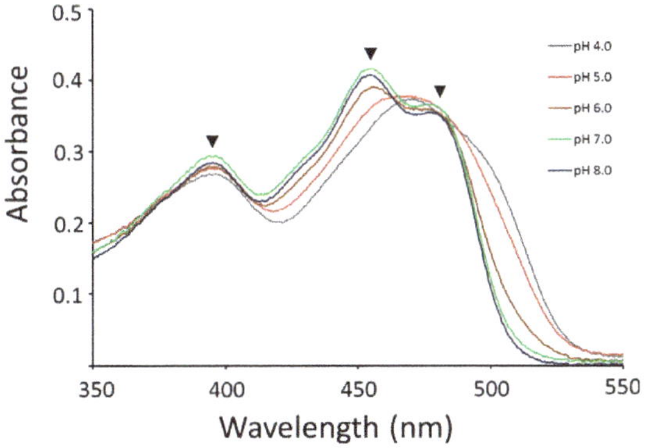

Figure 6. UV–visible absorption spectra of oxidized MMOR-FAD depending on the pH. Optical spectra were measured of purified MMOR-FAD (97–343 residues of MMOR from *M. sporium* 5) at different acidity values. Filled triangles represent flavin absorptions at 396, 456, and 476 nm.

2.5. Oxidation of Alkane, Halogen, and Aromatic Compounds

One interesting aspect of sMMO of the BMM superfamily members is the oxidation ability of various substrates, including alkanes, alkenes, aromatics, heterocyclics, and chlorinated compounds [4,14,50–52]. Dalton et al. showed that cell extracts of *M. capsulatus* Bath can hydroxylate C_1–C_8 alkanes and aromatic compounds, such as benzene and toluene [42]. Hydroxylation ability of purified sMMO components from *M. sporium* 5 was measured with alkanes, including pentane and heptane (Figure 7A). SEA results showed that pentane (102 ± 4.6 mU/mg) and heptane (109 ± 9.2 mU/mg) were hydroxylated by sMMO, and the overall activity toward heptane was slightly higher, indicating that the activity of sMMO toward *n*-alkanes decreased sharply from pentane to octane owing to the substrate size. The rate of substrate access to the diiron active site for hydroxylation is lowered because this substrate moves along the proposed pathways from cavity 3 to cavity 1 [23,53]. These results also suggested that non-specific hydroxylation of substrates occurs in *n*-alkanes because oxidation of alkanes through sMMO generates different hydroxylated products. *n*-Alkanes can be oxidized at the primary and secondary alkyl with different product ratios, although secondary alkyl hydroxylation is the major product in most cases [42]. Optimal enzyme activity was observed at pH 7.5 for most substrates in this study, but heptane showed slightly better activity at pH 7.5 (128 ± 4.1 mU/mg) compared with that at pH 7.0 (109 ± 9.2 mU/mg). These results also support that sMMO is a non-specific enzyme and that oxidation is not tightly regulated because hydroxylation of substrates produces different products. sMMO extracts from *M. capsulatus* Bath hydroxylate pentane to produce pentan-1-ol (27%), pentan-2-ol (69%), and pentan-3-ol (<3%), and other alkanes, including hexane, heptane, and octane, to yield non-specific hydroxylated products [42]. Complex formation among MMOH, MMOB, and MMOR may occur slightly differently based on the acidity of the solution, which would affect product formation and turnover number.

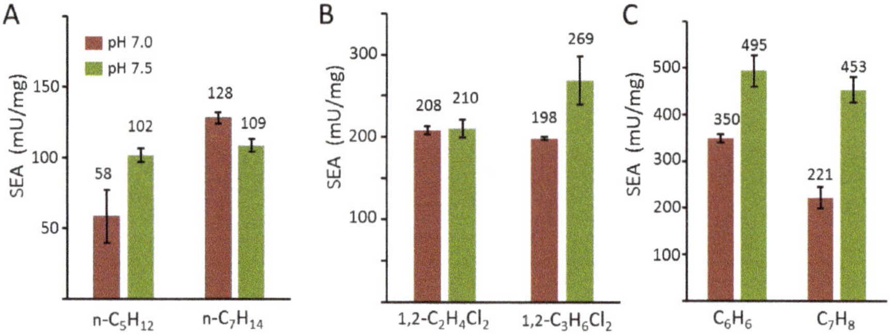

Figure 7. Soluble MMO from *M. sporium* 5 oxidized various substrates, including alkane, halogen, and aromatic compounds. (**A**) Normal pentane (n-C_5H_{12}) and normal heptane (n-C_7H_{14}) harbor more carbon atoms than propylene. The activities were measured at 340 nm by measuring NADH degradation. (**B**) Halogen compounds, including 1,2-dichloroethane (1,2-$C_2H_4Cl_2$) and 1,2-dichloropropane (1,2-$C_3H_6Cl_2$), were oxidized by sMMO. (**C**) Measurement of specific enzyme activity with benzene (C_6H_6) and toluene (C_7H_8). NADH consumption was not observed in the absence of substrates. SEA represents specific enzyme activity (n = 3, avg ± s.d.).

Hydroxylation of chlorinated or aromatic compounds is a fascinating ability because toxic and environmentally hazardous materials are converted to non-toxic or less-toxic products [14,42]. Owing to these reasons, sMMO is investigated for application in the field of environmental enzymology for hydroxylation of chlorinated and aromatic compounds. Chlorinated compounds, including 1,1-dichloroethane (1,1-$C_2H_4Cl_2$) and 1,1-dichloropropane (1,1-$C_3H_6Cl_2$), and aromatic compounds, including benzene and toluene, were incubated with the sMMO system (Figure 7B,C). These results showed that sMMO from *M. sporium* 5 did not exhibit substrate specificity like other sMMOs from different methanotrophs. The chlorinated compounds had lower SEA compared with propylene epoxidation but benzene and toluene showed almost similar activities to that of propylene at pH 7.5. These results suggest that sMMO from *M. sporium* 5 has lower substrate specificity to aromatic compounds. There is no direct evidence on how these large substrates get access to the diiron active sites; however, previous results demonstrated that a different molar ratio of MMOB affects the product distributions [22]. The cavities of sMMO allow small substrate transport pathways, such as methane, when crystals were compressed with Xe and large hydrophobic cavities were not developed like other BMM superfamily members. The relatively good enzyme activities in benzene by sMMO may be due to the structural flexibility near the diiron active sites in *M. sporium* 5 compared to other organisms.

3. Material and Methods

3.1. General Materials and Chemicals

M. sporium 5 (ATCC35069) was obtained from the American Type Culture Collection (ATCC), and nitrate mineral salts (NMS) media (ATCC 1306) were prepared based on the manufacturer's protocol. All chemicals included in NMS media were purchased from Sigma-Aldrich (St. Louis, MO, USA). pET-30a(+) and BL21(DE3) were purchased from EMD Millipore, and DH5α(DE3) was purchased from New England Biolabs. Synthetic nucleotides (*mmoC* and *mmoB* from *M. sporium* 5 and *M. species* M and primers (Table S1) were purchased and sequenced from Cosmogenetech. DNA extraction and purification were performed with a FavorPrep kit (Favorgen), and images were obtained from an E-Gel Imager System with UV Light Base (Thermo Fisher). Luria Broth (Lennox, powder microbial growth medium) was purchased from Sigma-Aldrich. Methane (99.9%), argon (99.9%), and propylene (99.5%) gases were purchased from Hankook Special Gases. All columns were attached to an ATKA Pure 25L, fast protein liquid chromatograph (GE Healthcare Life Science) to purify proteins. All chemicals were purchased from Sigma-Aldrich (St. Louis, MO, USA) and Alfa Aesar unless indicated otherwise.

3.2. Culturing M. sporium *5 and MMOH Purification*

M. sporium 5 was incubated in copper-limited NMS media at 30 °C in a 4-L rubber cap-sealed baffled Erlenmeyer flask at 220 rpm to achieve an OD_{600} of 8–10 with a methane:air (v/v) ratio of 1:5. pH was adjusted to 7.0 with 100 mM NaOH and 100 mM HCl. The optical density of grown cultures was measured at 600 nm using a UV–visible spectrometer (Cary 60 Agilent). Normal NMS medium contained 20.4 mg/mL iron during the cell culture, and additional iron was added to the iron-rich NMS media using $(NH_4)_2Fe(SO_4)_2{\cdot}6H_2O$ to attain final concentrations of 40.8, 61.2, or 122.4 mg/mL of iron. The images of media with different iron concentrations were obtained when OD_{600} was 8.5 (Figure 2B).

M. sporium 5 cells were harvested by centrifugation (11,300× *g*, 20 min; Supra 22K model, Hanil Science) at 4 °C. Cell pellet from 5 L of culture was suspended in 25 mM MOPS, 25 mM NaCl, 8 mM sodium thioglycolate, 2 mM L-cysteine, 200 μM $(NH_4)_2Fe(SO_4)_2{\cdot}6H_2O$, 5 mM $MgCl_2$, 0.25 μL/mL DNase, and 0.04 mg/mL phenylmethane sulfonyl fluoride (PMSF) at pH 6.5. The dissolved suspension was disrupted by sonication at 4 °C (CV334 model, Sonics). The cell lysate was centrifuged at 30,000× *g* for 45 min at 4 °C (Combi 514R model, Hanil Science), and the supernatant was carefully decanted and filtered through a 0.22-μm membrane (Merck Millipore).

The filtrate was loaded onto a DEAE sepharose fast flow column (GE Healthcare) packed in an XK 50/60 column (GE Healthcare) and equilibrated with Buffer A (25 mM MOPS at pH 6.5 containing 8 mM sodium thioglycolate, 2 mM L-cysteine, 200 μM $(NH_4)_2Fe(SO_4)_2{\cdot}6H_2O$, 5 mM $MgCl_2$, and 50 mM NaCl). After loading the supernatant, the column was equilibrated with 1000 mL of buffer A and eluted with 2000 mL of buffer A in a linear gradient from 0–500 mM NaCl at a flow rate of 1.0 mL/min. MMOH-containing fractions (Buffer A contained around 250 mM NaCl) were resolved by SDS-PAGE (Bio-Rad) and stained with Coomassie Brilliant Blue to confirm the presence of MMOH. MMOH-containing fractions were concentrated with a molecular weight cut-off membrane (Amicon ultracentrifugal filter 30 kDa; Merck Millipore). The concentrated MMOH was loaded onto a Superdex 200 column (GE Healthcare) packed in an XK26/70 column (GE Healthcare), equilibrated with 25 mM MOPS, containing 8 mM sodium thioglycolate, 2 mM L-cysteine, 200 μM $(NH_4)_2Fe(SO_4)_2 \cdot 6H_2O$, 200 mM NaCl, 1 mM DTT, and 5% glycerol. MMOH-containing fractions were resolved with SDS-PAGE and concentrated with a molecular weight cut-off membrane (30 kDa). The concentrated sample was loaded onto a Q sepharose fast flow column packed in an XK 26/40 column (GE Healthcare) with Buffer B (25 mM MOPS containing 1 mM DTT, 50 mM NaCl, and 10% glycerol). MMOH was eluted with Buffer B with a linear gradient from 0 to 500 mM NaCl. Selected fractions after SDS-PAGE were concentrated through an Amicon ultracentrifugal filter to obtain MMOH with >95% purity and stored at −88 °C before application for further experiments.

3.3. Construction of mmoB and mmoC from M. sporium *5 and M. species M*

A synthetic *mmoB* nucleotide sequence in a pUC18 vector (*pUC18-mmoB*, Cosmogentech) was amplified in DH5α (DE3) and then extracted using a QIAprep Spin Miniprep kit (QIAGEN). *mmoB* (417 bp) was amplified from *pUC18-mmoB* by PCR using two oligonucleotide-primers, *mmoB-M-Fd* and *mmoB-M-Rv* (Table S1). Amplified *mmoB* was digested with *Nde*I and *Hind*III (Takara) and was then inserted into pET30a(+) to construct *pET30a-mmoB* (Figure S6). *mmoB* from *M. species* M was constructed using the same protocols with two nucleotide primers, *mmoB-MM-Fd* and *mmoB-MM-Rv* (Table S1).

A synthetic *mmoC* nucleotide sequence in a pUC18 vector (*pUC18-mmoC*, Cosmogentech) was amplified in DH5α (DE3) and extracted using a QIAprep Spin Miniprep kit. *mmoC* (1,032 bp) was amplified in *pUC18-mmoC* by PCR using two oligonucleotide primers, *mmoC-M-Fd* and *mmoC-M-Rv* (Table S1). Amplified *mmoC* was digested with *Nde*I and *Hind*III and was then inserted into pET30a(+) to construct *pET30a-mmoC* (Figure S7). *mmoC* from *M. species* M was constructed using the same protocol with two nucleotide primers, *mmoC-MM-Fd* and *mmoC-MM-Rv* (Table S1).

mmoC-FAD (744 bp) was amplified in *pUC18-mmoC* by PCR using two oligonucleotide primers, *mmoC-FAD-M-Fd* and *mmoC-FAD-M-Rv* (Table S1). Amplified *mmoC* was digested with *Nde*I and *Hind*III and was then inserted into pET30a(+) to construct *pET30a-mmoC* (Figure S9).

3.4. Expression and Purification of MMOB

The constructed *pET30a-mmoB* was transformed into BL21 (DE3) cells (Novagen) and cultured in LB medium containing 50 μg/mL kanamycin at 37 °C until mid-log phase. Cell cultures were cultured until 5 h post-induction with 1.0 mM isopropyl β-D-1-thiogalactopyranoside (IPTG) before harvesting by centrifugation (11,300× *g* for 20 min at 4 °C). The cell pellets were suspended in 25 mM phosphate at pH 6.0, 75 mM NaCl, 1 mM Na_2-EDTA, 1 mM DTT, 5 mM $MgCl_2$, 0.25 μL/mL DNase I, and 0.04 mg/mL PMSF, and then suspensions were disrupted by sonication at 4 °C. The lysed cells were centrifuged at 26,000× *g* at 4 °C for 60 min, and supernatant was carefully decanted and filtered through a 0.22-μm membrane. The filtrate was loaded onto a Q sepharose fast flow column packed in a XK 26/40 column equilibrated in 25 mM phosphate (pH 6.0), containing 75 mM NaCl, 1 mM Na_2-EDTA, 1 mM DTT, and 5% glycerol (Buffer C). After loading the supernatant, the column was washed with 700 mL of buffer C with a linear gradient from 0 to 500 mM NaCl at a flow rate of 1.0 mL/min. MMOB-containing fractions were collected and analyzed by SDS-PAGE and were then concentrated using a molecular weight cut-off membrane (10 kDa). Concentrated MMOB was loaded onto a Superdex 75 column (GE Healthcare) packed in a XK16/70 column equilibrated in Buffer C. MMOB-containing fractions were collected and analyzed by SDS-PAGE and were then concentrated using a cut-off membrane to obtain MMOB (colorless solution) with >95% purity, which was stored at −88 °C before application for further experiments.

3.5. Expression and Purification of MMOR and MMOR-FAD

The constructed *pET30a-mmoC or pET30a-mmoC-FAD* was transformed into BL21 (DE3) cells (Novagen) and cultured in LB medium, containing 50 μg/mL kanamycin and 500 μM $(NH_4)_2Fe(SO4)_2{\cdot}6H_2O$, at 37 °C until mid-log phase. Cell cultures were incubated for 12 h post-induction with 0.5 mM IPTG at 25 °C at 200 rpm before harvesting by centrifugation (11,300× *g* for 20 min at 4 °C). Cell pellets were suspended in 25 mM MOPS at pH 6.5, containing 25 mM NaCl, 8 mM sodium thioglycolate, 2 mM L-cysteine, 200 μM $(NH_4)_2Fe(SO_4)_2{\cdot}6H_2O$, 5 mM $MgCl_2$, 1 unit/mL DNase I, and 0.2 mM PMSF, and were disrupted by sonication. The lysed cells were centrifuged at 26,000× *g* at 4 °C for 40 min, and the supernatant was carefully decanted and filtered through a 0.22 μm membrane. The filtrate was loaded onto a DEAE sepharose fast flow column packed in XK 50/40 column equilibrated in 25 mM MOPS (pH 6.5), containing 8 mM sodium thioglycolate, 2 mM L-cysteine, 200 μM $(NH_4)_2Fe(SO_4)_2{\cdot}6H_2O$, and 25 mM NaCl (Buffer D). After loading the supernatant, the column was washed with 1 L of buffer D with a linear gradient from 0 to 600 mM NaCl at a flow rate of 1.5 mL/min. MMOR-containing fractions were collected and analyzed by SDS-PAGE and were then concentrated using a molecular weight cut-off membrane (10 kDa). Concentrated MMOR was loaded onto a Q sepharose column (GE Healthcare) packed in XK26/40 column equilibrated in 25 mM MOPS (pH 6.5), containing 50 mM NaCl, 1 mM DTT, and 10% glycerol (Buffer E). After loading the supernatant, the column was washed with 1.2 L of buffer E with a linear gradient from 0 to 500 mM NaCl at a flow rate of 1.5 mL/min. MMOR-containing fractions were collected and analyzed by SDS-PAGE and were then concentrated using a cut-off membrane to obtain MMOR (red–brown solution) with >95% purity, which was stored at −88 °C before application for further experiments.

3.6. Measurement of SEA with Various Substrates

MMOH (1.0 μM), MMOB (2.0 μM), and 1.0 μM MMOR were added to 25 mM phosphate and 10 mM NaCl (Buffer F) and bubbled with propylene (Hankook Special Gas) for 20 min. Other substrates—including pentane, heptane, 1,2-dichloroethane, 1,2-dichloropropane, benzene, and toluene—were added to buffer F to obtain a final concentration of 20 mM. The whole mixture was incubated at 30 °C at the indicated pH values (4.0–8.0) for acidity-dependent experiments. Steady-state kinetics were measured using a Cary 60 UV–visible spectrometer at 340 nm (ε_{340} = 6220 $cm^{-1}M^{-1}$). The temperature was regulated by a water bath at 30 °C. Products were confirmed by gas chromatography (YL 6500GC system) using an Agilent HP-PLOT/Q stationery column (30 m × 0.535 mm × 40.00 μm). All experiments were performed in the absence of substrates as a negative control, which did not exhibit SEA.

SEA were measured in *M. sporium* 5 and *M. species* M using propylene. System I measured the activity by incubation with MMOH (1.0 mol equiv. from *M. sporium* 5), MMOB (2.0 mol equiv. from *Methylocystis sp.* M), and MMOR (0.5 mol equiv. from *M. sporium* 5). System II was incubated with MMOH (1.0 mol equiv. from *M. sporium* 5), MMOB (2.0 mol equiv. from *M. sporium* 5), and MMOR (0.5 mol equiv. from *M. species* M).

4. Conclusions

To elucidate the catalytic mechanisms of sMMO, *M. sporium* 5 can be used for isolating intermediates. Oxidation rates were considerably different depending on the concentration and proportion of MMOB, and it was observed that more than 2 mol equiv. of MMOB may retard electron transfer by blocking MMOR. The methanotrophic bacterium *M. sporium* 5 has been investigated for more than four decades, but its physicochemical properties are not intensively understood compared with those of *M. capsulatus* Bath and *M. trichosporium* OB3b.

One of the interesting aspects of sMMO is its broad oxygenase ability, and this study confirmed that sMMO from *M. sporium* 5 can hydroxylate diverse substrates and epoxidize propylene. The enzyme activity of sMMO from *M. sporium* 5 was comparable to that of other sMMOs and showed relatively high enzyme activity for aromatic substrates. Furthermore, it was observed that substitution with MMOB and MMOR from another species, *M. species* M, retains its enzyme activity (Figure S10). The amino acid sequences of MMOB and MMOR between *M. sporium* 5 and *M. species* M showed 99% and 94% identity, respectively (Figure 1). MMOB and MMOR of *M. species* M were expressed and purified to measure SEA, and these results showed that substituting with MMOB and MMOR from other species restored activities (Figure S10). System I represents MMOB substitution from *M. species* M, and System II represents MMOR substitution from the same species. Enzymatic activities proved that MMOB and MMOR showed cross-reactivity between *M. sporium* 5 and *M. species* M because System I and II yielded SEA values of 527 ± 8.3 mU/mg and 519 ± 28.2 mU/mg, respectively. The crossreactivity from *M. capsulatus* Bath and *M. sporium* 5 was tested, and results proved that enzymatic activities were unchanged [54]. These results prove that MMOB and MMOR retain their structural and functional roles with hydroxylase from other species. This fundamental study on purified enzymes of *M. sporium* 5 can provide another platform for further studying these complicated catalytic cycles.

Supplementary Materials: The following are available online at http://www.mdpi.com/2073-4344/8/12/582/s1, Figure S1. Multiple sequence amino acid alignment of sMMO from *Methylosinus sporium* 5 (type II), *Methylosinus trichosporium* OB3b (type II), *Methylocystis* species M (type II), and *Methylococcus capsulatus* Bath (type X). (A) Alignment of amino acids from the α-subunit of MMOH (GenBank accession number: ABD46892). Gray and red colors represent identical and different amino acids, respectively. Blue represents the identical residues among type II methanotrophs. (B) Alignment of amino acids from the β-subunit of MMOH (GenBank accession number: ABD46893). Gray and red colors represent identical and different amino acids, respectively. Blue represents the identical residues among type II methanotrophs. (C) Alignment of amino acids from the γ-subunit of MMOH (GenBank accession number: ABD46895). Gray and red colors represent identical and different amino acids, respectively. Blue represents the identical sequences among type II methanotrophs. Figure S2. Multiple amino acid alignment of MMOB (GenBank accession: ABD46894) from *Methylosinus sporium* 5 (type II), *Methylosinus trichosporium* OB3b (type II), *Methylocystis species* M (type II), and *Methylococcus capsulatus* Bath (type X). Figure S3. Multiple amino acid alignment of MMOR (GenBank accession: ABD46897) from *Methylosinus sporium* 5 (type II), *Methylosinus trichosporium* OB3b (type II), *Methylocystis species* M (type II), and *Methylococcus capsulatus* Bath (type X). Figure S4. Growth and pigmentation profile of sMMO from *M. sporium* 5 based on optical density at 600 nm (OD_{600}). (A) Cell growth at OD_{600} of 3.0 with one-time methane feeding (once/day). (B) Cell growth at OD_{600} of 8.2 with two-time methane supply (twice/day). Cultures A and B were proliferated for a week in media with the same iron concentration (20.4 mg/mL). Figure S5. Ferrozine assay of MMOH showed iron contents in MMOH (3.8–4.1 Fe/MMOH). The maximum wavelength was monitored at 562 nm from the iron–ferrozine complex. Heavy and light lines represent denatured MMOH and standard solutions, respectively. All experiments were performed in triplicate with R^2 values > 0.999. Figure S6. Constructed pET30a-*mmoB* plasmids and UV–visible spectrum of MMOB after purification. (A) Lane 1 represents the control ladder, lane 2 represents the pET30a-*mmoB* construct, and lane 3 represents nucleotides after *Nde*I and *Hind*III double digestion. (B) UV–vis spectrum of MMOB after construction. Figure S7. Constructed *pET30a-mmoC* plasmid and ferrozine assays from MMOR (A) Lane 1 represents the control ladder, lane 2 represents the *pET30a-mmoC* construct, and lane 3 represents the *pET30a-mmoC* after *Nde*I and *Hind*III double digestion. (B) Ferrozine assay of MMOR for demonstrating the iron content in MMOR (2.03–2.14 Fe/MMOR). The maximum wavelength was monitored at 562 nm from the iron–ferrozine complex. Heavy and light lines represent iron solutions from denatured MMOR and standard solutions, respectively. All experiments were performed in triplicate with R^2 values > 0.998. Figure S8. Specific enzyme activity based on the ratio of MMOR to MMOH in 2 mol equivalents of MMOB at pH 7.5. Figure S9. Constructed *pET30a-mmoC-FAD* plasmids. Lane 1 represents the control ladder, lane 2 represents the *pET30a-mmoC* construct, and lane 3 represents nucleotides after *Nde*I and *Hind*III double digestion. Figure S10. Measurement of crossreactivity in MMOB and MMOR. System I consists of MMOH (*M. sporium* 5), MMOB (*M. species* M), and MMOR (*M. sporium* 5). System II consists of MMOH (*M. sporium* 5), MMOB (*M. sporium* 5), and MMOR (*M. species* M). Specific enzymatic activities were measured at 30 °C and pH 7.5. Table S1. Primers used in this study.

Author Contributions: D.-H.L. and S.J.L designed the experiments; Y.R.P, M.Y.S., and H.S.Y. performed the gene cloning, expression, and purification of enzymes and activity tests. Y.R.P. and S.J.L. characterize enzymes and activity tests; S.J.L wrote the manuscript.

Funding: This research was supported by the C1 Gas Refinery Program through the National Research Foundation of Korea (NRF-2015M3D3A1A01064876) and Basic Science Research Program through the National Research Foundation of Korea (NRF) funded by the Ministry of Education (2017R1A6A1A03015876).

Conflicts of Interest: The authors declare no conflicts of interest.

References

1. Hanson, R.S.; Hanson, T.E. Methanotrophic bacteria. *Microbiol. Rev.* **1996**, *60*, 439–471. [PubMed]
2. Lawton, T.J.; Rosenzweig, A.C. Methane-oxidizing enzymes: An upstream problem in biological gas-to-liquids conversion. *J. Am. Chem. Soc.* **2016**, *138*, 9327–9340. [CrossRef] [PubMed]
3. Semrau, J.D.; DiSpirito, A.A.; Yoon, S. Methanotrophs and copper. *FEMS. Microbiol. Rev.* **2010**, *34*, 496–531. [CrossRef] [PubMed]
4. Lieberman, R.L.; Rosenzweig, A.C. Biological methane oxidation: Regulation, biochemistry, and active site structure of particulate methane monooxygenase. *Crit. Rev. Biochem. Mol.* **2004**, *39*, 147–164. [CrossRef] [PubMed]
5. Sullivan, J.P.; Dickinson, D.; Chase, H.A. Methanotrophs, *Methylosinus trichosporium* OB3b, sMMO, and their application to bioremediation. *Crit. Rev. Microbiol.* **1998**, *24*, 335–373. [CrossRef] [PubMed]
6. Leahy, J.G.; Batchelor, P.J.; Morcomb, S.M. Evolution of the soluble diiron monooxygenases. *FEMS Microbiol. Rev.* **2003**, *27*, 449–479. [CrossRef]
7. Warmuzinski, K. Harnessing methane emissions from coal mining. *Process Saf. Environ.* **2008**, *86*, 315–320. [CrossRef]
8. Haynes, C.A.; Gonzalez, R. Rethinking biological activation of methane and conversion to liquid fuels. *Nat. Chem. Biol.* **2014**, *10*, 331–339. [CrossRef] [PubMed]
9. Sirajuddin, S.; Barupala, D.; Helling, S.; Marcus, K.; Stemmler, T.L.; Rosenzweig, A.C. Effects of zinc on particulate methane monooxygenase activity and structure. *J. Biol. Chem.* **2014**, *289*, 21782–21794. [CrossRef] [PubMed]
10. Sirajuddin, S.; Rosenzweig, A.C. Enzymatic oxidation of methane. *Biochemistry* **2015**, *54*, 2283–2294. [CrossRef] [PubMed]
11. Wang, V.C.C.; Maji, S.; Chen, P.R.Y.; Lee, H.K.; Yu, S.S.F.; Chan, S.I. Alkane oxidation: Methane monooxygenases, related enzymes, and their biomimetics. *Chem. Rev.* **2017**, *117*, 8574–8621. [CrossRef] [PubMed]
12. Hakemian, A.S.; Rosenzweig, A.C. The biochemistry of methane oxidation. *Annu. Rev. Biochem.* **2007**, *76*, 223–241. [CrossRef] [PubMed]
13. Gassner, G.T.; Lippard, S.J. Component interactions in the soluble methane monooxygenase system from *Methylococcus capsulatus* (Bath). *Biochemistry* **1999**, *38*, 12768–12785. [CrossRef] [PubMed]
14. Merkx, M.; Kopp, D.A.; Sazinsky, M.H.; Blazyk, J.L.; Muller, J.; Lippard, S.J. Dioxygen activation and methane hydroxylation by soluble methane monooxygenase: A tale of two irons and three proteins. *Angew. Chem. Int. Ed.* **2001**, *40*, 2782–2807. [CrossRef]
15. Rosenzweig, A.C.; Nordlund, P.; Takahara, P.M.; Frederick, C.A.; Lippard, S.J. Geometry of the soluble methane monooxygenase catalytic diiron center in two oxidation states. *Chem. Biol.* **1995**, *2*, 409–418. [CrossRef]
16. Sazinsky, M.H.; Lippard, S.J. Correlating structure with function in bacterial multicomponent Monooxygenases and related diiron proteins. *Acct. Chem. Res.* **2006**, *39*, 558–566. [CrossRef] [PubMed]
17. Fox, B.G.; Froland, W.A.; Dege, J.E.; Lipscomb, J.D. Methane monooxygenase from *Methylosinus trichosporium* OB3b. *J. Biol. Chem.* **1989**, *264*, 10023–10033. [PubMed]
18. Hanson, R.S.; Wattenberg, E.V. Ecology of methylotrophic bacteria. *Biotechnology* **1991**, *18*, 325–348. [PubMed]
19. Banerjee, R.; Proshlyakov, Y.; Lipscomb, J.D.; Proshlyakov, D.A. Structure of the key species in the enzymatic oxidation of methane to methanol. *Nature* **2015**, *518*, 431–434. [CrossRef] [PubMed]
20. Elango, N.; Radhakrishnan, R.; Froland, W.A.; Wallar, B.J.; Earhart, C.A.; Lipscomb, J.D.; Ohlendorf, D.H. Crystal structure of the hydroxylase component of methane monooxygenase from *Methylosinus trichosporium* OB3b. *Protein Sci.* **1997**, *6*, 556–568. [CrossRef] [PubMed]
21. Rosenzweig, A.C.; Frederick, C.A.; Lippard, S.J.; Nordlund, P. Crystal structure of a bacterial nonheme iron hydroxylase that catalyzes the biological oxidation of methane. *Nature* **1993**, *366*, 537–543. [CrossRef] [PubMed]
22. Froland, W.A.; Andersson, K.K.; Lee, S.K.; Liu, Y.; Lipscomb, J.D. Methane monooxygenase component B and reductase alter the regioselectivity of the hydroxylase component-catalyzed reactions. A novel role for protein-protein interactions in an oxygenase mechanism. *J. Biol. Chem.* **1992**, *267*, 17588–17597. [PubMed]

23. Lee, S.J.; McCormick, M.S.; Lippard, S.J.; Cho, U.S. Control of substrate access to the active site in methane monooxygenase. *Nature* **2013**, *494*, 380–384. [CrossRef] [PubMed]
24. Ali, H.; Scanlan, J.; Dumont, M.G.; Murrell, J.C. Duplication of the *mmoX* gene in *Methylosinus sporium*: Cloning, sequencing and mutational analysis. *Microbiology* **2006**, *152*, 2931–2942. [CrossRef] [PubMed]
25. Nakajima, T.; Uchiyama, H.; Yagi, O.; Nakahara, T. Purification and properties of a aoluble methane monooxygenase from *Methylocystis* sp. M. *Biosci. Biotechnol. Biochem.* **1992**, *56*, 736–740. [CrossRef] [PubMed]
26. Pilkington, S.J.; Dalton, H. Purification and characterization of the soluble methane monooxygenase from *Methylosinus sporium* 5 demonstrates the highly conserved nature of this enzyme in methanotrophs. *FEMS Microbiol. Lett.* **1991**, *78*, 103–108. [CrossRef]
27. Whittenbury, R.; Phillips, K.C.; Wilkinson, J.F. Enrichment, isolation and some properties of methane utilizing bacteria. *J. Gen. Microbiol.* **1970**, *61*, 205–218. [CrossRef] [PubMed]
28. Blazyk, J.L.; Gassner, G.T.; Lippard, S.J. Intermolecular electron transfer reactions in soluble methane monooxygenase: A role for hysteresis in protein function. *J. Am. Chem. Soc.* **2005**, *127*, 17364–17376. [CrossRef] [PubMed]
29. Blazyk, J.L.; Lippard, S.J. Expression and characterization of ferredoxin and flavin adenine dinucleotide binding domains of the reductase component of soluble methane monooxygenase from *Methylococcus capsulatus* (Bath). *Biochemistry* **2002**, *41*, 15780–15794. [CrossRef] [PubMed]
30. Liu, Y.; Nesheim, J.C.; Paulsen, K.E.; Stankovich, M.T.; Lipscomb, J.D. Roles of the methane monooxygenase reductase component in the regulation of catalysis. *Biochemistry* **1997**, *36*, 5223–5233. [CrossRef] [PubMed]
31. Tinberg, C.E.; Lippard, S.J. Revisiting the mechanism of dioxygen activation in soluble methane monooxygenase from *M. capsulatus* (Bath): Evidence for a multi-step, proton-dependent reaction pathway. *Biochemistry* **2009**, *48*, 12145–12158. [CrossRef] [PubMed]
32. Wang, W.X.; Iacob, R.E.; Luoh, R.P.; Engen, J.R.; Lippard, S.J. Electron transfer control in soluble methane monooxygenase. *J. Am. Chem. Soc.* **2014**, *136*, 9754–9762. [CrossRef] [PubMed]
33. Matsen, J.B.; Yang, S.; Stein, L.Y.; Beck, D.; Kalyuzhnaya, M.G. Global molecular analyses of methane metabolism in methanotrophic alphaproteobacterium, *Methylosinus trichosporium* OB3b. Part I: Transcriptomic study. *Front. Microbiol.* **2013**, *4*, 40. [CrossRef] [PubMed]
34. Semrau, J.D.; Jagadevan, S.; DiSpirito, A.A.; Khalifa, A.; Scanlan, J.; Bergman, B.H.; Freemeier, B.C.; Baral, B.S.; Bandow, N.L.; Vorobev, A.; et al. Methanobactin and MmoD work in concert to act as the 'copper-switch' in methanotrophs. *Environ. Microbiol.* **2013**, *15*, 3077–3086. [CrossRef] [PubMed]
35. Grosse, S.; Laramee, L.; Wendlandt, K.D.; McDonald, I.R.; Miguez, C.B.; Kleber, H.P. Purification and characterization of the soluble methane monooxygenase of the type II methanotrophic bacterium *Methylocystis sp* strain WI 14. *Appl. Environ. Microbiol.* **1999**, *65*, 3929–3935. [PubMed]
36. Merkx, M.; Lippard, S.J. Why *OrfY*? Characterization of MMOD, a long overlooked component of the soluble methane monooxygenase from *Methylococcus capsulatus* (Bath). *J. Biol. Chem.* **2002**, *277*, 5858–5865. [CrossRef] [PubMed]
37. Wang, W.X.; Liang, A.D.; Lippard, S.J. Coupling oxygen consumption with hydrocarbon oxidation in bacterial multicomponent monooxygenases. *Acct. Chem. Res.* **2015**, *48*, 2632–2639. [CrossRef] [PubMed]
38. Shaofeng, H.; Shuben, L.; Jiayin, X.; Jianzhong, N.; Chungu, X.; Haidong, T.; Wei, T. Purification and biochemical characterization of soluble methane monooxygenase hydroxylase from *Methylosinus trichosporium* IMV 3011. *Biosci. Biotechnol. Biochem.* **2007**, *71*, 122–129. [CrossRef] [PubMed]
39. Rosenzweig, A.C.; Brandstetter, H.; Whittington, D.A.; Nordlund, P.; Lippard, S.J.; Frederick, C.A. Crystal structures of the methane monooxygenase hydroxylase from *Methylococcus capsulatus* (Bath): Implications for substrate gating and component interactions. *Proteins Struct. Funct. Genet.* **1997**, *29*, 141–152. [CrossRef]
40. Atkin, C.L.; Thelande, L.; Reichard, P.; Lang, G. Iron and free-radical in ribonucleotide reductase—Exchange of iron and Mössbauer spectroscopy of protein B2 subunit of *Escherichia coli* Enzyme. *J. Biol. Chem.* **1973**, *248*, 7464–7472. [PubMed]
41. Brandstetter, H.; Whittington, D.A.; Lippard, S.J.; Frederick, C.A. Mutational and structural analyses of the regulatory protein B of soluble methane monooxygenase from *Methylococcus capsulatus* (Bath). *Chem. Biol.* **1999**, *6*, 441–449. [CrossRef]
42. Colby, J.; Stirling, D.I.; Dalton, H. The soluble methane mono-oxygenase of *Methylococcus capsulatus* (Bath): Its ability to oxygenate *n*-alkanes, ethers, and alicyclic, aromatic and heterocyclic compounds. *Biochem. J.* **1977**, *165*, 395–402. [CrossRef] [PubMed]

43. Walters, K.J.; Gassner, G.T.; Lippard, S.J.; Wagner, G. Structure of the soluble methane monooxygenase regulatory protein B. *Proc. Natl. Acad. Sci. USA* **1999**, *96*, 7877–7882. [CrossRef] [PubMed]
44. Lund, J.; Dalton, H. Further Characterization of the FAD and Fe_2S_2 redox centers of component C, the NADH—Acceptor reductase of the soluble methane monooxygenase of *Methylococcus capsulatus* (Bath). *Eur. J. Biochem.* **1985**, *147*, 291–296. [CrossRef] [PubMed]
45. Lund, J.; Woodland, M.P.; Dalton, H. Electron transfer reactions in the soluble methane monooxygenase of *Methylococcus capsulatus* (Bath). *Eur. J. Biochem.* **1985**, *147*, 297–305. [CrossRef] [PubMed]
46. Kopp, D.A.; Gassner, G.T.; Blazyk, J.L.; Lippard, S.J. Electron-transfer reactions of the reductase component of soluble methane monooxygenase from *Methylococcus capsulatus* (Bath). *Biochemistry* **2001**, *40*, 14932–14941. [CrossRef] [PubMed]
47. Fox, B.G.; Liu, Y.; Dege, J.E.; Lipscomb, J.D. Complex formation between the protein components of methane monooxygenase from *Methylosinus trichosporium* Ob3b. Identification of sites of component interaction. *J. Biol. Chem.* **1991**, *266*, 540–550. [PubMed]
48. Liu, K.E.; Lippard, S.J. Redox properties of the hydroxylase component of methane monooxygenase from *Methylococcus capsulatus* (Bath). Effects of protein B, reductase, and substrate. *J. Biol. Chem.* **1991**, *266*, 12836–12839. [PubMed]
49. Green, J.; Prior, S.D.; Dalton, H. Copper ions as inhibitors of protein C of soluble methane monooxygenase of *Methylococcus capsulatus* (Bath). *Eur. J. Biochem.* **1985**, *153*, 137–144. [CrossRef] [PubMed]
50. Lipscomb, J.D. Biochemistry of the soluble methane monooxygenase. *Annu. Rev. Microbiol.* **1994**, *48*, 371–399. [CrossRef] [PubMed]
51. Murrell, J.C.; Gilbert, B.; McDonald, I.R. Molecular biology and regulation of methane monooxygenase. *Arch. Microbiol.* **2000**, *173*, 325–332. [CrossRef] [PubMed]
52. Pulver, S.; Froland, W.A.; Fox, B.G.; Lipscomb, J.D.; Solomon, E.I. Spectroscopic studies of the coupled binuclear non-heme iron active site in the fully reduced hydroxylase component of methane monooxygenase: Comparison to deoxy and deoxy-azide hemerythrin. *J. Am. Chem. Soc.* **1994**, *116*, 4529–4529. [CrossRef]
53. McCormick, M.S.; Lippard, S.J. Analysis of substrate access to active sites in bacterial multicomponent monooxygenase hydroxylases: X-ray crystal structure of xenon-pressurized phenol hydroxylase from *Pseudomonas* sp. OX1. *Biochemistry* **2011**, *50*, 11058–11069. [CrossRef] [PubMed]
54. Stirling, D.I.; Dalton, H. Properties of the methane monooxygenase from extracts of *Methylosinus trichosporium* OB3b and evidence for its similarity to the enzyme from *Methylococcus capsulatus* (Bath). *Eur. J. Biochem.* **1979**, *96*, 205–212. [CrossRef] [PubMed]

catalysts

MDPI

Article

Specific Immobilization of *Escherichia coli* Expressing Recombinant Glycerol Dehydrogenase on Mannose-Functionalized Magnetic Nanoparticles

Fei-Long Li [1,†], Meng-Yao Zhuang [1,†], Jia-Jia Shen [1], Xiao-Man Fan [1], Hyunsoo Choi [2], Jung-Kul Lee [2,*] and Ye-Wang Zhang [1,3,*]

1 School of Pharmacy, United Pharmaceutical Institute of Jiangsu University and Shandong Tianzhilvye Biotechnology Co. Ltd, Zhenjiang 212013, China; feilongli2018@outlook.com (F.-L.L.); zhuang_mengyao@outlook.com (M.-Y.Z.); shenjiajia20@163.com (J.-J.S.); fanxiaomanmail@yeah.net (X.-M.F.)

2 Department of Chemical Engineering, Konkuk University, 1 Hwayang Dong, Seoul 05029, Korea; hyunsoo4062@naver.com

3 College of Petroleum and Chemical Engineering, Beibu Gulf University, Qinzhou 535011, China

* Correspondence: jkrhee@konkuk.ac.kr (J.-K.L.); zhangyewang@ujs.edu.cn (Y.-W.Z.); Tel.: +82-458-3504 (J.-K.L.); +86-511-8503-8201 (Y.-W.Z.)

† These authors contributed equally to this work.

Received: 16 November 2018; Accepted: 13 December 2018; Published: 24 December 2018

Abstract: Mannose-functionalized magnetic nanoparticles were prepared for the immobilization of *Escherichia coli* cells harboring the recombinant glycerol dehydrogenase gene. Immobilization of whole *E. coli* cells on the carrier was carried out through specific binding between mannose on the nanoparticles and the FimH lectin on the *E. coli* cell surface via hydrogen bonds and hydrophobic interactions. The effects of various factors including cell concentration, pH, temperature, and buffer concentration were investigated. High degrees of immobilization (84%) and recovery of activity (82%) were obtained under the following conditions: cell/support 1.3 mg/mL, immobilization time 2 h, pH 8.0, temperature 4°C, and buffer concentration 50 mM. Compared with the free cells, the thermostability of the immobilized cells was improved 2.56-fold at 37 °C. More than 50% of the initial activity of the immobilized cells remained after 10 cycles. The immobilized cells were evaluated functionally by monitoring the catalytic conversion of glycerol to 1,3-dihydroxyacetone (DHA). After a 12 h reaction, the DHA produced by the immobilized cells was two-fold higher than that produced by the free cells. These results indicate that mannose-functionalized magnetic nanoparticles can be used for the specific recognition of gram-negative bacteria, which gives them great potential in applications such as the preparation of biocatalysts and biosensors and clinical diagnosis.

Keywords: mannose; magnetic nanoparticles; immobilization; whole cell; specific recognition

1. Introduction

As important biocatalysts, bacteria are widely used in sewage treatment, environmental remediation, and the production of drugs, pharmaceutical intermediates, food processing, and industrial chemical products [1,2]. However, free bacterial cells are unstable, and their reusability in culture medium or in a reaction solution requires critical reaction conditions. Compared with free cells, the immobilized cells can be easily re-used, and extremely high bacterial levels can be readily obtained in bioreactors using them [3]. Thus, immobilization of whole cells is receiving increased attention because of its potential in biocatalysis [4,5].

Generally, immobilization techniques have been proven to enhance the properties of biocatalysts, including whole cells and enzymes [6–10]. Nanostructure materials have been used as drug carriers and in other biological applications [11–17] and proved to have great potential in stabilizing cells, allowing their use for extended periods [18,19]. Among all nanomaterials, magnetic Fe_3O_4 nanoparticles have several advantages, including a large surface area, superparamagnetism, low toxicity, good biocompatibility, and ease of recycling [20,21]. Therefore, they are ideal for cell immobilization.

There are several widely used cell immobilization methods, including adsorption onto/into solid supports [22], entrapment in a matrix [23], cross-linking [24], and covalent binding to the carrier [25]. Of these methods, immobilization by adsorption has relatively slight influence on the biological activity of cells and often uses macroporous resins [26] or glass beads [27] as the immobilized carrier. However, weak binding and poor mechanical stability may cause cell leakage leading to a shortened utility lifespan [28]. Although the attachment of cells to supports using cross-linking or covalent binding reduces cell leakage, it is incompatible with cell viability since the cross-linking agents are highly toxic to microbial cells [29]. Recently, enzymes or active groups expressed on the cell surface have been reported [24,30]. However, it is difficult to express and control the orientation of heterologous proteins on the cell surface. Thus, the development of an appropriate immobilization method for bacterial cells remains an important goal.

It has been reported that gram-negative bacteria produce surface lectins which are usually presented in the form of filamentous pili, assembled from more than 1000 protein subunits [31,32]. As a typical gram-negative bacterial, *Escherichia coli* also produces cell surface lectins, and among these the most common is the FimH protein, which is comprised of 17 kDa subunits. These subunits have an extended trisaccharide binding site and preferentially bind oligomannose [33,34]. In the present study, we have developed a promising method for the immobilization of *E. coli* based on specific recognition between the FimH protein on the surface of bacteria and mannose-functionalized magnetic nanoparticles. These mannose-functionalized magnetic nanoparticles were fabricated (Scheme 1) and used as an effective carrier for the immobilization of *E. coli* cells expressing recombinant glycerol dehydrogenase, which was then used to synthesize 1,3-dihydroxyacetone (DHA) to investigate their biological activity (Scheme 2).

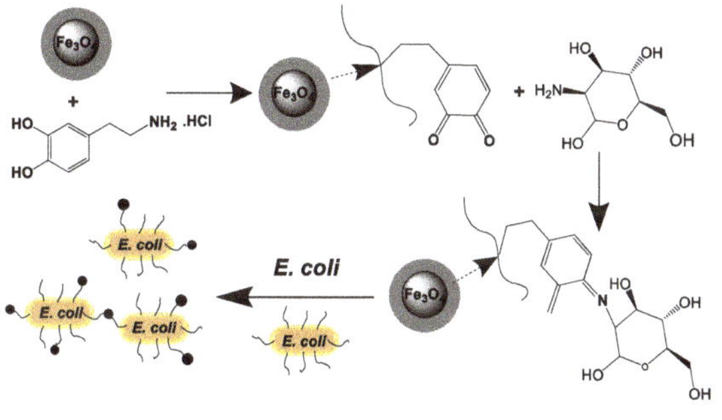

Scheme 1. Carrier preparation and *Escherichia coli* cell immobilization.

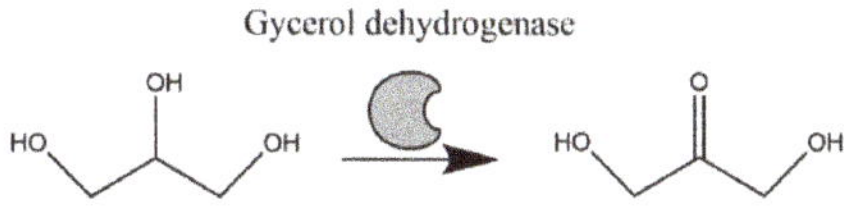

Scheme 2. Synthesis of 1,3-dihydroxyacetone (DHA) using *E. coli* expressing recombinant glycerol dehydrogenase.

2. Results and Discussion

2.1. Characterization of the Nanoparticles

The preparation of nanoparticles, including Fe_3O_4, Fe_3O_4@OA (oleic acid), and Fe_3O_4@OA@DP (dopamine), and the characterization of these nanoparticles using Fourier transform infrared spectroscopy (FTIR) and transmission electron microscopy (TEM) have been reported in our previous work [35]. To demonstrate that mannose was successfully anchored onto the surface of magnetic Fe3O4@OA@DP, the nanoparticles were analyzed using FTIR. As shown in Figure 1, in the Fe_3O_4 FTIR spectra, the band at 580 cm^{-1} was assigned to the Fe–O vibration. Compared with the Fe_3O_4@OA@DP nanoparticles, new bands at 1400 and 1643 cm^{-1} appeared in the Fe_3O_4@OA@DP–mannose nanoparticles spectrum. This was due to the fact that the residual quinone functional groups on the nanoparticles were reactive toward the nucleophilic amino groups in mannosamine and could be covalently coupled to mannosamine with polydopamine through a Michael addition and Schiff base formation [34]. These results demonstrated that the mannose was successfully functionalized onto the magnetic Fe_3O_4@OA@DP nanoparticles.

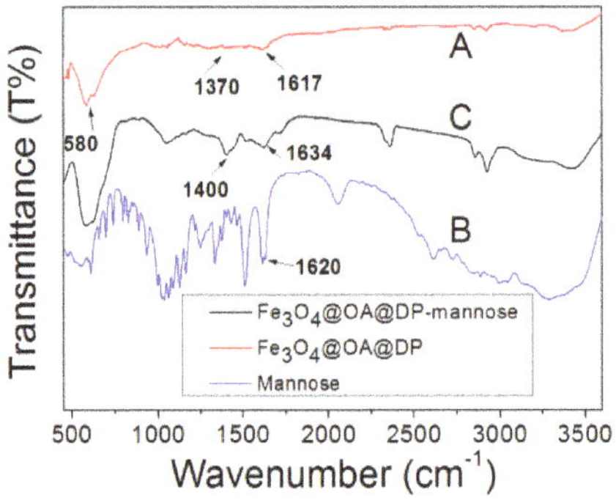

Figure 1. FTIR spectra of (**A**) Fe_3O_4@OA@DP, (**B**) mannose, and (**C**) Fe_3O_4@OA@DP–mannose.

The mannose-functionalized nanoparticles were then used to immobilize *E. coli* cells. To investigate the morphological changes and dispersional stabilization of the Fe_3O_4@OA@DP, Fe_3O_4@OA@DP–mannose, and Fe_3O_4@OA@DP–mannose–*E. coli* nanoparticles, transmission electron microscopy (TEM) was conducted (Figure 2). The average size of the Fe_3O_4@OA@DP nanoparticles (Figure 2A) was approximately 20 nm in diameter. After functionalization with mannose, the size of the Fe_3O_4@OA@DP–mannose (Figure 2B) particles became significantly larger than that of the Fe_3O_4@OA@DP nanoparticles, presenting an average diameter of 32 nm. From Figure 2C, it is clear that the *E. coli* cells were immobilized on the Fe_3O_4@OA@DP–mannose nanocarrier successfully. In previous reports, it was proven that there was no binding between mannose and *E. coli* lacking the FimH gene [36,37]. The specific recognition between the D-mannose-containing structure and *E. coli* containing the FimH gene was directly shown with receptor immuno-electron microscopy [38,39]. By adding a high concentration of D-mannose ammonium hydrochloride (100 mM), 95.2% of cells could be removed from the nanoparticles. This indicated that the interactions between FimH and mannose-functionalized nanoparticles were weaker than those between FimH and mannose monomers. This result is in agreement with the inhibition potency of mannosides with respect to the adhesion of *E. coli* to mannose-coated surfaces [40]. It has also been proven that the binding between mannose-containing magnetic nanoparticles is highly specific, which could be effective for the immobilization of *E. coli*.

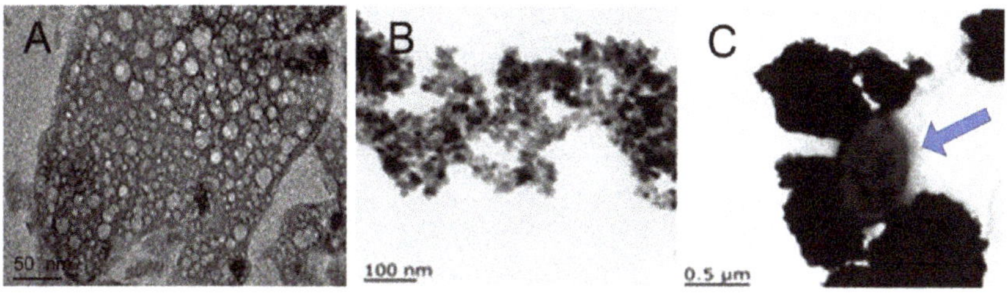

Figure 2. TEM of (**A**) Fe_3O_4@OA@DP, (**B**) Fe_3O_4@OA@DP-mannose, (**C**) Fe_3O_4@OA@DP–mannose–*E. coli*. The arrow indicates an immobilized *E. coli* cell.

2.2. Optimization of the Immobilization Conditions

In the present study, we utilized the specific binding between mannose and FimH lectin to immobilize *E. coli* cells expressing recombinant glycerol dehydrogenase onto magnetic nanoparticles. To achieve a high yield and biological activity of the cells, conditions such as pH, temperature, cell concentration, and buffer concentration were optimized. pH is an important factor for the immobilization of *E. coli* because the hydrogen bonds formed and the hydrophobic interactions involved in this process are highly dependent on it [33]. As shown in Figure 3A, the immobilization yield increased from 26.5% at pH 6.5 to the highest value of 82.4% at pH 8.0, but decreased sharply to 46.3% at pH 11. The recovery activity showed a similar tendency, which was 46.3% at pH 6, reaching the highest value of 85.1% at pH 9. Thereafter, it decreased to 73.6% at pH 10 and to 39.6% at pH 11. These results may be due to the fact that protons bind to the lone pair electrons on the amino N atom under acidic conditions. Under conditions of alkalinity (pH > 8.0), the protons on FimH lectin will likely be taken away by the hydroxide ions in the solution, leading to a destruction of the hydrogen bonds [41]. Thus, the optimal pH for immobilization was determined to be 8.0, on the basis of an evaluation of the immobilization yield and the recovery of activity.

The effect of cell concentration on the immobilization of *E. coli* was also examined (Figure 3B). The immobilization yield gradually decreased with the increase of cell concentration, with a yield as high as 100% at the lowest cell concentration of 4.55 mg/mL, whereas at the highest cell concentration of 25.5 mg/mL, the immobilization yield was only 64.33%. The reason for this might be that the number of nanoparticles was insufficient to immobilize all cells at high cell concentrations [42]. Unlike the immobilization yield, the activity recovery of the immobilized cells increased as cell concentration increased. At a cell concentration of 4.55 mg/mL, the activity recovery was only 31.45%, whereas it reached 92.74% when the cell concentration increased to 25.5 mg/mL. Based on the consideration of both the immobilization yield and activity recovery, a cell concentration of 13.45 mg/mL was selected for cell immobilization. The immobilization time was investigated from 0.25 to 2.0 h. After 2 h, the immobilization yield and the activity recovery reached their highest values at 82.4% and 82.5%, respectively. These values likely represent the dynamic balance between both adsorption and desorption [43,44].

Figure 3C shows the effect of temperature on the cells immobilized onto the functionalized Fe_3O_4@OA@DP nanoparticles. The optimal immobilization temperature was 4 °C. As the temperature increased, the immobilization yield and the activity recovery gradually decreased from 82.4 to 70.5% and 80 to 46%, respectively. When the temperature increases, secondary bonds within the enzyme structure are disrupted, which may lead to a change in the enzyme's conformation, resulting in a loss of cell activity. The Brownian motion between the molecules increases as well, which could further lead to the breakage of hydrogen bonds [45].

The effect of buffer concentration on the immobilization process was determined over the range 0.01 to 0.25 M (Figure 3D). The results indicated that the recovery of activity reached its highest value of 92% at a buffer concentration of 50 mM and then decreased with increasing buffer concentrations. Meanwhile, as the buffer concentration increased, the immobilization yield decreased.

The reason for this might be that a high buffer concentration affects the surface charge of the enzyme, removing its surface hydration layer, which would then affect the formation of hydrogen bonds [46]. The immobilization mechanism used here represents a highly specific recognition between mannose residues and the FimH protein. All the mannose hydroxyls, with the exception of the anomeric hydroxyl, interacted extensively with the binding domain residues in the FimH protein [47]. The amine groups at the N-terminal of the FimH protein interacted with the 2-OH, 6-OH, and the ring oxygen of the mannose moiety to form hydrogen bonds [48].

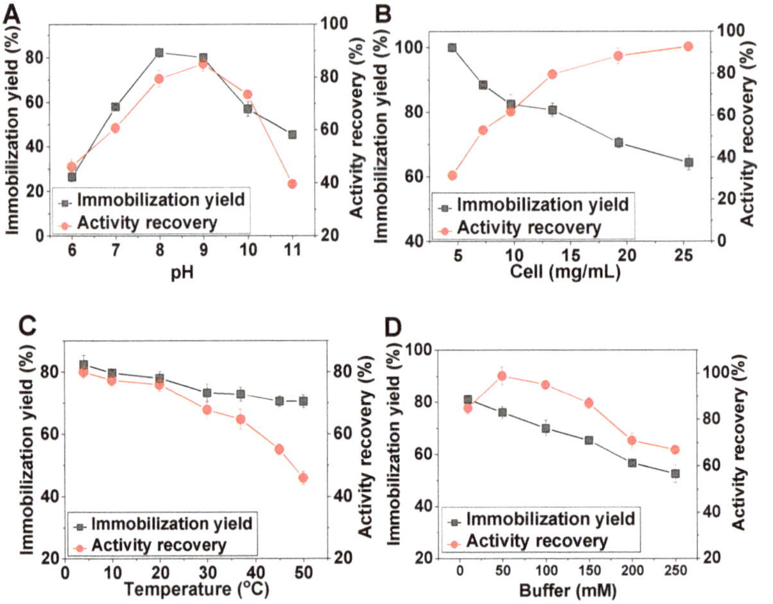

Figure 3. Effect of (**A**) pH, (**B**) cell concentration, (**C**) temperature, and (**D**) buffer concentration on the immobilization of *E. coli* cells. All the experiments were repeated three times, and the standard deviation was calculated and presented.

2.3. Effect of pH and Temperature on the Free and Immobilized Cells

The characteristics of the immobilized cells were then compared with those of the free cells. Figure 4A shows the effect of pH on the activity of the free and immobilized *E. coli* cells expressing recombinant glycerol dehydrogenase. The relative activity of the free cells was 75% at pH 6, with the highest value at pH 9.0; then, the activity decreased dramatically to 28.6% at pH 11. Although the immobilized cells showed a similar trend, the optimal pH was 8.0, and the activity slowly decreased to 81% at pH 11, which was 2.9-fold that of the free cells. Obviously, the immobilized cells displayed better activities at pH 10 and 11. The possible reason for the changes is that enhanced rigidity improved stability after immobilization onto the nanoparticles. Also, an alkaline medium (pH 9.0) exerts stress on the metabolic processes of cells. Moreover, the carrier for immobilization also weakens the mass transfer properties and decreases the permeability of *E. coli* cells [49,50]. Accordingly, the optimum pH for the immobilized cells shifted to 9.0. Additionally, the immobilized cells retained more than 60% residual activity at pH 6–11, whereas in comparison, the free cells exhibited only 14.4% activity at pH 11.0.

The effects of temperature on the activity of free and immobilized cells at pH 8.0 over the range 30 to 90 °C were checked. As shown in Figure 4B, the catalytic activities of the immobilized and free cells were both maximal at 50 °C. The reason for this might be that an increase in the temperature favored the accelerated movement of the cells, resulting in enhanced activity. However, as the temperature increased further, the secondary structures of the enzymes inside the cell could be disrupted, resulting in decreased activity. Compared with the free cells, immobilization restricted the inactivation of the cells. Thus, when the temperature increased to 90 °C, the free cells retained only 10.56% of their activity, while the immobilized cells retained more than 45% of their activity.

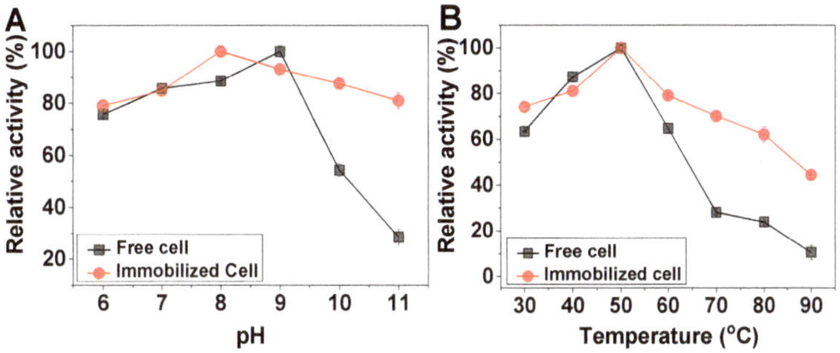

Figure 4. Effects of (**A**) pH and (**B**) temperature on the activities of the free and immobilized cells. All the experiments were repeated three times, and the standard deviation was calculated and presented as error bars.

2.4. Thermal Stability and Reusability of the Free and Immobilized Cells

The thermal stability of a cell is one of the most important properties for their industrial application [51]. The thermal stability of the free and immobilized cells was tested at 37 °C. It was clearly shown that the activity of the free cells decreased rapidly at 37 °C over a 0.5–3 h incubation, whereas the activity of the immobilized cells declined slowly. After incubation at 37 °C for 3 h, only 34% of the initial enzyme activity remained in the free cells, whereas the immobilized cells retained more than 87% of their initial enzyme activity. Moreover, the activity retention in the immobilized cell was 10 times higher than for the free cells after incubation at 37 °C for 3 h. These results demonstrated that after immobilization, the cells had better thermal stability. This may be due to solid-phase carriers protecting the enzyme molecules from external high temperatures, thereby enhancing the enzyme's thermal stability [52].

The reusability of the immobilized cells was also studied. As shown in Figure 5B, the immobilized cells had a higher degree of stability. After five cycles, the immobilized cells retained 70% of their initial activity, while the free cells retained only 10%. Moreover, the immobilized cells retained about 50% of their initial activity after 10 cycles. These results revealed that the immobilized cells had good operational stability and thermostability, which could lower the costs of industrial applications [53].

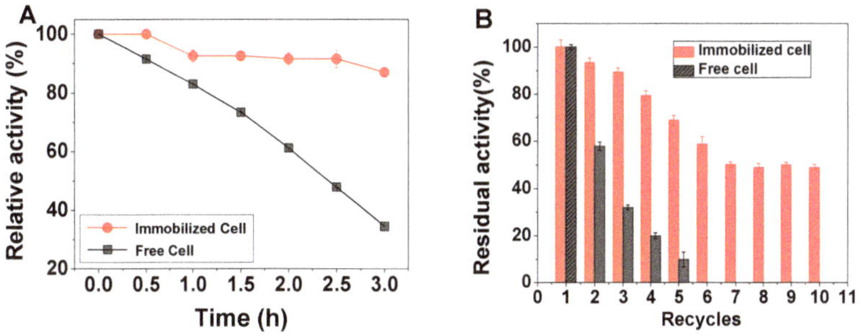

Figure 5. Thermal stability and reusability. (**A**) Thermal stability of the free and immobilized cells. (**B**) Reusability of the immobilized cells. All the experiments were repeated three times, and the standard deviation was calculated and presented as error bars.

2.5. Synthesis of 1,3-Dihydroxyacetone (DHA)

DHA is widely used in the cosmetic industry to produce artificial suntan lotion. The opportunity to produce DHA from glycerol is of particular interest due to the large surplus of glycerol that is formed as a by-product in the biodiesel industry [54]. To assess the possible industrial applications of the immobilized whole cells, the bioconversion of glycerol to DHA was carried out as a proof of concept. Immobilized cells harboring the glycerol dehydrogenase gene were selected for the synthesis of DHA. The time course of the conversion process is shown in Figure 6. After 12 h, approximately 7.5 mM DHA was produced by the immobilized cells, showing a two-fold improvement in the conversion rate compared with the free cells (3.48 mM DHA). Although the product inhibition still had an adverse effect on the system, the yield of DHA showed a relevant enhancement in comparison to previous studies [52]. Moreover, the immobilization method utilized in this study did not require enzyme purification and expensive cofactors. Therefore, it can reduce expenses significantly.

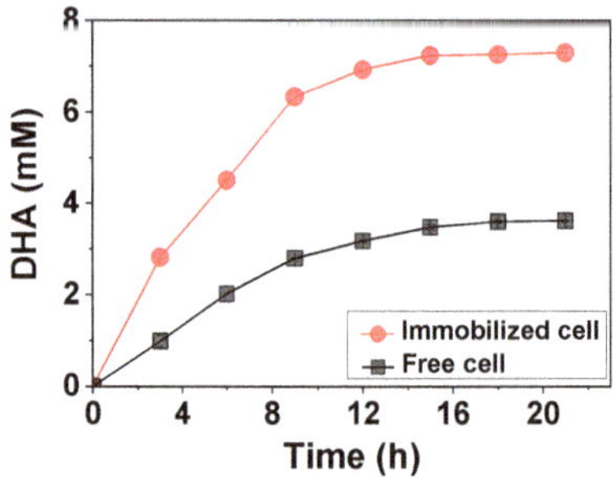

Figure 6. Time course of the synthesis 1,3-dihydroxyacetone (DHA) with immobilized cells. All the experiments were repeated three times, and the standard deviation was calculated and presented as error bars.

3. Materials and Methods

3.1. Materials

$FeSO_4{\cdot}7H_2O$, $FeCl_3{\cdot}6H_2O$, tryptone, oleic acid, glycerol sodium chloride, and yeast extract were from Sinopharm (Shanghai, China). Isopropyl-thio-β- D-galactoside (IPTG), kanamycin, and dopamine hydrochloride were all purchased from Sigma (St. Louis, MO, USA). D-mannosamine hydrochloride was obtained from Yuanye (Shanghai, China) Biotechnology Co. Ltd. Unless otherwise stated, all chemicals and reagents were commercial analytical- or biological-grade.

3.2. Preparation of Magnetic Nanocarriers Functionalized with Mannose

The preparation of magnetic nanocarrier Fe_3O_4@OA@DP was conducted in our previous report [3]. Mannose functionalization of this nanocarrier was performed using the following procedure: typically, 1.0 g of Fe_3O_4@OA@DP nanoparticles was dissolved in 100 mL of citrate buffer (10 mM, pH 4.0). Following this, D-mannosamine hydrochloride (0.23 g) was added to the solution, and the reaction mixture was stirred continuously for 3 h. The material was collected by magnetic separation and washed with deionized water. The product was dried for 12 h at 50 °C and is referred to as Fe_3O_4OA@DP-mannose.

3.3. Characterization of the Magnetic Nanoparticles

The morphologies of Fe_3O_4@OA@DP and Fe_3O_4@OA@DP–mannose nanoparticles were observed by TEM, (Tecnai 12, Philips FEI, Amsterdam, Netherlands). To do this, a few drops of a dilute sample of each of the nanoparticles were placed on a carbon-coated copper grid and dried overnight at room temperature before analysis.

The chemical structure and functional groups presented in the Fe_3O_4@OA@DP–mannose nanoparticles were examined using an FTIR spectrophotometer (PARAGON 500, Perkin Elmer, Waltham, MA, USA). The samples were dried at 80 °C and compressed with KBr. Spectra were recorded twice at a resolution of 4 cm^{-1}.

3.4. Immobilization of *E. coli* Cells onto the Fe_3O_4@OA@DP–Mannose Nanoparticles

E. coli cells expressing recombinant glycerol dehydrogenase were prepared in our previous work [55]. Immobilization of the *E. coli* cells was carried out as follows: the mannose-functionalized nanoparticles (30 mg) were added to PBS buffer (10 mM, pH 7.4) containing 1 mM Mn^{2+} and 1 mM Ca^{2+} and the *E. coli* cells. After incubation for 2 h at 4 °C with gentle shaking, the samples were separated by magnetic force and washed two times with deionized water. The immobilization yield (Y%) and the activity recovery (E%) were calculated as follows:

$$Y\% = 100 \times [(R0-R1-R2)/R0];\ E\% = 100 \times (A1/A0)$$

where R0 refers to the original optical density at 600 nm (OD_{600}) of the cell suspension, R1 refers to the OD_{600} of the suspension after immobilization and magnetic separation, R2 is the OD_{600} of the washing solution after magnetic separation, A0 is the total activity of the immobilized cells, and A1 is the total activity of the free cells.

3.5. Desorption of *E. coli* from the Fe_3O_4@OA@DP–Mannose Nanoparticles

Desorption of *E. coli* from the Fe_3O_4@OA@DP–mannose nanoparticles was performed as follows: the immobilized cells were collected using a magnetic field and added into PBS (pH 8.0) containing 1 mM Mn^{2+}, 1 mM Ca^{2+}, and 100 mM D-mannose. The mixture was incubated at 4 °C for 1 h with gentle shaking. The cells desorption rate was calculated by monitoring the increase of OD_{600} of the suspensions.

3.6. Activity Assay

The biological activity of *E. coli* was assessed using the harbored glycerol dehydrogenase, as previously described [3]. The activity was measured by the biocatalytic conversion of glycerol to 1,3-dihydroxyacetone (DHA). The reaction mixture (1 mL final volume) contained glycerol (100 mM, final concentration), 30 mg immobilized cells, or 2 mL (16 mg/mL) of fresh cell suspension, and PBS buffer (10 mM, pH 8.0). After a reaction lasting 5 min, the free or immobilized cells were separated by centrifugation respectively, and the DHA production was analyzed spectrophotometrically using the diphenylamine chromogenic method [56]. One unit of activity (U) was defined as the amount of DHA that can be released per gram of cells per minute under the assay conditions.

3.7. Effect of Buffer Concentration, pH, and Temperature on the Cells

The effect of buffer concentration on the activities of the free and immobilized cells was investigated over the concentration range 10 to 200 mM. All substrate mixtures used contained phosphate buffer at pH 8.0 at 25 °C. Every experiment was repeated in triplicate, and the standard deviation was also calculated. The activities of the immobilized and free cells were also measured at different pHs to determine the effect of pH on the cells. The buffers used in the reactions were the following: 0.2 M sodium carbonate buffer solution (pH 10.0 and 11.0), 0.2 M Tris-HCl solution (pH 8.0 and 9.0), and 0.2 M PBS (pH 6.0 and 7.0). In addition, the substrate was dissolved in 50 mM PBS at 25 °C. The effect of temperature on the activity of the free and immobilized cells was also investigated. The activities of the free and immobilized cells were determined over different temperatures ranging from 30 to 80 °C at intervals of 10 °C. The substrate solution was dissolved in 50 mM phosphate buffer (pH 8.0).

3.8. Thermal Stability and Reusability of the Immobilized Cells

Thermal stability was investigated by monitoring the remaining activity of the free and immobilized cells in a water bath at 37 °C for different times, ranging from 0.5 to 3.0 h, and then measuring the residual activity. The reusability of the immobilized cells was also investigated by the repeated use of the immobilized cells to catalyze the conversion of the substrate. The activity obtained after each round of use was compared with the initial activity to calculate the relative activity. Between two consecutive reactions, the cells were separated from the reaction media by centrifugation, washed with the 0.2 M PBS (pH 8.0), and then added to the new substrate solutions to start a new reaction. Each experiment was repeated in triplicate to minimize errors.

3.9. Synthesis of 1,3-Dihydroxyacetone (DHA) with the Immobilized Cells

The synthesis of DHA was performed with the immobilized cells using glycerol as the substrate. The total volume of the reaction system was set at 5 mL, and the final substrate concentration was 0.8 M. The reaction was carried out in phosphate buffer (pH 8.0) at 25 °C. DHA concentration was analyzed using the diphenylamine chromogenic method.

4. Conclusions

In summary, a unique method for the immobilization of *E. coli* cells expressing recombinant glycerol dehydrogenase on mannose-modified magnetic Fe_3O_4@OA@DP nanoparticles, based on the specific recognition of mannose by FimH, was developed. Because of the superparamagnetism of magnetic nanoparticles, the immobilized cells could be recycled easily under a magnetic field. The immobilized cells had a higher thermal stability and reusability than the free cells. Moreover, the immobilized cells demonstrate a promising future for producing 1,3-DHA; in fact, they produced 1,3-DHA at two times the rate of the free cells. Thus, mannose-functionalized magnetic nanoparticles have a promising industrial potential, and furthermore, this method could be used to recognize and label bacteria.

Author Contributions: F.-L.L. and M.-Y.Z. analyzed the data and wrote the manuscript; F.-L.L., M.-Y.Z., J.-J.S., X.-M.F. and H.C. performed the experiments; Y.-W.Z. and J.-K.L. designed the experiments and reviewed the manuscript. All authors read and approved the final manuscript.

Funding: This work was supported by the Ministry of Science, ICT and Future Planning, Republic of Korea (2013M3A6A8073184, NRF-2017R1A2B3011676, NRF-2017R1A4A1014806).

Conflicts of Interest: The authors declare that they have no competing interests.

References

1. Matsumoto, T.; Takahashi, S.; Kaieda, M.; Ueda, M.; Tanaka, A.; Fukuda, H.; Kondo, A. Yeast whole-cell biocatalyst constructed by intracellular overproduction of *Rhizopus oryzae* lipase is applicable to biodiesel fuel production. *Appl. Microbiol. Biotechnol.* **2001**, *57*, 515–520. [CrossRef] [PubMed]
2. Patel, S.K.S.; Kumar, V.; Mardina, P.; Li, J.; Lestari, R.; Kalia, V.C.; Lee, J.K. Methanol production from simulated biogas mixtures by co-immobilized *Methylomonas methanica* and *Methylocella tundrae*. *Bioresour. Technol.* **2018**, *263*, 25–32. [CrossRef] [PubMed]
3. Zhuang, M.Y.; Cong, W.; Xu, M.Q.; Ling, X.M.; Shen, J.J.; Zhang, Y.W. Using ConcanavalinA as a Spacer for Immobilization of *E. coli* onto Magnetic Nanoparticles. *Int. J. Biol. Macromol.* **2017**, *104*, 63–69. [CrossRef] [PubMed]
4. Krajčovič, T.; Bučko, M.; Vikartovská, A.; Lacík, I.; Uhelská, L.; Chorvát, D.; Neděla, V.; Tihlaříková, E.; Gericke, M.; Heinze, T.; et al. Polyelectrolyte Complex Beads by Novel Two-Step Process for Improved Performance of Viable Whole-Cell Baeyer-Villiger Monoxygenase by Immobilization. *Catalysts* **2017**, *7*, 353. [CrossRef]
5. Jyoti; Bhatia, K.; Chauhan, K.; Attri, C.; Seth, A. Improving stability and reusability of *Rhodococcus pyridinivorans* NIT-36 nitrilase by whole cell immobilization using chitosan. *Int. J. Biol. Macromol.* **2017**, *103*, 8–15. [CrossRef] [PubMed]

6. Patel, S.K.S.; Choi, S.H.; Kang, Y.C.; Lee, J.-K. Eco-Friendly Composite of Fe_3O_4-Reduced Graphene Oxide Particles for Efficient Enzyme Immobilization. *ACS Appl. Mater. Interfaces* **2017**, *9*, 2213–2222. [CrossRef] [PubMed]
7. Kim, T.-S.; Patel, S.K.S.; Selvaraj, C.; Jung, W.-S.; Pan, C.-H.; Kang, Y.C.; Lee, J.-K. A highly efficient sorbitol dehydrogenase from *Gluconobacter oxydans G624* and improvement of its stability through immobilization. *Sci. Rep.* **2016**, *6*, 33438. [CrossRef] [PubMed]
8. Patel, S.K.S.; Singh, R.K.; Kumar, A.; Jeong, J.-H.; Jeong, S.H.; Kalia, V.C.; Kim, I.-W.; Lee, J.-K. Biological methanol production by immobilized *Methylocella tundrae* using simulated biohythane as a feed. *Bioresour. Technol.* **2017**, *241*, 922–927. [CrossRef]
9. Patel, S.K.S.; Kondaveeti, S.; Otari, S.V.; Pagolu, R.T.; Jeong, S.H.; Kim, S.C.; Cho, B.-K.; Kang, Y.C.; Lee, J.-K. Repeated batch methanol production from a simulated biogas mixture using immobilized *Methylocystis bryophila*. *Energy* **2018**, *145*, 477–485. [CrossRef]
10. Patel, S.K.S.; Kumar, P.; Singh, M.; Lee, J.-K.; Kalia, V.C. Integrative approach to produce hydrogen and polyhydroxybutyrate from biowaste using defined bacterial cultures. *Bioresour. Technol.* **2015**, *176*, 136–141. [CrossRef]
11. Yang, L.; Xin, J.; Zhang, Z.; Yan, H.; Wang, J.; Sun, E.; Hou, J.; Jia, X.; Lv, H. TPGS-modified liposomes for the delivery of ginsenoside compound K against non-small cell lung cancer: Formulation design and its evaluation in vitro and in vivo. *J. Pharm. Pharmacol.* **2016**, *68*, 1109–1118. [CrossRef]
12. Wu, Y.-S.; Ngai, S.-C.; Goh, B.-H.; Chan, K.-G.; Lee, L.-H.; Chuah, L.-H. Anticancer Activities of Surfactin and Potential Application of Nanotechnology Assisted Surfactin Delivery. *Front. Pharmacol.* **2017**, *8*. [CrossRef] [PubMed]
13. Rui, M.; Xin, Y.; Li, R.; Ge, Y.; Feng, C.; Xu, X. Targeted Biomimetic Nanoparticles for Synergistic Combination Chemotherapy of Paclitaxel and Doxorubicin. *Mol. Pharm.* **2017**, *14*, 107–123. [CrossRef] [PubMed]
14. Ali, S.; Morsy, R.; El-Zawawy, N.; Fareed, M.; Bedaiwy, M. Synthesized zinc peroxide nanoparticles (ZnO_2-NPs): A novel antimicrobial, anti-elastase, anti-keratinase, and anti-inflammatory approach toward polymicrobial burn wounds. *Int. J. Nanomed.* **2017**, *12*, 6059–6073. [CrossRef]
15. Wang, G.; Wang, J.J.; Chen, X.L.; Du, L.; Li, F. Quercetin-loaded freeze-dried nanomicelles: Improving absorption and anti-glioma efficiency in vitro and in vivo. *J. Controll. Release* **2016**, *235*, 276–290. [CrossRef]
16. Shi, F.; Zhao, Y.; Firempong, C.K.; Xu, X. Preparation, characterization and pharmacokinetic studies of linalool-loaded nanostructured lipid carriers. *Pharm. Biol.* **2016**, *54*, 2320–2328. [CrossRef]
17. Peng, W.; Jiang, X.; Zhu, Y.; Omari-Siaw, E.; Deng, W.; Yu, J.; Xu, X.; Zhang, W. Oral delivery of capsaicin using MPEG-PCL nanoparticles. *Acta Pharmacol. Sin.* **2015**, *36*, 139–148. [CrossRef]
18. Zhang, H.; Firempong, C.K.; Wang, Y.; Xu, W.; Wang, M.; Cao, X.; Zhu, Y.; Tong, S.; Yu, J.; Xu, X. Ergosterol-loaded poly(lactide-co-glycolide) nanoparticles with enhanced in vitro antitumor activity and oral bioavailability. *Acta Pharmacol. Sin.* **2017**, *37*, 834–844. [CrossRef] [PubMed]
19. Kim, J.; Grate, J.W.; Wang, P. Nanostructures for enzyme stabilization. *Chem. Eng. Sci.* **2006**, *61*, 1017–1026. [CrossRef]
20. Xiong, F.; Hu, K.; Yu, H.; Zhou, L.; Song, L.; Zhang, Y.; Shan, X.; Liu, J.; Gu, N. A Functional Iron Oxide Nanoparticles Modified with PLA-PEG-DG as Tumor-Targeted MRI Contrast Agent. *Pharm. Res.* **2017**, *34*, 1683–1692. [CrossRef]
21. Shen, S.; Wu, L.; Liu, J.; Xie, M.; Shen, H.; Qi, X.; Yan, Y.; Ge, Y.; Jin, Y. Core–shell structured $Fe_3O_4@TiO_2$-doxorubicin nanoparticles for targeted chemo-sonodynamic therapy of cancer. *Int. J. Pharm.* **2015**, *486*, 380–388. [CrossRef] [PubMed]
22. Kilonzo, P.; Margaritis, A.; Bergougnou, M. Effects of surface treatment and process parameters on immobilization of recombinant yeast cells by adsorption to fibrous matrices. *Bioresour. Technol.* **2011**, *102*, 3662–3672. [CrossRef] [PubMed]
23. Zhang, Y.W.; Prabhu, P.; Lee, J.K. Alginate immobilization of recombinant *Escherichia coli* whole cells harboring L-arabinose isomerase for L-ribulose production. *Bioprocess Biosyst. Eng.* **2010**, *33*, 741–748. [CrossRef] [PubMed]
24. Ayer, M.; Klok, H.A. Cell-mediated delivery of synthetic nano- and microparticles. *J. Controll. Release* **2017**. [CrossRef] [PubMed]

25. Ni, K.; Lu, H.; Wang, C.; Black, K.C.; Wei, D.; Ren, Y.; Messersmith, P.B. A novel technique for *in situ* aggregation of *Gluconobacter oxydans* using bio-adhesive magnetic nanoparticles. *Biotechnol. Bioeng.* **2012**, *109*, 2970–2977. [CrossRef] [PubMed]
26. Zhang, Z.; Huang, J.; Jiang, S.; Liu, Z.; Gu, W.; Yu, H.; Li, Y. Porous starch based self-assembled nano-delivery system improves the oral absorption of lipophilic drug. *Int. J. Pharm.* **2013**, *444*, 162–168. [CrossRef] [PubMed]
27. Akashi, M.; Maruyama, I.; Fukudome, N.; Yashima, E. Immobilization of human thrombomodulin on glass beads and its anticoagulant activity. *Bioconj. Chem.* **1992**, *3*, 363. [CrossRef]
28. Guedri, H.; Durrieu, C. A self-assembled monolayers based conductometric algal whole cell biosensor for water monitoring. *Microchim. Acta* **2008**, *163*, 179–184. [CrossRef]
29. Chouteau, C.; Dzyadevych, S.; Durrieu, C.; Chovelon, J.M. A bi-enzymatic whole cell conductometric biosensor for heavy metal ions and pesticides detection in water samples. *Biosens. Bioelectron.* **2005**, *21*, 273–281. [CrossRef]
30. Lagarde, F.; Jaffrezic-Renault, N. Cell-based electrochemical biosensors for water quality assessment. *Anal. Bioanal. Chem.* **2011**, *400*, 947. [CrossRef]
31. Miller, E.; Garcia, T.; Hultgren, S.; Oberhauser, A.F. The mechanical properties of *E. coli* type 1 pili measured by atomic force microscopy techniques. *Biophys. J.* **2006**, *91*, 3848–3856. [CrossRef] [PubMed]
32. Li, H.; Dong, W.; Zhou, J.; Xu, X.; Li, F. Triggering effect of N-acetylglucosamine on retarded drug release from a lectin-anchored chitosan nanoparticles-in-microparticles system. *Int. J. Pharm.* **2013**, *449*, 37–43. [CrossRef]
33. Yazgan, I.; Noah, N.M.; Toure, O.; Zhang, S.; Sadik, O.A. Biosensor for selective detection of *E. coli* in spinach using the strong affinity of derivatized mannose with fimbrial lectin. *Biosens. Bioelectron.* **2014**, *61*, 266–273. [CrossRef]
34. Yang, W.; Pan, C.Y.; Luo, M.D.; Zhang, H.B. Fluorescent mannose-functionalized hyperbranched poly(amido amine)s: Synthesis and interaction with *E. coli*. *Biomacromolecules* **2010**, *11*, 1840–1846. [CrossRef] [PubMed]
35. Zhuang, M.Y.; Zhou, Q.L.; Wang, X.Y.; Zhang, J.X.; Xue, L.; Wang, R.; Zhang, J.X.; Zhang, Y.W. Immobilization of Lipase Onto Dopamine Functionalized Magnetic Nanoparticles. *Nanosci. Nanotechnol. Lett.* **2016**, *8*, 251–254. [CrossRef]
36. Klemm, P.; Jørgensen, B.J.; Die, I.V.; Han, D.R.; Bergmans, H. The fim genes responsible for synthesis of type 1 fimbriae in *Escherichia coli*, cloning and genetic organization. *Mol. Gen. Genet. MGG* **1985**, *199*, 410. [CrossRef]
37. Klemm, P.; Christiansen, G. Three FIM genes required for the regulation of length and mediation of adhesion of *Escherichia coli* type 1 fimbriae. *Mol. Gen. Genet. MGG* **1987**, *208*, 439–445. [CrossRef] [PubMed]
38. Krogfelt, K.A.; Bergmans, H.; Klemm, P. Direct evidence that the FimH protein is the mannose-specific adhesin of *Escherichia coli* type 1 fimbriae. *Infect. Immun.* **1990**, *58*, 1995–1998. [CrossRef]
39. Aprikian, P.; Tchesnokova, V.; Kidd, B.; Yakovenko, O.; Yarov-Yarovoy, V.; Trinchina, E.; Vogel, V.; Thomas, W.; Sokurenko, E. Interdomain Interaction in the FimH Adhesin of *Escherichia coli* Regulates the Affinity to Mannose. *J. Biol. Chem.* **2007**, *282*, 23437. [CrossRef] [PubMed]
40. Möckl, L.; Fessele, C.; Despras, G.; Bräuchle, C.; Lindhorst, T.K. En route from artificial to natural: Evaluation of inhibitors of mannose-specific adhesion of *E. coli* under flow. *Biochim. Biophys. Acta* **2016**, *1860*, 2031–2036. [CrossRef] [PubMed]
41. Liu, W.; Zhou, F.; Zhang, X.-Y.; Li, Y.; Wang, X.-Y.; Xu, X.-M.; Zhang, Y.-W. Preparation of Magnetic Fe_3O_4@SiO_2 Nanoparticles for Immobilization of Lipase. *J. Nanosci. Nanotechnol.* **2014**, *14*, 3068–3072. [CrossRef]
42. Liu, C.H.; Li, X.Q.; Jiang, X.P.; Zhuang, M.Y.; Zhang, J.X.; Bao, C.H.; Zhang, Y.W. Preparation of Functionalized Graphene Oxide Nanocomposites for Covalent Immobilization of NADH Oxidase. *Nanosci. Nanotechnol. Lett.* **2016**, *8*, 164–167. [CrossRef]
43. Tao, Q.-L.; Li, Y.; Shi, Y.; Liu, R.-J.; Zhang, Y.-W.; Guo, J. Application of Molecular Imprinted Magnetic Fe_3O_4@SiO_2 Nanoparticles for Selective Immobilization of Cellulase. *J. Nanosci. Nanotechnol.* **2016**, *16*, 6055–6060. [CrossRef] [PubMed]
44. Shi, Y.; Liu, W.; Tao, Q.-L.; Jiang, X.P.; Liu, C.-H.; Zeng, S.; Zhang, Y.-W. Immobilization of Lipase by Adsorption onto Magnetic Nanoparticles in Organic Solvents. *J. Nanosci. Nanotechnol.* **2016**, *16*, 601–607. [CrossRef] [PubMed]

45. Rambaud, C.; Oppenländer, A.; Trommsdorff, H.P.; Vial, J.C. Tunneling dynamics of delocalized protons in hydrogen bonds at low temperatures. *J. Lumin.* **1990**, *45*, 310–312. [CrossRef]
46. Liu, F.; Kozlovskaya, V.; Zavgorodnya, O.; Martinezlopez, C.; Catledge, S.; Kharlampieva, E. Encapsulation of anticancer drug by hydrogen-bonded multilayers of tannic acid. *Soft Matter* **2014**, *10*, 9237. [CrossRef]
47. Sharon, N. Carbohydrates as future anti-adhesion drugs for infectious diseases. *Biochim. Biophys. Acta BBA Gen. Subj.* **2006**, *1760*, 527–537. [CrossRef]
48. Hung, C.-S.; Bouckaert, J.; Hung, D.; Pinkner, J.; Widberg, C.; DeFusco, A.; Auguste, C.G.; Strouse, R.; Langermann, S.; Waksman, G.; et al. Structural basis of tropism of *Escherichia coli* to the bladder during urinary tract infection: FimH mannose-binding pocket. *Mol. Microbiol.* **2002**, *44*, 903–915. [CrossRef]
49. Yang, H.; Shen, Y.; Xu, Y.; Maqueda, A.S.; Zheng, J.; Wu, Q.; Tam, J. A novel strategy for the discrimination of gelatinous Chinese medicines based on enzymatic digestion followed by nano-flow liquid chromatography in tandem with orbitrap mass spectrum detection. *Int. J. Nanomed.* **2015**, 4947–4955. [CrossRef]
50. Man, R.C.; Ismail, A.F.; Fuzi, S.F.Z.M.; Ghazali, N.F.; Illias, R.M. Effects of Culture Conditions of Immobilized Recombinant *Escherichia coli* on Cyclodextrin Glucanotransferase (CGTase) Excretion and Cell Stability. *Process Biochem.* **2016**, *51*, 474–483. [CrossRef]
51. Zhang, Y.-W.; Jeya, M.; Lee, J.-K. L-Ribulose production by an *Escherichia coli* harboring l-arabinose isomerase from *Bacillus licheniformis*. *Appl. Microbiol. Biotechnol.* **2010**, *87*, 1993–1999. [CrossRef]
52. Zhang, Y.; Gao, F.; Zhang, S.P.; Su, Z.G.; Ma, G.H.; Wang, P. Simultaneous production of 1,3-dihydroxyacetone and xylitol from glycerol and xylose using a nanoparticle-supported multi-enzyme system with in situ cofactor regeneration. *Bioresour. Technol.* **2011**, *102*, 1837–1843. [CrossRef] [PubMed]
53. Singh, R.K.; Zhang, Y.W.; Nguyen, N.P.T.; Jeya, M.; Lee, J.K. Covalent immobilization of β-1,4-glucosidase from *Agaricus arvensis* onto functionalized silicon oxide nanoparticles. *Appl. Microbiol. Biotechnol.* **2011**, *89*, 337–344. [CrossRef] [PubMed]
54. Jiang, X.P.; Lu, T.T.; Liu, C.H.; Ling, X.M.; Zhuang, M.Y.; Zhang, J.X.; Zhang, Y.W. Immobilization of dehydrogenase onto epoxy-functionalized nanoparticles for synthesis of (R)-mandelic acid. *Int. J. Biol. Macromol.* **2016**, *88*, 9–17. [CrossRef] [PubMed]
55. Zhuang, M.Y.; Jiang, X.P.; Ling, X.-M.; Xu, M.-Q.; Zhu, Y.-H.; Zhang, Y.W. Immobilization of glycerol dehydrogenase and NADH oxidase for enzymatic synthesis of 1,3-dihydroxyacetone with *in situ* cofactor regeneration. *J. Chem. Technol. Biotechnol.* **2018**, *93*, 2351–2358. [CrossRef]
56. Zhou, Y.J.; Yang, W.; Wang, L.; Zhu, Z.; Zhang, S.; Zhao, Z.K. Engineering NAD^+ availability for *Escherichia coli* whole-cell biocatalysis: A case study for dihydroxyacetone production. *Microb. Cell Fact.* **2013**, *12*, 103. [CrossRef]

catalysts

MDPI

Article

Fluorescein Diacetate Hydrolysis Using the Whole Biofilm as a Sensitive Tool to Evaluate the Physiological State of Immobilized Bacterial Cells

Anna Dzionek [1], Jolanta Dzik [2], Danuta Wojcieszyńska [1,*] and Urszula Guzik [1]

[1] Department of Biochemistry, Faculty of Biology and Environmental Protection, University of Silesia in Katowice, Jagiellońska 28, 40-032 Katowice, Poland; adzionek@us.edu.pl (A.D.); urszula.guzik@us.edu.pl (U.G.)

[2] Institute of Technology and Mechatronics, University of Silesia in Katowice, Żytnia 12, 41-200 Sosnowiec, Poland; jolanta.dzik@us.edu.pl

* Correspondence: danuta.wojcieszynska@us.edu.pl; Tel.: +32-2009-567

Received: 16 August 2018; Accepted: 30 September 2018; Published: 2 October 2018

Abstract: Due to the increasing interest and the use of immobilized biocatalysts in bioremediation studies, there is a need for the development of an assay for quick and reliable measurements of their overall enzymatic activity. Fluorescein diacetate (FDA) hydrolysis is a widely used assay for measuring total enzymatic activity (TEA) in various environmental samples or in monoculture researches. However, standard FDA assays for TEA measurements in immobilized samples include performing an assay on cells detached from the carrier. This causes an error, because it is not possible to release all cells from the carrier without affecting their metabolic activity. In this study, we developed and optimized a procedure for TEA quantification in the whole biofilm formed on the carrier without disturbing it. The optimized method involves pre-incubation of immobilized carrier in phosphate buffer (pH 7.6) on the orbital shaker for 15 min, slow injection of FDA directly into the middle of the immobilized carrier, and incubation on the orbital shaker (130 rpm, 30 °C) for 1 h. Biofilm dry mass was obtained by comparing the dried weight of the immobilized carrier with that of the unimmobilized carrier. The improved protocol provides a simple, quick, and more reliable quantification of TEA during the development of immobilized biocatalysts compared to the original method.

Keywords: immobilization; fluorescein diacetate; polyurethane foam; biofilm; total enzymatic activity

1. Introduction

Increasing technological and civilization progress resulted in the level of anthropogenic pollution (e.g. pesticides, heavy metals, pharmaceuticals, dyes) in the natural environment increasing significantly in recent years. However, scientific progress made it possible to cheaply and effectively reduce the amount of these pollutants in the environment through bioremediation. This process is based on microorganisms equipped with systems of enzymes that allow them to obtain carbon and energy from xenobiotics [1–3].

An important attribute of stable bioremediation systems is their well-shaped microflora. For that reason, introduction of new microorganisms into the bioremediation systems very often ends, however, with their quick removal by the microflora present in the system. One of the common methods used to increase the chance of survival upon introducing microorganisms into the new system is their immobilization. In addition, immobilized biocatalysts bring certain advantages into bioremediation studies, such as reducing costs, ensuring a stable microenvironment for cells and their enzymes, and increasing the efficiency and resistance of biocatalysts to adverse environmental conditions and high pollutant concentration. Immobilized biocatalysts were extensively examined in the treatment of wastewaters contaminated with various pollutants, and their potential is promising [2,4–7].

Among various immobilization techniques, particular attention in bioremediation studies is paid to the ability of some bacterial strains to form biofilms on various materials. This technique is simple, fast, cheap, and non-toxic for cells and the environment. One of the most important advantages of this method, considering bioremediation systems, is also the spread of the introduced cells within the system, caused by the detachment of external parts of the biofilm in one of its growth phases. The necessary condition, in this technique, to receive a stable and efficient immobilized biocatalyst, is the development of a biofilm strongly attached to the surface of the carrier [4,8–10]. To obtain this kind of biofilm, it is necessary to optimize conditions of the immobilization process for each strain and the carrier [4,11].

Currently, the most commonly method used to determine the efficiency of immobilization is the plate method which relies on plating and subsequent counting of colony-forming units (CFUs) released from the carriers [12] or determination of dry weight of the immobilized biomass [13]. However, none of the above methods determine the physiological state of immobilized cells, which is significantly affected by the quality of the formed biofilm. An indirect method allowing determination of immobilization efficiency is to conduct pollution degradation tests for which an immobilized biocatalyst was developed [14]. However, with multifactor optimization, determining the immobilization efficiency using this method is problematic, especially in the case of hardly biodegradable pollutants that are decomposed over a long period of time. In such cases, enzymatic determination of the metabolic activity of microbial cells may be the solution.

Fluorescein diacetate (3′,6′-diacetyl-fluorescein; FDA) is a prefluorophore, which can be hydrolyzed by a wide spectrum of non-specific extracellular enzymes and membrane-bound enzymes like proteases, lipases, and esterases. Fluorescein, which is a product of hydrolysis, has a yellow-green color and is characterized by strong light absorption at 490 nm. For this reason, the concentration of fluorescein after enzymatic reactions can be easily measured spectrophotometrically. Moreover, measurements of enzymatic activity using FDA hydrolysis correlate with other parameters, such as biomass, ATP content, oxygen consumption, or optical density, and therefore, are often expressed as the total enzymatic activity (TEA) [15–17].

Despite its simplicity, determination of enzymatic activity of immobilized bacterial cells with FDA was presented so far in only one study [18]. The method proposed by Liang et al. [18] assumes the determination of FDA-hydrolyzing enzyme activity of cells that are detached from the carrier. A measurement of enzymatic activity performed in this way carries an error for two very important reasons. Firstly, it is impossible to detach the entire biofilm from the carrier in a non-toxic way because of the biofilm binding strength [19]. On the other hand, bacterial cells at different depths of the biofilm are characterized by different enzymatic activities [11,20]. Therefore, depending on the biofilm binding strength, its various layers with different enzymatic activities can be released and assumed as a total activity. In this study, we made an attempt to apply an appropriate modification to this method to eliminate the mentioned errors. The most important modification was to skip the step of cell removal from the biofilm and to conduct the FDA assay on the entire biofilm with the carrier. To achieve a reliable and reproductive assay, tests were started by determining the ability of carrier to adsorb the product of FDA hydrolysis. We also examined the influence of shaking, and determined which of the substrate application methods resulted in the highest FDA hydrolysis efficiency and the lowest coefficient of variation. Due to the fact that the repeatability and sensitivity of methods based on enzymatic activity depend on the operational conditions [21,22], the optimization of conditions such as pH and incubation time was performed. As a result, a sensitive and reproducible method was developed to determine the total enzymatic activity (TEA) of the entire biofilm formed on the carrier without disturbing it. Using this method, it is possible to determine the efficiency of immobilization during the optimization of its conditions quickly and precisely.

2. Results and Discussion

2.1. Fluorescein Adsorption by Polyurethane Foam (PUR)

After the decision to carry out the enzymatic assay on the biofilm along with the carrier, particular attention should be paid to the possible interaction of the reaction product with the carrier. Fluorescein, without ionic functional groups (e.g., COO^-), is characterized by very limited solubility in water. As the ionization increases, the interaction of the dye with oppositely charged functional groups of the carrier will also increase due to ion exchange. For this reason, the sorption of fluorescent dyes depends on both the pH and the functional groups of the carrier. Due to the presence of two negatively charged groups, and the absence of positive charges, fluorescein is much better adsorbed by positively charged surfaces than by negative ones [23–25].

In this study, the immobilization of the naproxen- and ibuprofen-degrading bacterium *Bacillus thuringiensis* B1 (2015b) [26] was conducted on PUR as a carrier. This is one of the most commonly used materials for microorganism immobilization, and it is characterized by good mechanical strength, non-toxicity, large surface area, and low price [8,27]. It was also shown that polyurethane foam, due to the presence of neutral carbamate groups, is a good sorbent of hydrophobic compounds [8,28]. Therefore, since fluorescein exhibits hydrophobic characteristics [24], its adsorption by polyurethane foam was investigated.

Sterile PUR cubes were incubated for 1 h with fluorescein formed during the hydrolysis of fluorescein diacetate to test the adsorption capacity of PUR. Conducted tests showed that, in the analyzed range of fluorescein concentrations (0.5–5 μg/mL), its adsorption by PUR did not exceed 9% of the dye, but the adsorption value depended on the initial concentration of fluorescein (Figure 1). The average value of adsorbed fluorescein in concentrations below 2.5 μg/mL was equal to 3.8 ± 1.6%, which was a statistically insignificant result (*t*-test; $p \geq 0.05$). However, when the initial fluorescein concentration was higher than 2.5 μg/mL, 7.7 ± 1.12% of dye adsorption was observed. Due to the fact that this result was a statistically significant difference (*t*-test, $p \leq 0.05$), in this study, when the obtained concentration was in the range of 2.5–5 μg/mL, the adsorption of fluorescein by PUR was included in the final concentration.

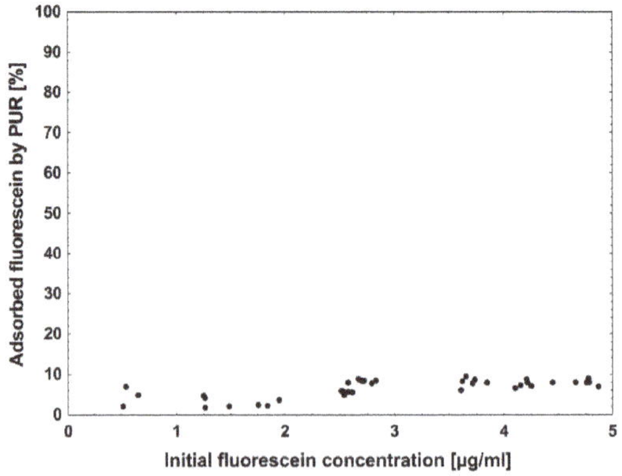

Figure 1. Fluorescein adsorption by sterile polyurethane foam (PUR) cubes depending on the initial concentration of the dye after 1 h of incubation.

Adsorption of fluorescein by materials used as carriers for immobilization is not yet extensively described. A good sorbent material, zeolite, was shown to adsorb 17% of the dye during overnight incubation [23]. However, the material which did not adsorb fluorescein, due to the negative charge of its surface, was silica gel [24].

The adsorption capacity of the carrier can be one of the most important factors which significantly affects the reliability of the FDA assay. For this reason, the adsorption test should be performed for each carrier at the beginning of the optimization of immobilization.

2.2. Fluorescein Diacetate Application and Impact of Shaking

Depending on bacterial strain, the physiological condition of the cells, and the environmental conditions, biofilms can be flat or consist of numerous water channels and extensive structures. They may contain a small number of cells and a rich matrix, or be very densely packed with cells. The structure of the biofilm and the condition of the cells at different depths differ significantly. However, the transport of water, metabolites, or nutrients in any type of biofilm is conducted in the same way. Mass transfer in the biofilm follows the principles of diffusion (in the biofilm matrix) and advection (in the water channels). Because mass transport in the biofilm is limited in its deeper layers, due to the slower diffusion through the matrix, a chemical gradient is created that affects the physiological state of cells at different heights of the biofilm [4,11,20,29]. For that reason, an examination of the physiological state of the biofilm should concern each of its layers. However, this causes technical complications that must be investigated to correctly perform the enzymatic assay and obtain reliable results.

In order to check whether the method of application of fluorescein diacetate would affect the reproducibility and efficiency of its hydrolysis, FDA was applied to the buffer solution or injected directly into the immobilized PUR cube and incubated for 1 h. Depending on the site of FDA application, a different hydrolysis efficiency and coefficient of variation was observed (Table 1). The most reproducible and efficient result was obtained when the substrate was applied directly into the center of immobilized PUR cubes (262 ± 18 µg/g dry mass per h). Addition of FDA to the phosphate buffer caused a large discrepancy in the obtained results (210 ± 48 µg/g dry mass per h).

Table 1. Reproducibility of the method for determining fluorescein diacetate (FDA) hydrolytic activity depending on the method of FDA application. TEA—total enzymatic activity; SD—standard deviation; CV—coefficient of variation.

Location of Application	Biofilm Dry Mass (g)	Fluorescein Concentration (µg/mL)	TEA (µg/g dry mass per h)	Mean	SD	CV (%)
Solution	0.0082–0.0089	1.35–2.30	157–267	210	48	23
Carrier	0.0084–0.0086	2.02–2.40	238–283	262	18	7

The immobilization of bacterial cells on polyurethane foam often results in the formation of a very abundant biofilm, both on its surface and inside the pores. As a result, a high cell density can be obtained in a small volume of the carrier, but also with limited mass transfer to the internal parts of the carrier [20,30]. For that reason, the application of FDA to the buffer solution could cause the adsorption of the FDA to only occur due to a biofilm located on the outer parts of the PUR. Therefore, different amounts of substrate could penetrate into the PUR interior, causing divergences. Nevertheless, it should also be taken into account that the release of fluorescein from the biofilm, especially from the internal parts of the carrier, may be slower due to the limited mass transfer and electrostatic repulsion with amino acids present in the biofilm matrix [31]. In order to achieve results with the smallest error, the final procedure assumes injecting the FDA directly into immobilized carriers placed in a phosphate buffer.

To evaluate the impact of agitation on the efficiency and reproducibility of FDA assay with immobilized B1 (2015b) cells on PUR, hydrolytic activity was measured after 1 h in static conditions, and upon subjection to a rotation rate of 130 rpm (Table 2). Under static conditions, a higher concentration of fluorescein (275 µg/g dry mass per h) was observed in comparison to assays conducted with shaking (249 µg/g dry mass per h). However, this result was the least reproducible as confirmed by the obtained coefficient of variation (46%). Results obtained during assays shaken at 130 rpm proved to be the most reproducible with the smallest coefficient of variation (8%).

Table 2. Impact of shaking on the reproducibility of FDA assay.

Agitation	Dry Biofilm Mass (g)	Fluorescein Concentration (µg/mL)	TEA (µg/g dry mass per h)	Mean	SD	CV (%)
With	0.0079–0.0086	1.36–3.89	164–469	275	126	46
Without	0.0081–0.0089	1.92–2.32	226–273	249	21	8

In enzymatic assays, proper mixing is necessary to ensure sufficient substrate contact with enzyme active sites. However, excessive shaking, due to the shear forces, can deactivate the enzymes and reduce the efficiency of enzymatic reactions [32]. On the other hand, in static incubation, FDA will be rapidly hydrolyzed near the biofilm, while the rest of the FDA may not be transferred to the biofilm surface and matrix [17]. However, it was shown that shaking at below 200 rpm does not damage the enzymes and provides the best efficiency of enzymatic reactions in soils [17,33]. For that reason, in the final method, samples were incubated with shaking at 130 rpm.

2.3. *pH Optimization*

One of the crucial factors influencing enzyme activity is the pH of the assay mixture. Therefore, each enzyme is characterized by a specific pH value at which it works most efficiently. At the optimal pH, the active site of the enzyme is properly spatially shaped. This behavior is related to the proper protonation of amino acids included in the active site. However, due to the fact that FDA hydrolysis is carried out by many different enzymes, determining the optimum for the reaction involves determining the optimum of the enzyme group. It should also be noted that one of the FDA hydrolysis products is acetic acid; therefore, it is necessary to perform the assay in a buffer with an appropriate buffering capacity [17,34]. The temperature of the assay mixture also affects it pH value. Thus, to best assess the physiological state of the analyzed bacterial cells, the assay was carried out at the optimal temperature for their growth (30 °C).

In order to select the optimal pH of phosphate buffer, the hydrolysis of fluorescein diacetate in pH-buffering solutions ranging from 6.8 to 7.6 was examined.

Conducted assays showed significant differences in FDA hydrolysis at different pH levels (Figure 2). Incubation of the immobilized B1 (2015b) strain with FDA in the buffer with the lowest pH (6.8) resulted in the smallest amount of released fluorescein in 1.5 h (54 ± 6 µg/g dry mass per h). As the pH of the buffer increased, hydrolytic activity also increased. Maximum FDA hydrolysis was observed at pH 7.4–7.6 (138 ± 7 µg/g dry mass per h to 128 ± 5 µg/g dry mass per h). According to Guilbault and Kramer [35], FDA-hydrolyzing enzymes exhibit the highest activity at a pH from 7 to 8. However, most researchers use pH 7.6, which is very beneficial [22], mainly because of the fact that abiotic FDA hydrolysis is statistically significant at higher pH values. For this reason, pH above 7.6 was not examined during evaluation.

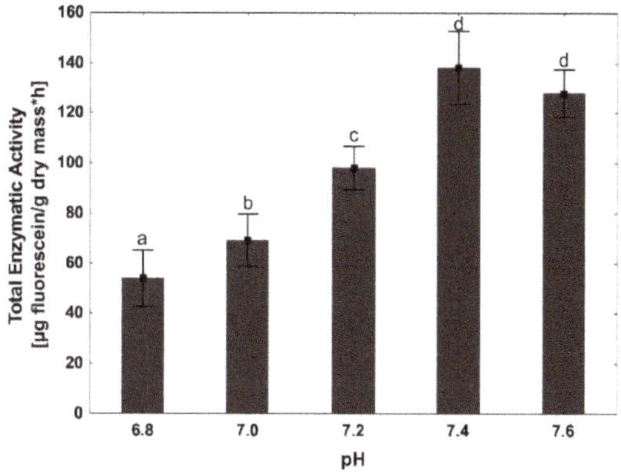

Figure 2. Effect of pH on the enzymatic hydrolysis of fluorescein diacetate (FDA) by *Bacillus thuringiensis* B1 (2015b) cells immobilized onto PUR. Error bars were obtained based on the standard deviation. Statistically significant differences are marked with letters (post hoc, $p \leq 0.05$).

Depending on the type of the carrier and its ionization, pH of the environment may influence abiotic degradation of FDA [17]. In this study, abiotic and spontaneous FDA hydrolysis in the presence of PUR in the analyzed pH range was not statistically significant (data not shown).

The lack of background in quantification of microbial enzymatic activity is undoubtedly an advantage. However, not every carrier will affect the abiotic FDA degradation; therefore, the above test should be performed in phosphate buffer (pH 7.6) before the FDA assay with immobilized cells.

2.4. Incubation Time

The biofilm matrix is a complex mixture of many compounds such as polymers, proteins, polysaccharides, and nucleic acids. Other important components of the biofilm matrix are also cellular elements, including enzymes. They may come from autolysed cells or may be secreted by viable cells to facilitate degradation of macromolecular substances adsorbed by extracellular polymeric substances (EPS) [11,20]. Frølund et al. [36] also demonstrated the presence of enzymes responsible for the hydrolysis of FDA in the biofilm matrix. They observed much greater enzymatic activity per cell in activated sludge flocs than in sludge cultures. Jørgensen et al. [37] also noted that they may be responsible for 20–30% of FDA hydrolysis reactions from samples. However, due to the anionic nature of the biofilm, accumulation of negatively charged fluorescein in the biofilm matrix after 1.5 h of incubation was not statistically significant (data not shown).

In this study, we investigated the temporal variation of fluorescein release from B1 (2015b) cells immobilized onto PUR during 1.5 h of incubation. A linear relationship was observed throughout all analyzed times of incubation with the maximum amount of released fluorescein after 1.5 h of incubation (128 ± 5 µg/g dry mass per h, Figure 3). This result show that FDA hydrolysis was not limited by substrate concentration over the analyzed period of time. Due to the fact that the assay was conducted at a favorable temperature for bacterial cell proliferation (30 °C) [38], it was suggested that a long-term incubation could lead to a result that would not reflect the enzymatic activity of the original sample [17,22,39]. Adam and Duncan [39] also pointed out that it is more important to estimate the hydrolytic potential of the samples than to obtain the highest concentration of fluorescein; therefore, they recommend that incubation last not longer than one hour. On the other hand, Green et al. [22] recommended that incubation last longer than 2 h for soil samples, thereby allowing better differentiation of the results.

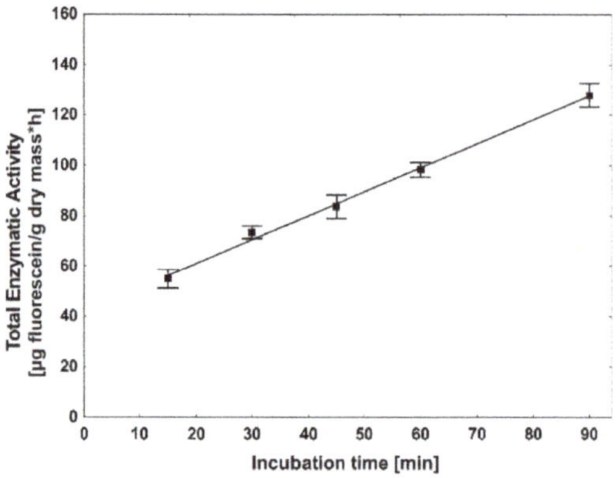

Figure 3. Fluorescein release over time during FDA hydrolysis by immobilized B1 (2015b) cells onto PUR. Error bars were obtained based on the standard deviation.

In the analyzed period of time, errors associated with the growth of microorganisms were eliminated. However, to allow a longer differentiation of samples, an incubation time of 1 h was chosen for the final procedure.

2.5. Sensitivity Assay—Carbon Starvation

The incubation time during immobilization is an extremely important parameter which determines the formation of a stable and strong biofilm, which, after reaching maturity, will be fully resistant to adverse environmental conditions and will be able to degrade higher concentrations of impurities. However, to produce biofilms, bacterial cells must be metabolically active. One of the basic factors affecting metabolic activity is the availability of easily assimilable carbon sources. During a shortage of carbon sources, bacterial cells will reduce their size, and very often, their shape as well (they become more round). If, however, nutrition level drops to a minimum, the response to these conditions involves limiting endogenous metabolism to such a level that they will not be able to reproduce, but will remain active [40–42]. Under these conditions, the vegetative bacterial cells can survive, depending on the strain, from a few to even 100 days (e.g., *Arthrobacter crystallopoietes*) [43].

In order to determine sensitivity of the optimized method, it was observed how the total enzymatic activity (TEA) of bacterial cells in the developing biofilm decreased to the point of minimal endogenous metabolism under starvation during the immobilization process. Seventy-two hours of incubation without a carbon source in the medium resulted in a gradual decrease in TEA (Table 3). After 24 h of incubation, when immobilized B1 (2015b) cells were using accumulated sources of energy, the highest enzymatic activity (360 ± 24 µg/g dry mass per h) was observed. Along with the progressing starvation, after 72 h, a nearly twofold reduction in mean TEA was observed (170 ± 7 µg/g dry mass per h), which indicates the exhaustion of energy reserves and restriction of metabolic activity. The obtained results agree with those obtained by Gengenbacher et al. [44] and Voelker et al. [45], in which a significant decrease in the amount of ATP was demonstrated in nutrient-starved *Bacillus subtilis* and *Mycobacterium tuberculosis*, which indicates a reduction of metabolic activity. It should be also noted that the obtained fluorescein concentration after the analyzed period of time does not differ despite increasing biofilm mass (Table 3). This result can be caused by continuous EPS production by bacterial cells without progressing colonization of the carrier. However, monitoring of the changes in the optical density (OD_{600}) of the medium during immobilization reveal that, with progressing incubation, more cells migrated from the medium. After 24, 48, and 72 h, reductions in the initial OD_{600} value were observed to be 39.5 ± 0.9, 48.8 ± 3.5, and $54.8 \pm 9.1\%$, respectively. It was observed also that, despite the increasing amount of EPS during incubation, it did not exceed 13–15% of the biofilm mass (Table 3). These results clearly indicate that the drop in TEA was caused by the decreasing activity of newly colonizing bacterial cells, instead of the increasing amount of EPS.

Table 3. Impact of carbon starvation on the TEA of the immobilized *Bacillus thuringiensis* B1 (2015b) strain during the immobilization process. EPS—extracellular polymeric substances.

Incubation time (h)	Biofilm dry mass (g)	Dry EPS mass(g)	Fluorescein concentration (μg/ml)	TEA (μg/g dry mass per h)	Mean	SD	CV (%)
24	0.0034–0.0049	0.0005–0.0007	1.21–1.66	326–383	360	24	7
48	0.0064–0.0078	0.0008–0.0010	1.63–2.45	255–325	287	28	10
72	0.0080–0.0091	0.0010–0.0013	1.43–1.50	166–180	170	7	4

In comparison to the TEA values obtained from planktonic B1 (2015b) cells present in the medium, cells immobilized in the biofilm were characterized by better resistance to starvation. After 48 h of incubation, unimmobilized B1 (2015b) cells showed the lowest TEA value (161 $\pm$ 17 μg/g dry mass per h), which was maintained until the end of the analyzed period of time. This result shows that the TEA in the range of 160–170 μg/g dry mass per h indicated that the B1 cells (2015b) were limited to endogenous metabolism.

It was noticed that starvation of *Bacillus thuringiensis* B1 (2015b) cells promoted their immobilization on polyurethane foam. To observe the progress of immobilization, SEM micrographs were prepared after 24, 48, and 72 h of incubation of polyurethane foam with *Bacillus thuringiensis* B1 (2015b) cells (Figure 4).

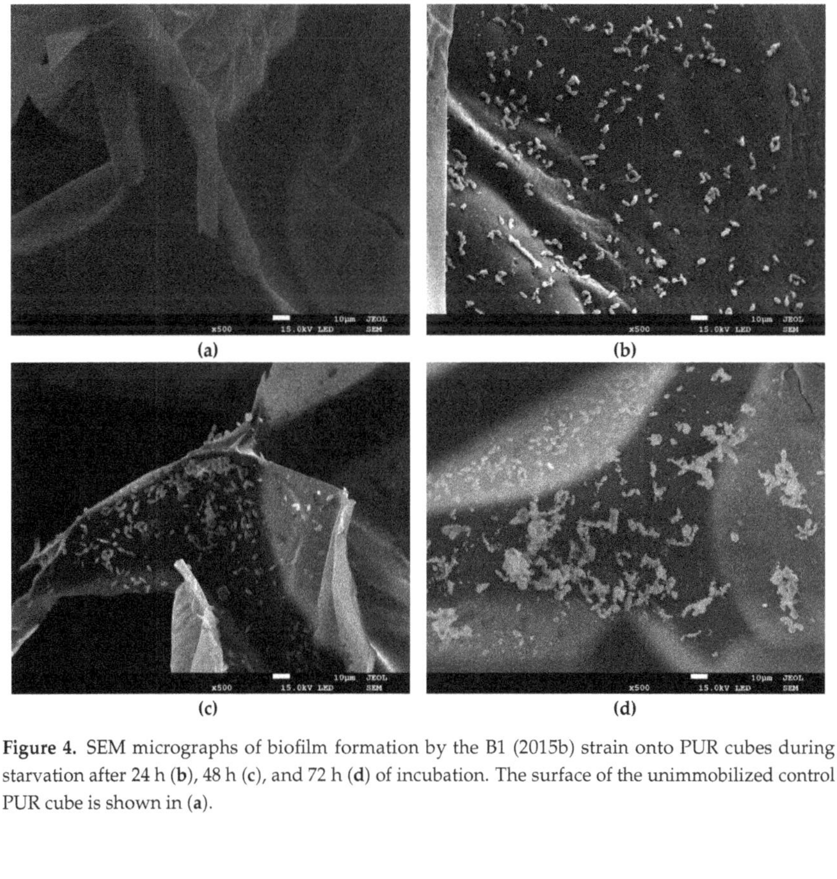

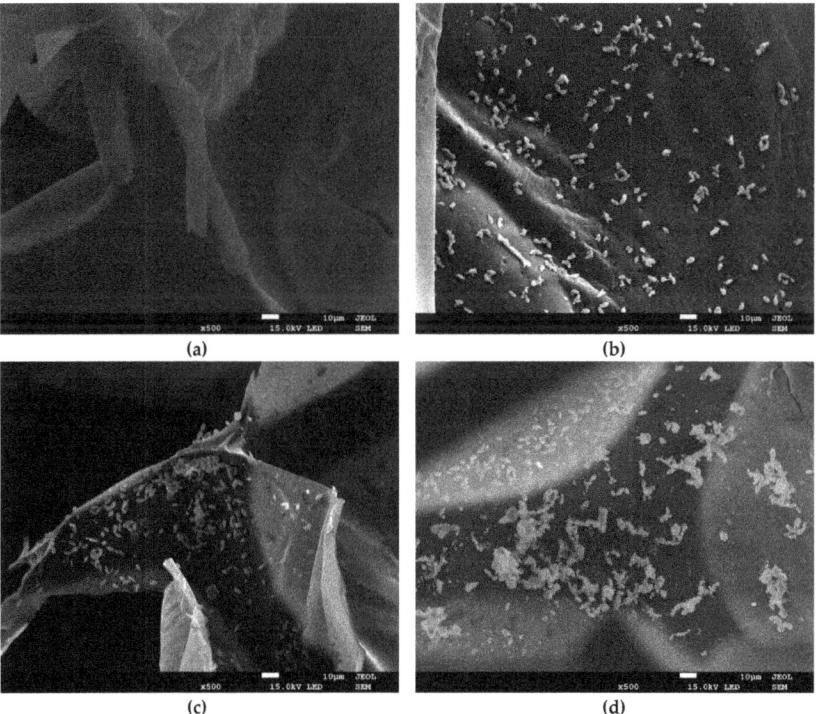

Figure 4. SEM micrographs of biofilm formation by the B1 (2015b) strain onto PUR cubes during starvation after 24 h (**b**), 48 h (**c**), and 72 h (**d**) of incubation. The surface of the unimmobilized control PUR cube is shown in (**a**).

As can be seen in Figure 4b, the colonization of polyurethane foam by B1 (2015b) cells was already evident after 24 h of immobilization. Adsorption of bacterial cells onto the surface of polyurethane foam was observed, which indicates the start of the biofilm formation process. Over time, the bacterial cells began forming microcolonies and secreting extracellular polymeric substances (EPS; Figure 4c). After 72 h, accumulated cells and extracellular matrices in the form of connected aggregates were observed on the PUR surface (Figure 4d). As is known, the limitation of nutrients such as carbon, nitrogen, or phosphorus in the medium is an inducer of sporulation in *Bacillus subtilis* and *Bacillus cereus* cells. The transcription factor *Spo0A* is activated, which, apart from participating in the production of spores, also promotes the formation of biofilm by induction of EPS production. This is one of the defense mechanisms of this genus during the absence of nutrients [46,47]. Due to the fact that the *Spo0A* gene was found in the genome of *Bacillus thuringiensis* [48,49], the mechanism of biofilm induction during starvation in *B. thuringiensis* B1 (2015b) used in this study could be similar. As a result, the decrease in total enzymatic activity of immobilized B1 (2015b) cells was caused by spending energy reserves on EPS synthesis and biofilm formation.

During estimation of the immobilization efficiency, only by determining the dry mass of immobilized bacterial cells could 72 h of incubation be considered optimal. However, after examination of the enzymatic activity, it is shown that the bacterial cells were weakened. The biodegradation tests in this case could be significantly prolonged due to the time when the biofilm would be regenerating. If, however, contamination indicates toxic effects on the strain, it could even lead to their death. With the developed method, it is possible to examine the physiological state of the biofilm formed on the carrier, thereby optimizing the immobilization process, allowing one to obtain a biofilm with the highest enzymatic activity.

2.6. Comparision of the Modified FDA Method with Oxygen Consumption

Oxygen is the key substrate conditioning the metabolism of aerobic organisms. It is necessary for ATP synthesis, and therefore, for the growth, proliferation, and synthesis of various cellular elements. However, due to its poor solubility in water, cultures of aerobic microorganisms must be constantly mixed to ensure its transition from the gas phase. For this reason, oxygen availability in the medium can be a growth-limiting factor for cell cultures [50]. Because of the unique mass transfer properties of biofilms, they are able to adsorb oxygen even at low concentrations. As a result, biofilms ensure, in the top layers, a constant amount of oxygen depending on the cell's oxygen demand, which, in turn, results from the physiological state of bacterial cells [51]. Recent studies showed that differences in oxygen concentration in the horizontal direction at the same depth of aerobic biofilms are statistically insignificant; however, like the previously mentioned nutrients, they are significant in the vertical direction [52,53].

The oxygen uptake rate (OUR), due to its good correlation with metabolic activity, provides valuable information on the physiological state of microbial cells. Due to its relatively simple calculation, OUR is often used to characterize activated sludge [54], production processes [55], and bioremediation [56,57].

To test whether the optimized method would present the same relationships as those used to assess the physiological state of microbial cells, the oxygen uptake rate (OUR) was measured during the starvation assay described above. Table 4 summarizes the obtained values of OUR and TEA after 24, 48, and 72 h of incubation of B1 (2015b) cells with PUR in a medium without carbon sources. The trend in oxygen uptake rate showed a very good correlation with the decrease in total enzymatic activity. After 72 h, immobilized cells showed a nearly twofold reduction in OUR (70 ± 4 μg/g dry mass per h) compared to values obtained after 24 h of incubation (176 ± 13 μg/g dry mass per h). A similar decrease in OUR value in a trickling filter biofilm was also observed by Cox et al. [41] in the absence of toluene, which was the only carbon source in the experiment.

Table 4. Comparison of TEA and oxygen uptake rate (OUR) during carbon starvation of immobilized B1 (2015b) cells during the immobilization process. Data are presented as means ± standard deviation of three replicates.

Incubation time (h)	TEA (μg/g dry mass per h)	OUR (μg O_2/g dry mass per h)
24	360 ± 24	176 ± 13
48	287 ± 28	120 ± 9
72	170 ± 7	70 ± 4

A comparison of the proposed and optimized method in this study for evaluating the physiological state of immobilized cells in biofilms and the method based on the determination of oxygen uptake requires a consideration of the advantages and disadvantages of each technique. OUR is well-established method that allows an indirect estimation of metabolic activity. Its biggest advantage is the duration of the measurement, because results can be obtained after 10 minutes. On the other hand, this method requires experience, due to the fact that, depending on the method of flask sealing and oxygen removal from the gas phase, the result may be burdened with various errors caused by the transfer of oxygen from the air. The determination of the physiological state by direct analysis of the activity of non-specific enzymes responsible for organic matter degradation proposed herein does not require any technical steps that may disturb the final results.

To conclude, in this study, a modification of the FDA assay was optimized in a way which allows results which are reproducible and have a low coefficient of variation. The result also implies the diversity of activities resulting from the heterogeneity of the biofilm. However, due to the possible fluorescein adsorption by the carrier, it is necessary to carry out adsorption tests. With the proposed method, it is possible to monitor changes in the physiological state of the biofilm formed on the carrier through optimization of the immobilization process. By conducting the optimization in this way, the development of an immobilized biocatalyst was possible with the highest enzymatic activity, and thus, with the highest biodegradation capacity or resistance to harsh environmental conditions.

3. Materials and Methods

3.1. Materials

Polyurethane foam (PUR) used in this study is a commonly used material to protect packages during transport (Instapak®, Charlotte, NY, USA). The carrier was trimmed into 1 × 1 × 1 cm cubes with a weight of 10 ± 5 mg, and was washed two times with distilled water to remove impurities, before being autoclaved (121 °C, 1.2 atm, 20 min). All the chemicals were purchased from Sigma-Aldrich (St. Louis, MO, USA).

3.2. Bacterial Strains and Growth Conditions

Bacterial strain *Bacillus thuringiensis* B1 (2015b) isolated from the soil of the chemical factory "Organika-Azot" in Jaworzno (Poland) was used for immobilization [26]. Strain B1 (2015b) was grown in the nutrient broth (BBL) at 30 °C on a rotary shaker at 130 rpm for 24 h. After cultivation, cells were harvested by centrifugation (5000 rpm, 15 min), washed twice with a sterile mineral salt medium according to Greń et al. [58], and re-suspended in the same medium. A bacterial suspension at a final concentration corresponding to an optical density (OD_{600}) of 0.8 was used for immobilization.

3.3. Immobilization Procedure

Each Erlenmeyer flask (250 mL) containing sterile carrier material (0.1 g) was inoculated with the bacterial cell suspension (100 mL). The mineral salt medium [58] in which the immobilization process was conducted did not contain any carbon sources. The immobilization process was carried out on the orbital shaker (130 rpm) at 30 °C for 72 h. After incubation, the medium was removed and immobilized PURs were suspended in NaCl solution (0.9%), centrifuged at 500 rpm for 2 min to remove unbound microorganisms, rinsed with 0.9% NaCl, and used for further analysis.

3.4. Standard Method of Non-Specific Esterase Activity with FDA Assay

The physiological state of the bacterial cells was determined via measurements of non-specific esterase activity with fluorescein diacetate (FDA) as a substrate. The original method [18] includes detaching microorganisms from the carriers by shaking (5 g) in 100 mL of distilled water (200 rpm, 30 min). In the next step, 2 mL of the microorganism suspension was added to 8 mL of phosphate buffer (pH 7.0) and incubated for 15 min with shaking at 30 °C. After pre-incubation, 0.1 mL of FDA stock solution (4.8 mM, dissolved in acetone) was added to each sample and incubated for 2 h. Fluorescein concentration was measured spectrophotometrically (Genesys 20, Thermo Fisher Scientific, Inc., Rochester, NY, USA) at 490 nm and was calculated on the basis of a standard curve.

3.5. Abiotic Controls for FDA Assay

To examine fluorescein adsorption by PUR, sterile carrier cubes (one cube per assay) were placed in solutions with different concentrations (0.5–5 µg/mL) of sterile fluorescein suspended in phosphate buffer (pH 7.0) and incubated in the dark on the orbital shaker (130 rpm, 30 °C). After 1 h of incubation, absorbance (λ = 490 nm) was measured. Additionally, control samples were prepared in the case of FDA autohydrolysis and the natural coloration of the sample with and without sterile carriers.

3.6. Optimization Procedure

The main aim of the optimization procedure was to skip the step of detachment of microorganisms from the carrier in such a way to allow testing of the enzymatic activity of the entire biofilm formed on the carrier without disturbing it. For the best reproducibility, the impacts of substrate application method (FDA added to the liquid or into the carrier) and agitation (with or without) were examined. In order to maximize the activity of non-specific esterases, optimizations of the pH (6.8–7.6) and of the incubation time of immobilized strain B1 (2015b) (15–90 min) with FDA solution were also performed.

3.7. Modified Method of Non-Specific Esterase Activity with FDA Assay

The final methodology is defined as follows: an immobilized PUR cube was placed into 8 mL of phosphate buffer (pH 7.6) and incubated for 15 min on the orbital shaker. In the next step, 0.1 mL of FDA solution in acetone (4.8 mM) was slowly injected directly into the middle of the carrier and incubated on the orbital shaker (130 rpm, 30 °C) for 1 h. Fluorescein concentration was measured as described in Section 3.4.

3.8. Sensitivity Assay—Carbon Starvation

To determine the sensitivity of the method, the impact of carbon starvation on the metabolic response of bacterial cells and the immobilization process was monitored and expressed as total enzymatic activity (TEA). In this test, bacterial cells were immobilized onto PUR as described in Section 3.3 with the incubation time varied to 24, 48, or 72 h. After incubation, the FDA hydrolysis potential of the immobilized bacterial cells was examined. The biofilm's dry mass was calculated by comparing the dried weight of the immobilized carrier (dried at 105 °C for 2 h and stored in a desiccator) with that of the unimmobilized carriers incubated and dried under the same conditions. TEA was expressed in µg of fluorescein obtained from 1 g of biofilm dry mass for 1 h [59]. TEA values for unimmobilized cells of B1 (2015b) were obtained in the same way as for immobilized cells, except that 2 mL of the culture was added to the phosphate buffer (pH 7.6) and, after 1 h of incubation with FDA, bacterial cells were collected through filtration on 0.2-µm Nuclepore filters [15,17,59]. Migration of the bacterial cells from the medium was determined using spectrophotometry (OD_{600}; Genesys 20, Thermo Fisher Scientific, Inc., Rochester, NY, USA). EPS extraction from the immobilized PUR cubes was conducted according to the protocol proposed by Subramanian et al. [60] with some modifications. The PUR cube after 24, 48, or 72 h of incubation was transferred from the medium into 20 mL of distilled water, centrifuged (500 rpm for 2 min) to remove unbound microorganisms, and re-suspended in the same volume of Milli-Q water (Burlington, MA, USA). In the next step, the sample was ultrasonically treated three times for 15 s with a time interval of 10 s, and centrifuged (without carrier, 14000 rpm for 20 min at 4 °C). The collected supernatant containing EPS was precipitated with 2.2 volumes of absolute chilled ethanol through incubation of the mixture at 20 °C for 1 h, and was separated by centrifugation at 6000 rpm for 15 min at 4 °C. The dry EPS mass was obtained by drying the pellet at room temperature and overnight storage in the desiccator.

3.9. Scanning Electron Microscopy

Scanning electron microscopy (SEM) was used to illustrate biofilm formation onto a carrier during starvation. For this purpose, immobilized carrier cubes were fixed in 3% glutaraldehyde and 1% osmium tetroxide, dehydrated with ethanol (30, 50, 70, 80, 90, 95, and 100%, each for 10 min), dried by lyophilization, covered with gold, and observed with a high-resolution electron microscope JSM-7100F TTL LV (JEOL, Tokio, Japan).

3.10. Oxygen Consumption

Oxygen uptake rate (OUR) was determined using an Elmetron multiparameter equipped with a Clark electrode. One immobilized PUR cube was introduced into a flask containing 15 mL of oxygen-saturated phosphate buffer (pH 7.6, 20 °C). To minimize the measurement error, the vessels were placed on a magnetic stirrer and sealed. The decrease in oxygen concentration was registered every 30 sec for 10 min. Oxygen uptake rate was calculated from the slope of a linear regression line through the obtained results and expressed as OUR (µg of consumed O_2 by 1 g of biofilm dry mass during 1 h) [54,61].

3.11. Statistical Analysis

All experiments were performed in at least three replicates. The values of the efficiency of immobilization and enzyme activities were analyzed by one-way ANOVA ($p \leq 0.05$ was considered significant) using the STATISTICA 12 PL software package (StatSoft Inc., Kraków, Poland). A post hoc test was applied to assay the differences between the treatments. To express the repeatability and precision of conducted assays, the coefficient of variation (CV) was calculated as the quotient of the standard deviation and the mean of the obtained TEA from each flask.

Author Contributions: Conceptualization, A.D. and U.G. Data curation, A.D. Formal analysis, D.W. Investigation, A.D. Methodology, A.D. Supervision, U.G. Visualization, J.D. Writing—original draft, A.D., D.W., and U.G. Writing—review and editing, A.D., D.W., and U.G.

Funding: This work was financed by the National Science Centre (Poland), granted on the basis of decision DEC-2017/25/N/NZ9/00422.

Acknowledgments: We are grateful to Justyna Michalska for help with the oxygen uptake rate assay and Małgorzata Adamczyk-Habrajska for help with the preparation of SEM micrographs.

Conflicts of Interest: The authors declare no conflicts of interest.

References

1. Wojcieszyńska, D.; Domaradzka, D.; Hupert-Kocurek, K.; Guzik, U. Enzymes Involved in Naproxen Degradation by *Planococcus* sp. S5. *Pol. J. Microbiol.* **2016**, *65*, 177–182. [CrossRef] [PubMed]
2. Dzionek, A.; Wojcieszyńska, D.; Guzik, U. Natural carriers in bioremediation: A review. *Electron. J. Biotechnol.* **2016**, *19*, 28–36. [CrossRef]
3. Bayat, Z.; Hassanshahian, M.; Cappello, S. Immobilization of microbes for bioremediation of crude oil polluted environments: A mini review. *Open Microbiol. J.* **2015**, *9*, 48–54. [PubMed]

4. Dzionek, A.; Wojcieszyńska, D.; Hupert-Kocurek, K.; Adamczyk-Habrajska, M.; Guzik, U. Immobilization of *Planococcus* sp. S5 strain on the loofah sponge and its application in naproxen removal. *Catalysts* **2018**, *8*, 176. [CrossRef]
5. Partovinia, A.; Rasekh, B. Review of the immobilized microbial cell systems for bioremediation of petroleum hydrocarbons polluted environments. *Crit. Rev. Environ. Sci. Technol.* **2018**, *48*, 1–38. [CrossRef]
6. Sarioglu, O.F.; Celebioglu, A.; Tekinay, T.; Uyar, T. Evaluation of fiber diameter and morphology differences for electrospun fibers on bacterial immobilization and bioremediation performance. *Int. Biodeterior. Biodegradation* **2017**, *120*, 66–70. [CrossRef]
7. Mrozik, A.; Piotrowska-Seget, Z. Bioaugmentation as a strategy for cleaning up of soils contaminated with aromatic compounds. *Microbiol. Res.* **2010**, *165*, 363–375. [CrossRef] [PubMed]
8. Alessandrello, M.J.; Parellada, E.A.; Juárez Tomás, M.S.; Neske, A.; Vullo, D.L.; Ferrero, M.A. Polycyclic aromatic hydrocarbons removal by immobilized bacterial cells using annonaceous acetogenins for biofilm formation stimulation on polyurethane foam. *J. Environ. Chem. Eng.* **2017**, *5*, 189–195. [CrossRef]
9. Ohashi, A.; Harada, H. Adhesion strength of biofilm developed in an attached-growth reactor. *Water Sci. Technol.* **1994**, *29*, 281–288. [CrossRef]
10. Stanley, P.M. Factors affecting the irreversible attachment of *Pseudomonas aeruginosa* to stainless steel. *Can. J. Microbiol.* **1983**, *29*, 1493–1499. [CrossRef] [PubMed]
11. Sutherland, I.W. The biofilm matrix—An immobilized but dynamic microbial environment. *Trends Microbiol.* **2001**, *9*, 222–227. [CrossRef]
12. Niknezhad, S.V.; Asadollahi, M.A.; Zamani, A.; Biria, D. Production of xanthan gum by free and immobilized cells of *xanthomonas campestris* and *xanthomonas pelargonii*. *Int. J. Biological Macromol.* **2016**, *82*, 751–756. [CrossRef] [PubMed]
13. Nie, M.; Nie, H.; He, M.; Lin, Y.; Wang, L.; Jin, P.; Zhang, S. Immobilization of biofilms of *pseudomonas aeruginosa* ny3 and their application in the removal of hydrocarbons from highly concentrated oil-containing wastewater on the laboratory scale. *J. Environ. Manag.* **2016**, *173*, 34–40. [CrossRef] [PubMed]
14. Ferreira, L.; Rosales, E.; Sanromán, M.A.; Pazos, M. Preliminary testing and design of permeable bioreactive barrier for phenanthrene degradation by *pseudomonas stutzeri* CECT 930 immobilized in hydrogel matrices. *J. Chem. Technol. Biotechnol.* **2014**, *90*, 500–506. [CrossRef]
15. Swisher, R.; Carroll, G.C. Fluorescein diacetate hydrolysis as an estimator of microbial biomass on coniferous needle surfaces. *Microb. Ecol.* **1980**, *6*, 217–226. [CrossRef] [PubMed]
16. Fontvieille, D.A.; Outaguerouine, A.; Thevenot, D.R. Fluorescein diacetate hydrolysis as a measure of microbial activity in aquatic systems: Application to activated sludges. *Environ. Technol.* **1992**, *13*, 531–540. [CrossRef]
17. Jiang, S.; Huang, J.; Lu, H.; Liu, J.; Yan, C. Optimisation for assay of fluorescein diacetate hydrolytic activity as a sensitive tool to evaluate impacts of pollutants and nutrients on microbial activity in coastal sediments. *Mar. Pollut. Bull.* **2016**, *110*, 424–431. [CrossRef] [PubMed]
18. Liang, Y.; Zhang, X.; Dai, D.; Li, G. Porous biocarrier-enhanced biodegradation of crude oil contaminated soil. *Int. Biodeterior. Biodegradation* **2009**, *63*, 80–87. [CrossRef]
19. Picioreanu, C.; van Loosdrecht, M.C.M.; Heijnen, J.J. Two-dimensional model of biofilm detachment caused by internal stress from liquid flow. *Biotechnol. Bioeng.* **2001**, *72*, 205–218. [CrossRef]
20. Flemming, H.-C.; Wingender, J.; Szewzyk, U.; Steinberg, P.; Rice, S.A.; Kjelleberg, S. Biofilms: An emergent form of bacterial life. *Nat. Rev. Microbiol.* **2016**, *14*, 563–575. [CrossRef] [PubMed]
21. Hupert-Kocurek, K.; Guzik, U.; Wojcieszyńska, D. Characterization of catechol 2,3-dioxygenase from *Planococcus* sp. strain S5 induced by high phenol concentration. *Acta Biochem. Pol.* **2012**, *59*, 345–351.
22. Green, V.S.; Stott, D.E.; Diack, M. Assay for fluorescein diacetate hydrolytic activity: optimization for soil samples. *Soil Biol. Biochem.* **2006**, *38*, 693–701. [CrossRef]
23. Fisher, K.A.; Huddersman, K.D.; Taylor, M.J. Comparison of micro- and mesoporous inorganic materials in the uptake and release of the drug model fluorescein and its analogues. *Chem. - Eur. J.* **2003**, *9*, 5873–5878. [CrossRef] [PubMed]
24. Kasnavia, T.; Vu, D.; Sabatini, D.A. Fluorescent dye and media properties affecting sorption and tracer selection. *Groundwater* **1999**, *37*, 376–381. [CrossRef]
25. Sabatini, D.A.; Austin, T.A. Characteristics of rhodamine WT and fluorescein as adsorbing ground-water tracers. *Groundwater* **1991**, *29*, 341–349. [CrossRef]

26. Marchlewicz, A.; Domaradzka, D.; Guzik, U.; Wojcieszyńska, D. *Bacillus thuringiensis* b1 (2015b) is a gram-positive bacteria able to degrade naproxen and ibuprofen. *Water, Air, Soil Pollut.* **2016**, *227*, 1–8. [CrossRef] [PubMed]
27. Manohar, S.; Kim, C.K.; Karegoudar, T.B. Enhanced degradation of naphthalene by immobilization of *pseudomonas* sp. strain NGK1 in polyurethane foam. *Appl. Microbiol. Biotechnol.* **2001**, *55*, 311–316. [CrossRef] [PubMed]
28. Jain, P.; Pradeep, T. Potential of silver nanoparticle-coated polyurethane foam as an antibacterial water filter. *Biotechnol. Bioeng.* **2005**, *90*, 59–63. [CrossRef] [PubMed]
29. Phoenix, V.R.; Holmes, W.M. Magnetic resonance imaging of structure, diffusivity, and copper immobilization in a phototrophic biofilm. *Appl. Environ. Microbiol.* **2008**, *74*, 4934–4943. [CrossRef] [PubMed]
30. Dikshit, P.K.; Moholkar, V.S. Kinetic analysis of dihydroxyacetone production from crude glycerol by immobilized cells of *gluconobacter oxydans* MTCC 904. *Bioresour. Technol.* **2016**, *216*, 948–957. [CrossRef] [PubMed]
31. Karunakaran, E.; Biggs, C.A. Mechanisms of *bacillus cereus* biofilm formation: An investigation of the physicochemical characteristics of cell surfaces and extracellular proteins. *Appl. Microbiol. Biotechnol.* **2010**, *89*, 1161–1175. [CrossRef] [PubMed]
32. Ingesson, H.; Zacchi, G.; Yang, B.; Esteghlalian, A.R.; Saddler, J.N. The effect of shaking regime on the rate and extent of enzymatic hydrolysis of cellulose. *J. Biotechnol.* **2001**, *88*, 177–182. [CrossRef]
33. Sirisha, E.; Rajasekar, N.; Narasu, M.L. Isolation and optimization of lipase producing bacteria from oil contaminated soils. *Adv. Biol. Res.* **2010**, *4*, 249–252.
34. Bisswanger, H. Enzyme assays. *Perspect. Sci.* **2014**, *1*, 41–55. [CrossRef]
35. Guilbault, G.G.; Kramer, D.N. Fluorometric determination of lipase, acylase, alpha-, and gamma-chymotrypsin and inhibitors of these enzymes. *Anal. Chem.* **1964**, *36*, 409–412. [CrossRef]
36. Frølund, B.; Griebe, T.; Nielsen, P.H. Enzymatic activity in the activated-sludge floc matrix. *Appl. Microbiol. Biotechnol.* **1995**, *43*, 755–761. [CrossRef] [PubMed]
37. Jørgensen, P.E.; Eriksen, T.; Jensen, B.K. Estimation of viable biomass in wastewater and activated sludge by determination of ATP, oxygen utilization rate and FDA hydrolysis. *Water Res.* **1992**, *26*, 1495–1501. [CrossRef]
38. Pietikäinen, J.; Pettersson, M.; Bååth, E. Comparison of temperature effects on soil respiration and bacterial and fungal growth rates. *FEMS Microbiol. Ecol.* **2005**, *52*, 49–58. [CrossRef] [PubMed]
39. Adam, G.; Duncan, H. Development of a sensitive and rapid method for the measurement of total microbial activity using fluorescein diacetate (FDA) in a range of soils. *Soil Biol. Biochem.* **2001**, *33*, 943–951. [CrossRef]
40. Sebastián, M.; Auguet, J.-C.; Restrepo-Ortiz, C.X.; Sala, M.M.; Marrasé, C.; Gasol, J.M. Deep ocean prokaryotic communities are remarkably malleable when facing long-term starvation. *Environ. Microbiol.* **2018**, *20*, 713–723. [CrossRef] [PubMed]
41. Cox, H.H.J.; Deshusses, M.A. Effect of starvation on the performance and re-acclimation of biotrickling filters for air pollution control. *Environ. Sci. Technol.* **2002**, *36*, 3069–3073. [CrossRef] [PubMed]
42. Roszak, D.B.; Colwell, R.R. Survival strategies of bacteria in the natural environment. *Microbiol. Rev.* **1987**, *51*, 365–379. [PubMed]
43. Ensign, J.C. Long-term starvation survival of rod and spherical cells of *Arthrobacter crystallopoietes*. *J. Bacteriol.* **1970**, *103*, 569–577. [PubMed]
44. Gengenbacher, M.; Rao, S.P.S.; Pethe, K.; Dick, T. Nutrient-starved, non-replicating mycobacterium tuberculosis requires respiration, ATP synthase and isocitrate lyase for maintenance of ATP homeostasis and viability. *Microbiology* **2009**, *156*, 81–87. [CrossRef] [PubMed]
45. Voelker, U.; Voelker, A.; Maul, B.; Hecker, M.; Dufour, A.; Haldenwang, W.G. Separate mechanisms activate sigma B of *Bacillus subtilis* in response to environmental and metabolic stresses. *J. Bacteriol.* **1995**, *177*, 3771–3780. [CrossRef] [PubMed]
46. Mielich-Süss, B.; Lopez, D. Molecular mechanisms involved in *Bacillus subtilis* biofilm formation. *Environ. Microbiol.* **2015**, *17*, 555–565. [CrossRef] [PubMed]
47. Sonenshein, A.L. Control of sporulation initiation in *Bacillus subtilis*. *Curr. Opin. Microbiol.* **2000**, *3*, 561–566. [CrossRef]
48. Majed, R.; Faille, C.; Kallassy, M.; Gohar, M. *Bacillus cereus* biofilms—same, only different. *Front. Microbiol.* **2016**, *7*, 1054. [CrossRef] [PubMed]
49. Correction: SinR controls enterotoxin expression in *bacillus thuringiensis* biofilms. *PLoS ONE* **2014**, *9*, e96707.

50. Garcia-Ochoa, F.; Gomez, E.; Santos, V.E.; Merchuk, J.C. Oxygen uptake rate in microbial processes: An overview. *Biochem. Eng. J.* **2010**, *49*, 289–307. [CrossRef]
51. Norsker, N.H.; Nielsen, P.H.; Hvitved-Jacobsen, T. Influence of oxygen on biofilm growth and potential sulfate reduction in gravity sewer biofilm. *Water Sci. Tech.* **1995**, *31*, 159–167. [CrossRef]
52. Zhou, X.-H.; Qiu, Y.-Q.; Shi, H.-C.; Yu, T.; He, M.; Cai, Q. A new approach to quantify spatial distribution of biofilm kinetic parameters by in situ determination of oxygen uptake rate (our). *Environ. Sci. Technol.* **2009**, *43*, 757–763. [CrossRef] [PubMed]
53. De Beer, D.; Stoodley, P.; Roe, F.; Lewandowski, Z. Effects of biofilm structures on oxygen distribution and mass transport. *Biotechnol. Bioeng.* **1994**, *43*, 1131–1138. [CrossRef] [PubMed]
54. Zhang, L.; Wu, W.; Wang, J. Immobilization of activated sludge using improved polyvinyl alcohol (PVA) gel. *J. Environ. Sci.* **2007**, *19*, 1293–1297. [CrossRef]
55. Bandaiphet, C.; Prasertsan, P. Effect of aeration and agitation rates and scale-up on oxygen transfer coefficient, kla in exopolysaccharide production from *enterobacter cloacae* WD7. *Carbohydr. Polym.* **2006**, *66*, 216–228. [CrossRef]
56. Ganesh, R.; Balaji, G.; Ramanujam, R.A. Biodegradation of tannery wastewater using sequencing batch reactor—respirometric assessment. *Bioresour. Technol.* **2006**, *97*, 1815–1821. [CrossRef] [PubMed]
57. Schäfer, S.; Schrader, J.; Sell, D. Oxygen uptake rate measurements to monitor the activity of terpene transforming fungi. *Process Biochem.* **2004**, *39*, 2221–2228. [CrossRef]
58. Greń, I.; Wojcieszyńska, D.; Guzik, U.; Perkosz, M.; Hupert-Kocurek, K. Enhanced biotransformation of mononitrophenols by *stenotrophomonas maltophilia* kb2 in the presence of aromatic compounds of plant origin. *World J. Microbiol. Biotechnol.* **2009**, *26*, 289–295. [CrossRef]
59. Battin, T.J. Assessment of fluorescein diacetate hydrolysis as a measure of total esterase activity in natural stream sediment biofilms. *Sci. Total. Environ.* **1997**, *198*, 51–60. [CrossRef]
60. Subramanian, S.B.; Yan, S.; Tyagi, R.D.; Surampalli, R.Y. Extracellular polymeric substances (EPS) producing bacterial strains of municipal wastewater sludge: Isolation, molecular identification, EPS characterization and performance for sludge settling and dewatering. *Water Res.* **2010**, *44*, 2253–2266. [CrossRef] [PubMed]
61. Amon, R.M.W.; Benner, R. Photochemical and microbial consumption of dissolved organic carbon and dissolved oxygen in the Amazon River system. *Geochim. Cosmochim. Acta* **1996**, *60*, 1783–1792. [CrossRef]

catalysts

MDPI

Article

Immobilization of Eversa Lipase on Octyl Agarose Beads and Preliminary Characterization of Stability and Activity Features

Sara Arana-Peña †, Yuliya Lokha † and Roberto Fernández-Lafuente *

Departamento de Biocatálisis, Instituto de Catálisis-CSIC, Campus UAM-CSIC, 28049 Madrid, Spain; sara_arana@hotmail.com (S.A.-P.); yuliyalokha@gmail.com (Y.L.)

* Correspondence: rfl@icp.csic.es

† Both authors have evenly contributed to this paper.

Received: 11 October 2018; Accepted: 31 October 2018; Published: 2 November 2018

Abstract: Eversa is an enzyme recently launched by Novozymes to be used in a free form as biocatalyst in biodiesel production. This paper shows for first time the immobilization of Eversa (a commercial lipase) on octyl and aminated agarose beads and the comparison of the enzyme properties to those of the most used lipase, the isoform B from *Candida antarctica* (CALB) immobilized on octyl agarose beads. Immobilization on octyl and aminated supports of Eversa has not had a significant effect on enzyme activity versus p-nitrophenyl butyrate (pNPB) under standard conditions (pH 7), but immobilization on octyl agarose beads greatly enhanced the stability of the enzyme under all studied conditions, much more than immobilization on aminated support. Octyl-Eversa was much more stable than octyl-CALB at pH 9, but it was less stable at pH 5. In the presence of 90% acetonitrile or dioxane, octyl-Eversa maintained the activity (even increased the activity) after 45 days of incubation in a similar way to octyl-CALB, but in 90% of methanol, results are much worse, and octyl-CALB became much more stable than Eversa. Coating with PEI has not a clear effect on octyl-Eversa stability, although it affected enzyme specificity and activity response to the changes in the pH. Eversa immobilized octyl supports was more active than CALB versus triacetin or pNPB, but much less active versus methyl mandelate esters. On the other hand, Eversa specificity and response to changes in the medium were greatly modulated by the immobilization protocol or by the coating of the immobilized enzyme with PEI. Thus, Eversa may be a promising biocatalyst for many processes different to the biodiesel production and its properties may be greatly improved following a suitable immobilization protocol, and in some cases is more stable and active than CALB.

Keywords: Eversa; interfacial activation; lipase immobilization; enzyme stabilization; enzyme modulation

1. Introduction

Lipases are very interesting enzymes in biocatalysis due to their broad substrate specificity, their selectivity and high stability, combined in many instances with a very high enantio or regio selectivity and specificity [1–10]. Moreover, they can perform their function in a wide diversity of reaction media [11–13].

One of the hottest topics in lipase application is the production of biodiesel [14–16]. Although, currently, chemical catalysis is used in most factories, the use of lipases may have interest to save energy, reduce contaminants and side-products and also to prevent pretreatment of the oils (as acid oils are not compatible with alkali catalysis) [17,18].

Lipases catalytic mechanism is quite peculiar, as the enzyme is in a conformational equilibrium [19–21]. In the closed form, a polypeptide (named lid) blocks the active center, making the enzyme inactive in many cases. In the open form, the lid is shifted and exposes to the medium a large hydrophobic pocket containing the active center. This pocket may become adsorbed and stabilized to any hydrophobic surface, like drops of oils (the so-called interfacial activation) [21,22], but also to hydrophobic proteins [23,24], another open form of a lipase [25,26] or a hydrophobic support [27].

Thus, hydrophobic supports can promote the one step immobilization/stabilization/purification and hyperactivation of lipases [28], because the open form of the lipase becomes fixed after immobilization [29]. Moreover, the lipase is in monomeric form [25–29], ensuring that all lipases have a similar conformation and state.

As a result of the interest for new lipases, the most important enzyme producing companies have developed an intense search for lipases directed to some specific uses. Among them, Novozymes has launched a lipase preparation called Eversa. In 2014, the enzyme was announced as "Novozymes Eversa®, the first commercially available enzymatic solution to make biodiesel from waste oils" [30]. Since then, the enzyme has been mainly used in biodiesel production [31–38]. In all these papers, the enzyme has been used in free form as recommended by the supplier. Thus, there is not a proper characterization of the enzyme in terms of stability, activity, etc. in other reactions. An exception to the use of free enzyme is the very recent paper from Remonatto et al. that used immobilized Eversa, although again without a characterization of the immobilized enzyme properties and focused on biodiesel production [39].

Even though this enzyme may be used in free form, their immobilization may enhance many properties. A proper immobilization may improve enzyme stability, activity, selectivity, or specificity, resistance to inhibitors or chemicals, even purity [40–47]. This improvement has already been shown in biodiesel production using different hydrophobic supports [48–50].

On the other hand, following the concept of combilipase, it is difficult to believe that a single lipase may be the ideal one for all oils, and for each of the possible substrates contained in a heterofunctional product like a vegetal oil [51,52]. Therefore, it seems sensitive to evaluate in a more global form the properties of this new lipase.

Thus, in this paper we have intended to immobilize and to perform a first characterization of some features of Eversa (e.g., stability and activity at different pH values versus different substrates, stability in the presence of different organic solvents). To this goal, the enzyme has been immobilized on octyl agarose, via interfacial activation as described above [29]. The main problem of this immobilization protocol is the possibility of enzyme desorption during operation. This may occur at high temperature or in the presence of high concentrations of organic solvents [53], also if the substrate or the product has detergent properties [54–56]. To avoid enzyme release, the employment of heterofunctional supports (having acyl groups plus some chemically reactive groups) has proved to be a valid solution [53,57,58]. Another simpler alternative is the chemical or physical crosslinking of the enzyme molecules, which reduces the risks of enzyme release from the support [59]. The coating of immobilized lipases with PEI is one of the most used proposals [60–62], as this reagent may have many positive impacts in enzyme properties. [63]. Moreover, lipases coated with PEI have been proposed as a first step in the building of coimmobilized enzymes enabling the reuse of the least stable one [64,65]. Finally, this polymer has been also proposed to coimmobilize enzymes and cofactors [66].

The new Eversa enzyme has been compared with the lipase B from *Candida antarctica* (CALB), perhaps the most popular lipase in biocatalysis, immobilized also in octyl agarose [67,68].

Thus, the main objective of the current paper is to show a first evaluation of the properties of Eversa, immobilized on octyl agarose beads and treated with PEI, under different conditions, which may be expected that will have a likely impressive impact in the next future due to some of their properties [31–39]. Eversa has been also ionically exchanged in aminated agarose for comparison with octyl agarose.

To this goal, Eversa immobilized via interfacial activation or ion exchange has been compared to CALB immobilized via interfacial in terms of activity versus different substrates and under different conditions as well as their stabilities in different media and conditions, to analyze the properties of the new enzyme compared to CALB and the effects of the immobilization of the enzyme, even though the manufacturer claims that this is an enzyme to be used in free form.

2. Results

2.1. Immobilization of CALB and Eversa

Figure 1 shows the immobilization courses of CALB and Eversa on octyl agarose beads and in the case of Eversa, on monoaminoethyl-N-ethyl-agarose (MANAE) agarose beads. CALB slightly decreased its activity after immobilization on octyl agarose, while the activity of Eversa remained almost intact, being immobilization marginally more rapid, using both enzymes immobilization yield is over 95%. Eversa on MANAE also maintains full activity, but immobilization yield is around 80%.

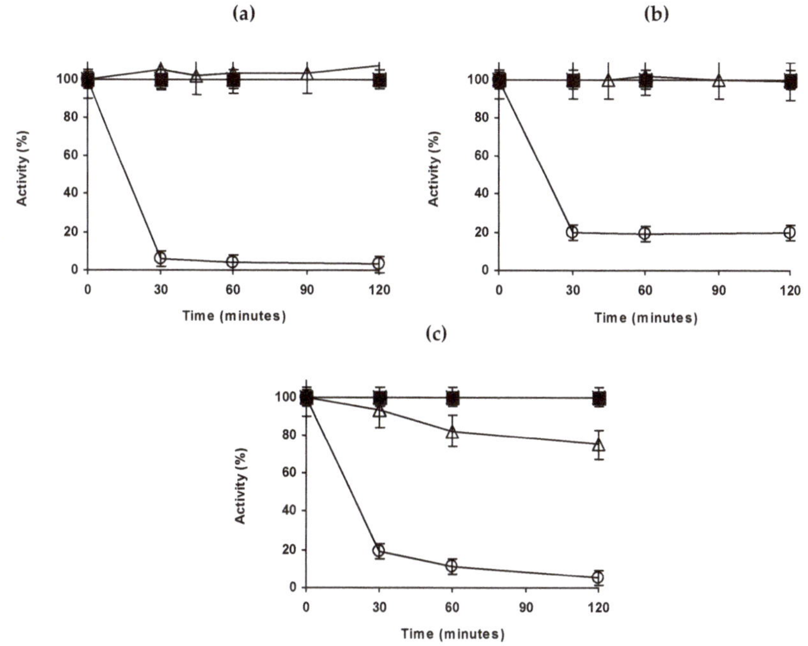

Figure 1. Immobilization courses of Eversa on octyl (**a**) or MANAE (**b**) agarose beads and CALB on octyl (**c**) agarose beads. Experiments were performed as described in methods section. Solid squares: reference, circles: supernatant; empty triangles: suspension.

Figure S1 shows the SDS-PAGE of Eversa and CALB, both in free and immobilized on octyl agarose. Both enzymes are quite pure, purification after immobilization is therefore negligible.

Most lipases increase its activity after immobilization on octyl agarose [27], CALB is a known exception due to the small lid that not fully isolated the active center [53], Eversa has a similar behavior.

2.2. Modification of Octyl-Eversa with PEI

The modification of octyl-Eversa con PEI did not have a significant effect on enzyme the standard activity assay (recovered activity versus pNPB was 105% after PEI coating).

2.3. Thermal Stability of the Different Lipase Biocatalysts

The three immobilized Eversa preparations (enzyme immobilized on octyl, enzyme immobilized on octyl and then modified with PEI and the enzyme immobilized on MANAE agarose) and the free Eversa were inactivated at different pH values, and compared to octyl-CALB (Figure 2).

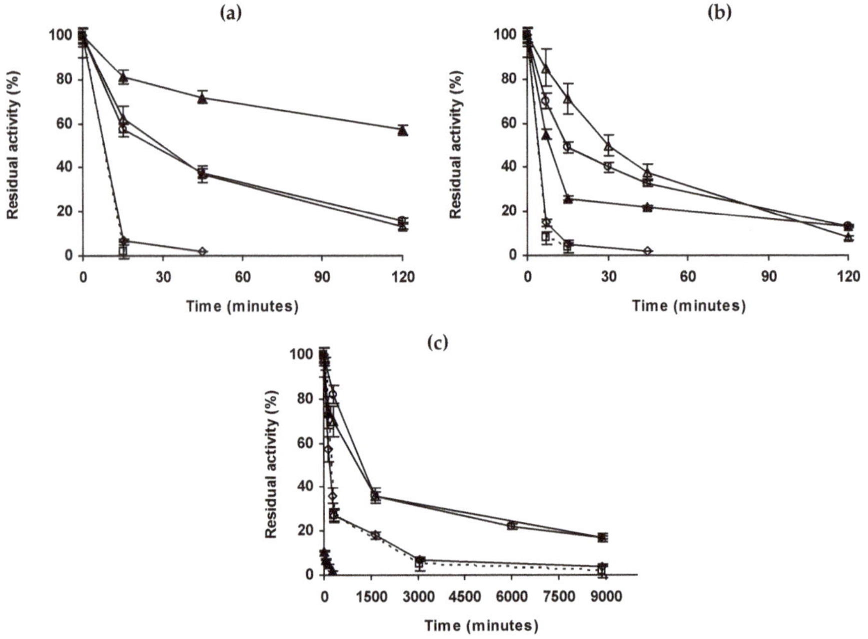

Figure 2. Thermal inactivations of different lipase preparations at pH 5 (**a**) and 7 (**b**) (both at 80 °C) or 9 (**c**) (65 °C). Other specifications are described in methods section. Solid triangles: octyl-CALB, empty squares, dotted line, free Eversa; empty circles: Octyl-Eversa; empty triangles: Octyl-Eversa-PEI; empty rhombus: MANAE-Eversa.

At pH 5 and 80 °C (Figure 2a), immobilized CALB is clearly the most stable preparation, retaining almost 60% of the initial activity after 2 h while the most stable immobilized Eversa preparation retained only around 15%. Octyl-Eversa modification with PEI did not result in a significant stabilization, but both preparations were much more stable than the enzyme immobilized on MANAE or the free enzyme, both almost fully inactive in the first measure.

At pH 7 and 80 °C (Figure 2b), octyl-Eversa presented a stability similar to that at pH 5, while CALB become more unstable. That way, octyl-Eversa become more stable than octyl-CALB in the first inactivation steps, but later CALB retained similar activity. The coating with PEI of octyl-Eversa also offered some stabilization in the first inactivation steps, but this effect is not so clear later. Again, free enzyme and MANAE-Eversa were almost fully inactive in the first activity determination point.

Finally, at pH 9 (temperature was decreased to 65 °C because of the lower stability of all preparations) all Eversa preparations, including the free enzyme, became more stable than octyl-CALB (only 15% of residual activity after 15 min while the less stable Eversa preparation presented 60% after 2 h). Immobilization on octyl greatly improved the stability of Eversa, while PEI coating has no effect except in the first steps of the inactivation.

The results fit the expect ones: immobilization on octyl-agarose should provide certain stabilization because the open and stabilized form of the lipase has some thermodynamic advantages [69,70]. That way this preparation interfacially activated versus an octyl layer is even more stable than multipoint covalently attached preparations [71], and that way it is expected that this can be more stable than the ionically adsorbed enzyme, where stabilization is hard to expect [72]. CALB is more stable than Eversa at pH 5 while at pH 7 differences are short and at pH 9, CALB is much more unstable than Eversa. This should be associated to differences on enzymes structures.

2.4. Inactivation of the Enzyme Preparations in Organic Solvents

The four enzyme immobilized preparations were incubated in 90% of different solvents at pH 7 and 25 °C (Figure 3). Using acetonitrile, Eversa preparations were more stable than octyl-CALB. Even there was a clear hyperactivation of the Eversa biocatalysts. For example, MANAE-Eversa after 45 days gave a 60% of hyperactivation. PEI coating of octyl-Eversa had not a positive effect. In dioxane, all preparations maintained their activity almost intact after 45 days, octyl-Eversa-PEI showing the highest activity after 45 days (120%), shortly followed by octyl-CALB.

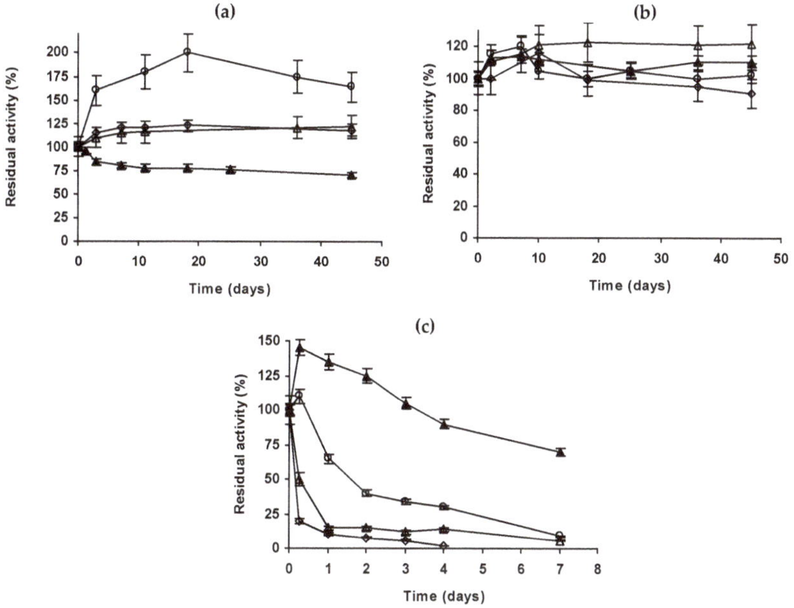

Figure 3. Inactivation in 90% of acetonitrile (**a**), dioxane (**b**) or methanol (**c**) of different preparations of lipases. Other specifications are described in methods section. Solid triangles: octyl-CALB, empty circles: Octyl-Eversa; empty triangles: Octyl-Eversa-PEI; empty rhombus: MANAE-Eversa.

Differences become clearer using methanol. In this case, Eversa become clearly less stable than CALB, and octyl-Eversa is the most stable preparation. In this case, MANEA- or PEI-coated preparations were much less stable than the octyl preparation.

The solvent inactivation of octyl lipase preparations is associated to the enzyme release from the support [53–56,60–62] and this will not occur in MANAE. Apparently, Eversa became so strongly adsorbed on octyl agarose that this effect was not very significant. Furthermore, the coating of the enzymes with PEI has prevented enzyme leakage and the promotion of a certain solvent partition as main functions [60–62]. Using octyl-Eversa, only using dioxane the effect is somehow positive, with acetonitrile and methanol the PEI coating effect is even negative. The presence of a cationic polymer or surface near the enzyme may stabilize incorrect structures, with negative effects on enzyme stability [73].

Considering that this enzyme has been designed to produce biodiesel [31–39], it is surprising that methanol, a substrate in this reaction, is the one with a higher negative effect on enzyme stability when compared to CALB. However, it may be expected that in anhydrous media the effect of methanol may be different, but this relatively poor stability of the Eversa enzyme in methanol should be considered in the biodiesel reaction design.

2.5. Effect of pH on Lipase Activity versus pNPB

Figure 4 shows the specific activity of Eversa and CALB at different pH values after immobilization on different supports. First point that is evident is that all Eversa preparations were much more active than CALB versus this substrate, except MANAE and free enzyme at pH 4.

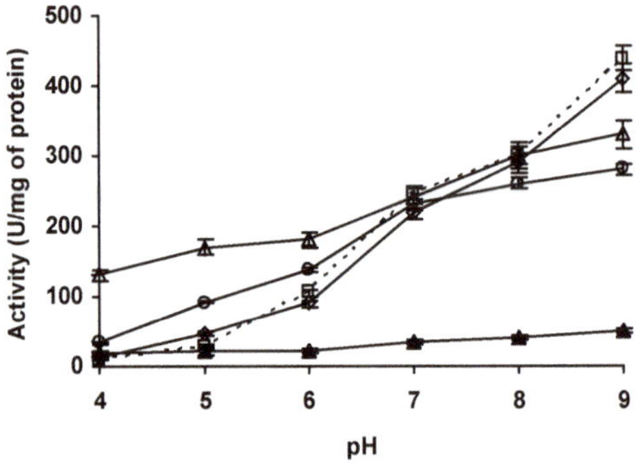

Figure 4. Effect of pH on the activity of different preparations of lipases versus pNPB. Experiments were performed as described in Methods. Solid triangles: octyl-CALB, empty squares, dotted line, free Eversa; empty circles: Octyl-Eversa; empty triangles: Octyl-Eversa-PEI; empty rhombus: MANAE-Eversa.

The maximum activity was observed in all cases at the highest studied pH, pH 9. At higher pH the low stability of pNPB reduces the reliability of the results. Free Eversa activity greatly increased when increasing the pH value, while the other preparations increase of activity is in a milder way. The largest difference when comparing the activities at pH 4 and 9 was that between octyl coated with PEI and the free enzyme. Using octyl-Eversa- PEI the increment in activity from pH 4 to pH 9 is 2.5 times while for the free enzyme is 48 folds. For octyl-Eversa, it is eight-fold and for MANAE-Eversa, it is 34.

After immobilization on octyl agarose beads, the activity almost remained unaltered at pH 7, however greatly increased at acid pH value decreased the enzyme activity at alkaline pH value (mainly at pH 9). The coating of octyl-Eversa with PEI has no significant effect at pH 7, but at all other pH values produced a very significant increase in activity (e.g., at pH 4 from 33 to 130 and at pH 9 from 240 to 330). Immobilization of Eversa on MANAE gave very low activity at pH 4 (just over the free enzyme), at pH 7 the activity is similar to all other preparations, but at pH 8 and 9 it behaves more similarly to the free enzyme and gave the highest activity among the immobilized preparations.

This great difference of the effects of pH value on the activity of different enzyme preparations agrees with previous reports [74] and can be explained by the complexity of the lipase activity: we are measuring in some of them the activity of the active form and the percentage of molecules on the open form, the strength of the interaction between the enzyme and cationic (MANAE, PEI) or even anionic groups in the support [64], and the activity of the structure of the "open form" resulting in each condition. These results exemplify that a lipase cannot be discarded by a negative result in a specific condition, neither a positive one should be expected to be extrapolated to any other condition [74].

2.6. Effect of Some Ions on the Lipase Stability

It has been recently described that some anions, like phosphate, have a negative effect on the stability of many lipases [75]. However, these effects were only on stability at high temperature or very high buffer concentration, the activity remained unaltered when changing the buffer [75]. On the other hand, some cations (e.g., Ca^{2+}) have shown to have positive effects on some lipase stabilities [76,77]. Curiously, this effect depends on the pH and immobilization lipase form [76,77].

Thus, we have decided to analyze the effect of $CaCl_2$ and sodium phosphate on the stability of the different Eversa forms at pH 7, compared to octyl-CALB. Figure 5 shows the results.

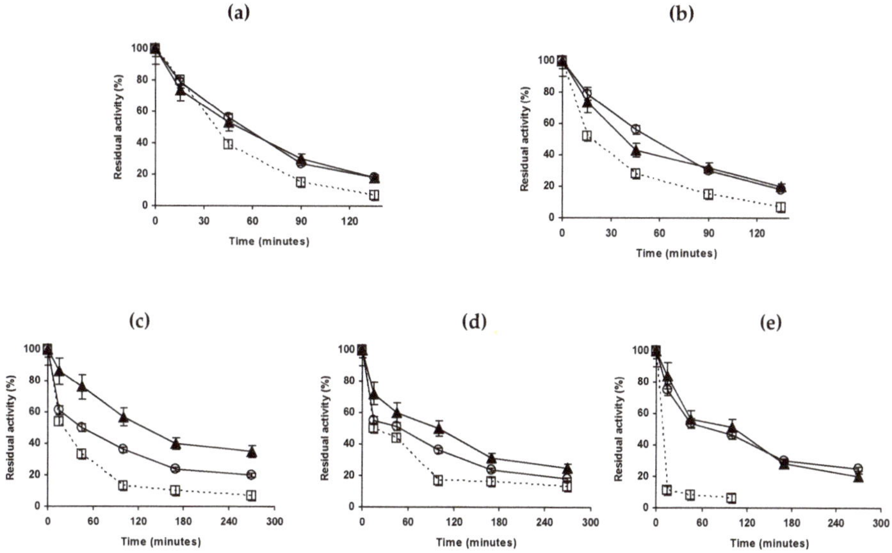

Figure 5. Effect of different ions on the stability of diverse lipase preparations an pH 7. (**a**): free Eversa, (72 °C); (**b**): MANAE-Eversa (72 °C); (**c**): Octyl-Eversa (78 °C), (**d**): Octyl-Eversa-PEI (78 °C), (**e**): Octyl-CALB (78 °C). Inactivation in 100 mM phosphate: empty squares; inactivation in 100 mM Tris: empty circles; inactivation in 100 mM Tris/10 mM $CaCl_2$: full triangles.

Using free Eversa, the effects of the additives are not very significant. Stability in phosphate is slightly lower than in Tris, and $CaCl_2$ has no effect. Using MANAE, the negative effect of phosphate is clear, while the effect of $CaCl_2$ is negligible. In these cases the inactivations were performed at 72 °C because at 78 °C the inactivation was too rapid to be able to perform a reliable comparison. The other biocatalysts were inactivated at 78 °C. The situation using octyl-Eversa was different; here a positive effect of $CaCl_2$ and a negative effect of phosphate were evident. The coating with PEI reduces the effect of the ions, but that were still clear. Octyl-CALB stability is strongly reduced in 100 mM sodium phosphate, while the addition of 10 mM $CaCl_2$ has not an evident effect (slightly higher stability in initial steps of the inactivation).

Thus, phosphate negative effect on lipase stability previously described [75] is also found using Eversa, although the effect is maximized in the octyl preparation. In any case, this effect is much lower than the one found using octyl-CALB, where phosphate produces a drastic reduction on enzyme stability, while using octyl-Eversa is reduced just by 2–3 folds. The positive effect of $CaCl_2$ on lipase stability is only found using octyl-Eversa, in a similar way to the previous results, [76] but it is not very significant (a 2–3 fold factors) compared to the values described with other enzymes [76].

Differences between CALB and Eversa should be based in differences in enzyme structure, in the case of MANAE and octyl-Eversa, this should be based in the fact that while in one case the open structure is stabilized [29], in the ion exchanged enzyme the conformational equilibrium remains. Apparently, the immobilization of Eversa on octyl agarose beads permits to maximize the positive or negative effects of the studied ions.

2.7. Enzyme Specificity

To investigate of the immobilization of Eversa may alter the enzyme specificity as described with other lipases [74,78], several substrates have been used. Table 1 shows the activity of different immobilized enzyme preparations versus p-NPB, R or S methyl mandelate and triacetin.

Table 1. Activity of different lipase biocatalyst versus different substrates at pH 7 and 25 °C. Other reaction conditions are described in Methods section. Activity is given as micromoles/min/mg of enzyme.

Biocatalyst	pH	Substrate			
		pNPB	(R)-Methyl Mandelate	Enantiomer Activity Ratio (VR/VS)	Triacetin
Octyl-Eversa	5	92 ± 5	0.0112 ± 0.0011	0.76	25.7 ± 0.8
	7	230 ± 15	0.0103 ± 0.0009	0.9	21.6 ± 0.6
	9	280 ± 18	0.0050 ± 0.0007	0.94	20 ± 0.9
Octyl-Eversa treated with PEI	5	170 ± 10	0.0083 ± 0.0007	0.64	28.3 ± 1.0
	7	240 ± 13	0.0073 ± 0.0004	0.8	22.3 ± 0.6
	9	330 ± 15	0.0032 ± 0.0005	0.77	25.0 ± 1.2
MANAE-Eversa	5	47 ± 4	0.0024 ± 0.0003	0.51	2.95 ± 0.08
	7	220 ± 12	0.0091 ± 0.0008	1	4.8 ± 0.3
	9	410 ± 21	0.0040 ± 0.0004	0.74	11.0 ± 0.6
Octyl-CALB	5	21 ± 2	49.3 ± 3.1	12.0	9.6 ± 1.0
	7	35 ± 2	81.1 ± 3.8	8.4	19.9 ± 1.6
	9	50 ± 3	63.3 ± 4.6	11.3	10.3 ± 0.5

As described above, Eversa is more active versus pNPB than CALB at all pHs studied. In all cases, the activity increases with pH. Coating of octyl-Eversa with PEI increases the activity mainly at pH 5 and 9, while it had a marginal effect at pH 7. MANAE-Eversa is the least active Eversa immobilized biocatalyst at pH 5 and the most active at pH 9.

Focusing on mandelate esters, the situation is quite different. Octyl-CALB is much more active than all Eversa preparations, preferring the R isomer, with a ratio in the rate of hydrolysis of the R isomer versus S enantiomer near 10 in all range of studied pH values. Activity is higher at pH 7 than at pH 5 or 9, but the enantiomeric ratio is lower at that pH. This enzyme is particularly active versus aromatic acyl donors. Although Eversa is hundreds fold less active; the most active Eversa preparation is octyl-Eversa. This enzyme prefers the hydrolysis of the S isomer, but with very low differences (maximum, a factor of less than 2). The effect of the pH on the enzyme activity depends on the preparation. Octyl-Eversa has the maximum activity versus methyl mandelate at pH 5 (in opposition with the behavior using pNPB), with a significant drop at pH 9. The picture is similar after coating with PEI, although activity is 20–40% lower. MANAE-Eversa has optimal activity at pH 7, with higher activity at pH 9 than at pH 5 in opposition to the octyl preparations. This preparation offered the highest (a factor of 2 at pH 5) and the lowest (at pH 7, with identical activities) differences in the reaction with both enantiomers.

Using triacetin, octyl-Eversa is quite more active than octyl-CALB, mainly at alkaline and acid pH values. While CALB maintains a maximum activity at pH 7 (less than 20 U/mg) with around a 50% decrease a pH 5 or 9, octyl-Eversa maintains a similar level of activity (from around 25 U/mg at pH 5 to around 21 U/mg at pH 9), although the maximum activity is found at pH 5 and the minimum at pH 9 (similar to mandelate esters, and opposite to pNPB). The coating with PEI produced a slightly increase in the activity (similar to using pNPB and opposite to the results using mandelate esters), and a change in the effect of the pH: now there a valley at pH 7 (22.3 U/mg) and at pH 5 (28.3 U/mg) and pH 9 (25 U/mg) the activities were higher. MANAE-Eversa have lower activity than even octyl-CALB, but it presented a clear maximum at pH 9, at this pH value the activity is higher than the activity of octyl-CAB. Minimal activity is found at pH 5.

The results above newly exemplify how immobilization of an enzyme on different supports or the physical treatment of the immobilized enzymes may greatly alter enzyme specificity and response to the changes in the medium [74,78]. Results suggested that Eversa is specially indicated to reactions where the acyl donor is aliphatic, if the acyl door is aromatic the activity greatly decrease.

Octyl-Eversa was reused seven times in hydrolysis of (R) or (S)-methyl mandelate at pH 7 without any significant change in enzyme activity (results not shown).

3. Materials and Methods

3.1. Materials

Commercial solutions of lipase B from *Candida antartica* (CALB) (7.3 mg of protein per mL) and lipase Eversa Transform 2.0 FG (55.4 mg of protein per mL) were kindly gifted by Novozymes (Alcobendas, Spain). p-nitrophenyl butyrate (p-NPB), branched polyethyleneimine (PEI), (MW 25,000), octyl Cl-4B Sepharose beads, R- and S-methyl mandelate, triacetin, 1,4-dioxane and acetonitrile were from Sigma Aldrich (Alcobendas, Spain). Low molecular weight (LMW) calibration kit for SDS electrophoresis (14.4–97 kDa) was from GE Healthcare (Madrid, Spain), and 4% BCL agarose beads standard from Agarose Bead Technologies (Lisbon, Portugal). All other reagents were of analytical grade. Protein concentration was determined using Bradford's method [79] with bovine serum albumin as a standard. Monoaminoethyl-N-ethyl-agarose (MANAE-agarose) was prepared as previously described [80,81].

3.2. Determination of Enzyme Activity

Enzyme activity was measured by recording the increase in absorbance at 348 nm produced by the release p-nitrophenol in the hydrolysis of 0.4 mM p-NPB in 25 mM sodium phosphate buffer at pH 7.0 and 25 °C (ε under these conditions is 5150 M^{-1} cm^{-1}), between 50 and 200 µL of enzyme solution or suspension were added to 2.5 mL of substrate solution to commence the reaction.

3.3. Immobilization of Enzymes

The enzymes were insolubilized employing 1 mg or 0.5 mg of protein per g of wet support, for CALB and Eversa respectively. First, the commercial solutions of enzymes were diluted in the appropriate volume of 5 mM sodium phosphate at pH 7. Then, the support was added, 1 g of wet support per 10 mL of enzyme solution and left under mild stirring. The activities of suspension and supernatant were measured employing pNPB. After full enzyme immobilization, the suspensions were filtered and immobilized lipase biocatalysts were washed several times with distilled water.

3.4. Coating of Immobilized Eversa by Polyethylenimine

To 100 mL of 10% PEI solution at pH 7, 10 g of wet immobilized Eversa were added and left under gentle stirring for 18 h at 4 °C. Then, the suspension was filtered and washed several times with distilled water.

3.5. Thermal Stability

Biocatalysts were incubated at pH 9, 7, or 5, using 50 mM sodium carbonate, tris or sodium acetate, respectively, at different temperatures. The activity was measured periodically using pNPB and residual activity was expressed as percentage of initial activity (initial enzyme). In some cases some additives were added (100 mM phosphate, 10 mM $CaCl_2$).

3.6. *Stability of the Lipase Biocatalysts in the Presence of Organic Co-Solvents*

Immobilized enzymes were incubated in mixtures of 90% (*v*/*v*) 1,4-dioxane, methanol or acetonitrile: 10% (*v*/*v*) 100 mM Tris-HCL (pH 7) at 25 °C. Periodically, samples were withdrawn, and their residual activities were determined with pNPB. Residual activity was expressed as percentage of initial activity.

3.7. *Effect of pH on Enzyme Activity*

To study the effect of pH on enzyme activity, biocatalysts were resuspended in 10 mL of 25 mM Tris-HCl buffer pH 7.0 and activity was measured using pNPB as described previously but using 25 mM buffers at different pH (sodium acetate at pH 4–6, sodium phosphate at pH 6–8 and sodium carbonate at pHs 8 and over). In some instances, the activity was measured when the enzyme was added to the buffer and 10 min after to ensure internal and external pH equilibration [74].

3.8. *SDS-PAGE Analysis*

SDS-polyacrylamide gel electrophoresis was performed on a 5% polyacrylamide stacking gel and a 12% polyacrylamide resolving gel according to Laemmli [82] to analyze the amount of proteins that immobilize on a support. In case of free enzymes, the solutions were diluted in rupture buffer (4% SDS and 10% mercaptoethanol) to the final concentration of 1 mg of protein/mL. Using octyl-CALB, 100 mg were re-suspended in 200 μL of rupture buffer and for immobilized EVERSA 100 mg were re-suspended in 100 μL of rupture buffer. The suspensions were boiled for 8 min and 15 μL aliquots of supernatant were loaded. The samples were run at 100 V. Gels were stained with Coomassie brilliant blue.

3.9. *Hydrolisis of Triacetin*

Triacetin was diluted to 50 mM in 50 mM sodium phosphate pH 7.0. Biocatalyst samples between 0.15–0.45 g were added to 5 mL of triacetin solution and the reaction suspensions were kept under gentle stirring at 25 °C. The concentrations of reaction products were determined by RP-HPLC (Jasco PU-2085) coupled with a UV-1575 Intelligent UV–VIS detector. A 20 μL sample of proper dilution of the reaction mixture was injected and separation was performed on a Kromasil C18 column (15 cm × 0.46 cm) using 15% acetonitrile-water (*v*/*v*) as mobile phase at a flow rate of 1 mL/min at 25 °C. Detection was set at 230 nm. Retention times were 25 min for triacetin and 5 min for diacetin. Activity was determined with a maximum triacetin hydrolysis of 15% to prevent hydrolysis of 1,3 diacetin [83]. One unit of enzyme activity was determined as the amount of enzyme necessary to produce 1 μmol of diacetin per minute under the conditions described above.

3.10. *Hydrolysis of Methyl Mandelate*

A total of 0.05–0.45 g of wet biocatalyst were added to 5 mL of 50 mM R- or S-methyl mandelate in 50 mM sodium phosphate buffer at pH 7.0 and 25 °C. The conversion degree was determined using HPLC as in above point, employing acetonitrile/10 mM ammonium acetate (35:65, *v*/*v*) at pH 2.8 as mobile phase. Retention times were 2.4 min for acid and 4.2 min for ester.

4. Conclusions

Eversa properties such as stability or activity may be enhanced after a proper immobilization, although the supplier recommends use in free form. Immobilization of the enzyme on octyl agarose beads has permitted to improve the enzyme stability under a wide range of conditions. The enzyme immobilized on octyl agarose is quite sensible to the presence of phosphate (negative effect) or Ca^{2+} (positive effect). The enzyme is extremely stable in acetonitrile or dioxane, but not so stable in the presence of methanol. Compared to CALB, Eversa is more thermostable at pH 9. At pH 7, only octyl-Eversa is slightly more stable than octyl-CALB, octyl-CALB is the most stable preparation while at pH 5. Immobilization on octyl agarose gave higher thermal stability at Eversa in all conditions than immobilization on MANAE agarose, but the coating with PEI has an unclear effect (perhaps due to the low loading of enzyme). Curiously for an enzyme designed to be used in biodiesel production, methanol presented a very negative effect on enzyme stability compared to acetonitrile or dioxane, being this effect more deleterious than for CALB. In any case, Eversa may be an interesting biocatalyst for many reactions different to biodiesel production and its immobilization on octyl agarose impairs many positive effects.

Supplementary Materials: The following are available online at http://www.mdpi.com/2073-4344/8/11/511/s1, Figure S1. SDS-PAGE of different lipase preparations. Experiments were performed as described in Methods. Lane 1: Molecular markers, Lane 2: Octyl-CALB, Lane 3: CALB extract, Lane 4: Octyl-Eversa, Lane 5: Eversa extract.

Author Contributions: R.F.-L. conceived, designed the experiments, and wrote the paper; S.A.-P. and Y.L. performed the experiments and helped in writing the paper.

Funding: We gratefully recognize the support from the MINECO from Spanish Government, (project number CTQ2017-86170-R) and Colciencias (Colombia) (project number FP 44842-076-2016).

Acknowledgments: We gratefully recognize the support from the MINECO from Spanish Government, (project number CTQ2017-86170-R) and Colciencias (Colombia) (project number FP 44842-076-2016). The authors wish to thank Ramiro Martínez (Novozymes, Spain) for kindly supplying the enzymes used in this research. The suggestions of Berenguer during the writing of this paper are also gratefully recognized.

Conflicts of Interest: The authors declare no conflict of interest.

References

1. Seddigi, Z.S.; Malik, M.S.; Ahmed, S.A.; Babalghith, A.O.; Kamal, A. Lipases in asymmetric transformations: Recent advances in classical kinetic resolution and lipase—Metal combinations for dynamic processes. *Coord. Chem. Rev.* **2017**, *348*, 54–70. [CrossRef]
2. Bansode, S.R.; Rathod, V.K. An investigation of lipase catalysed sonochemical synthesis: A review. *Ultrason. Sonochem.* **2017**, *38*, 503–529. [CrossRef] [PubMed]
3. Angajala, G.; Pavan, P.; Subashini, R. Lipases: An overview of its current challenges and prospectives in the revolution of biocatalysis. *Biocatal. Agric. Biotechnol.* **2016**, *7*, 257–270. [CrossRef]
4. Kumar, A.; Dhar, K.; Kanwar, S.S.; Arora, P.K. Lipase catalysis in organic solvents: Advantages and applications. *Biol. Proced. Online* **2016**, *18*, 2. [CrossRef] [PubMed]
5. Carvalho, A.C.L.M.; Fonseca, T.S.; De Mattos, M.C.; De Oliveira, M.C.F.; De Lemos, T.L.G.; Molinari, F.; Romano, D.; Serra, I. Recent advances in lipase-mediated preparation of pharmaceuticals and their intermediates. *Int. J. Mol. Sci.* **2015**, *16*, 29682–29716. [CrossRef] [PubMed]
6. Borrelli, G.M.; Trono, D. Recombinant lipases and phospholipases and their use as biocatalysts for industrial applications. *Int. J. Mol. Sci.* **2015**, *16*, 20774–20840. [CrossRef] [PubMed]
7. de Miranda, A.S.; Miranda, L.S.M.; de Souza, R.O.M.A. Lipases: Valuable catalysts for dynamic kinetic resolutions. *Biotechnol. Adv.* **2015**, *33*, 372–393. [CrossRef] [PubMed]
8. Yang, Y.; Zhang, J.; Wu, D.; Xing, Z.; Zhou, Y.; Shi, W.; Li, Q. Chemoenzymatic synthesis of polymeric materials using lipases as catalysts: A review. *Biotechnol. Adv.* **2014**, *32*, 642–651. [CrossRef] [PubMed]
9. Zhang, J.; Shi, H.; Wu, D.; Xing, Z.; Zhang, A.; Yang, Y.; Li, Q. Recent developments in lipase-catalyzed synthesis of polymeric materials. *Process Biochem.* **2014**, *49*, 797–806. [CrossRef]
10. Stergiou, P.-Y.; Foukis, A.; Filippou, M.; Koukouritaki, M.; Parapouli, M.; Theodorou, L.G.; Hatziloukas, E.; Afendra, A.; Pandey, A.; Papamichael, E.M. Advances in lipase-catalyzed esterification reactions. *Biotechnol. Adv.* **2013**, *31*, 1846–1859. [CrossRef] [PubMed]
11. García-Verdugo, E.; Altava, B.; Burguete, M.I.; Lozano, P.; Luis, S.V. Ionic liquids and continuous flow processes: A good marriage to design sustainable processes. *Green Chem.* **2015**, *17*, 2693–2713. [CrossRef]
12. Lozano, P.; Bernal, J.M.; Garcia-Verdugo, E.; Sanchez-Gomez, G.; Vaultier, M.; Burguete, M.I.; Luis, S.V. Sponge-like ionic liquids: A new platform for green biocatalytic chemical processes. *Green Chem.* **2015**, *17*, 3706–3717. [CrossRef]
13. Lozano, P.; Nieto, S.; Serrano, J.L.; Perez, J.; Sánchez-Gomez, G.; García-Verdugo, E.; Luis, S.V. Flow biocatalytic processes in ionic liquids and supercritical fluids. *Mini-Rev. Org. Chem.* **2017**, *14*, 65–74. [CrossRef]

14. Hama, S.; Noda, H.; Kondo, A. How lipase technology contributes to evolution of biodiesel production using multiple feedstocks. *Curr. Opin. Biotechnol.* **2018**, *50*, 57–64. [CrossRef] [PubMed]
15. Lozano, P.; Bernal, J.M.; Sánchez-Gómez, G.; López-López, G.; Vaultier, M. How to produce biodiesel easily using a green biocatalytic approach in sponge-like ionic liquids. *Energy Environ. Sci.* **2013**, *6*, 1328–1338. [CrossRef]
16. Sankaran, R.; Show, P.L.; Chang, J.-S. Biodiesel production using immobilized lipase: Feasibility and challenges. *Biofuels Bioprod. Biorefin.* **2016**, *10*, 896–916. [CrossRef]
17. Aransiola, E.F.; Ojumu, T.V.; Oyekola, O.O.; Madzimbamuto, T.F.; Ikhu-Omoregbe, D.I.O. A review of current technology for biodiesel production: State of the art. *Biomass Bioenergy* **2014**, *61*, 276–297. [CrossRef]
18. Christopher, L.P.; Kumar, H.; Zambare, V.P. Enzymatic biodiesel: Challenges and opportunities. *Appl. Energy* **2014**, *119*, 497–520. [CrossRef]
19. Brzozowski, A.M.; Derewenda, U.; Derewenda, Z.S.; Dodson, G.G.; Lawson, D.M.; Turkenburg, J.P.; Bjorkling, F.; Huge-Jensen, B.; Patkar, S.A.; Thim, L. A model for interfacial activation in lipases from the structure of a fungal lipase-inhibitor complex. *Nature* **1991**, *351*, 491–494. [CrossRef] [PubMed]
20. Jaeger, K.-E.; Dijkstra, B.W.; Reetz, M.T. Bacterial biocatalysts: Molecular biology, three-dimensional structures, and biotechnological applications of lipases. *Annu. Rev. Microbiol.* **1999**, *53*, 315–351. [CrossRef] [PubMed]
21. Grochulski, P.; Li, Y.; Schrag, J.D.; Bouthillier, F.; Smith, P.; Harrison, D.; Rubin, B.; Cygler, M. Insights into interfacial activation from an open structure of *Candida rugosa* lipase. *J. Biol. Chem.* **1993**, *268*, 12843–12847. [PubMed]
22. Verger, R. 'Interfacial activation' of lipases: Facts and artifacts. *Trends Biotechnol.* **1997**, *15*, 32–38. [CrossRef]
23. Wang, P.; He, J.; Sun, Y.; Reynolds, M.; Zhang, L.; Han, S.; Liang, S.; Sui, H.; Lin, Y. Display of fungal hydrophobin on the *Pichia pastoris* cell surface and its influence on Candida antarctica lipase B. *Appl. Microbiol. Biotechnol.* **2016**, *100*, 5883–5895. [CrossRef] [PubMed]
24. Palomo, J.M.; Peñas, M.M.; Fernández-Lorente, G.; Mateo, C.; Pisabarro, A.G.; Fernández-Lafuente, R.; Ramírez, L.; Guisán, J.M. Solid-phase handling of hydrophobins: Immobilized hydrophobins as a new tool to study lipases. *Biomacromolecules* **2003**, *4*, 204–210. [CrossRef] [PubMed]
25. Palomo, J.M.; Ortiz, C.; Fuentes, M.; Fernandez-Lorente, G.; Guisan, J.M.; Fernandez-Lafuente, R. Use of immobilized lipases for lipase purification via specific lipase-lipase interactions. *J. Chromatogr. A* **2004**, *1038*, 267–273. [CrossRef] [PubMed]
26. Palomo, J.M.; Ortiz, C.; Fernández-Lorente, G.; Fuentes, M.; Guisán, J.M.; Fernández-Lafuente, R. Lipase-lipase interactions as a new tool to immobilize and modulate the lipase properties. *Enzyme Microb. Technol.* **2005**, *36*, 447–454. [CrossRef]
27. Fernandez-Lafuente, R.; Armisén, P.; Sabuquillo, P.; Fernández-Lorente, G.; Guisán, J.M. Immobilization of lipases by selective adsorption on hydrophobic supports. *Chem. Phys. Lipids* **1998**, *93*, 185–197. [CrossRef]
28. Palomo, J.M.; Muoz, G.; Fernández-Lorente, G.; Mateo, C.; Fernández-Lafuente, R.; Guisán, J.M. Interfacial adsorption of lipases on very hydrophobic support (octadecyl-Sepabeads): Immobilization, hyperactivation and stabilization of the open form of lipases. *J. Mol. Catal. B Enzym.* **2002**, *19*, 279–286. [CrossRef]
29. Manoel, E.A.; dos Santos, J.C.S.; Freire, D.M.G.; Rueda, N.; Fernandez-Lafuente, R. Immobilization of lipases on hydrophobic supports involves the open form of the enzyme. *Enzyme Microb. Technol.* **2015**, *71*, 53–57. [CrossRef] [PubMed]
30. New Enzyme Technology Converts Waste Oils into Biodiesel. Available online: https://www.novozymes.com/es/news/news-archive/2014/12/new-enzyme-technology-converts-waste-oil-into-biodiesel (accessed on 11 October 2018).
31. Adewale, P.; Vithanage, L.N.; Christopher, L. Optimization of enzyme-catalyzed biodiesel production from crude tall oil using Taguchi method. *Energy Convers. Manag.* **2017**, *154*, 81–91. [CrossRef]
32. Andrade, T.A.; Errico, M.; Christensen, K.V. Transesterification of castor oil catalyzed by liquid enzymes: Optimization of reaction conditions. *Comput. Aided Chem. Eng.* **2017**, *40*, 2863–2868. [CrossRef]
33. Andrade, T.A.; Errico, M.; Christensen, K.V. Evaluation of Reaction Mechanisms and Kinetic Parameters for the Transesterification of Castor Oil by Liquid Enzymes. *Ind. Eng. Chem. Res.* **2017**, *56*, 9478–9488. [CrossRef]
34. Nguyen, H.C.; Huong, D.T.M.; Juan, H.-Y.; Su, C.-H.; Chien, C.-C. Liquid lipase-catalyzed esterification of oleic acid with methanol for biodiesel production in the presence of superabsorbent polymer: Optimization by using response surface methodology. *Energies* **2018**, *11*, 1085. [CrossRef]

35. He, Y.; Li, J.; Kodali, S.; Balle, T.; Chen, B.; Guo, Z. Liquid lipases for enzymatic concentration of n-3 polyunsaturated fatty acids in monoacylglycerols via ethanolysis: Catalytic specificity and parameterization. *Bioresour. Technol.* **2017**, *224*, 445–456. [CrossRef] [PubMed]
36. Andrade, T.A.; Errico, M.; Christensen, K.V. Castor oil transesterification catalysed by liquid enzymes: Feasibility of reuse under various reaction conditions. *Chem. Eng. Trans.* **2017**, *57*, 913–918. [CrossRef]
37. Uliana, N.R.; Polloni, A.; Paliga, M.; Veneral, J.G.; Quadri, M.B.; Oliveira, J.V. Acidity reduction of enzymatic biodiesel using alkaline washing. *Renew. Energy* **2017**, *113*, 393–396. [CrossRef]
38. Remonatto, D.; Santin, C.M.T.; De Oliveira, D.; Di Luccio, M.; De Oliveira, J.V. FAME Production from waste oils through commercial soluble lipase Eversa® catalysis. *Ind. Biotechnol.* **2016**, *12*, 254–262. [CrossRef]
39. Remonatto, D.; de Oliveira, J.V.; Manuel Guisan, J.; de Oliveira, D.; Ninow, J.; Fernandez-Lorente, G. Production of FAME and FAEE via alcoholysis of sunflower oil by Eversa lipases immobilized on hydrophobic supports. *Appl. Biochem. Biotechnol.* **2018**, *185*, 705–716. [CrossRef] [PubMed]
40. Barbosa, O.; Ortiz, C.; Berenguer-Murcia, Á.; Torres, R.; Rodrigues, R.C.; Fernandez-Lafuente, R. Strategies for the one-step immobilization-purification of enzymes as industrial biocatalysts. *Biotechnol. Adv.* **2015**, *33*, 435–456. [CrossRef] [PubMed]
41. Garcia-Galan, C.; Berenguer-Murcia, A.; Fernandez-Lafuente, R.; Rodrigues, R.C. Potential of different enzyme immobilization strategies to improve enzyme performance. *Adv. Synth. Catal.* **2011**, *353*, 2885–2904. [CrossRef]
42. Mateo, C.; Palomo, J.M.; Fernandez-Lorente, G.; Guisan, J.M.; Fernandez-Lafuente, R. Improvement of enzyme activity, stability and selectivity via immobilization techniques. *Enzyme Microb. Technol.* **2007**, *40*, 1451–1463. [CrossRef]
43. Bilal, M.; Rasheed, T.; Zhao, Y.; Iqbal, H.M.N.; Cui, J. "Smart" chemistry and its application in peroxidase immobilization using different support materials. *Int. J. Biol. Macromol.* **2018**, *119*, 278–290. [CrossRef] [PubMed]
44. Guzik, U.; Hupert-Kocurek, K.; Wojcieszynska, D. Immobilization as a strategy for improving enzyme properties- Application to oxidoreductases. *Molecules* **2014**, *19*, 8995–9018. [CrossRef] [PubMed]
45. Di Cosimo, R.; Mc Auliffe, J.; Poulose, A.J.; Bohlmann, G. Industrial use of immobilized enzymes. *Chem. Soc. Rev.* **2013**, *42*, 6437–6474. [CrossRef] [PubMed]
46. Cantone, S.; Ferrario, V.; Corici, L.; Ebert, C.; Fattor, D.; Spizzo, P.; Gardossi, L. Efficient immobilisation of industrial biocatalysts: Criteria and constraints for the selection of organic polymeric carriers and immobilisation methods. *Chem. Soc. Rev.* **2013**, *42*, 6262–6276. [CrossRef] [PubMed]
47. Sheldon, R.A.; van Pelt, S. Enzyme immobilisation in biocatalysis: Why, what and how. *Chem. Soc. Rev.* **2013**, *42*, 6223–6235. [CrossRef] [PubMed]
48. Friedrich, J.L.R.; Peña, F.P.; Garcia-Galan, C.; Fernandez-Lafuente, R.; Ayub, M.A.Z.; Rodrigues, R.C. Effect of immobilization protocol on optimal conditions of ethyl butyrate synthesis catalyzed by lipase B from *Candida antarctica*. *J. Chem. Technol. Biotechnol.* **2013**, *88*, 1089–1095. [CrossRef]
49. Séverac, E.; Galy, O.; Turon, F.; Pantel, C.A.; Condoret, J.-S.; Monsan, P.; Marty, A. Selection of CalB immobilization method to be used in continuous oil transesterification: Analysis of the economical impact. *Enzyme Microb. Technol.* **2011**, *48*, 61–70. [CrossRef] [PubMed]
50. Tacias-Pascacio, V.G.; Virgen-Ortïz, J.J.; Jimenez -Perez, M.; Yates, M.; Torrestiana-Sanchez, B.; Rosales-Quintero, A.; Fernandez-Lafuente, R. Evaluation of different lipase biocatalysts in the production of biodiesel from used cooking oil: Critical role of the immobilization support. *Fuel* **2017**, *200*, 1–10. [CrossRef]
51. Poppe, J.K.; Matte, C.R.; Do Carmo Ruaro Peralba, M.; Fernandez-Lafuente, R.; Rodrigues, R.C.; Ayub, M.A.Z. Optimization of ethyl ester production from olive and palm oils using mixtures of immobilized lipases. *Appl. Catal. A-Gen.* **2015**, *490*, 50–56. [CrossRef]
52. Alves, J.S.; Vieira, N.S.; Cunha, A.S.; Silva, A.M.; Záchia Ayub, M.A.; Fernandez-Lafuente, R.; Rodrigues, R.C. Combi-lipase for heterogeneous substrates: A new approach for hydrolysis of soybean oil using mixtures of biocatalysts. *RSC Adv.* **2014**, *4*, 6863–6868. [CrossRef]
53. Rueda, N.; Dos Santos, J.C.S.; Torres, R.; Ortiz, C.; Barbosa, O.; Fernandez-Lafuente, R. Improved performance of lipases immobilized on heterofunctional octyl-glyoxyl agarose beads. *RSC Adv.* **2015**, *5*, 11212–11222. [CrossRef]

54. Virgen-Ortíz, J.J.; Tacias-Pascacio, V.G.; Hirata, D.B.; Torrestiana-Sanchez, B.; Rosales-Quintero, A.; Fernandez-Lafuente, R. Relevance of substrates and products on the desorption of lipases physically adsorbed on hydrophobic supports. *Enzyme Microb. Technol.* **2017**, *96*, 30–35. [CrossRef] [PubMed]
55. Hirata, D.B.; Albuquerque, T.L.; Rueda, N.; Virgen-Ortíz, J.J.; Tacias-Pascacio, V.G.; Fernandez-Lafuente, R. Evaluation of different immobilized lipases in transesterification reactions using tributyrin: Advantages of the heterofunctional octyl agarose beads. *J. Mol. Catal. B Enzym.* **2016**, *133*, 117–123. [CrossRef]
56. Hirata, D.B.; Albuquerque, T.L.; Rueda, N.; Sánchez-Montero, J.M.; Garcia-Verdugo, E.; Porcar, R.; Fernandez-Lafuente, R. Advantages of heterofunctional octyl supports: Production of 1,2-dibutyrin by specific and selective hydrolysis of tributyrin catalyzed by immobilized lipases. *ChemistrySelect* **2016**, *1*, 3259–3270. [CrossRef]
57. Bernal, C.; Illanes, A.; Wilson, L. Heterofunctional hydrophilic-hydrophobic porous silica as support for multipoint covalent immobilization of lipases: Application to lactulose palmitate synthesis. *Langmuir* **2014**, *30*, 3557–3566. [CrossRef] [PubMed]
58. Guajardo, N.; Bernal, C.; Wilson, L.; Cabrera, Z. Selectivity of R-α-monobenzoate glycerol synthesis catalyzed by Candida antarctica lipase B immobilized on heterofunctional supports. *Process Biochem.* **2015**, *50*, 1870–1877. [CrossRef]
59. Fernandez-Lorente, G.; Filice, M.; Lopez-Vela, D.; Pizarro, C.; Wilson, L.; Betancor, L.; Avila, Y.; Guisan, J.M. Cross-linking of lipases adsorbed on hydrophobic supports: Highly selective hydrolysis of fish oil catalyzed by RML. JAOCS. *J. Am. Oil. Chem. Soc.* **2011**, *88*, 801–807. [CrossRef]
60. Fernandez-Lopez, L.; Pedrero, S.G.; Lopez-Carrobles, N.; Virgen-Ortíz, J.J.; Gorines, B.C.; Otero, C.; Fernandez-Lafuente, R. Physical crosslinking of lipase from *Rhizomucor miehei* immobilized on octyl agarose via coating with ionic polymers: Avoiding enzyme release from the support. *Process Biochem.* **2017**, *54*, 81–88. [CrossRef]
61. Fernandez-Lopez, L.; Virgen-Ortíz, J.J.; Pedrero, S.G.; Lopez-Carrobles, N.; Gorines, B.C.; Otero, C.; Fernandez-Lafuente, R. Optimization of the coating of octyl-CALB with ionic polymers to improve stability and decrease enzyme leakage. *Biocatal. Biotransf.* **2018**, *36*, 47–56. [CrossRef]
62. Zaak, H.; Fernandez-Lopez, L.; Otero, C.; Sassi, M.; Fernandez-Lafuente, R. Improved stability of immobilized lipases via modification with polyethylenimine and glutaraldehyde. *Enzyme Microb. Technol.* **2017**, *106*, 67–74. [CrossRef] [PubMed]
63. Virgen-Ortíz, J.J.; Dos Santos, J.C.S.; Berenguer-Murcia, Á.; Barbosa, O.; Rodrigues, R.C.; Fernandez-Lafuente, R. Polyethylenimine: A very useful ionic polymer in the design of immobilized enzyme biocatalysts. *J. Mater. Chem. B* **2017**, *5*, 7461–7490. [CrossRef]
64. Peirce, S.; Virgen-Ortíz, J.J.; Tacias-Pascacio, V.G.; Rueda, N.; Bartolome-Cabrero, R.; Fernandez-Lopez, L.; Russo, M.E.; Marzocchella, A.; Fernandez-Lafuente, R. Development of simple protocols to solve the problems of enzyme coimmobilization. Application to coimmobilize a lipase and a β-galactosidase. *RSC Adv.* **2016**, *6*, 61707–61715. [CrossRef]
65. Zaak, H.; Kornecki, J.F.; Siar, E.-H.; Fernandez-Lopez, L.; Corberán, V.C.; Sassi, M.; Fernandez-Lafuente, R. Coimmobilization of enzymes in bilayers using PEI as a glue to reuse the most stable enzyme: Preventing pei release during inactivated enzyme desorption. *Process Biochem.* **2017**, *61*, 95–101. [CrossRef]
66. Velasco-Lozano, S.; Benítez-Mateos, A.I.; López-Gallego, F. Co-immobilized phosphorylated cofactors and enzymes as self-sufficient heterogeneous biocatalysts for chemical processes. *Angew. Chem. Int. Ed.* **2017**, *56*, 771–775. [CrossRef] [PubMed]
67. Gotor-Fernández, V.; Busto, E.; Gotor, V. *Candida antarctica* lipase B: An ideal biocatalyst for the preparation of nitrogenated organic compounds. *Adv. Synth. Catal.* **2006**, *348*, 797–812. [CrossRef]
68. Anderson, E.M.; Larsson, K.M.; Kirk, O. One biocatalyst—Many applications: The use of *Candida antarctica* B-lipase in organic synthesis. *Biocatal. Biotransf.* **1998**, *16*, 181–204. [CrossRef]
69. Jaeger, K.-E.; Ransac, S.; Koch, H.B.; Ferrato, F.; Dijkstra, B.W. Topological characterization and modeling of the 3D structure of lipase from *Pseudomonas aeruginosa*. *FEBS Lett.* **1993**, *332*, 143–149. [CrossRef]
70. Cygler, M.; Schrag, J.D. Structure and conformational flexibility of *Candida rugosa* lipase. *Biochim. Biophys. Acta Mol. Cell Biol. Lipids* **1999**, *1441*, 205–214. [CrossRef]
71. Dos Santos, J.C.S.; Rueda, N.; Torres, R.; Barbosa, O.; Gonçalves, L.R.B.; Fernandez-Lafuente, R. Evaluation of divinylsulfone activated agarose to immobilize lipases and to tune their catalytic properties. *Process Biochem.* **2015**, *50*, 918–927. [CrossRef]

72. Santos, J.C.S.D.; Barbosa, O.; Ortiz, C.; Berenguer-Murcia, A.; Rodrigues, R.C.; Fernandez-Lafuente, R. Importance of the support properties for immobilization or purification of enzymes. *ChemCatChem* **2015**, *7*, 2413–2432. [CrossRef]
73. Virgen-Ortíz, J.J.; Peirce, S.; Tacias-Pascacio, V.G.; Cortes-Corberan, V.; Marzocchella, A.; Russo, M.E.; Fernandez-Lafuente, R. Reuse of anion exchangers as supports for enzyme immobilization: Reinforcement of the enzyme-support multiinteraction after enzyme inactivation. *Process Biochem.* **2016**, *51*, 1391–1396. [CrossRef]
74. Rodrigues, R.C.; Ortiz, C.; Berenguer-Murcia, A.; Torres, R.; Fernández-Lafuente, R. Modifying enzyme activity and selectivity by immobilization. *Chem. Soc. Rev.* **2013**, *42*, 6290–6307. [CrossRef] [PubMed]
75. Zaak, H.; Fernandez-Lopez, L.; Velasco-Lozano, S.; Alcaraz-Fructuoso, M.T.; Sassi, M.; Lopez-Gallego, F.; Fernandez-Lafuente, R. Effect of high salt concentrations on the stability of immobilized lipases: Dramatic deleterious effects of phosphate anions. *Process Biochem.* **2017**, *62*, 128–134. [CrossRef]
76. Fernandez-Lopez, L.; Bartolome-Cabrero, R.; Rodriguez, M.D.; Dos Santos, C.S.; Rueda, N.; Fernandez-Lafuente, R. Stabilizing effects of cations on lipases depend on the immobilization protocol. *RSC Adv.* **2015**, *5*, 83868–83875. [CrossRef]
77. Fernandez-Lopez, L.; Rueda, N.; Bartolome-Cabrero, R.; Rodriguez, M.D.; Albuquerque, T.L.; Dos Santos, J.C.S.; Barbosa, O.; Fernandez-Lafuente, R. Improved immobilization and stabilization of lipase from *Rhizomucor miehei* on octyl-glyoxyl agarose beads by using CaCl2. *Process Biochem.* **2016**, *51*, 48–52. [CrossRef]
78. Palomo, J.M.; Fernandez-Lorente, G.; Mateo, C.; Ortiz, C.; Fernandez-Lafuente, R.; Guisan, J.M. Modulation of the enantioselectivity of lipases via controlled immobilization and medium engineering: Hydrolytic resolution of mandelic acid esters. *Enzyme Microb. Technol.* **2002**, *31*, 775–783. [CrossRef]
79. Bradford, M.M. A rapid and sensitive method for the quantitation of microgram quantities of protein utilizing the principle of protein-dye binding. *Anal. Biochem.* **1976**, *72*, 248–254. [CrossRef]
80. Fernandez-Lafuente, R.; Rosell, C.M.; Rodriguez, V.; Santana, C.; Soler, G.; Bastida, A.; Guisán, J.M. Preparation of activated supports containing low pK amino groups. A new tool for protein immobilization via the carboxyl coupling method. *Enzyme Microb. Technol.* **1993**, *15*, 546–550. [CrossRef]
81. Pereira, M.G.; Facchini, F.D.A.; Filó, L.E.C.; Polizeli, A.M.; Vici, A.C.; Jorge, J.A.; Fernandez-Lorente, G.; Pessela, B.C.; Guisan, J.M.; Polizeli, M.L.T.M. Immobilized lipase from *Hypocrea pseudokoningii* on hydrophobic and ionic supports: Determination of thermal and organic solvent stabilities for applications in the oleochemical industry. *Process Biochem.* **2015**, *50*, 561–570. [CrossRef]
82. Laemmli, U.K. Cleavage of structural proteins during the assembly of the head of bacteriophage T4. *Nature* **1970**, *227*, 680–685. [CrossRef] [PubMed]
83. Hernandez, K.; Garcia-Verdugo, E.; Porcar, R.; Fernandez-Lafuente, R. Hydrolysis of triacetin catalyzed by immobilized lipases: Effect of the immobilization protocol and experimental conditions on diacetin yield. *Enzyme Microb. Technol.* **2011**, *48*, 510–517. [CrossRef] [PubMed]

catalysts

Article

Preparation of Magnetic Cross-Linked Amyloglucosidase Aggregates: Solving Some Activity Problems

Murilo Amaral-Fonseca [1], Willian Kopp [2], Raquel de Lima Camargo Giordano [1], Roberto Fernández-Lafuente [3,*] and Paulo Waldir Tardioli [1,*]

1 Postgraduate Program in Chemical Engineering, Department of Chemical Engineering, Federal University of São Carlos (PPGEQ/UFSCar), Rod. Washington Luiz, km 235, São Carlos 13565-905, SP, Brazil; muriloaf88@hotmail.com (M.A.-F.); raquel@ufscar.br (R.d.L.C.G.)

2 Kopp Technologies (KTech), Rua Alfredo Lopes, 1717, Jardim Macarengo, São Carlos 13560-460, SP, Brazil; solutions@kopptechnologies.com

3 Departamento de Biocatális, ICP-CSIC, Campus UAM-CSIC, 28040 Madrid, Spain

* Correspondence: rfl@icp.csic.es (R.F.-L.); pwtardioli@ufscar.br (P.W.T.); Tel.: +34-915954941 (R.F.-L.); +55-16-3351-9362 (P.W.T.)

Received: 8 October 2018; Accepted: 24 October 2018; Published: 26 October 2018

Abstract: The preparation of Cross-Linked Enzyme Aggregates (CLEAs) is a simple and cost-effective technique capable of generating insoluble biocatalysts with high volumetric activity and improved stability. The standard CLEA preparation consists of the aggregation of the enzyme and its further crosslinking, usually with glutaraldehyde. However, some enzymes have too low a content of surface lysine groups to permit effective crosslinking with glutaraldehyde, requiring co-aggregation with feeders rich in amino groups to aid the formation of CLEAs. The co-aggregation with magnetic particles makes their handling easier. In this work, CLEAs of a commercial amyloglucosidase (AMG) produced by *Aspergillus niger* were prepared by co-aggregation in the presence of polyethyleneimine (PEI) or starch with aminated magnetic nanoparticles (MNPs) or bovine serum albumin (BSA). First, CLEAs were prepared only with MNPs at different glutaraldehyde concentrations, yielding a recovered activity of around 20%. The addition of starch during the precipitation and crosslinking steps nearly doubled the recovered activity. Similar recovered activity (around 40%) was achieved when changing starch by PEI. Moreover, under the same conditions, AMG co-aggregated with BSA was also synthesized, yielding CLEAs with very similar recovered activity. Both CLEAs (co-aggregated with MNPs or BSA) were four times more stable than the soluble enzyme. These CLEAs were evaluated in the hydrolysis of starch at typical industrial conditions, achieving more than 95% starch-to-glucose conversion, measured as Dextrose Equivalent (DE). Moreover, both CLEAS could be reused for five cycles, maintaining a DE of around 90%. Although both CLEAs had good properties, magnetic CLEAs could be more attractive for industrial purposes because of their easy separation by an external magnetic field, avoiding the formation of clusters during the filtration or centrifugation recovery methods usually used.

Keywords: cross-linked enzyme aggregate; amyloglucosidase; magnetic nanoparticles; bovine serum albumin; polyethyleneimine; starch hydrolysis

1. Introduction

Amyloglucosidase (1,4-α-D-glucan hydrolase, EC 3.2.1.3) is an enzyme that catalyzes the release of glucose from the non-reducing ends of glucose polysaccharides. In addition to selectively hydrolyzing α-1,4-glycosidic bonds, it is also capable of hydrolyzing starch branches (α-1,6-glycosidic bonds), but in a slower manner [1,2]. Fungal amyloglucosidases, like the ones produced by *Aspergillus niger* (AMG), may present more than one form, with different molecular weights [3–5]. *Aspergillus niger* produces two isoforms, namely G1 isoform, which corresponds to a protein having a catalytic domain (structure on the left in Figure 1), and a starch-binding domain (structure on the right in Figure 1) with a total of 640 amino acid residues and Mw around 68 kDa, and G2 isoform, which contains only the catalytic domain with 470 amino acid residues and Mw around 50 kDa [4,5]. Both isoforms are highly glycosylated by both N-linked and O-linked carbohydrates. These isoforms are derived from the same genetic material, but differ because of a different RNA splicing after transcription [6].

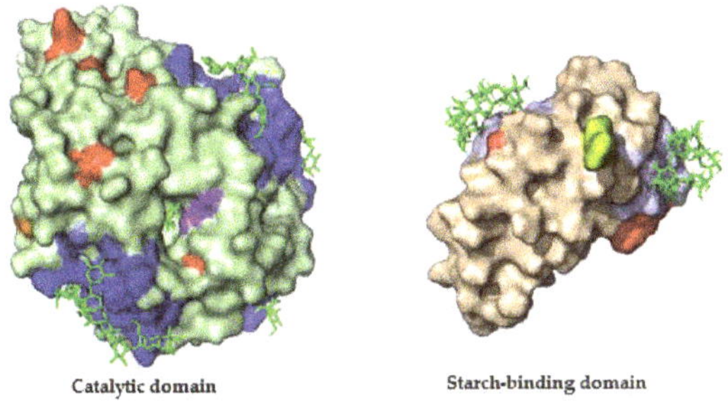

Figure 1. Three-dimensional structures of *A. niger* amyloglucosidase, showing the catalytic domain on the left and the starch-binding domain on the right (3eqa and 1ac0 PDB structures [4,5], respectively). Color patterns: red (lysine residues), purple (active site), orange (N-terminal residue), lemon (C-terminal residue), blue and light blue (glycosylated regions in the catalytic and starch binding domains, respectively). The figures were generated using PyMol (The PyMol Molecular Graphics System; Version 2.1.0; Schrödinger, LLC).

AMG from *A. niger* is extensively used in industrial starch saccharification, almost completely hydrolyzing maltodextrins, amylose, and amylopectins to produce glucose syrup, which serves as a substrate in the production of other syrups for application in the beverage and food industries [2,6,7].

The saccharification of the starch is usually carried out by the enzyme in the soluble form. Because of this, the enzyme is neither recovered nor reused [8]. This drawback can be overcome by the use of enzyme immobilization techniques that allow for enzyme recovery, reuse, and often improvement of the operational stability [9–11]. These advantages could contribute to reduced processing costs and, therefore, a lower final price of the product. Several studies reported the immobilization of amyloglucosidase by different protocols [12–16].

An alternative to the immobilization of enzymes on solid supports is carrier-free immobilization, such as the crosslinked enzyme aggregates (CLEAs) proposed by Sheldon [17–19]. This methodology involves the precipitation of the proteins by the addition of precipitating agents (e.g., salts, organic solvents or nonionic polymers), where the enzymes associate as insoluble aggregates [20]. Thereafter, the formed aggregates are cross-linked using bifunctional (usually glutaraldehyde) or polyfunctional (dextran polyaldehyde, for example) agents [21]. CLEAs of many different enzymes have been reported [18,22–26], including amylolytic enzymes, e.g., β-amylase [27] and amyloglucosidases [28,29].

The preparation of CLEAs may lead to some problems. At first glance, it should be relatively simple to find a precipitant that permits the recovery of high levels of enzyme activity. However, the crosslinking step may be problematic if the enzyme is poor in external free primary amino groups. This problem has been solved using Lys-rich proteins as protein feeders [22,24,27,30], co-immobilizing enzymes and a primary amino rich polymer (e.g., polyethylenimine, PEI) [31] or even enriching the enzyme in primary amino groups via chemical modification [32]. AMG has 13 Lys, but only nine are exposed to the medium [3,5], two of which are located in the starch-binding domain and one close to the active site (see Figure 1). Because of this, the crosslinking of the enzyme with bifunctional agents may be poor, resulting in a CLEA with low mechanical stability and allowing leaching of the enzyme in the reaction medium [33–36]. In some instances, the crosslinking with glutaraldehyde may lead to enzyme inactivation by altering the active center; in these cases large aldehyde polymers (e.g., aldehyde dextran) or other crosslinkers have been proposed [21,37].

The use of CLEAs also poses some problems. First, CLEAs are mechanically fragile, complicating their recovery. Second, pore sizes may be small and this can lead to high diffusional limitations [38].

The first problem may be solved using magnetic nanoparticles (MNPs) that are co-aggregated with the enzyme to permit magnetic handling of the final CLEAs [39]. The size and functionalization of the MNPs can determine the final properties of the magnetic CLEA [40]. Immobilization of different amylase-related enzymes using CLEA technology associated with MNPs functionalized with 3-aminopropyltriethoxysilane (APTES) has already been reported, showing that its application confers mechanical stability and efficient magnetic separation of CLEAs [28,29,41].

The second problem may be reduced if strategies to enlarge the pores of the CLEA are utilized. The co-aggregation of PEI with the enzyme can significantly improve the crosslinking efficiency, preventing leakage and promoting the generation of a hydrophilic microenvironment that protects the enzyme from organic solvents [20,34–36,42], but that may also result in enlarged pore sizes. Similarly, the use of polymers like starch during the aggregation and crosslinking steps may facilitate the formation of large pores during CLEA production (and perhaps, as it is a substrate of the enzyme, it may protect the active center of the enzyme during the different steps of the CLEA production). As at the final CLEA preparation the starch is degraded by α-amylase and washed away from the CLEA, this strategy may reduce internal mass-transfer limitations and, thus, increase the catalytic efficiency [43,44].

In this context, this paper intends the co-aggregation of commercial AMG with MNPs to get a magnetic CLEA. A set of parameters was evaluated to prepare CLEAs of AMG with high activity, reduced leaching, and high thermal and operational stabilities, such as type and amount of precipitants, concentration of glutaraldehyde, time of glutaraldehyde treatment, stirring speed, use of PEI and starch during precipitation, and crosslinking steps to generate larger pores and, perhaps, to protect the active center of the enzyme. The addition of starch or PEI has been studied as a strategy for the formation of CLEAs with larger pores (among other likely effects). The catalytic properties of the most active and stable CLEAs were characterized (optimal pH and temperature for enzyme activity, thermal stability and performance in the hydrolysis of starch at high starch concentration (typical industrial conditions)).

2. Results and Discussion

2.1. Precipitant Selection

Because the biochemical and structural properties differ from one enzyme to another, a screening of precipitants should be performed for the preparation of CLEAs of a particular enzyme [20,45]. Thus, in this work, a screening of five precipitants (acetone, ethanol, iso-propanol, ammonium sulfate and PEG) was carried out aiming at full protein precipitation and high recovered activity of the re-dissolved precipitate. Commercial AMG 300L is relatively pure (Figure 2), exhibiting two main bands (around 70 and 100 kDa), probably corresponding to the G1 and G2 isoforms [3].

Figure 3 shows that acetone, ethanol, and isopropanol were capable of precipitating around 80% of proteins, retaining high activity of the re-dissolved precipitates (around 90%), while ammonium sulfate precipitated only around 50% of the proteins and polyethylene glycol (PEG) did not have precipitation action on AMG (data not shown). The different precipitation yields may be explained by the different mechanisms of aggregation of each precipitant (changes in the hydration state of the molecules, or changes in the dielectric constant of the solution) [45].

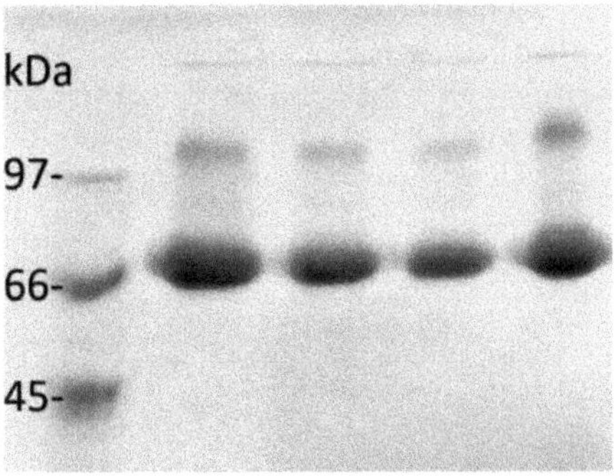

Figure 2. Electrophoresis gel (10% SDS-PAGE) of commercial amyloglucosidase (AMG, 300L).

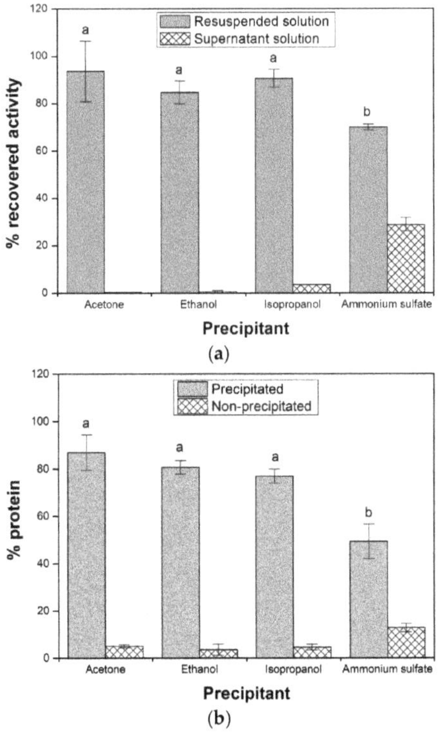

Figure 3. Screen of precipitants for amyloglucosidase (AMG 300L), the percentage of recovered activity (**a**) and the percentage of precipitated protein (**b**). Precipitation conditions: precipitant/enzyme solution volume ratio of 9:1, at 4 °C, 60 min precipitation under 150 rpm shaking, and enzyme solution prepared in 50 mM sodium citrate buffer pH 4.5. Note: Values are shown as the mean of triplicate measurements ± s.d. Means followed by the same letter are not statistically different by Tukey's test ($p < 0.05$). Percentage of protein was calculated taken the initial protein as 100%.

Ethanol exhibited good performance at precipitating AMG and is economically and environmentally more friendly (low-cost and -toxicity, and renewable) than the other precipitants evaluated; therefore, it was selected as a precipitant of AMG for the preparation of CLEAs.

Figure 4 shows the influence of the precipitant/enzyme solution volume ratio on the protein precipitation yields and recovered activity of AMG in the re-dissolved precipitate. It can be observed that for ethanol concentrations from 60% to 90% (*v*/*v*), protein precipitation yields were not statistically different. In terms of recovered activity, 90% (*v*/*v*) ethanol exhibited a small improvement (around 85% recovered activity). Thus, a volume ratio of 9:1 (precipitant/enzyme solution) was selected for the further assays.

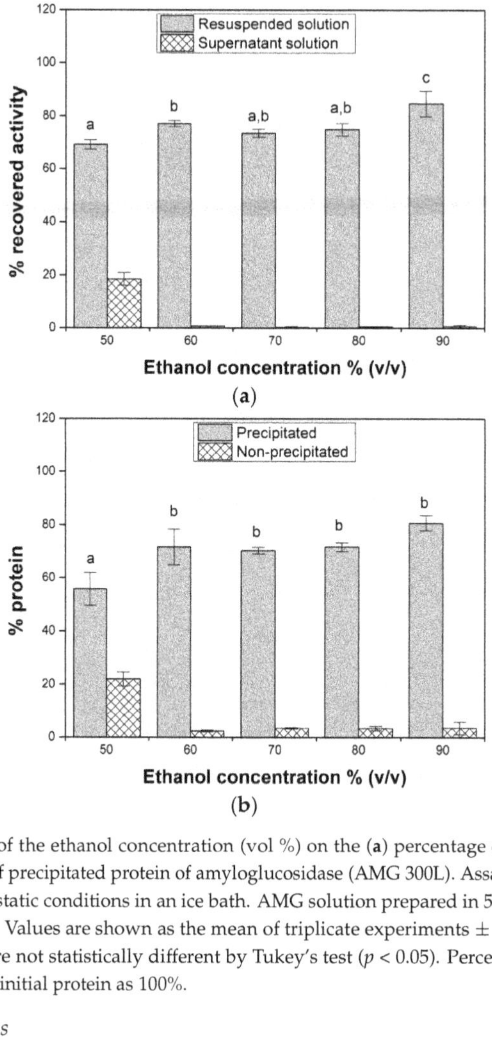

Figure 4. Influence of the ethanol concentration (vol %) on the (**a**) percentage of recovered activity, and (**b**) percentage of precipitated protein of amyloglucosidase (AMG 300L). Assay conditions: 30 min precipitation under static conditions in an ice bath. AMG solution prepared in 50 mM sodium citrate buffer pH 4.5. Note: Values are shown as the mean of triplicate experiments ± s.d. Means followed by the same letter are not statistically different by Tukey's test ($p < 0.05$). Percentage of protein was calculated taken the initial protein as 100%.

2.2. Preparation of CLEAs

Initially, CLEAs were prepared without co-feeders or any other aid. Figure 5 shows that CLEAs could be only formed using 500 mM glutaraldehyde in the crosslinking step. But the recovered activity was lower than 20%. Using 100 or 300 mM glutaraldehyde, the aggregates were re-dissolved due to inefficient crosslinking.

The co-aggregation with PEI allowed the formation of CLEAs at 100 mM glutaraldehyde (recovered activity around 25%). The increase in the glutaraldehyde concentration led to CLEAs with lower recovered activity, probably due to excessive enzyme modification. The co-aggregation with starch produced better results (recovered activity around 35%), but required higher glutaraldehyde concentration in the crosslinking step. The combined use of PEI and starch did not improve the recovered activity.

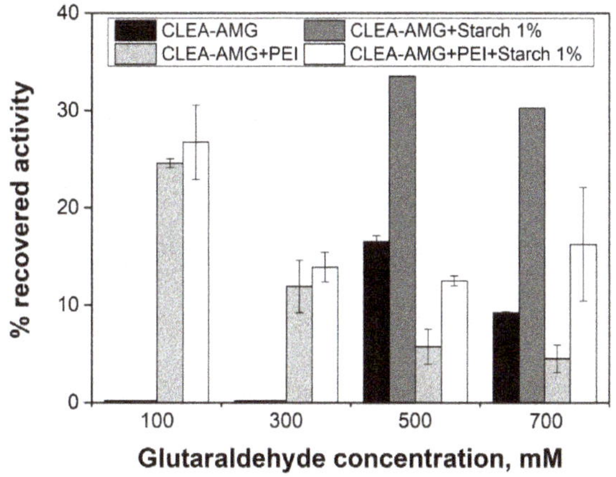

Figure 5. CLEAs of amyloglucosidase (AMG 300L) prepared using only enzyme, co-aggregation with polyethyleneimine (protein/PEI mass ratio of 1:1) and/or starch (1%, *w*/*v*). Assay conditions: ethanol as precipitant (volume ratio of 1:9, enzyme solution pH 7.0 or 10.0/ethanol), 30 min precipitation in an ice bath, glutaraldehyde as a cross-linker (100–700 mM), 16 h crosslinking under gently stirring at 4 °C. Note: Values are shown as the mean of duplicate experiments ± s.d.

2.3. *Preparation of Magnetic AMG CLEAs*

In order to prepare easily manageable CLEAs, MNPs were co-aggregated with AMG. A set of commercial MNPs functionalized with amino or amino/hydrophobic groups (Table 1) was evaluated.

Table 1. Content of amino and hydrophobic (octyl or octadecyl) groups in the magnetic nanoparticles supplied by Koop Technologies (São Carlos, SP, Brazil).

Magnetic Nanoparticle	$-NH_2$ Content (µmol/g)		-C8 or -C18 (µmol/g)
NP-N-1	1013.1 [1]		Not applicable
NP-N-2	265.5 [1]		
DCNP-N	522.1 [1]		
N (100%)	275.7 [1]	348 ± 15 [2]	
N (75%) C8 (25%)		136 ± 10 [2]	282 ± 19 [2]
N (75%) C18 (25%)		310 ± 39 [2]	479 ± 21 [2]

[1] Amino content quantified according to TNBS method [46]. [2] Amino content quantified according to CNHS method. Note: Column 1 lists the names of commercial magnetic nanoparticles (MNPs) as provided by the manufacturer. N, C8 and C18 indicate chemicals used for the synthesis of the MNPs, such as (3-aminopropyl)triethoxysilane, triethoxy(octyl)silane and n-octadecyltriethoxysilane, respectively.

Figure 6 shows the recovered activity of CLEAs prepared by co-aggregation with MNPs in a AMG/MNPs mass ratio of 1:1 [39]. As described above, using a glutaraldehyde concentration below 500 mM did not form CLEAs using only AMG, which was fully leached after resuspension in buffer or washing steps. However, using MNPs CLEAs were formed for all glutaraldehyde concentrations evaluated, despite the low recovered activity (less than 20%). The magnetic nanoparticle DCNP-N, containing 522 µmol of amino groups per gram, did give better results even at the lowest glutaraldehyde concentration; therefore, it was selected for further experiments.

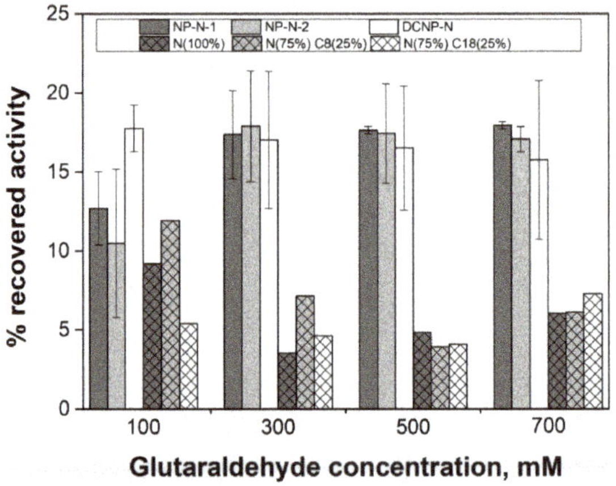

Figure 6. Effect of different commercial aminated magnetic nanoparticles (NP-N-1, NP-N-2, N(100%), N(75%)-C8(25%), N(75%)-C18(25%)) on the recovered activity of CLEAs of amyloglucosidase (AMG 300L). Synthesis conditions: enzyme/magnetic nanoparticle mass ratio of 1:1, ethanol/enzyme solution volume ratio of 9:1 (in 100 mM carbonate buffer pH 10.0), glutaraldehyde concentration ranging from 100 to 700 mM, precipitation and crosslinking under static conditions in an ice bath. Activity of AMG CLEAs was measured with 1% (*w*/*v*) dextrin in 50 mM sodium citrate buffer pH 4.5–10 min reaction at 55 °C under 900 rpm stirring. Values are shown as the mean of triplicate experiments ± s.d.

2.4. Evaluation of Glutaraldehyde Effect on Enzyme Activity

In order to investigate the probable deleterious effect of glutaraldehyde on the AMG, the activity of free AMG was measured after incubation of the enzyme with glutaraldehyde at different concentrations in the same conditions used in the preparation of CLEAs (16 h at 4 °C), but in the absence of ethanol. Starch was also added to the enzyme/glutaraldehyde solutions to evaluate whether the active site would be protected in the presence of a natural substrate. Table 2 shows that more than 75% of activity was lost when free AMG (without starch) was incubated in the presence of glutaraldehyde. On the other hand, when starch was added to the enzyme solution, slight protection was observed. The interaction of the starch with the amino acid residues at the active site could avoid distorting changes in the tertiary structure of AMG promoted by excessive cross-links with glutaraldehyde and/or could prevent the establishment of a covalent link between glutaraldehyde and the ε-amino group from the lysine residue located close to the active site. The increase of the starch concentration from 1% to 5% (*w*/*v*) did not cause an improvement in the residual activity, probably due to the high viscosity of the medium at low temperature (i.e., 4 °C), preventing the formation of a homogeneous mixture [43].

Table 2. Effect of glutaraldehyde on the free amyloglucosidase activity in the presence or absence of starch. Residual activity was measured after 16 h incubation at 4 °C and the initial activity was taken to be 100%.

Glutaraldehyde Concentration	Residual Activity (%)			
	Without Starch	Starch 1%	Starch 2.5%	Starch 5%
100 mM	34.46	48.12	43.18	36.37
300 mM	20.25	30.53	24.25	23.22
500 mM	19.48	29.63	26.80	24.40
700 mM	17.33	24.92	25.73	20.29

Figure 7 shows that the inactivation of AMG is very fast in the presence of glutaraldehyde, mainly at high glutaraldehyde concentrations. For short time periods and low glutaraldehyde concentrations, the residual activity is high, but these conditions do not favor the formation of stable CLEAs [28,47,48]. Thus, 16 h was kept as the glutaraldehyde treatment time in the preparation of CLEAs in the presence of 1% (*w*/*v*) starch as a pore-making agent. In the CLEA preparation (at 4 °C), the starch hydrolysis rate catalyzed by AMG is very low, thus preserving the starch molecules large enough to serve as pore-making agents.

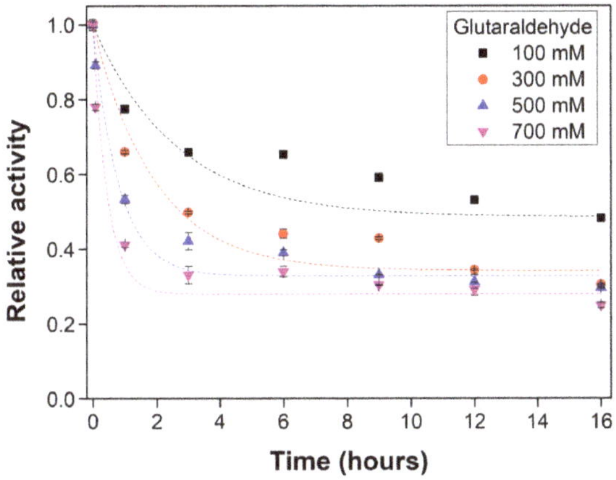

Figure 7. Profile of enzymatic inactivation as a function of glutaraldehyde concentration in the presence of 1% (*w*/*v*) starch. Assay conditions: water/enzyme solution volume ratio of 9:1 (in 100 mM carbonate buffer pH 10.0), glutaraldehyde concentration ranging from 100 to 700 mM, incubated statically in a refrigerator at 4 °C. Activity of amyloglucosidase was measured with 1% (*w*/*v*) dextrin in 50 mM sodium citrate buffer pH 4.5, 10 min reaction at 55 °C under 900 rpm stirring. Values correspond to a single assay with activity values measured in duplicate.

Table 3 shows that CLEAs prepared in the presence of 1% (*w*/*v*) starch (with and without the aminated magnetic nanoparticles DCNP-N) retained 2-3 times more activity, suggesting that the control of the pore size may play an important role in the enhanced activity during CLEA preparation. The combination of starch and DCNP-N yielded CLEAs with around 40% recovered activity using 500 mM glutaraldehyde. Although in the absence of DCNP-N the recovered activity was only a little lower (around 30%), the co-aggregation of AMG with DCNP-N has the advantage of easy separation of CLEAs without the formation of large clusters.

Table 3. Comparison of the effect of glutaraldehyde concentration on the recovered activity of CLEAs (without and with DCNP-N [a]) synthesized in the presence of 1% (*w*/*v*) starch. Synthesis conditions: 16 h precipitation/crosslinking under static conditions in an ice bath, mass ratio protein/DCNP-N of 1:1, volume ratio enzyme-DCNP-N suspension/ethanol of 1:9.

Glutaraldehyde Concentration	Recovered Activity (%)			
	Without Starch	Starch 1%	DCNP-N	DCNP-N+Starch 1%
100 mM	-	-	17.74 ± 1.47	18.72 ± 0.29
300 mM	-	-	17.03 ± 4.34	18.81 ± 0.31
500 mM	16.53 ± 0.64	33.53	16.52 ± 3.93	39.40 ± 1.44
700 mM	9.20 ± 0.11	30.28	15.76 ± 5.03	29.75 ± 1.15

[a] DCNP-N refer to magnetic nanoparticles functionalized with amino groups.

2.5. Study of the Co-Aggregation of AMG and Polyethyleneimine (PEI)

In order to improve the recovered activity of CLEAs of AMG, PEI was co-aggregated with the enzyme and the MNPs DCNP-N. Figure 8 shows that CLEAs could be formed even at a low glutaraldehyde concentration, yielding higher recovered activity compared to the CLEAs prepared without MNPs (Figure 5).

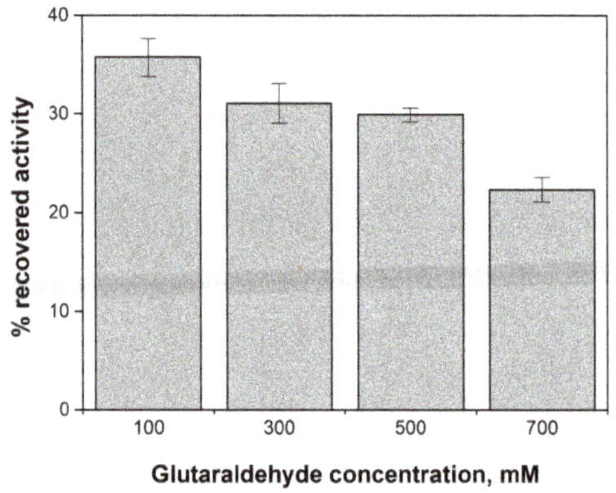

Figure 8. Effect of glutaraldehyde concentration on the recovered activity of CLEAs synthesized in the presence of polyethyleneimine (PEI). Synthesis conditions: PEI/enzyme mass ratio of 1:1 (30 min of gently stirring at 25 °C), addition of the magnetic nanoparticles DCNP-N to a DCNP-N/amyloglucosidase mass ratio of 1:1, precipitation with ethanol (volume ratio enzyme solution in 100 mM phosphate buffer pH 7.00 to ethanol of 1:9), crosslinking with glutaraldehyde concentrations ranging from 100 to 700 mM. Precipitation and crosslinking steps were performed under static conditions in an ice bath. Activity of immobilized amyloglucosidase was measured with 1% (*w*/*v*) dextrin in 50 mM sodium citrate buffer pH 4.5, 10 min reaction at 55 °C under 900 rpm stirring. Values are shown as the mean of duplicate experiments ± s.d.

Despite the improvement obtained with the addition of PEI, the recovered activity of magnetic CLEAs (maximum around 35% in Figure 8) were lower than that achieved using starch as protective additive (maximum around 40% in Table 3). Therefore, the synthesis of CLEAs of AMG co-aggregated with the magnetic nanoparticles DCNP-N in the presence of both PEI and 1% (*w*/*v*) starch was also evaluated.

Figure 9 shows that the presence of PEI and starch enabled the formation of CLEAs even at low concentrations of glutaraldehyde (12.5 mM). However, below 100 mM glutaraldehyde the immobilization yields were very low, probably because of insufficient cross-links to form stable structures [33,49]. On the other hand, when the glutaraldehyde concentrations ranged from 300 to 700 mM the immobilization yields were close to 100%, but the recovered activity was very low, probably due to the excessive crosslinking and/or deleterious effect of glutaraldehyde on the tertiary structure of AMG, as previously discussed. Therefore, a minimum concentration of 300 mM glutaraldehyde was selected for future experiments.

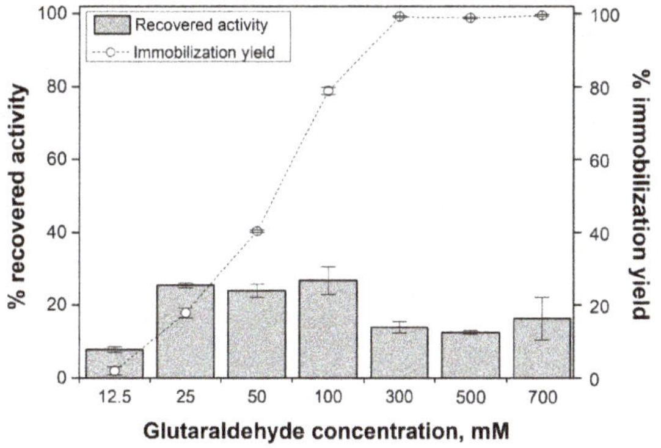

Figure 9. Effect of glutaraldehyde concentration on CLEA of amyloglucosidase (AMG) synthesized in the presence of polyethyleneimine (PEI) and co-aggregated with the magnetic nanoparticles DCNP-N and starch 1% (*w*/*v*). Synthesis conditions: treatment of AMG with PEI/enzyme mass ratio of 1:1 (30 min under gently stirring at 25 °C), co-aggregation with DCNP-N/enzyme mass ratio of 1:1, ethanol/enzyme solution (in 100 mM phosphate buffer pH 7.0) volume ratio of 9:1, crosslinking with glutaraldehyde concentrations ranging from 100 to 700 mM, precipitation and crosslinking under static conditions in an ice bath. Activity of immobilized AMG was measured with 1% (*w*/*v*) dextrin in 50 mM sodium citrate buffer pH 4.5, 10 min reaction at 55 °C under 900 rpm stirring. Values are shown as the mean of duplicate experiments ± s.d.

2.6. *Influence of Agitation, Glutaraldehyde Treatment Time, Cross-Linker Concentration, and Co-Feeders in the Activity of AMG-CLEAs*

It has been reported that at 4 °C, glutaraldehyde treatment time from 6 to 16 h is required to prepare stable CLEAs [28,47,48]. Thus, AMG CLEAs production was evaluated at 4 °C under stirring for 4 and 16 h, and under static conditions for 16 h.

Figure 10 shows that the glutaraldehyde treatment time was a key factor only for CLEAs prepared by co-aggregation with the magnetic nanoparticles DCNP-N and 1% (*w*/*v*) starch and crosslinking with 500 mM glutaraldehyde. Under these conditions, the immobilization yield increased from 45.3% to 70% and the recovered activity increased around twofold by increasing the time from 4 to 16 h. When PEI was added, no significant difference was observed. For the CLEAs prepared in the presence of PEI, the immobilization yields were higher than 95% for all conditions, but the recovered activity were higher when CLEAs were prepared under gently stirring in 3D Platform Shaker. The gentle stirring promoted the formation of homogeneous CLEAs, which visibly reduced the particle size, thereby probably minimizing mass transfer problems [20].

CLEAs previously prepared with 1% (*w*/*v*) starch, polyethyleneimine, and combinations thereof were synthesized again under stirring for 16 h at 4 °C; the co-aggregation with the aminated magnetic nanoparticles DCNP-N was compared with the most used co-feeder (bovine serum albumin), and two concentrations of glutaraldehyde (300 and 500 mM) were evaluated. These CLEAs were compared for their recovered activity as well as for their performance in the starch hydrolysis. The hydrolysis conditions were 55 °C under 300 rpm shaking for 6 h, using as a substrate 3% (*w*/*v*) starch prepared in 50 mM sodium citrate buffer pH 4.5 and pre-hydrolyzed with α-amylase. At the end of the reaction, the Dextrose Equivalent (DE) of the hydrolyzed starch and residual activity of the CLEA were measured.

Figure 11a shows that magnetic CLEAs prepared in the presence of starch (DCNP-N+Starch 1%) yielded the highest DE (83.2), but its residual activity was very low (3.5%). Although the recovered activity was high (around 40%), these CLEAs were not sufficiently stable in starch hydrolysis conditions, even using a co-feeder rich in amino groups and high glutaraldehyde concentration in their preparation. On the other hand, magnetic CLEAs prepared in the presence of polyethyleneimine were also active in the starch hydrolyses (DE above 60) and stable (residual activiy around 40% and 60% for 300 and

500 mM glutaraldehyde, respectively). The addition of 1% (*w*/*v*) starch in the preparation of these CLEAs was not advantageous from a stability point of view.

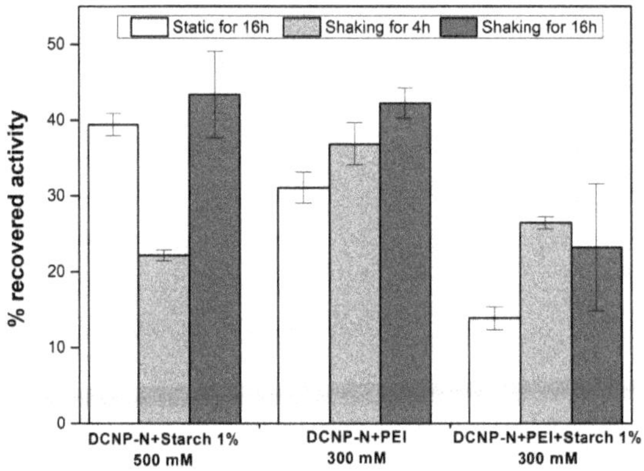

Figure 10. Comparison of the effect of agitation and crosslinking time on the recovered activity of CLEAs of AMG prepared as following: co-aggregation with the magnetic nanoparticles DCNP-N and 1% (*w*/*v*) starch and crosslinking with 500 mM glutaraldehyde (DCNP-N+Starch 1%-500 mM); co-aggregation with the magnetic nanoparticles DCNP-N and polyethyleneimine (PEI) and crosslinking with 300 mM glutaraldehyde (DCNP-N+PEI-300 mM); co-aggregation with the magnetic nanoparticles DCNP-N, 1% (*w*/*v*) starch and PEI and crosslinking with 300 mM glutaraldehyde (DCNP-N+PEI+Starch 1%-300 mM). All CLEAs were prepared at 4 °C, mass ratio protein/DCNP-N of 1:1, precipitation with ethanol at volume ratio 1:9 (enzyme solution:ethanol). Values are shown as the mean of duplicate experiments ± s.d.

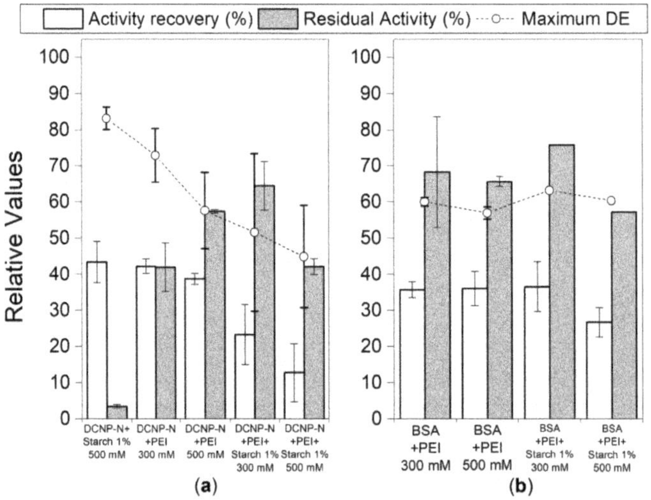

Figure 11. Recovered activity and performance of CLEAs of amyloglucosidase (AMG) in the hydrolysis of 3% (*w*/*v*) starch (Dextrose Equivalent and residual activity). CLEAs were prepared by co-aggregation with the magnetic nanoparticles DCNP-N or bovine serum albumin (BSA) in the presence of 1% (*w*/*v*) starch or polyethyleneimine (PEI) or both. A volume of 20 mL of starch (3%, *w*/*v*) solution was hydrolyzed by 20 U of AMG CLEAs at 55 °C and pH 4.5 for 6 h under 300 rpm stirring. Values are shown as the mean of duplicate experiments ± s.d.

CLEAs prepared under the same conditions, replacing magnetic nanoparticles with bovine serum albumin as the co-feeder, showed good performance (Figure 11b) regarding the recovered activity (around 35%), DE in the starch hydrolysis (around 60), and residual activity (above 60%). Although this co-feeder is widely used in the preparation of CLEAs [20,26,27,30,50], the replacement by magnetic nanoparticles is advantageous because of the ease of capture by an external magnetic field [40], avoiding the formation of clusters usually observed in the separation of CLEAs co-aggregated with bovine serum albumin by centrifugation [45]. Moreover, magnetic CLEAs prepared in the presence of polyethyleneimine and crosslinking with 500 mM glutaraldehyde (CLEA DCNP-N+PEI) showed similar performance (recovered activity around 40%, DE in the starch hydrolysis around 60, and residual activity around 60%), so were selected to be kinetically characterized and used in the hydrolysis of starch under industrial conditions. For comparison, CLEAs of AMG prepared by co-aggregation with bovine serum albumin and crosslinking with 500 mM glutaraldehyde (CLEA BSA+PEI) were also selected.

2.7. Characterization of the CLEAs of AMG

The CLEAs AMG+DCNP-N+PEI and AMG+BSA+PEI were characterized regarding the activity as a function of pH and temperature, thermal stability at 55 °C and pH 4.5, and their performance in the hydrolysis of starch under industrial conditions (high starch concentration, i.e., 35%, *w*/*v*).

Figure 12 shows that the activity profiles as a function of pH for CLEAs of AMG shifted the maximum activity from pH 4.5 to 3.0 compared to the free AMG. This can be associated with a higher stability of immobilized enzyme at this drastic pH.

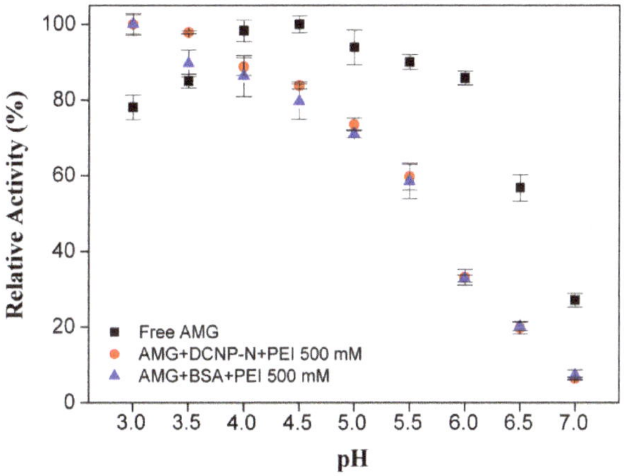

Figure 12. Effect of pH on the activity of free amyloglucosidase (AMG) and AMG CLEAs prepared by enzyme treatment with polyethyleneimine (PEI) and by co-aggregation with the magnetic nanoparticles DCNP-N or bovine serum albumin (BSA). Activity assay conditions: 1% (*w*/*v*) dextrin solution in 50 mM buffer, 10 min reaction at 55 °C under 900 rpm stirring. Values are shown as the mean of triplicate experiments ± s.d.

Figure 13 shows that the maximum activity of the CLEAs was lowered from 75 °C (for free AMG) to 65 °C. This suggests that the immobilized enzyme was less stable at this high temperature, or that the CLEA structure may change and in that way alter the diffusional problems. However, at 55 °C and pH 4.5 the CLEA activity corresponds to around 80% of the maximum activity, while for soluble AMG the activity at these conditions corresponds to around 50% of the maximum value. Thus, the CLEAs of AMG could be more advantageous from an industrial point of view, because the typical industrial conditions of the saccharification of starch are pH 4.0–4.5 and temperature 55–60 °C [51–53].

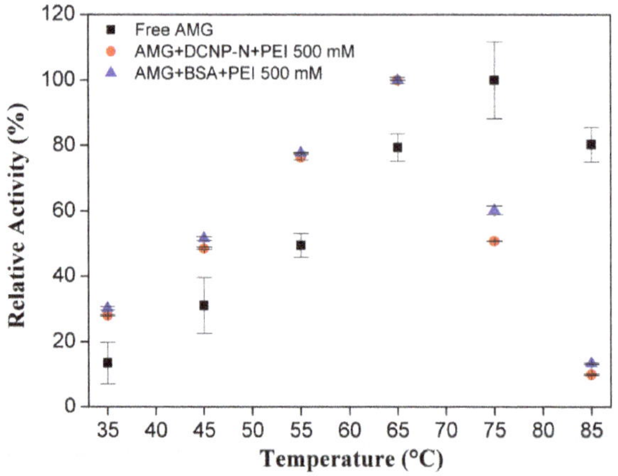

Figure 13. Effect of the temperature on the activity of free amyloglucosidase (AMG) and AMG CLEAs prepared by enzyme treatment with polyethyleneimine (PEI) and co-aggregation with the magnetic nanoparticles DCNP-N or bovine serum albumin (BSA). Activity assays conditions: 1% (*w*/*v*) dextrin solution in 50 mM sodium citrate buffer pH 4.5, 10 min reaction under 900 rpm stirring. Values are shown as the mean of triplicate experiments ± s.d.

Figure 14 shows the thermal inactivation profiles of free (45 and 55 °C and pH 4.5) and immobilized AMG at these typical industrial conditions (55 °C and pH 4.5) [51]. Table 4 shows the parameters of the Sadana and Henley model [54] fitted to the experimental data. It can be observed that CLEAs of AMG (with both magnetic nanoparticles and BSA as co-feeders) were around 4 times more stable than the free AMG at 55 °C, having a fraction of immobilized molecules more resistant to the inactivation as indicated by the parameter α (0.36–0.46 for CLEAs and 0.23 for free AMG). This higher stability is indicative of effective crosslinking with glutaraldehyde, aided by polyethyleneimine in the complex and rigid structure of the CLEAs (enzyme and co-feeders). It has been reported that the immobilization tends to restrict the conformational flexibility of the enzyme, which prevents conformational changes [11,38,55].

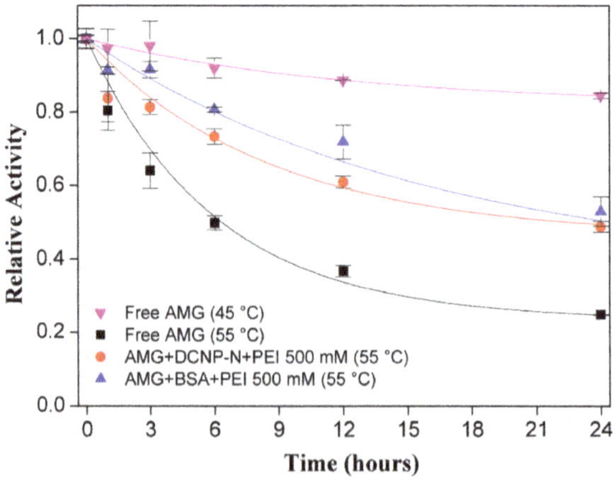

Figure 14. Profile of thermal inactivation of free amyloglucosidase (AMG) and AMG CLEAs prepared by enzyme treatment with polyethyleneimine (PEI) and by co-aggregation with the magnetic nanoparticles DCNP-N or bovine serum albumin (BSA) at 45 and 55 °C and pH 4.5 (50 mM sodium citrate buffer). Values are shown as the mean of triplicate experiments ± s.d.

Table 4. Half-life times ($t_{1/2}$) and stability factor (SF) for amyloglucosidase (free and immobilized AMG) at 55 °C and pH 4.5.

Biocatalyst	$t_{1/2}$ (h)	SF	k_d	α	Adj. R^2
Free AMG	6.34	1.00	0.167 ± 0.014	0.234 ± 0.005	0.99
AMG+DCNP-N+PEI 500 mM	22.67	3.58	0.117 ± 0.016	0.462 ± 0.027	0.98
AMG+BSA+PEI 500 mM	24.60	3.88	0.062 ± 0.019	0.361 ± 0.119	0.95

Note: SF is the ratio between the half-life of the immobilized enzyme and of the free enzyme; the inactivation parameters (k_d and α) were estimated using the Sadana and Henley model [54]; DCNP-N is magnetic nanoparticles functionalized with amino groups; PEI is polyethyleneimine; BSA is bovine serum albumin. All CLEAs were prepared by crosslinking with 500 mM glutaraldehyde.

Figure 15 shows the profiles of starch hydrolysis catalyzed by free and immobilized AMG at typical industrial conditions (35% starch solution and 2 mL of soluble AMG/kg of starch). When CLEA was used as the catalyst, the amount of CLEA was equivalent to 2 mL of soluble AMG in terms of activity. Because at 45 °C the free AMG is highly stable (more than 80% residual activity after 24 h incubation), starch was also hydrolyzed at 45 °C using free AMG, maintaining the other conditions.

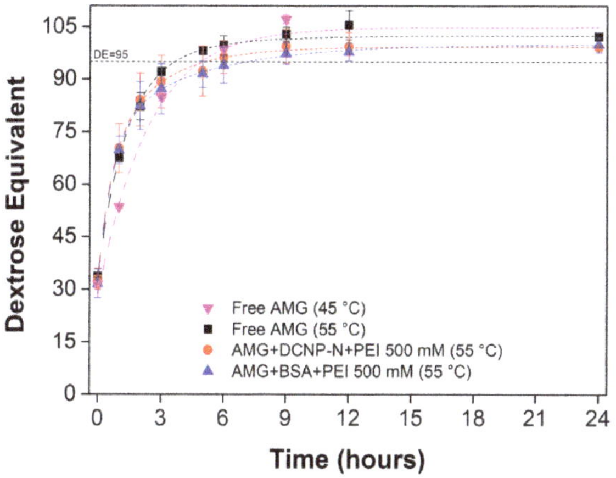

Figure 15. Profile of hydrolysis of 35% (*w*/*v*) starch (pre-hydrolyzed with α-amylase) catalyzed by free amyloglucosidase (AMG) and AMG CLEAs prepared by enzyme treatment with polyethyleneimine (PEI) and by co-aggregation with the magnetic nanoparticles DCNP-N or bovine serum albumin (BSA) at 45 and 55 °C and pH 4.5 under 900 rpm stirring. Amount of AMG: 2 mL of free enzyme/kg of starch (37,820 U/kg of starch) or equivalent units to the CLEAs. The reaction conversion (starch-to-glucose) was monitored by Dextrose Equivalent (DE) measuring reducing sugars by DNS method. Values are shown as the mean of duplicate experiments ± s.d.

It can be observed that the hydrolysis profiles for all biocatalysts (free and immobilized enzymes) were very closed, achieving a Dextrose Equivalent (DE) around 95 after a 6-h reaction, even at 45 °C using free AMG, which shows that the enzyme load is high enough to guarantee high hydrolysis rates. The increase in DE for larger reaction times is very low, not justifying longer hydrolysis because the productivity of the process drops a lot (in $g_{glucose}$ $L^{-1} \cdot h^{-1}$): 6.6 for 6 h, 2.7 for 9 h, and 1.35 for 12 h. Maximum conversions of starch hydrolysis by AMG ranging from 90 to 98% [6] had been previously reported, which shows the excellent performance of CLEAs of AMG prepared in this work.

2.8. Reuse Assays

Because of the similar behavior of hydrolysis at 45 and 55 °C, achieving a DE around 95 after 6 h, the reuse assays were performed at 45 °C because of the high stability of AMG. Figure 16 shows that after five 6-h batches, DE is maintained around 85% using both biocatalysts (CLEAs of AMG prepared with BSA or MNP in presence of PEI). Handling and recovery of the CLEA were easy when MNP was used applying an external magnetic field (Figure 17).

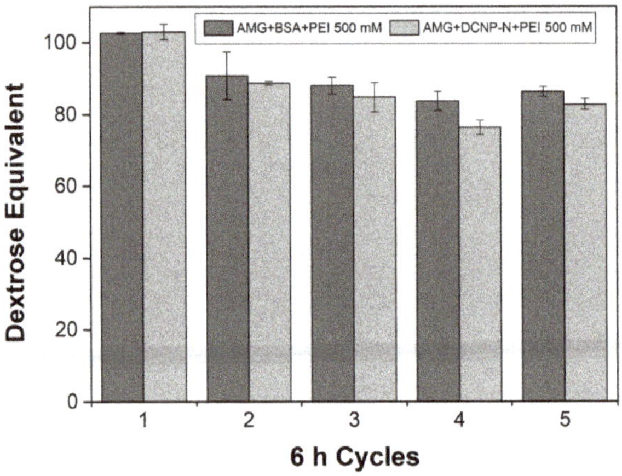

Figure 16. Hydrolysis of pre-hydrolyzed starch with α-amylase (35%, *w*/*v*) catalyzed by CLEAs of amyloglucosidase (AMG) prepared by enzyme treatment with polyethyleneimine (PEI) and by co-aggregation with the magnetic nanoparticles DCNP-N or bovine serum albumin (BSA) at 45 °C and pH 4.5 for 6 h reaction. Values are shown as the mean of duplicate experiments ± s.d.

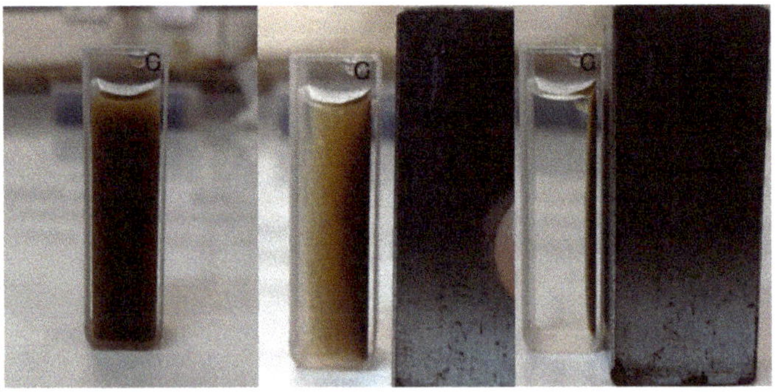

Figure 17. Separation of the magnetic CLEAs of amyloglucosidase by the action of an external magnetic field.

3. Materials and Methods

3.1. Materials

AMG 300L from *Aspergillus niger* (EC 3.2.1.3; glucoamylase; amyloglucosidase) and α-amylase BAN 480L were from Novozymes A/S (Bagsværd, Denmark), dextrin 10, bovine serum albumin (BSA), and polyethylenimine (oligomer mixture, Mn ~423) were purchased from Sigma-Aldrich (St. Louis, MO, USA), Soluble starch and glutaraldehyde 25% (*v*/*v*) aqueous solution from Vetec (Duque de Caxias, RJ, Brazil). Mono reagent colorimetric enzymatic (GOD-POD) obtained from Gold Analisa (Belo Horizonte, MG, Brazil). The magnetic nanoparticles were purchased from Kopp Technologies (São Carlos, SP, Brazil). All other chemicals (analytical grade) were purchased from Synth (Diadema, SP, Brazil) and Vetec (Duque de Caxias, RJ, Brazil).

3.2. Precipitant Screening

The precipitation of amyloglucosidase was carried out by adding the protein precipitant (acetone, ethanol, isopropanol or saturated ammonium sulfate solution) to an enzyme solution (protein concentration of 20 mg·mL^{-1} in 50 mM carbonate buffer pH 4.5) in a volume ratio of 9:1. The mixture was incubated in shaker under 150 rpm at 4 °C for 60 min, the precipitate was recovered by centrifugation (1500× *g* at 4 °C for 5 min) and resuspended in 1 mL of 50 mM sodium citrate buffer pH 4.5. Enzyme activity and protein content were measured in the supernatant and in the re-suspended precipitate as described in Sections 3.7 and 3.8, respectively. For the precipitant chosen based on the highest yields of activity and protein, the volume ratio of precipitant and enzyme solution was also evaluated, but at static conditions in an ice bath for 30 min. Within the volume ratios evaluated, the better one was chosen for further assays.

3.3. General Crosslinking Procedure Using Glutaraldehyde

Ethanol was added to an enzyme solution prepared in 100 mM carbonate buffer pH 10.0 to a volume ratio of 1:9 (enzyme/ethanol). After 30 min precipitation in an ice bath, volumes of glutaraldehyde 25% (v/v) to final concentrations of 100, 300, 500 and 700 mM were added to the precipitated enzyme (crosslinking step). The aggregate suspension was homogenized and incubated statically in a refrigerator at 4 °C for 16 h or under gently stirring in a three-dimensional laboratory agitator (KASVI, K45-4020) at 4 °C for 4 or 16 h. After the incubation period, the suspension was centrifuged (1500× *g* for 5 min at 4 °C) and the precipitate was washed twice with 50 mM sodium citrate buffer pH 4.5, and finally re-suspended in the same buffer. Then, the activity recovery was calculated by Equation (1):

$$Activity\ recovery\ (\%) = \frac{Total\ activity\ in\ the\ CLEA\ suspension}{Total\ activity\ offered\ initially} \times 100. \quad (1)$$

The same procedure described above was used for evaluation of co-aggregation of AMG with magnetic nanoparticles (MNPs) (characteristics described in Table 1) or bovine serum albumin (BSA) as additive or co-feeder, respectively. In this case, the MNPs or BSA was added to the enzyme solution (in 100 mM carbonate buffer pH 10) to a final protein/MNPs (or BSA) mass ratio of 1:1 (20 mg total mL^{-1}). When BSA was used, the CLEAs were separated by centrifugation (1500× *g* for 5 min at 4 °C); on the other hand, when MNPs were used, the CLEAs were separated by applying an external magnetic field.

Starch was also evaluated as a protector additive of the enzyme active site. In this case, a 1% (w/v) starch solution was prepared in a 100 mM carbonate buffer pH 10 and was used to prepare the mixture of enzyme and other additives in the CLEA preparation. The other steps remained unaltered.

Polyethyleneimine (PEI) was also evaluated as a crosslinking aid. In this case, an aqueous solution of PEI was prepared and the pH was adjusted to 7.0. This solution was mixed with the enzyme solution (protein concentration of 20 mg mL^{-1} in 100 mM sodium phosphate buffer pH 7.0) to a protein/PEI mass ratio of 1:1, as described by López-Gallego et al. [34]. After 30 min stirring in a tridimensional laboratory agitator at room temperature, the precipitant was added, and the protocol followed the procedure described above.

3.4. Characterization of the Biocatalysts

The activity of soluble and immobilized AMG was measured as a function of the pH (at 55 °C) in the range from 3.0 to 7.0, and as a function of the temperature (at pH 4.5) in the range from 35 to 85 °C. For pH values from 3.0 to 5.5, 50 mM sodium citrate and acetate buffers were used, and for pH values from 5.5 to 7.0 a 50 mM sodium phosphate buffer was used.

Thermal inactivation assays were carried out at 55 °C and pH 4.5 (50 mM sodium citrate buffer). Enzyme activity was measured at regular time intervals until 24 h. The model of Sadana and

Henley [54] was fitted to the experimental data to determine the half-life. The stabilization factor (SF) was calculated as the ratio between the half-life of the immobilized enzyme and that of the free enzyme.

3.5. Hydrolysis of Starch

A solution of starch (35%, *w*/*v*) was prepared in 50 mM citrate buffer (pH 4.5). The starch was pre-hydrolyzed with α-amylase (BAN 480L, 3 mL/kg starch) at 60 °C for 20 min. The temperature was fitted to 55 °C and soluble AMG (2 mL/kg starch) or CLEAs of AMG (37,820 U/kg starch, equivalent to the amount of soluble enzyme) was added. The reaction was carried out for 24 h in a batch reactor under 900 rpm stirring using a cylindrical impeller without blades to prevent loss of biocatalyst by shearing. Samples of the reaction medium were withdrawn at regular time intervals to measure reducing sugars by the DNS method [56]. Dextrose Equivalent (DE) [57] was calculated by Equation (2) to construct the hydrolysis time profile.

$$Dextrose\ Equivalent = \frac{Amount\ of\ reducing\ sugar\ (expressed\ as\ glucose)}{Starch\ dry\ mass} \times 100 \quad (2)$$

In the selection of CLEAs, their performance in the hydrolysis of starch was evaluated using 3% (*w*/*v*) starch. The hydrolysis reactions were performed at 55 °C and pH 4.5 for 6 h under 300 rpm stirring. Reducing sugars and Dextrose Equivalent were determined as described above, only at the final stage of the reaction.

3.6. Reuse Assays

The reusability assays of amyloglucosidase CLEA were performed at 45 °C under 300 rpm shaking with a solution of pre-hydrolyzed starch (35%, *w*/*v*) prepared in 50 mM citrate buffer pH 4.5. After each cycle of 6 h reaction time, the conversion of starch (DE) was determined and the CLEA was recovered by applying an external magnetic field and/or by centrifugation (1500× *g* for 5 min at 4 °C), washed with citrate buffer (50 mM, pH 4.5) and re-suspended in a fresh substrate solution.

3.7. Enzymatic Activity Assay

Enzymatic activity of amyloglucosidase (free and immobilized as CLEAs) was determined by calculating the initial velocity of glucose formation catalyzed by a known amount of enzyme. The standard substrate was dextrin 1% (*w*/*v*, in 50 mM sodium citrate buffer at pH 4.5). The reaction was carried out at 55 °C for 10 min under 900 rpm stirring. Samples were withdrawn every 2.5 min, the reaction was quenched with 1 M HCl, and glucose was measured as described in Section 3.8. One unit of enzyme activity (U) was defined as the amount of enzyme required to release 1 μmol of glucose per minute under the conditions above.

3.8. Determination of Glucose and Protein Concentration

Glucose was determined spectrophotometrically at 505 nm by glucose oxidase and peroxidase colorimetric enzymatic test (GOD-POD). The GOD-POP solution (1 mL) was added to the sample (10 μL) and incubated at 37 °C for 10 min [58]. The concentration of glucose was quantified using a glucose concentration vs. absorbance curve constructed with glucose as standard.

The protein content was determined spectrophotometrically at 595 nm by the Bradford method, using bovine serum albumin as the standard protein [59].

3.9. SDS-PAGE Electrophoresis

The commercial AMG was characterized by polyacrylamide gel electrophoresis (10% SDS-PAGE) using Coomassie Brilliant Blue for staining [60].

4. Conclusions

This study showed that promising CLEAs of amyloglucosidase could be synthesized by co-aggregation with aminated magnetic nanoparticles (MNPs) or bovine serum albumin (BSA), using polyethyleneimine as an aid in the crosslinking step with glutaraldehyde. Both CLEAs showed around 40% of the offered activity, thermal stability approximately 4 times higher than the soluble enzyme, and small changes in the catalytic properties. Moreover, the starch saccharification at typical industrial conditions, i.e., 35% (w/v) pre-hydrolyzed starch with α-amylase, 55 °C and pH 4.5, catalyzed by the CLEAs (co-aggregated with MNPs or BSA) showed similar behavior to the soluble enzyme, yielding a Dextrose Equivalent around 95 after a 6-h reaction. In addition, both CLEAS could be reused in five 6-h cycles at 45 °C and maintain a DE above 85. These findings could be attractive to the amylolytic industry because highly concentrated starch solutions may be processed by immobilized AMG as well as the soluble enzyme (including 10 °C below the conventionally used temperature), having the additional advantage of being easily separated from the reaction medium and reused in the process when MNPs are used instead of the protein co-feeders commonly used in the CLEA methodology, such as BSA, whose cost is too high for the synthesis of CLEAs for large-scale applications.

Author Contributions: M.A.-F. performed all experimental assays; W.K. synthesized the magnetic nanoparticles and characterized them regarding the amino and hydrophobic groups; R.d.L.C.G., R.F.-L., and P.W.T. designed and supervised all experiments, and wrote/revised the manuscript with the help of M.A.-F. as part of his Doctorate in Chemical Engineering. All authors have given approval to the final version of the manuscript.

Funding: This research was funded in part by the Conselho Nacional de Desenvolvimento Científico e Tecnológico, grant #142107/2015-8, Coordenação de Aperfeiçoamento de Pessoal de Nível Superior (CAPES), Finance Code 001, and Fundação de Amparo à Pesquisa do Estado de São Paulo (FAPESP), grant #2016/10636-8.

Acknowledgments: The authors thank LNF Latino Americana (Bento Gonçalves, Brazil) for providing the α-amylase BAN 480L and Novozymes A/S (Araucária, Brazil) for the donation of amyloglucosidase AMG 300L. We gratefully recognize financial support from Project CTQ2017-86170-R (MINECO, Spain).

Conflicts of Interest: The authors declare no conflict of interest.

References

1. Parkin, K.L. Enzymes. In *Fennema's Food Chemistry*; Damodaran, S., Parkin, K.L., Fennema, O.R., Eds.; CRC Press: Boca Raton, FL, USA, 2007; pp. 331–435, ISBN 978-0849392726.
2. Crabb, W.D.; Mitchinson, C. Enzymes involved in the processing of starch to sugars. *Trends Biotechnol.* **1997**, *15*, 349–352. [CrossRef]
3. Svensson, B.; Larsen, K.; Svendsen, I.; Boel, E. The complete amino acid sequence of the glycoprotein, glucoamylase G1, from *Aspergillus niger*. *Carlsberg Res. Commun.* **1983**, *48*, 529–544. [CrossRef]
4. Lee, J.; Paetzel, M. Structure of the catalytic domain of glucoamylase from *Aspergillus niger*. *Acta Crystallogr. Sect. F Struct. Biol. Cryst. Commun.* **2011**, *67*, 188–192. [CrossRef] [PubMed]
5. Sorimachi, K.; Gal-Coëffet, M.-F.L.; Williamson, G.; Archer, D.B.; Williamson, M.P. Solution structure of the granular starch binding domain of *Aspergillus niger* glucoamylase bound to β-cyclodextrin. *Structure* **1997**, *5*, 647–661. [CrossRef]
6. Tomasik, P.; Horton, D. Enzymatic conversions of starch. In *Advances in Carbohydrate Chemistry and Biochemistry*; Academic Press: Oxford, UK, 2012; Volume 68, pp. 59–436, ISBN 9780123965233.
7. Pazur, J.H.; Ando, T. The action of an amyloglucosidase of *Aspergillus niger* on starch and malto-oligosaccharides. *J. Biol. Chem.* **1959**, *234*, 1966–1970. [PubMed]
8. Parker, K.; Salas, M.; Nwosu, V.C. High fructose corn syrup: Production, uses and public health concerns. *Biotechnol. Mol. Biol. Rev.* **2010**, *5*, 71–78.
9. Kennedy, J.F.; Cabral, J.M. Enzyme imobilization. In *Biotechnology*; Rehm, H.J., Reed, G., Eds.; VCH: Weinheim, Germany, 1987; pp. 347–404.
10. Homaei, A.A.; Sariri, R.; Vianello, F.; Stevanato, R. Enzyme immobilization: An update. *J. Chem. Biol.* **2013**, *6*, 185–205. [CrossRef] [PubMed]

11. Mateo, C.; Palomo, J.M.; Fernandez-Lorente, G.; Guisan, J.M.; Fernandez-Lafuente, R. Improvement of enzyme activity, stability and selectivity via immobilization techniques. *Enzyme Microb. Technol.* **2007**, *40*, 1451–1463. [CrossRef]
12. Silva, R.N.; Asquieri, E.R.; Fernandes, K.F. Immobilization of *Aspergillus niger* glucoamylase onto a polyaniline polymer. *Process Biochem.* **2005**, *40*, 1155–1159. [CrossRef]
13. Bryjak, J. Glucoamylase, α-amylase and β-amylase immobilisation on acrylic carriers. *Biochem. Eng. J.* **2003**, *16*, 347–355. [CrossRef]
14. Shah, C.; Sellappan, S.; Madamwar, D. Entrapment of enzyme in water-restricted microenvironment—Amyloglucosidase in reverse micelles. *Process Biochem.* **2000**, *35*, 971–975. [CrossRef]
15. Tardioli, P.W.; Vieira, M.F.; Vieira, A.M.S.; Zanin, G.M.; Betancor, L.; Mateo, C.; Fernández-Lorente, G.; Guisán, J.M. Immobilization-stabilization of glucoamylase: Chemical modification of the enzyme surface followed by covalent attachment on highly activated glyoxyl-agarose supports. *Process Biochem.* **2011**, *46*, 409–412. [CrossRef]
16. Wang, J.; Zhao, G.; Li, Y.; Liu, X.; Hou, P. Reversible immobilization of glucoamylase onto magnetic chitosan nanocarriers. *Appl. Microbiol. Biotechnol.* **2013**, *97*, 681–692. [CrossRef] [PubMed]
17. Sheldon, R.A. Cross-linked enzyme aggregates (CLEA®s): Stable and recyclable biocatalysts. *Biochem. Soc. Trans.* **2007**, *35*, 1583–1587. [CrossRef] [PubMed]
18. Schoevaart, R.; Wolbers, M.W.; Golubovic, M.; Ottens, M.; Kieboom, A.P.G.; van Rantwijk, F.; van der Wielen, L.A.M.; Sheldon, R.A. Preparation, optimization, and structures of cross-linked enzyme aggregates (CLEAs). *Biotechnol. Bioeng.* **2004**, *87*, 754–762. [CrossRef] [PubMed]
19. Cao, L.; van Rantwijk, F.; Sheldon, R.A. Cross-linked enzyme aggregates: A simple and effective method for the immobilization of penicillin acylase. *Org. Lett.* **2000**, *2*, 1361–1364. [CrossRef] [PubMed]
20. Talekar, S.; Joshi, A.; Joshi, G.; Kamat, P.; Haripurkar, R.; Kambale, S. Parameters in preparation and characterization of cross linked enzyme aggregates (CLEAs). *RSC Adv.* **2013**, *3*, 12485–12511. [CrossRef]
21. Mateo, C.; Palomo, J.M.; van Langen, L.M.; van Rantwijk, F.; Sheldon, R.A. A new, mild cross-linking methodology to prepare cross-linked enzyme aggregates. *Biotechnol. Bioeng.* **2004**, *86*, 273–276. [CrossRef] [PubMed]
22. Mafra, A.C.O.; Beltrame, M.B.; Ulrich, L.G.; Giordano, R.D.L.C.; Ribeiro, M.P.D.A.; Tardioli, P.W. Combined CLEAs of invertase and soy protein for economically feasible conversion of sucrose in a fed-batch reactor. *Food Bioprod. Process.* **2018**, *110*, 145–157. [CrossRef]
23. Talekar, S.; Ghodake, V.; Kate, A.; Samant, N.; Kumar, C.; Gadagkar, S. Preparation and Characterization of Cross-linked Enzyme Aggregates of *Saccharomyces cerevisiae* Invertase. *Aust. J. Basic Appl. Sci.* **2010**, *4*, 4760–4765.
24. Ramos, M.D.; Miranda, L.P.; Giordano, R.L.C.; Fernandez-Lafuente, R.; Kopp, W.; Tardioli, P.W. 1,3-Regiospecific ethanolysis of soybean oil catalyzed by crosslinked porcine pancreas lipase aggregates. *Biotechnol. Prog.* **2018**. [CrossRef] [PubMed]
25. Wilson, L.; Betancor, L.; Fernández-Lorente, G.; Fuentes, M.; Hidalgo, A.; Guisán, J.M.; Pessela, B.C.C.; Fernández-Lafuente, R. Cross-Linked Aggregates of Multimeric Enzymes: A Simple and Efficient Methodology to Stabilize Their Quaternary Structure. *Biomacromolecules* **2004**, *5*, 814–817. [CrossRef] [PubMed]
26. Mafra, A.C.O.; Kopp, W.; Beltrame, M.B.; de Lima Camargo Giordano, R.; de Arruda Ribeiro, M.P.; Tardioli, P.W. Diffusion effects of bovine serum albumin on cross-linked aggregates of catalase. *J. Mol. Catal. B Enzym.* **2016**, *133*, 107–116. [CrossRef]
27. Araujo-Silva, R.; Mafra, A.C.O.; Rojas, M.J.; Kopp, W.; Giordano, R.D.C.; Fernandez-Lafuente, R.; Tardioli, P.W. Maltose Production Using Starch from Cassava Bagasse Catalyzed by Cross-Linked β-Amylase Aggregates. *Catalysts* **2018**, *8*, 170. [CrossRef]
28. Nadar, S.S.; Rathod, V.K. Magnetic macromolecular cross linked enzyme aggregates (CLEAs) of glucoamylase. *Enzyme Microb. Technol.* **2016**, *83*, 78–87. [CrossRef] [PubMed]
29. Gupta, K.; Jana, A.K.; Kumar, S.; Maiti, M. Immobilization of amyloglucosidase from SSF of *Aspergillus niger* by crosslinked enzyme aggregate onto magnetic nanoparticles using minimum amount of carrier and characterizations. *J. Mol. Catal. B Enzym.* **2013**, *98*, 30–36. [CrossRef]
30. Shah, S.; Sharma, A.; Gupta, M.N. Preparation of cross-linked enzyme aggregates by using bovine serum albumin as a proteic feeder. *Anal. Biochem.* **2006**, *351*, 207–213. [CrossRef] [PubMed]

31. Wilson, L.; Fernández-Lorente, G.; Fernández-Lafuente, R.; Illanes, A.; Guisán, J.M.; Palomo, J.M. CLEAs of lipases and poly-ionic polymers: A simple way of preparing stable biocatalysts with improved properties. *Enzyme Microb. Technol.* **2006**, *39*, 750–755. [CrossRef]
32. Galvis, M.; Barbosa, O.; Ruiz, M.; Cruz, J.; Ortiz, C.; Torres, R.; Fernandez-Lafuente, R. Chemical amination of lipase B from *Candida antarctica* is an efficient solution for the preparation of crosslinked enzyme aggregates. *Process Biochem.* **2012**, *47*, 2373–2378. [CrossRef]
33. Sheldon, R.A. Cross-linked enzyme aggregates as industrial biocatalysts. *Org. Process Res. Dev.* **2011**, *15*, 213–223. [CrossRef]
34. López-Gallego, F.; Betancor, L.; Hidalgo, A.; Alonso, N.; Fernández-Lafuente, R.; Guisán, J.M. Co-aggregation of enzymes and polyethyleneimine: A simple method to prepare stable and immobilized derivatives of glutaryl acylase. *Biomacromolecules* **2005**, *6*, 1839–1842. [CrossRef] [PubMed]
35. Zheng, J.; Chen, Y.; Yang, L.; Li, M.; Zhang, J. Preparation of Cross-Linked Enzyme Aggregates of Trehalose Synthase via Co-aggregation with Polyethyleneimine. *Appl. Biochem. Biotechnol.* **2014**, *174*, 2067–2078. [CrossRef] [PubMed]
36. Virgen-Ortíz, J.J.; Dos Santos, J.C.; Berenguer-Murcia, Á.; Barbosa, O.; Rodrigues, R.C.; Fernandez-Lafuente, R. Polyethylenimine: A very useful ionic polymer in the design of immobilized enzyme biocatalysts. *J. Mater. Chem. B* **2017**, 7461–7490. [CrossRef]
37. Velasco-Lozano, S.; López-Gallego, F.; Vázquez-Duhalt, R.; Mateos-Díaz, J.C.; Guisán, J.M.; Favela-Torres, E. Carrier-free immobilization of lipase from candida rugosa with polyethyleneimines by carboxyl-activated cross-linking. *Biomacromolecules* **2014**, *15*, 1896–1903. [CrossRef] [PubMed]
38. Garcia-Galan, C.; Berenguer-Murcia, Á.; Fernandez-Lafuente, R.; Rodrigues, R.C. Potential of Different Enzyme Immobilization Strategies to Improve Enzyme Performance. *Adv. Synth. Catal.* **2011**, *353*, 2885–2904. [CrossRef]
39. Talekar, S.; Ghodake, V.; Ghotage, T.; Rathod, P.; Deshmukh, P.; Nadar, S.; Mulla, M.; Ladole, M. Novel magnetic cross-linked enzyme aggregates (magnetic CLEAs) of alpha amylase. *Bioresour. Technol.* **2012**, *123*, 542–547. [CrossRef] [PubMed]
40. Kopp, W.; Silva, F.A.; Lima, L.N.; Masunaga, S.H.; Tardioli, P.W.; Giordano, R.C.; Araújo-Moreira, F.M.; Giordano, R.L.C. Synthesis and characterization of robust magnetic carriers for bioprocess applications. *Mater. Sci. Eng. B* **2015**, *193*, 217–228. [CrossRef]
41. Panek, A.; Pietrow, O.; Synowiecki, J. Characterization of glucoamylase immobilized on magnetic nanoparticles. *Starch/Staerke* **2012**, *64*, 1003–1008. [CrossRef]
42. Mateo, C.; Abian, O.; Fernandez-Lafuente, R.; Guisan, J.M. Reversible enzyme immobilization via a very strong and nondistorting ionic adsorption on support-polyethylenimine composites. *Biotechnol. Bioeng.* **2000**, *68*, 98–105. [CrossRef]
43. Wang, M.; Jia, C.; Qi, W.; Yu, Q.; Peng, X.; Su, R.; He, Z. Porous-CLEAs of papain: Application to enzymatic hydrolysis of macromolecules. *Bioresour. Technol.* **2011**, *102*, 3541–3545. [CrossRef] [PubMed]
44. Talekar, S.; Shah, V.; Patil, S.; Nimbalkar, M. Porous cross linked enzyme aggregates (p-CLEAs) of *Saccharomyces cerevisiae* invertase. *Catal. Sci. Technol.* **2012**, *2*, 1575–1579. [CrossRef]
45. Cui, J.D.; Jia, S.R. Optimization protocols and improved strategies of cross-linked enzyme aggregates technology: Current development and future challenges. *Crit. Rev. Biotechnol.* **2015**, *35*, 15–28. [CrossRef] [PubMed]
46. Snyder, S.L.; Sobocinski, P.Z. An improved 2,4,6-trinitrobenzenesulfonic acid method for the determination of amines. *Anal. Biochem.* **1975**, *64*, 284–288. [CrossRef]
47. Talekar, S.; Nadar, S.; Joshi, A.; Joshi, G. Pectin cross-linked enzyme aggregates (pectin-CLEAs) of glucoamylase. *RSC Adv.* **2014**, *4*, 59444–59453. [CrossRef]
48. Migneault, I.; Dartiguenave, C.; Bertrand, M.J.; Waldron, K.C. Glutaraldehyde: Behavior in aqueous solution, reaction with proteins, and application to enzyme crosslinking. *Biotechniques* **2004**, *37*, 790–802. [CrossRef] [PubMed]
49. Sheldon, R.A.; van Pelt, S. Enzyme immobilisation in biocatalysis: Why, what and how. *Chem. Soc. Rev.* **2013**, *42*, 6223–6235. [CrossRef] [PubMed]
50. Dal Magro, L.; Hertz, P.F.; Fernandez-Lafuente, R.; Klein, M.P.; Rodrigues, R.C. Preparation and characterization of a Combi-CLEAs from pectinases and cellulases: A potential biocatalyst for grape juice clarification. *RSC Adv.* **2016**, *6*, 27242–27251. [CrossRef]

51. Hobbs, L. Sweeteners from Starch: Production, Properties and Uses. In *Starch*; BeMiller, J., Whistler, R., Eds.; Academic Press: Cambridge, UK, 2009; pp. 797–832, ISBN 9780127462752.
52. Synowiecki, J. The Use of Starch Processing Enzymes in the Food Industry. In *Industrial Enzymes*; Springer: Dordrecht, The Netherlands, 2007; pp. 19–34.
53. Guzmán-Maldonado, H.; Paredes-López, O.; Biliaderis, C.G. Amylolytic enzymes and products derived from starch: A review. *Crit. Rev. Food Sci. Nutr.* **1995**, *35*, 373–403. [CrossRef] [PubMed]
54. Sadana, A.; Henley, J.P. Single-step unimolecular non-first-order enzyme deactivation kinetics. *Biotechnol. Bioeng.* **1987**, *30*, 717–723. [CrossRef] [PubMed]
55. Rodrigues, R.C.; Ortiz, C.; Berenguer-Murcia, Á.; Torres, R.; Fernández-Lafuente, R. Modifying enzyme activity and selectivity by immobilization. *Chem. Soc. Rev.* **2013**, *42*, 6290–6307. [CrossRef] [PubMed]
56. Miller, G.L. Use of Dinitrosalicylic Acid Reagent for Determination of Reducing Sugar. *Anal. Chem.* **1959**, *31*, 426–428. [CrossRef]
57. Kearsley, M.W.; Dziedzic, S.Z. Physical and chemical properties of glucose syrups. In *Handbook of Starch Hydrolysis Products and Their Derivatives*; Springer: Boston, MA, USA, 1995; pp. 129–154.
58. Trinder, P. Determination of blood glucose using an oxidase-peroxidase system with a non-carcinogenic chromogen. *J. Clin. Pathol.* **1969**, *22*, 158–161. [CrossRef] [PubMed]
59. Bradford, M.M. A rapid and sensitive method for the quantitation of microgram quantities of protein utilizing the principle of protein-dye binding. *Anal. Biochem.* **1976**, *72*, 248–254. [CrossRef]
60. Laemmli, U.K. Cleavage of structural proteins during the assembly of the head of bacteriophage T4. *Nature* **1970**, *227*, 680–685. [CrossRef] [PubMed]

catalysts

Review

Combined Cross-Linked Enzyme Aggregates as Biocatalysts

Meng-Qiu Xu [1], Shuang-Shuang Wang [2], Li-Na Li [1], Jian Gao [2,*] and Ye-Wang Zhang [1,2,*]

1 School of Pharmacy, United Pharmaceutical Institute of Jiangsu University and Shandong Tianzhilvye Biotechnology Co. Ltd., Jiangsu University, Zhenjiang 212013, China; xumenqiu@outlook.com (M.-Q.X.); lilina20182018@outlook.com (L.-N.L.)

2 College of Petroleum and Chemical Engineering, Qinzhou University, Qinzhou 535011, China; doublewang123@163.com

* Correspondence: jgao12@163.com (J.G.); zhangyewang@ujs.edu.cn (Y.-W.Z.); Tel.: +86-511-8503-8201 (Y.-W.Z.)

Received: 17 September 2018; Accepted: 14 October 2018; Published: 17 October 2018

Abstract: Enzymes are efficient biocatalysts providing an important tool in many industrial biocatalytic processes. Currently, the immobilized enzymes prepared by the cross-linked enzyme aggregates (CLEAs) have drawn much attention due to their simple preparation and high catalytic efficiency. Combined cross-linked enzyme aggregates (combi-CLEAs) including multiple enzymes have significant advantages for practical applications. In this review, the conditions or factors for the preparation of combi-CLEAs such as the proportion of enzymes, the type of cross-linker, and coupling temperature were discussed based on the reaction mechanism. The recent applications of combi-CLEAs were also reviewed.

Keywords: biocatalysis; Combi-CLEAs; cascade reactions; immobilization

1. Introduction

With a high catalytic efficiency, chemo-, region-, and stereoselectivities, enzyme-mediated biocatalytic reactions have a wide range of applications in the fermentation, chemical, food industry and environmental management [1–5]. These reactions usually involve one enzyme or multi-enzymes as catalysts. Compared with single enzymes, multi-enzymes can produce more valuable products although the composition and preparation are complicated. A lot of research has employed soluble enzymes to catalyze reactions [6,7]. The poor operational stability and difficult enzyme separation hampered the development of multiple enzymes as biocatalysts [8]. These limitations could be overcome by membrane reactors [9]. However, the applications of such membrane bioreactors are relatively expensive because of membrane fouling [10], high energy consumption [11], and difficulty to separate macromolecular substrate and enzymes from the reaction mixture. To lower the cost of biocatalysts, the most practical option is enzyme immobilization [12,13].

Enzyme immobilization offers a considerable prospect of reusability and increases the enzyme stability, such as the improvement of organic solvents resistance, pH tolerance, and thermal stability [14–17]. Immobilization of enzymes onto solid carriers is an effective method for stabilizing enzymes because of the superior physical stability and easier recovery [18]. Till today, many carriers including nanomaterials [19–29], magnetic materials [30–35], and graphene carriers [36,37] are widely used for drug delivery, food preparation, and the immobilization of enzymes. Most of these carriers are metal particles [38–42], and some of them as are biological macromolecules [43–47]. Unfortunately, extra carrier leads to the dilution of activity of immobilized enzymes because of the introducing of a large portion of the non-catalytic mass. This will result in lower space-time yields, volumetric activity, and productivity in the enzymatic reactions [48,49]. Furthermore, the preparation and modification of carrier significantly increase additional costs of the biocatalysts. To overcome these shortcomings, carrier-free techniques for immobilization of enzymes have received much interest since they were developed.

Carrier-free immobilized enzymes were developed since the middle of the 20th century [48]. As the most typical carrier-free immobilized enzyme, cross-linked enzyme crystal (CLEC) is chemically cross-linked between enzyme crystals [50]. It was firstly reported by Quiocho et al. in 1964 [51]. In their research, carboxypeptidase A crystals were obtained from pure enzyme protein by adjusting the pH of carboxypeptidase A containing 1 M NaCl solution and dialyzing to reduce the salt concentration. CLEC was found to have advantages such as the controllable particle size, high recycling rate and favorable stability, and enhanced tolerance to organic reagent and extreme pH. Because of these advantages, it has been successfully used in chromatography, drug release, chiral synthesis, and other fields. However, it is difficult to prepare CLECs in the industrial scale because of the critical conditions for protein crystallization. These conditions include suitable ionic species and strength, proper seed loading, cooling rate, and temperature [52,53]. The crystal size and shape were affected directly by these factors, and they will determinate the activity of CLEC [54–56]. These conditions are hardly controlled simultaneously in industrial production, and the preparation of high purity protein is another burden for production of CLEC. To overcome CLEC's drawbacks, cross-linked enzyme aggregate (CLEA) was then developed. This technique does not require highly purified enzymes but it could be performed starting from crude enzyme preparations.

2. Combi-CLEA

2.1. Advantages of CLEAs and Combi-CLEAs

Developed from CLEC, CLEA also has all its advantages like a high specific activity, high resistance to pH organic solvent, improved thermal stability. Additionally, it also has a good stability under operation and storage conditions which CLEC process could be achieved by CLEA because there is no dilution caused by the insertion of solid carriers [57,58]. The CLEA displays high resistance to organic solvents, extreme pH, and high temperatures. This is largely due to the fact that immobilization decreases the enzyme flexibility and suppresses enzymes towards unfolding tertiary structures necessary for catalytic activity [59–61]. Thus, CLEAs would not lose much activity after several re-utilizations. For example, the thermal stability of CLEAs subtilisin was more improved than that of its soluble counterpart and the immobilized enzyme kept 45% of the catalytic activity after 10 reuses [62]. The investigation of sucrose phosphorylase CLEAs found that the optimum temperature of the immobilized enzyme was increased by 17 °C and had a broader pH range [63]. Epoxide hydrolases CLEAs could protect enzymes from hydrophobic organic solvents as the activity of CLEAs was 21.5% higher than that of the free enzymes in *n*-hexane [64]. The storage stability of immobilized enzymes was also improved; CLEAs kept about 67% and the free enzymes remained less than 35% of the initial activity. These properties of CLEAs will undoubtedly promote the development of novel processes of industrial applications. Compared with CLEC, the advantages of CLEAs are that the highly pure enzyme is not required in the process of immobilization and CLEA can be prepared directly from a crude enzyme. CLEAs of hydroxynitrile lyase were prepared from crude enzyme precipitation resulting in an obvious improvement of the enzyme stability in acidic conditions [53]. The general procedure for CLEA preparation includes precipitation and cross-linking. Firstly, adding salts or water-miscible organic solvents or nonionic polymers to precipitate the enzyme to obtain physical aggregates from aqueous solutions [54,65]. Physical aggregates keep most of the enzyme activity because the interaction between the molecules of aggregates is non-covalent, which makes the protein form insoluble structures and will not destroy their tertiary structure [59,61]. Precipitation is also a purification process of enzymes. Thus, the CLEA combines purification and immobilization into one operational unit. In principle, CLEAs can be prepared from the crude enzyme directly [66,67]. The simple preparation process of CLEAs indicates CLEAs are much economical and practical in the industry. Cross-linking with bifunctional reagents will make the immobilized enzyme aggregates much more stable due to the fixation of molecular conformation. Therefore, the immobilized enzyme could be used for multiple cycles while maintaining most of its initial activity.

Normally, biocatalytic reactions in vivo involve more than one enzyme and thus keep a high efficiency of the life cycle. To achieve such a goal of the multiple-enzymes catalysis, combined cross-linked enzyme aggregates (combi-CLEAs) were developed based on the CLEAs. Combi-CLEAs include two or more immobilized enzymes that can catalyze cascade or parallel reactions in one reaction system. For example, combi-CLEAs of xylanase, cellulase and β-1,3-glucanase was used to achieve one-pot bioconversion of lignocellulosic biomass to fermentable sugars [8]. To make the combi-CLEAs highly efficient, the reaction conditions such as enzymes ratio, the pH of the reaction medium, and the temperature of preparation should be finely optimized [68,69].

2.2. Factors Influencing CLEAs and Combi-CLEAs Preparation

In order to prepare highly efficient and stable industrial biocatalysts by combi-CLEAs, their preparative conditions are important in affecting catalytic properties that include the activity, stability, and kinetic parameters [69,70]. Many factors such as the ratio of enzymes, the type of precipitants, the crosslinkers, the cross-linking time, and the pH of the reaction system influence the preparation of combi-CLEAs are discussed below (Table 1).

2.2.1. Proportion of Enzymes

The proportion of every enzyme involved in the fabrication of combi-CLEAs affects the catalytic efficiency. Thus, it is necessary to determine the optimal ratio of enzymes because each enzyme has its own catalytic rate under the reaction conditions. For example, in glucose oxidase (GOD) and versatile peroxidase (VP) CLEAs the glucose can be catalyzed by GOD to D-glucono-δ-lactone and hydrogen peroxide which is the substrate of VP. Excessive GOD catalyzes the production of superfluous hydrogen peroxide that would cause the inactivation of the VP, and a low concentration of GOD would limit the apparent rate of the combi-CLEAs. When the ratio of VP and GOD is 10:7, the maximal apparent rate of combi-CLEAs was observed, which was about 2-fold higher than that without optimization. The reason might be that the produced hydrogen peroxide could be consumed completely by VP [71]. Perfectly setting the right ratio of enzymes in the preparation process of combi-CLEAs is inevitable fundamental in order to make the catalysts economical, have a relative high productivity, with high reaction rates, and more stable in practical applications. Combi-CLEAs of amylosucrase (AS), maltooligosyltrehalose synthase (MTS), and maltooligosyltrehalose trehalohydrolase (MTH) were prepared to produce trehalose from sucrose. AS catalyzes sucrose to form maltooligosaccharides that are substrates for MTS and MTH. The ratio of MTS and MTH was investigated and it was found that the yield of trehalose was similar in the range from 14:1 to 1:14 of MTS:MTH. The results indicated that MTS and MTH were not rate-limiting enzymes in the cascade reactions. The same quantity of MTS and MTH was arbitrarily used in further optimization reactions. The ratios of AS:(MTS and MTH) from 0.25:1 to 8:1 were further investigated. When the ratio was set as 8:1, the trehalose yield was 13 times higher than that of 1:1 [72]. It could be concluded that AS catalyzed the formation of sufficient maltooligosaccharides and long-glucan polymers led to more efficient MTS and MTH catalyzed reactions.

2.2.2. The Precipitants

Adding neutral salts or water-miscible organic solvents or nonionic polymers to free enzymes could induce the physical aggregation and precipitation of protein molecules [73]. These additives could change the hydration state of the enzyme molecule or change the dielectric constant of the solution. Meanwhile, the supramolecular structure is formed in aggregates by non-covalent bonds, the original tertiary structure of enzymes is not destroyed, and the structure may collapse in an aqueous medium. Every enzyme has a unique primary sequence and quaternary structure, so the optimal precipitation condition varies from one enzyme to another. Different precipitant induces distinct conformations of enzyme aggregates which affect the catalytic properties of CLEAs. In the screening of precipitants, the precipitation efficiency of the enzyme is usually determined, then the physical aggregates are re-dissolved in an appropriate buffer and evaluated with their activities [65]. To choose suitable precipitation agents will benefit the preparation of combi-CLEAs. Manganese peroxidase was precipitated in the form of CLEAs using acetone, ammonium sulfate, ethanol, 2-propanol, and *tert*-butanol, followed by glutaraldehyde cross-linking for 3 h at 4 °C. The results showed that the activity recovery and aggregation yield of acetone were at a maximum, reached 31.26 and 73.46%. The CLEAs recovered 47.57% of the initial activity following cross-linking with glutaraldehyde [74]. The aggregation yield and activity recovery of combi-CLEAs of xylanase and mannanase treated by acetone were both 1.2-folds than those of treated by ammonium sulfate. After cross-linking with 125 mM glutaraldehyde, combi-CLEAs precipitated with 80% acetone retained a higher activity than that precipitated with 80% ammonium sulfate [14]. Four ice-cold organic solvents (acetone, acetonitrile, ethanol, and 2-propanol) were used as the precipitant agents for the preparation of peroxidase CLEAs. The results showed that the activity recoveries of CLEAs were less than 10% when ethanol and 2-propanol were served as precipitants. When acetone was utilized as the precipitant, a 28% activity recovery and 78% aggregation yield of CLEAs were obtained. After incubating at 70 °C for 15 min, the CLEAs precipitated with acetone remained about 37% activity while the free counterpart was found totally inactivated [75]. Compared with acetone and *t*-butanol, dimethoxyethane was found to be the best precipitant for the preparation of combi-CLEAs including lipase, α-amylase, and phospholipase A_2. The preparation parameters included 5 mL of pre-cooling dimethoxyethane, 20 mM of glutaraldehyde, 4 h of cross-linking time and a 4 °C cross-linking temperature. The obtained combi-CLEAs could keep most initial activities after 3 cycles [76]. Generally, the best precipitant for enzymes could be variable, and consequently, the selection of precipitation parameters is a critical step in the preparation of CLEAs with a high recovery of enzyme activity. Additionally, a suitable precipitant should be inexpensive and commercially available. More importantly, it should be aqueous soluble and not react with enzymes and buffers.

2.2.3. The Cross-Linker

The cross-linker is a bifunctional agent that can covalently link the amino acid residues of enzyme surfaces [77,78]. Glutaraldehyde has been used as a cross-linker for years in the preparation of CLEAs since it is inexpensive and readily available in commercial quantities [79]. CLEAs were prepared by reacting ε-amino groups of lysine residues on the surface of neighboring enzyme molecules with glutaraldehyde to form inter- and intramolecular aldol condensates [80]. The end products obtained under alkaline or acidic conditions were different. Under alkaline conditions, the end product formed a Schiff base with an amino group from one protein molecule and a C–N bond by Michael addition to the β-carbon in the amino group from another protein molecule. Under acidic conditions, the end product formed a Schiff base with an amino group from one protein molecule and a C–N bond by anti-Markownikoff addition to the α-carbon in another amino group from a neighboring protein (Figure 1) [81]. However, there it was reported that alcohol dehydrogenase and nitrilase lost most activities after cross-linking with glutaraldehyde [82]. Glutaraldehyde has a relatively small size, and it could occupy the binding sites of substrates or even block the entry of macromolecular substrates into the catalytic center which could inactivate the enzymes [80]. To prevent this, cross-linkers with large sizes like dextran polyaldehyde were explored. When β-mannanase aggregate was cross-linked with 0.01 mL of 20% glutaraldehyde or 0.1 mL of 3% dextran polyaldehyde for 16 h at 4 °C, it was found that the CLEAs prepared with linear dextran polyaldehyde showed a higher activity to macromolecular substrates, which was 16 times higher than that prepared with glutaraldehyde. It could be explained that such CLEAs contained a porous structure with low steric hindrance [83]. CLEAs of lipase were prepared with the *p*-benzoquinone as cross-linker and it was observed that this biocatalyst retained 75.18% of their initial activity. Additionally, after heat treatment for 96 h at 50 °C, the residual activity of CLEAs prepared using *p*-benzoquinone was 5.01-fold higher than that of the CLEAs cross-linked with glutaraldehyde [84]. In all the above-mentioned reports, the reaction mechanism is based on the formation of the covalent bonds between the aldehyde group or ketone of cross-linkers and lysine residues of enzymes to form a Schiff base. Another reaction model was developed with new cross-linkers. As well as lysine amino groups, free carboxyl groups from aspartic and glutamic acids are also potentially reactive and could be used to increase the number of possible cross-linking sites. The free carboxyl groups could be cross-linked with amino-containing polymers to form CLEAs. Several interesting cross-linking agents containing amino groups have been successfully developed. Polyethylenimines was used for the production of a stable *Candida rugose* lipase CLEAs by the cross-linking of carboxyl groups activated with carbodiimide. Compared with the amino-CLEAs of lipase cross-linked with glutaraldehyde, carboxyl-CLEAs cross-linked with polyethylenimines had a higher activity and thermostability [85]. Laccase CLEAs were prepared by cross-linking the carboxyl groups activated with carbodiimide and the amino groups of chitosan to form amide bonds at 20 °C for 4 h. The CLEAs retained 65% of their initial activity while the free laccase was completely denatured after 12 h of thermal denaturation. The three-dimensional structure of laccase molecules was strengthened by the covalent linking enzyme with chitosan, preventing the unfolding of laccase under heat stress conditions [86]. The improved thermos-stability would make the biocatalyst suitable for applications in industrial processes carried out at high temperatures. The enzymes should be stable in the pH range which chitosan could be dissolved when using chitosan as the linker. Otherwise, the enzyme would be inactivated during the cross-linking reaction. The selection of the cross-linking agent should be based on the number of positive or negative amino acids on the enzyme surface and the location of the catalytic center. Amine-rich precipitators (such as polyethylenimines) should be selected if the negative residues (aspartic or glutamic acid) are more abundant than the positive residues on the protein surface. Conversely, cross-linking agents with aldehyde groups (such as glutaraldehyde) or ketone groups (such as *p*-benzoquinone) should be chosen to prepare CLEAs with more positive residues on the surface. Otherwise, macromolecular cross-linking agents should be adopted because small size cross-linking molecules might occupy the reactive center and interrupt the normal enzymatic function when the catalytic center located at the protein surface.

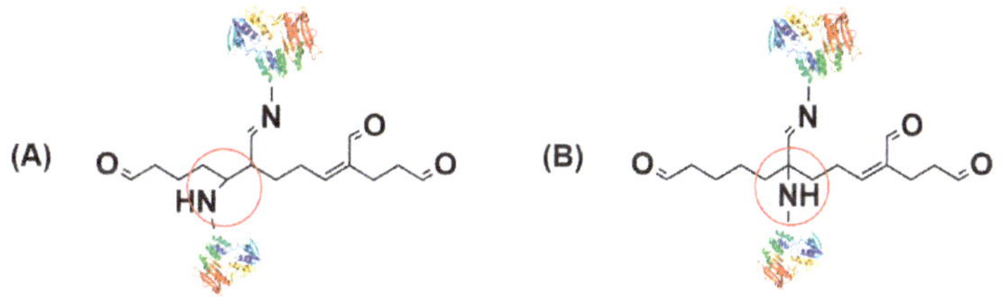

Figure 1. The reaction of polymeric glutaraldehyde with lysine residues protein in alkaline conditions (**A**) and acidic conditions (**B**).

The ratio of cross-linker and protein also should be considered because the cross-linker influences the activity and morphology of the resulting CLEAs. As a result, a suitable quantity of cross-linking agent is essential for CLEA. According to the previous reports, the residual activity of CLEA has a maximum value while changing concentrations of cross-linker [87–90]. Therefore, an appropriate cross-linking agent should be quantitatively used according to the active residues on the enzyme surface. Additionally, a series of proportions of enzyme and cross-linkers should be investigated to obtain the maximum value of enzyme activity retention. As reported in the literature, insufficient bonds were formed at a lower concentration of the crosslinker. On the contrary, a high concentration of cross-linker will be harmful to the CLEA because it can damage the flexibility of the enzyme and result in a change of rigidity. The rigidification of enzymes could prevent the substrate from reaching the active sites and increase the internal mass transfer limitations, consequently decrease the catalytic activity of CLEAs [91–93]. Since enzymes might be susceptible to high concentrations of glutaraldehyde, adding bovine serum albumin (BSA), which owns a lot of amine groups, could avoid the excessive cross-linking of enzymes [84,94]. CLEAs of halohydrin dehalogenase were obtained by slowly adding 0.75% (*v*/*v*) glutaraldehyde for 6 h under stirring. The activity retention of CLEAs (91.2%) was highest when the cross-linking was carried out with 0.75% (*v*/*v*) of glutaraldehyde. When the concentration of glutaraldehyde was increased, the enzyme activity of CLEAs decreased, which might be attributed to mass transfer limitations caused by excessive cross-linking [94].

2.2.4. Effect of Temperature on the Cross-Linking

CLEAs' preparation depends on the accessibility of the cross-linker with protein residues, and the reaction rate also depends on the reaction temperature [95,96]. At lower temperatures, cross-linking reaction with low reaction rate requires longer reaction time while it will finish in a shorter time at a higher temperature. At higher temperatures, the possibility of enzyme irreversible denaturation tends to increase, but at low temperatures, the enzyme is stable and maintain good catalytic activity [97]. There are several reports that the CLEAs preparation carried out in 4 °C could maintain higher specific activity [74,75,98,99]. It is mainly because enzymes undergo partial unfolding and the heat disrupts the intramolecular bonds in the tertiary and quaternary structure [100].

Table 1. The factors for preparing CLEAs.

Enzymes	Factors				References
	Enzymes Proportion	Precipitants	Cross-Linker	Cross-Linking Temperature	
Glucose oxidase and versatile peroxidase	10:7 (mass)	Polyethylene glycol	Glutaraldehyde	30 °C	[71]
Amylosucrase, maltooligosyltrehalose synthase and maltooligosyltrehalose trehalohydrolase	16:1:1 (mass)	Acetone	Glutaraldehyde	4 °C	[72]
Xylanase and mannanase	1:1 (mass)	Acetone	Glutaraldehyde	37 °C	[14]
Lipase, α-amylase and phospholipase A_2	NA	Dimethoxyethane	Glutaraldehyde	4 °C	[76]
Amylase, glucoamylase and pullulanase	3:3:1 (activity)	Ammonium sulfate	Glutaraldehyde	35 °C	[101]
α-L-arabinosidase and β-D-glucosidase	NA	Ammonium sulfate	Glutaraldehyde	4 °C	[102]
X-prolyl-dipeptidyl aminopeptidase and general aminopeptidase N	1:1 (mass)	Ammonium sulfate	Glutaraldehyde	Ice	[103]
Lipase and protease	NA	Ammonium sulfate	Glutaraldehyde	4 °C	[104]
Eductases and glucose dehydrogenase	1.1:5 (activity)	Ammonium sulfate	Oxidized dextran	4 °C	[105]
Glucose oxidase and horseradish peroxidase	150:1 (mass)	Acetonitrile	Glutaraldehyde	NA	[106]
Ketoreductase and D-glucose dehydrogenase	1:1(mass)	1,2-Dimethoxyethane	Glutaraldehyde	20 °C	[107]

NA: Not available.

3. Applications of Combined CLEAs

3.1. Amylosucrase, Maltooligosyltrehalose Synthase, and Maltooligosyltrehalose

Amylosucrase (AS), maltooligosyltrehalose synthase (MTS), and maltooligosyltrehalose trehalohydrolase (MTH) were co-immobilized as combi-CLEAs to one-pot bioconversion of sucrose to trehalose [72] (shown in Figure 2). Physical aggregates obtained by adding 3.6 mL of cold acetone to a 3.6 mL of a mixture solution containing 4 mg of AS, 0.25 mg of MTS, 0.25 mg of MTH, and 9 mg of BSA for 30 min at 4 °C. Then the physical aggregates reacted with 10 mM of glutaraldehyde at 4 °C for 4 h, and the combi-CLEAs were harvested by centrifugation. In this multiple-enzymes catalyzed cascade system, AS is responsible for the bioconversion of sucrose to maltooligosaccharides which are substrates of MTS and transformed into maltooligosyltrehalose. The produced maltooligosyltrehalose was then cleaved by MTH to get trehalose and a shorter maltooligosaccharide. Their experiments showed that combi-CLEA could be used at high substrate concentrations (up to 400 mM). It is found that the activity of combi-CLEAs was well maintained after five cycles. The combi-CLEAs catalyzed multi-step bioconversions into a single reaction system, which brought many advantages such as the low cost of the substrate, handling simplicity, and reusability.

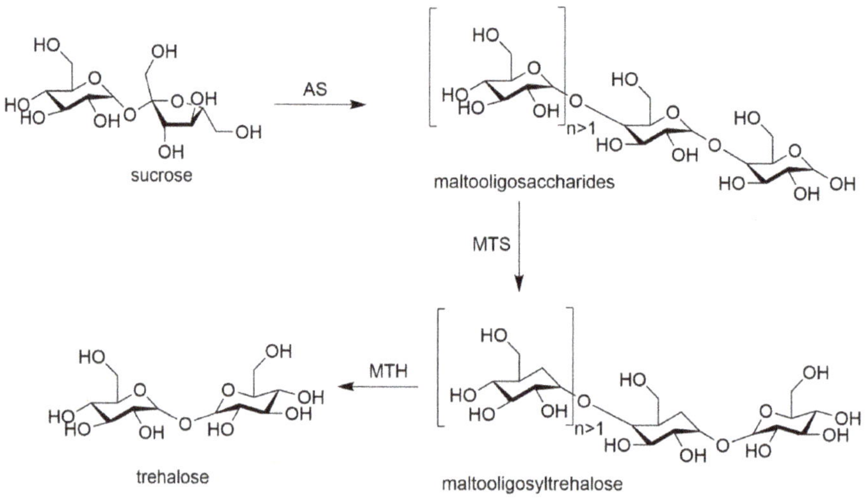

Figure 2. The one-step bioconversion of sucrose to trehalose with combi-CLEAs of AS, MTS, and MTH.

3.2. Hydroxynitrile Lyase and Nitrilase

Combi-CLEAs of hydroxynitrile lyase (HnL) and enantioselective nitrilase (NLase) were prepared to synthesize enantiomerically pure (*S*)-mandelic acid [100]. HnL catalyzed the conversion of aldehyde into the corresponding nitrile which was a substrate of NLase. The enantiomeric excess value of (*S*)-mandelic acid synthetized by the mixture of HnL-CLEAs and NLase-CLEAs was 94%. Combi-CLEA resulted in a further improvement and 98% enantiomerically pure (*S*)-mandelic acid was obtained (Figure 3A). It could be explained that the nitrile intermediate was immediately hydrolyzed in the combi-CLEA particles, which suppressed nitrile diffusion into the water phase and possible racemization. Therefore, it is concluded that combi-CLEAs could improve its stereoselectivity. On the contrary, the chemical process for the production of tmandelic acid includes tedious four-steps. Firstly, a mixture comprising benzaldehyde, trimethylsilyl cyanide, and ZnI_2 was stirred for 24 h, followed by chemical hydrolysis in the presence of concentrated hydrochloric acid for 24 h. Then the mixture was boiled to remove water and hydrochloric acid. Finally, the racemic mandelic acid was extracted and recrystallized from benzene (Figure 3B) [108]. Compared with the chemical method, the combi-CLEAs mediated synthesis of mandelic acid has significant advantages such as having fewer unit operations, a smaller reactor volume and solvent, less waste generation, good stereo-selectivity, and it is less time-consuming [11,102]. Additionally, there is no need to isolate intermediates which brings many potential economic and environmental advantages. Furthermore, in combi-CLEAs catalyzed cascade reactions, the equilibrium of the reaction could be driven to the target product, thereby improving the catalytic yield [103,104].

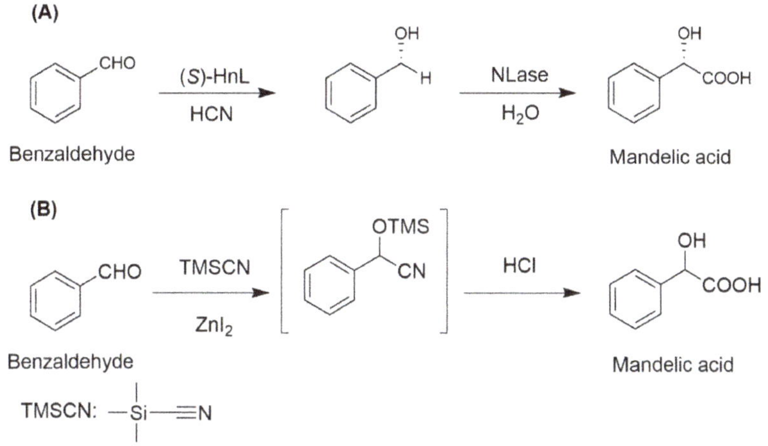

Figure 3. The Biocatalytic (**A**) and Chemical (**B**) synthesis of mandelic acid from benzaldehyde.

3.3. *Amylase, Glucoamylase, and Pullulanase*

Combi-CLEAs of amylase, glucoamylase, and pullulanase were obtained for one-pot hydrolysis of starch (Figure 4) [101]. In this report, amylase (45 U), glucoamylase (45 U), and pullulanase (15 U) were precipitated with 10 mL of saturated ammonium sulfate for 0.5 h at 4 °C and cross-linked with 40 mM of glutaraldehyde for 4.5 h at 30 °C. The biocatalyst was harvested by centrifugation. The starch conversion of 100% was obtained by combi-CLEAs, whereas a 60% and 40% conversion were obtained by using CLEAs mixture and mixed free enzymes, respectively. The observed different starch conversion attributed to a lower thermal stability of free enzymes and enzymes in separated CLEAs at the reaction temperature compared to those in combi-CLEAs. The reason for a higher rate of starch conversion was that increased proximity of enzymes and reduced the diffusion limitation of the substrate from one enzyme to another by using combi-CLEAs. Moreover, combi-CLEAs showed the highest thermal stability at 55 °C and 75 °C. As immobilized enzymes, the combi-CLEAs have a more evident protection from thermal denaturation and require much more energy to break down the active conformation [38,109].

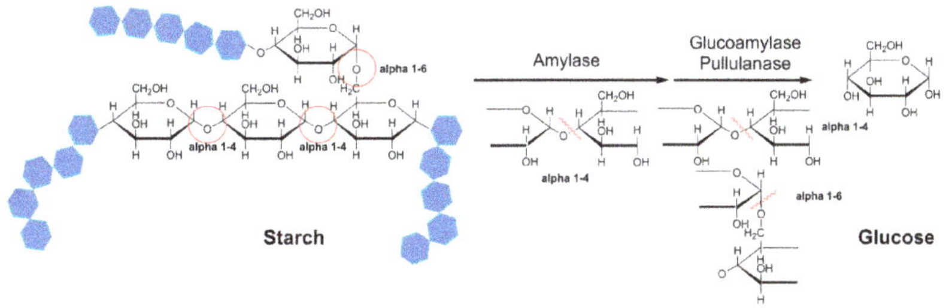

Figure 4. The combi-CLEAs of amylase, glucoamylase, and pullulanase for hydrolyzing of starch.

3.4. L-Arabinosidase and D-Glycosidase

α-L-arabinosidase (ARA) and β-D-glucosidase (βG) were co-immobilized to form combi-CLEAs [102]. Ammonium sulfate (40 mL) was added slowly to the mixture solution (10 mL) containing 20 mg of enzymes and 6.6 mg of BSA under stirring at 4 °C for 30 min. Then 1.06 mg glutaraldehyde was added slowly at 4 °C under stirring for 1 h. The biocatalyst was obtained by centrifugation from suspension. In the hydrolysis of diglycoside, the corresponding sugar and glucoside are released by ARA and then the glucoside is hydrolyzed by βG, liberating the aromatic compound. ARA catalyzes the dissociation of the monoterpenyl β-D-glycoside from its corresponding residual sugar. Then βG catalyzes substrate to release monoterpenes. The released volatile terpenes could enhance wine aroma (Figure 5) [110]. The results showed that the half-lives of βG and ARA in combi-CLEAs were 33.7 and 8.8 times higher than those of the soluble enzymes, respectively. The immobilized biocatalysts were more stable than the soluble enzymes and the higher stability of combi-CLEAs was explained by inter and intramolecular covalent cross-linking [36,111].

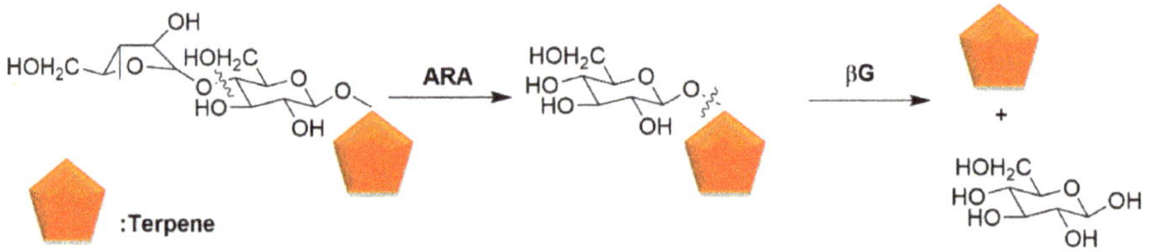

Figure 5. The combi-CLEAs of L-arabinose and β–glucosidase involved in aroma release in wine.

3.5. Aminopeptidase N and X-Prolyl-Dipeptidyl Aminopeptidase

Proline-specific X-prolyl-dipeptidyl aminopeptidase (PepX and aminopeptidase N (PepN) are used in the food industry for cheese-making, baking, and meat tenderization. The enzymatic hydrolysis of proteins can result in a bitter taste. A significant reduction in casein hydrolysate bitterness can be achieved by using PepX combined PepN. A total of 12.5 μg of PepX and 12.5 μg of (PepN) were co-immobilized through precipitation with pre-cooling 4 M ammonium sulfate for 15 min and cross-linked with 50 mM glutaraldehyde. The suspension was centrifuged to separate the combi-CLEAs [103]. The combi-CLEAs were applied to a pre-hydrolyzed casein solution (Figure 6). It could be concluded that, compared with the free PepX and PepN, the relative degree of hydrolysis of the combi-CLEAs of PepX and PepN was increased by approximately 52%. The hydrolysis of food proteins can result in an improved digestibility, modification of sensory quality such as texture or taste, improvement of antioxidant capability or reduction in allergenic compounds. This proved that the combi-CLEAs might have the potential for application in protein hydrolysis.

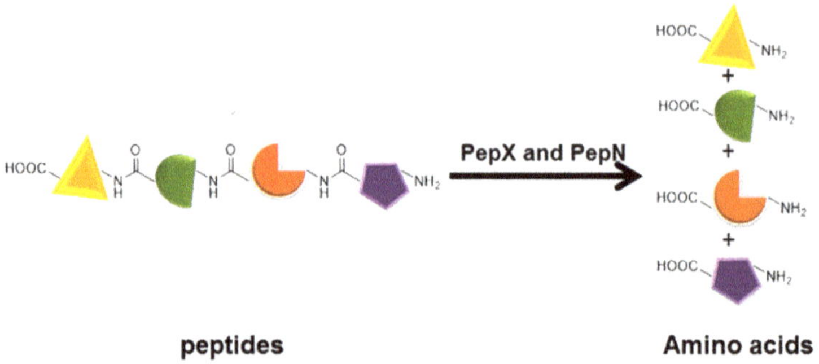

Figure 6. The combi-CLEAs of PepX and PepN.

3.6. Lipase and Protease

Protease and lipase are generally used to hydrolyze proteins and long-chain fatty acid esters, respectively. A novel combi-CLEAs comprising the enzymes lipase and protease from viscera were prepared by Mahmod et al. [104]. The optimum conditions for preparing combi-CLEAs included 65 mM glutaraldehyde, 55% (*w*/*v*) ammonium sulfate, and 0.113 mM BSA. Reactions catalyzed by these combi-CLEAs could facilitate the removal of different kinds of food stains and biodiesel production. The stain removal percentage was improved 67.78% when the combi-CLEAs was added to a commercial detergent. In addition, the combi-CLEAs were used to catalyze biodiesel production from vegetable oil with a percentage conversion of 51.7%. The study had presented that combi-CLEAs with a high activity could expand application in the washing process, as well as in catalyzing biodiesel production from vegetable oil.

3.7. Eductase and Glucose Dehydrogenase

Eductases (ERs) (11 U) and glucose dehydrogenase (GDH) (50 U) were employed to form combi-CLEAs by 1 h of precipitation with 4.0 M ammonium sulfate and 3 h cross-linking with 15% (*v*/*v*) glutaraldehyde or oxidized dextran [105]. Then the combi-CLEAs of ER and GDH were collected by centrifugation. The temperature of the whole process was controlled at 4 °C. In this combi-CLEA catalyzed system, ER reduce the C=C bond by using NADH as the cofactor, and GDH can in situ regenerate NADH and make the reaction efficient (Figure 7). Cofactor regeneration is an important issue for the biochemical or pharmaceutical process, and the stability of the biocatalysts is essential for the biotransformation. In order to confirm the thermal stability of the combi-CLEAs, the immobilized biocatalyst and free enzymes were investigated at 50 °C for 8 h. The ERs in combi-CLEAs could keep 65.2% of its initial activity, whereas free ER and GDH had approximately only maintained 9.2% and 19.4% of their initial activity, respectively. The reasons for the improved thermostability of immobilized enzymes might be the suitable microenvironment and the steric constrained structure created by the immobilization of the enzyme molecules. In addition, the activity of ERs in combi-CLEAs could maintain 110% of their initial activity after 14 cycles. The co-immobilized combi-CLEAs were successfully constructed and the stability of the enzymes was improved, which could be implemented in the cofactor regeneration.

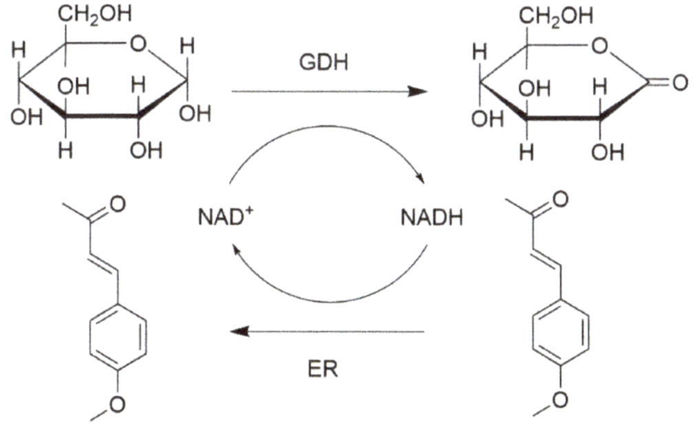

Figure 7. The combi-CLEAs of eductases and glucose dehydrogenase for cofactor regeneration system.

3.8. Peroxidase and Glucose Oxidase

Combi-CLEAs of Versatile peroxidase (VP) and glucose oxidase (GOD) were obtained by Taboada et al. [71]. In their report, the detailed conditions were 0.5 mg of VP, 0.35 mg of GOD, 900 μL of 70% polyethylene glycol, 72 mM glutaraldehyde and cross-linking for 21.5 h at 30 °C. The obtained biocatalysts were collected by centrifugation. The combi-CLEAs provided an integrated system in which GOD oxidized glucose in situ produced hydrogen peroxide required by VP. The activity of VP is easily affected by the concentration of hydrogen peroxide. However, the immobilization could increase the stability of VP against hydrogen peroxide. The combi-CLEAs was more stable than free VP, and their results showed that the residual activities of combi-CLEAs and free VP were 50% and 10% at 20 mM hydrogen peroxide, respectively. The improved stability of the immobilized enzymes could be concluded that the technique of cross-linking enzyme can protect the enzyme structure. In batch experiments, four endocrine disrupting chemicals (bisphenol A, nonylphenol, triclosan, 17α-ethinylestradiol, and the hormone 17β-estradiol) were eliminated by either combi-CLEAs or the free enzymes.

3.9. Glucose Oxidase and Horseradish Peroxidase

Glucose oxidase (GOx) and horseradish peroxidase (HRP) were combined into CLEAs [106]. This preparation was performed in a self-made millifluidic reactor that consisted of an inner capillary (ID 0.3 mm, OD 0.4 mm) and an outer capillary (ID 0.8 mm, OD 1.0 mm). The inner capillary was fixed by a T-shape connector at the center of the outer capillary. The distance from the confluence point to the outlet was 20 mm. The enzyme mixture flowed through inner capillary while acetonitrile containing 0.5 mM of glutaraldehyde was in the outer capillary. When the weight ratio of GOx/HRP was 150 and the flow rate of the inner and outer capillary was 20 μL/min, the combi-CLEAs could retain 96.5% of free enzyme activity. Glucose is the substrate of GOx and the product hydrogen peroxide is the substrate of HRP (Figure 8). The apparent K_m value of combi-CLEA (12.4 ± 0.03 mM) was lower than that of the free enzyme (19.3 ± 0.09 mM). The apparent $V_{\max}$ value of combi-CLEA (15.8 ± 0.05 μM/min) was closed to that of the free enzyme (15.6 ± 0.09 μM/min). Moreover, the catalytic efficiency (k_{cat}/K_m) of combi-CLEA was 1.47 times higher than the free enzyme. These results indicated that combi-CLEA led to a slightly higher reaction rate than free enzymes. The increased reaction rate could be attributed to the in situ fast consumption of hydrogen peroxide by HRP inside the combi-CLEAs.

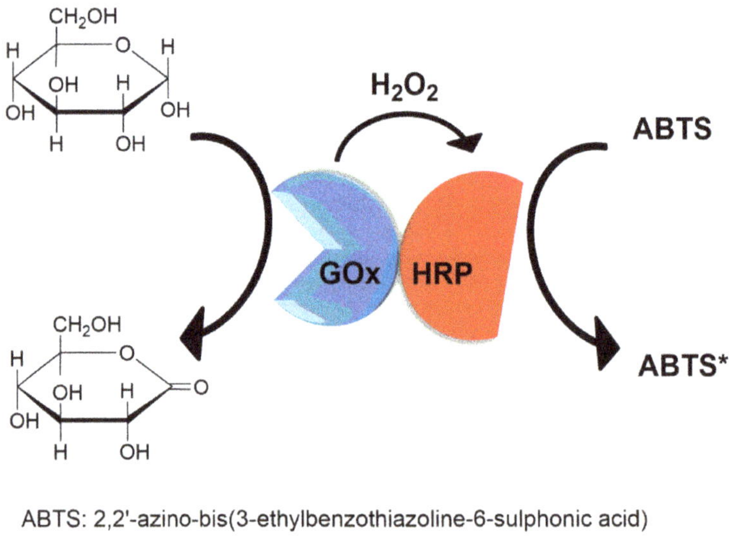

Figure 8. The combi-CLEAs of glucose oxidase and horseradish peroxidase.

3.10. Alcohol Dehydrogenase and Glucose Dehydrogenase

Another example for the cofactor regeneration system is combi-CLEAs of alcohol dehydrogenase (ADH) and glucose dehydrogenase (GDH) [89]. In this work, 90% acetone was added to the crude ADH and GDH solution for 30 min at 4 °C and then cross-linked with 2.5% glutaraldehyde (glutaraldehyde:total protein at 0.5:1) for 1.5 h at ambient temperature. The combi-CLEAs were collected by centrifugation. The obtained immobilized catalysts catalyze the reduction of 1-(2,6-dichloro-3-fluorophenyl) acetophenone and they retained 81.90% (ADH) and 40.29% (GDH) of the initial activity, respectively. The optimum pH was found to be 7 for free enzymes and combi-CLEAs. Meanwhile, combi-CLEAs showed a higher stability under acidic conditions for cofactor regeneration. Since the cofactor recycling will lead to the continuous better acidification of the reaction system, this result especially benefits the reductive coupling reactions of ADH and GDH. This allows the use of these biocatalysts more efficiently in a broader pH region compared to free ADH/GDH (Figure 9).

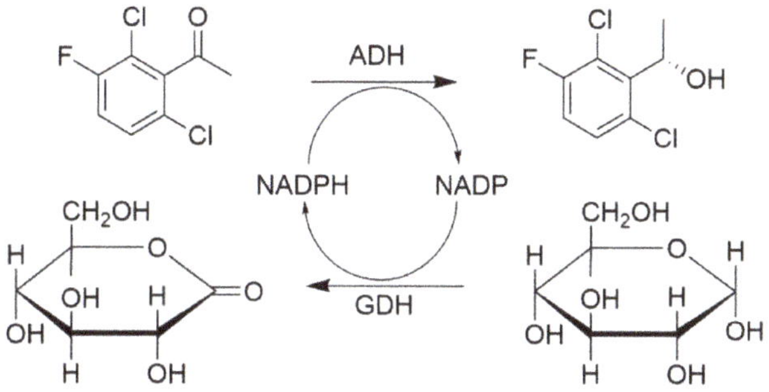

Figure 9. Combi-CLEAs of ADH and GDH for cofactor regeneration system.

3.11. Ketoreductase and Glucose Dehydrogenase

Combi-CLEAs composed of ketoreductase and D-glucose dehydrogenase were prepared according to the work of Ning et al. [107]. They added 1,2-Dimethoxyethane prechilled at 4 °C into the enzyme mixture to a final concentration of 90% (*v*/*v*) at 4 °C for 20 min, and the CLEAs were obtained by cross-linking with 0.2% (*w*/*v*) glutaraldehyde at 20 °C for 1 h. The whole process was under shaking at 200 rpm. These combi-CLEAs had been demonstrated to be a robust regeneration system for pyridine nucleotide cofactor (Figure 10). The results indicated it is an effective cofactor regeneration system and the optimal substrate concentration in a biphasic system of combi-CLEAs was 300 mM. The improvement of affinity for substrate suggesting that combi-CLEAs had a positive effect on volumetric productivity.

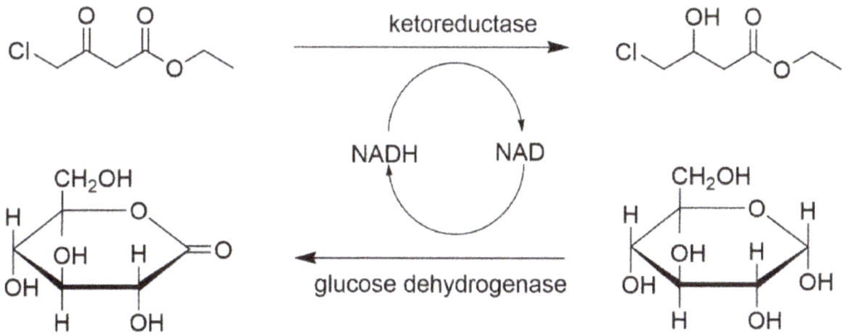

Figure 10. The combi-CLEAs of ketoreductase and D-glucose dehydrogenase for the cofactor regeneration system.

4. Conclusions and Prospects

Combi-CLEA includes multiple enzymes and it is a promising immobilization method without carriers that combined purification and immobilization into one step and do not require highly pure enzymes. However, enzymes from different sources require the optimization of the precipitation and cross-linking parameters. Although the precipitant and cross-linker are the priorities for the preparation of combi-CLEA, all the parameters should be considered carefully because of the interactions among them. Generally, combi-CLEAs can improve enzyme stability and apparent catalytic rate. These will definitely benefit the biotransformation in industrial applications, especially in the cofactor regeneration systems. When combi-CLEAs are used for cofactor regeneration, some advantages such as a high selectivity, high efficiency, and low diffusional limitation are obvious. Due to most oxidoreductases require stoichiometric amounts of expensive cofactors such as NAD^+, the Combi-CLEAs of oxidoreductases or can ensure the reuse of cofactors to reduce the cost. It is a robust regeneration system which could be prepared in a simple, rapid, and economical way.

With the development of biomolecular engineering, tailor-made enzymes with specific modifications are produced in a bench scale. Chemical or biological modifications of enzyme surface residues with the specific incorporation of unnatural or functional amino acids will make the preparation of combi-CLEA easier. Some unnatural amino acids with azide (e.g., L-azidohomoalanine, p-azidophenylalanine, and 5-azidopentanoic acid) and alkyne (e.g., L-homopropargylglycine) could be used for bioorthogonal reactions with "click" linking, so the cross-linking reaction could be achieved easily by coupling both the residue-specific and site-specific incorporation of unnatural amino acids into a single protein or different proteins. As a consequence, these enzymes could be conjugated directly to form combi-CLEAs without a cross-linker. If so, combi-CLEAs will be applied to a wide field of cascade processes of unprecedented complexity, efficiency, and elegance.

Author Contributions: M.-Q.X., and Y.-W.Z. prepared the manuscript. M.-Q.X., S.-S.W., and L.-N.L. collected the literature. J.G. and Y.-W.Z. aided in writing and revised the manuscript.

Funding: This research was funded by Guangxi Natural Science Foundation, grant number 2016GXNSFCA380011. The APC was funded by "Seagull Plan" from Qinzhou city.

Conflicts of Interest: All authors declare no conflict of interest.

References

1. Wen, Y.; Xu, L.; Chen, F.; Gao, J.; Li, J.; Hu, L.; Li, J. Discovery of a novel inhibitor of NAD(P)$^+$-dependent malic enzyme (ME2) by high-throughput screening. *Acta Pharmacol. Sin.* **2014**, *35*, 674–684. [CrossRef] [PubMed]
2. Feng, W.; Zhao, T.; Zhou, Y.; Li, F.; Zou, Y.; Bai, S.; Wang, W.; Yang, L.; Wu, X. Optimization of enzyme-assisted extraction and characterization of collagen from Chinese sturgeon (Acipenser sturio Linnaeus) skin. *Pharmacogn. Mag.* **2013**, *9*, 32–37. [CrossRef]
3. Yang, H.; Shen, Y.; Xu, Y.; Maqueda, A.S.; Zheng, J.; Wu, Q.; Tam, J.P. A novel strategy for the discrimination of gelatinous Chinese medicines based on enzymatic digestion followed by nano-flow liquid chromatography in tandem with orbitrap mass spectrum detection. *Int. J. Nanomed.* **2015**, *10*, 4947–4955. [CrossRef] [PubMed]
4. Magne, V.; Amounas, M.; Innocent, C.; Dejean, E.; Seta, P. Enzyme textile for removal of urea with coupling process: Enzymatic reaction and electrodialysis. *Desalination* **2018**, *144*, 163–166. [CrossRef]
5. Liu, L.; Zhang, R.; Deng, Y.; Zhang, Y.; Xiao, J.; Huang, F.; Wei, W.; Zhang, M. Fermentation and complex enzyme hydrolysis enhance total phenolics and antioxidant activity of aqueous solution from rice bran pretreated by steaming with α-amylase. *Food Chem.* **2017**, *221*, 636–643. [CrossRef] [PubMed]
6. Shah, S.; Agera, R.; Sharma, P.; Sunder, A.V.; Bajwa, H.; James, H.M.; Gaikaiwari, R.P.; Wangikar, P.P. Development of biotransformation process for asymmetric reduction with novel anti-Prelog NADH-dependent alcohol dehydrogenases. *Process Biochem.* **2018**. [CrossRef]
7. Zhang, L.; Singh, R.; Sivakumar, D.; Guo, Z.; Li, J.; Chen, F.; He, Y.; Guan, X.; Kang, Y.C.; Lee, J.K. An artificial synthetic pathway for acetoin, 2,3-butanediol, and 2-butanol production from ethanol using cell free multi-enzyme catalysis. *Green Chem.* **2017**, *221*, 636–643. [CrossRef]

8. Touahar, I.E.; Haroune, L.; Ba, S.; Bellenger, J.-P.; Cabana, H. Characterization of combined cross-linked enzyme aggregates from laccase, versatile peroxidase and glucose oxidase, and their utilization for the elimination of pharmaceuticals. *Sci. Total Environ.* **2014**, *481*, 90–99. [CrossRef] [PubMed]
9. Lloret, L.; Eibes, G.; Feijoo, G.; Moreira, M.T.; Lema, J.M. Degradation of estrogens by laccase from Myceliophthora thermophila in fed-batch and enzymatic membrane reactors. *J. Hazard. Mater.* **2012**, 175–183. [CrossRef] [PubMed]
10. Le-Clech, P.; Chen, V.; Fane, T.A.G. Fouling in membrane bioreactors used in wastewater treatment. *J. Membr. Sci.* **2006**, *284*, 17–53. [CrossRef]
11. Sheldon, R.A. E factors, green chemistry and catalysis: An odyssey. *Chem. Commun.* **2008**, *39*, 3352–3365. [CrossRef] [PubMed]
12. Zhuang, M.Y.; Jiang, X.P.; Ling, X.M.; Xu, M.Q.; Zhu, Y.H.; Zhang, Y.W. Immobilization of glycerol dehydrogenase and NADH oxidase for enzymatic synthesis of 1,3-dihydroxyacetone with in situ cofactor regeneration. *J. Chem. Technol. Biotechnol.* **2018**, *93*, 2351–2358. [CrossRef]
13. Shi, Y.; Liu, W.; Tao, Q.L.; Jiang, X.P.; Liu, C.H.; Zeng, S.; Zhang, Y.W. Immobilization of Lipase by Adsorption Onto Magnetic Nanoparticles in Organic Solvents. *J. Nanosci. Nanotechnol.* **2016**, *16*, 601–607. [CrossRef] [PubMed]
14. Bhattacharya, A.; Pletschke, B.I. Strategic optimization of xylanase–mannanase combi-CLEAs for synergistic and efficient hydrolysis of complex lignocellulosic substrates. *J. Mol. Catal. B Enzym.* **2015**, *115*, 140–150. [CrossRef]
15. Stepankova, V.; Bidmanova, S.; Koudelakova, T.; Prokop, Z.; Chaloupkova, R.; Damborsky, J. Strategies for Stabilization of Enzymes in Organic Solvents. *ACS Catal.* **2013**, *3*, 2823–2836. [CrossRef]
16. Guzik, U.; Hupertkocurek, K.; Wojcieszyńska, D. Immobilization as a strategy for improving enzyme properties-application to oxidoreductases. *Molecules* **2014**, *19*, 8995–9018. [CrossRef] [PubMed]
17. Hwang, E.T.; Gu, M.B. Enzyme stabilization by nano/microsized hybrid materials. *Eng. Life Sci.* **2013**, *13*, 49–61. [CrossRef]
18. Wang, F.; Chen, L.; Zhang, D.; Jiang, S.; Shi, K.; Huang, Y.; Li, R.; Xu, Q. Methazolamide-loaded solid lipid nanoparticles modified with low-molecular weight chitosan for the treatment of glaucoma: Vitro and vivo study. *J. Drug Target.* **2014**, *22*, 849–858. [CrossRef] [PubMed]
19. Chen, Y.; Yuan, L.; Zhou, L.; Zhang, Z.; Cao, W.; Wu, Q. Effect of cell-penetrating peptide-coated nanostructured lipid carriers on the oral absorption of tripterine. *Int. J. Nanomed.* **2012**, *7*, 4581–4591. [CrossRef]
20. Wang, M.; Qi, W.; Su, R.; He, Z. Advances in carrier-bound and carrier-free immobilized nanobiocatalysts. *Chem. Eng. Sci.* **2015**, *135*, 21–32. [CrossRef]
21. Xie, M.; Xu, Y.; Shen, H.; Shen, S.; Ge, Y.; Xie, J. Negative-charge-functionalized mesoporous silica nanoparticles as drug vehicles targeting hepatocellular carcinoma. *Int. J. Pharm.* **2014**, *474*, 223–231. [CrossRef] [PubMed]
22. Liu, H.; Shi, S.; Cao, J.; Ji, L.; He, Y.; Xi, J. Preparation and evaluation of a novel bioactive glass/lysozyme/PLGA composite microsphere. *Drug Dev. Ind. Pharm.* **2015**, *41*, 458–463. [CrossRef] [PubMed]
23. Shi, F.; Zhao, Y.; Firempong, C.K.; Xu, X. Preparation, characterization and pharmacokinetic studies of linalool-loaded nanostructured lipid carriers. *Pharm. Biol.* **2016**, *54*, 2320–2328. [CrossRef] [PubMed]
24. Wang, G.; Wang, J.-J.; Chen, X.-L.; Du, L.; Li, F. Quercetin-loaded freeze-dried nanomicelles: Improving absorption and anti-glioma efficiency in vitro and in vivo. *J. Control. Release* **2016**, *235*, 276–290. [CrossRef] [PubMed]
25. Zhu, T.; Tao, Z.; Jia, L.; Luo, Y.-F.; Xu, J.; Chen, R.-H.; Ge, Z.-J.; Ma, T.-L.; Chen, H. Multifunctional nanocomposite based on halloysite nanotubes for efficient luminescent bioimaging and magnetic resonance imaging. *Int. J. Nanomed.* **2016**, *11*, 4765–4776. [CrossRef] [PubMed]
26. Rui, M.; Xin, Y.; Li, R.; Ge, Y.; Feng, C.; Xu, X. Targeted Biomimetic Nanoparticles for Synergistic Combination Chemotherapy of Paclitaxel and Doxorubicin. *Mol. Pharm.* **2017**, *14*, 107–123. [CrossRef] [PubMed]
27. Zhang, H.; Firempong, C.K.; Wang, Y.; Xu, W.; Wang, M.; Cao, X.; Zhu, Y.; Tong, S.; Yu, J.; Xu, X. Ergosterol-loaded poly(lactide-*co*-glycolide) nanoparticles with enhanced in vitro antitumor activity and oral bioavailability. *Acta Pharmacol. Sin.* **2017**, *37*, 834–844. [CrossRef] [PubMed]

28. Wu, Y.-S.; Ngai, S.-C.; Goh, B.-H.; Chan, K.-G.; Lee, L.-H.; Chuah, L.-H. Anticancer Activities of Surfactin and Potential Application of Nanotechnology Assisted Surfactin Delivery. *Front. Pharmacol.* **2017**, *8*. [CrossRef] [PubMed]
29. Cao, X.; Deng, W.W.; Fu, M.; Wang, L.; Tong, S.S.; Wei, Y.W.; Xu, Y.; Su, W.Y.; Xu, X.M.; Yu, J.N. In vitro release and *in vitro–in vivo* correlation for silybin meglumine incorporated into hollow-type mesoporous silica nanoparticles. *Int. J. Nanomed.* **2012**, *753*. [CrossRef]
30. Peng, W.; Jiang, X.; Zhu, Y.; Omari-Siaw, E.; Deng, W.; Yu, J.; Xu, X.; Zhang, W. Oral delivery of capsaicin using MPEG-PCL nanoparticles. *Acta Pharmacol. Sin.* **2015**, *36*, 139–148. [CrossRef] [PubMed]
31. Shen, S.; Wu, L.; Liu, J.; Xie, M.; Shen, H.; Qi, X.; Yan, Y.; Ge, Y.; Jin, Y. Core–shell structured Fe_3O_4@TiO_2-doxorubicin nanoparticles for targeted chemo-sonodynamic therapy of cancer. *Int. J. Pharm.* **2015**, *486*, 380–388. [CrossRef] [PubMed]
32. Liu, W.; Zhou, F.; Zhang, X.-Y.; Li, Y.; Wang, X.-Y.; Xu, X.-M.; Zhang, Y.-W. Preparation of Magnetic Fe_3O_4@SiO_2 Nanoparticles for Immobilization of Lipase. *J. Nanosci. Nanotechnol.* **2014**, *14*, 3068–3072. [CrossRef] [PubMed]
33. Li, Y.; Wang, X.-Y.; Jiang, X.-P.; Ye, J.-J.; Zhang, Y.-W.; Zhang, X.-Y. Fabrication of graphene oxide decorated with Fe_3O_4@SiO_2 for immobilization of cellulase. *J. Nanopart. Res.* **2015**, *17*, 8. [CrossRef]
34. Wang, X.-Y.; Jiang, X.-P.; Li, Y.; Zeng, S.; Zhang, Y.-W. Preparation Fe_3O_4@chitosan magnetic particles for covalent immobilization of lipase from *Thermomyces lanuginosus*. *Int. J. Biol. Macromol.* **2015**, *75*, 44–50. [CrossRef] [PubMed]
35. Ling, X.-M.; Wang, X.-Y.; Ma, P.; Yang, Y.; Qin, J.-M.; Zhang, X.-J.; Zhang, Y.-W. Covalent Immobilization of Penicillin G Acylase onto Fe_3O_4@Chitosan Magnetic Nanoparticles. *J. Microbiol. Biotechnol.* **2016**, *26*, 829–836. [CrossRef] [PubMed]
36. Tao, Q.-L.; Li, Y.; Shi, Y.; Liu, R.-J.; Zhang, Y.-W.; Guo, J. Application of Molecular Imprinted Magnetic Fe_3O_4@SiO_2 Nanoparticles for Selective Immobilization of Cellulase. *J. Nanosci. Nanotechnol.* **2016**, *16*, 6055–6060. [CrossRef] [PubMed]
37. Liu, C.H.; Li, X.Q.; Jiang, X.P.; Zhuang, M.Y.; Zhang, J.X.; Bao, C.H.; Zhang, Y.W. Preparation of Functionalized Graphene Oxide Nanocomposites for Covalent Immobilization of NADH Oxidase. *Nanosci. Nanotechnol. Lett.* **2016**, *8*, 164–167. [CrossRef]
38. Gao, J.; Lu, C.-L.; Wang, Y.; Wang, S.-S.; Shen, J.-J.; Zhang, J.-X.; Zhang, Y.-W. Rapid Immobilization of Cellulase onto Graphene Oxide with a Hydrophobic Spacer. *Catalysts* **2018**, *8*, 180. [CrossRef]
39. Jiao, Z.; Chen, Y.; Wan, Y.; Zhang, H. Anticancer efficacy enhancement and attenuation of side effects of doxorubicin with titanium dioxide nanoparticles. *Int. J. Nanomed.* **2011**, *6*, 2321–2326. [CrossRef] [PubMed]
40. Xiong, F.; Hu, K.; Yu, H.; Zhou, L.; Song, L.; Zhang, Y.; Shan, X.; Liu, J.; Gu, N. A Functional Iron Oxide Nanoparticles Modified with PLA-PEG-DG as Tumor-Targeted MRI Contrast Agent. *Pharm. Res.* **2017**, *34*, 1683–1692. [CrossRef] [PubMed]
41. Yuan, L.; Geng, L.; Ge, L.; Yu, P.; Duan, X.; Chen, J.; Chang, Y. Effect of iron liposomes on anemia of inflammation. *Int. J. Pharm.* **2013**, *454*, 82–89. [CrossRef] [PubMed]
42. Du, F.; Lou, J.; Jiang, R.; Fang, Z.; Zhao, X.; Niu, Y.; Zou, S.; Zhang, M.; Gong, A.; Wu, C. Hyaluronic acid-functionalized bismuth oxide nanoparticles for computed tomography imaging-guided radiotherapy of tumor. *Int. J. Nanomed.* **2017**, *12*, 5973–5992. [CrossRef] [PubMed]
43. Ali, S.S.; Morsy, R.; Elzawawy, N.A.; Fareed, M.F.; Bedaiwy, M.Y. Synthesized zinc peroxide nanoparticles (ZnO_2-NPs): A novel antimicrobial, anti-elastase, anti-keratinase, and anti-inflammatory approach toward polymicrobial burn wounds. *Int. J. Nanomed.* **2017**, *12*, 6059–6073. [CrossRef] [PubMed]
44. Chen, Y.; Yuan, L.; Congyan, L.; Zhang, Z.; Zhou, L.; Qu, D. Antitumor activity of tripterine via cell-penetrating peptide-coated nanostructured lipid carriers in a prostate cancer model. *Int. J. Nanomed.* **2013**, 4339–4350. [CrossRef] [PubMed]
45. Zhang, Z.; Huang, J.; Jiang, S.; Liu, Z.; Gu, W.; Yu, H.; Li, Y. Porous starch based self-assembled nano-delivery system improves the oral absorption of lipophilic drug. *Int. J. Pharm.* **2013**, *444*, 162–168. [CrossRef] [PubMed]
46. Rui, M.; Qu, Y.; Gao, T.; Ge, Y.; Feng, C.; Xu, X. Simultaneous delivery of anti-miR21 with doxorubicin prodrug by mimetic lipoprotein nanoparticles for synergistic effect against drug resistance in cancer cells. *Int. J. Nanomed.* **2017**, *12*, 217–237. [CrossRef] [PubMed]

47. Li, H.; Dong, W.F.; Zhou, J.Y.; Xu, X.M.; Li, F.Q. Triggering effect of N-acetylglucosamine on retarded drug release from a lectin-anchored chitosan nanoparticles-in-microparticles system. *Int. J. Pharm.* **2013**, *449*, 37–43. [CrossRef] [PubMed]
48. Cao, L.; Lv, L.; Sheldon, R.A. Immobilised enzymes: Carrier-bound or carrier-free? *Curr. Opin. Biotechnol.* **2003**, *14*, 387–394. [CrossRef]
49. Tischer, W.; Kasche, V. Immobilized enzymes: Crystals or carriers? *Trends Biotechnol.* **1999**, *17*, 326–335. [CrossRef]
50. Jegan Roy, J.; Emilia Abraham, T. Strategies in Making Cross-Linked Enzyme Crystals. *Chem. Rev.* **2004**, *104*, 3705–3722. [CrossRef] [PubMed]
51. Quiocho, F.A.; Richards, F.M. Intermolecular cross linking of a protein in the crystallinestate: Carboxypeptidasw-A. *Proc. Natl. Acad. Sci. USA* **1964**, *52*, 833–839. [CrossRef] [PubMed]
52. Kwon, J.S.; Nayhouse, M.; Christofides, P.D.; Orkoulas, G. Protein Crystal Shape and Size Control in Batch Crystallization: Comparing Model Predictive Control with Conventional Operating Policies. *Ind. Eng. Chem. Res.* **2014**, *53*, 5002–5014. [CrossRef]
53. Liu, J.J.; Ma, C.Y.; Hu, Y.D.; Wang, X.Z. Effect of seed loading and cooling rate on crystal size and shape distributions in protein crystallization—A study using morphological population balance simulation. *Comput. Chem. Eng.* **2010**, *34*, 1945–1952. [CrossRef]
54. Velascolozano, S.; Lópezgallego, F.; Mateosdíaz, J.C.; Favelatorres, E. Cross-linked enzyme aggregates (CLEA) in enzymeimprovement—A review. *Biocatalysis* **2016**, *1*, 166–177. [CrossRef]
55. Clair, N.L.S.; Navia, M.A. Cross-linked enzyme crystals as robust biocatalysts. *J. Am. Chem. Soc.* **1992**, *114*, 7314–7316.
56. Zelinski, T.; Waldmann, H. Cross-Linked Enzyme Crystals(CLECs): Efficient and Stable Biocatalysts for Preparative Organic Chemistry. *Angew. Chem. Int. Ed. Engl.* **1997**, *36*, 722–724. [CrossRef]
57. Roessl, U.; Nahálka, J.; Nidetzky, B. Carrier-free immobilized enzymes for biocatalysis. *Biotechnol. Lett.* **2010**, *32*, 341–350. [CrossRef] [PubMed]
58. Liu, Y.; Guo, Y.L.; Chen, D.W.; Peng, C.; Yan, Y.J. Conformation and Activity of Sol-Gels Encapsulated Cross-Linked Enzyme Aggregates of Lipase from *Burkholderia cepacia*. *Adv. Mater. Res.* **2011**, *291–294*, 614–620. [CrossRef]
59. Goetze, D.; Foletto, E.F.; da Silva, H.B.; Silveira, V.C.C.; Dal Magro, L.; Rodrigues, R.C. Effect of feather meal as proteic feeder on combi-CLEAs preparation for grape juice clarification. *Process Biochem.* **2017**, *62*, 122–127. [CrossRef]
60. Min, H.K.; Park, S.; Yong, H.K.; Won, K.; Sang, H.L. Immobilization of formate dehydrogenase from *Candida boidinii* through cross-linked enzyme aggregates. *J. Mol. Catal. B Enzym.* **2013**, *97*, 209–214. [CrossRef]
61. Matijošytė, I.; Arends, I.W.C.E.; Vries, S.D.; Sheldon, R.A. Preparation and use of cross-linked enzyme aggregates (CLEAs) of laccases. *J. Mol. Catal. B Enzym.* **2010**, *62*, 142–148. [CrossRef]
62. Sangeetha, K.; Abraham, T.E. Preparation and characterization of cross-linked enzyme aggregates (CLEA) of Subtilisin for controlled release applications. *Int. J. Biol. Macromol.* **2008**, *43*, 314–319. [CrossRef] [PubMed]
63. Cerdobbel, A.; De, W.K.; Desmet, T.; Soetaert, W. Sucrose phosphorylase as cross-linked enzyme aggregate: Improved thermal stability for industrial applications. *Biotechnol. J.* **2010**, *5*, 1192–1197. [CrossRef] [PubMed]
64. Yu, C.Y.; Li, X.F.; Lou, W.Y.; Zong, M.H. Cross-linked enzyme aggregates of Mung bean epoxide hydrolases: A highly active, stable and recyclable biocatalyst for asymmetric hydrolysis of epoxides. *J. Biotechnol.* **2013**, *166*, 12–19. [CrossRef] [PubMed]
65. Lanfranchi, E.; Grill, B.; Raghoebar, Z.; Pelt, S.V.; Sheldon, R.; Steiner, K.; Glieder, A.; Winkler, M. Production of hydroxynitrile lyase from *D. tyermanii* (DtHNL) in Komagataella phaffii and its immobilization as CLEA to generate a robust biocatalyst. *Chembiochem* **2017**, *19*, 312–316. [CrossRef] [PubMed]
66. Mafra, A.C.O.; Kopp, W.; Beltrame, M.B.; Giordano, R.D.L.C.; Tardioli, P.W. Diffusion effects of bovine serum albumin on cross-linked aggregates of catalase. *J. Mol. Catal. B Enzym.* **2016**, *133*, 107–116. [CrossRef]
67. Talekar, S.; Joshi, A.; Joshi, G.; Kamat, P.; Haripurkar, R.; Kambale, S. Parameters in preparation and characterization of cross linked enzyme aggregates (CLEAs). *RSC Adv.* **2013**, *3*, 12485–12511. [CrossRef]
68. Periyasamy, K.; Santhalembi, L.; Mortha, G.; Aurousseau, M.; Subramanian, S. Carrier-free co-immobilization of xylanase, cellulase and β-1,3-glucanase as combined cross-linked enzyme aggregates (combi-CLEAs) for one-pot saccharification of sugarcane bagasse. *RSC Adv.* **2016**, *6*, 32849–32857. [CrossRef]

69. Hanefeld, U.; Gardossi, L.; Magner, E. Understanding enzyme immobilisation. *Chem. Soc. Rev.* **2009**, *38*, 453–468. [CrossRef] [PubMed]
70. Chen, Y.; Jiang, Q.; Sun, L.; Li, Q.; Zhou, L.; Chen, Q.; Li, S.; Yu, M.; Li, W. Magnetic Combined Cross-Linked Enzyme Aggregates of Ketoreductase and Alcohol Dehydrogenase: An Efficient and Stable Biocatalyst for Asymmetric Synthesis of (R)-3-Quinuclidinol with Regeneration of Coenzymes *In Situ*. *Catalysts* **2018**, *8*, 334. [CrossRef]
71. Taboada-Puig, R.; Junghanns, C.; Demarche, P.; Moreira, M.T.; Feijoo, G.; Lema, J.M.; Agathos, S.N. Combined cross-linked enzyme aggregates from versatile peroxidase and glucose oxidase: Production, partial characterization and application for the elimination of endocrine disruptors. *Bioresour. Technol.* **2011**, *102*, 6593–6599. [CrossRef] [PubMed]
72. Jung, D.-H.; Jung, J.-H.; Seo, D.-H.; Ha, S.-J.; Kweon, D.-K.; Park, C.-S. One-pot bioconversion of sucrose to trehalose using enzymatic sequential reactions in combined cross-linked enzyme aggregates. *Bioresour. Technol.* **2013**, *130*, 801–804. [CrossRef] [PubMed]
73. Jian, D.C.; Shi, R.J. Optimization protocols and improved strategies of cross-linked enzyme aggregates technology: Current development and future challenges. *Crit. Rev. Biotechnol.* **2015**, *35*, 15–28. [CrossRef]
74. Bilal, M.; Asgher, M.; Iqbal, H.M.N.; Hu, H.; Zhang, X. Bio-based degradation of emerging endocrine-disrupting and dye-based pollutants using cross-linked enzyme aggregates. *Environ. Sci. Pollut. Res.* **2017**, *24*, 7035–7041. [CrossRef] [PubMed]
75. Tandjaoui, N.; Tassist, A.; Abouseoud, M.; Couvert, A.; Amrane, A. Preparation and characterization of cross-linked enzyme aggregates (CLEAs) of *Brassica rapa* peroxidase. *Biocatal. Agric. Biotechnol.* **2015**, *4*, 208–213. [CrossRef]
76. Dalal, S.; Kapoor, M.; Gupta, M.N. Preparation and characterization of combi-CLEAs catalyzing multiple non-cascade reactions. *J. Mol. Catal. B Enzym.* **2007**, *44*, 128–132. [CrossRef]
77. Andreazza, R.; Pieniz, S.; Okeke, B.; Camargo, F.A. Evaluation of copper resistant bacteria from vineyard soils and mining waste for copper biosorption. *Braz. J. Microbiol.* **2011**, *42*, 66–74. [CrossRef] [PubMed]
78. Pchelintsev, N.A.; Youshko, M.I.; Švedas, V.K. Quantitative characteristic of the catalytic properties and microstructure of cross-linked enzyme aggregates of penicillin acylase. *J. Mol. Catal. B Enzym.* **2009**, *56*, 202–207. [CrossRef]
79. Richards, F.M.; Knowles, J.R. Glutaraldehyde as a protein cross-linkage reagent. *J. Mol. Biol.* **1968**, *37*, 231–233. [CrossRef]
80. Dartiguenave, M.C.; Bertrand, M.J.; Waldron, K.C. Glutaraldehyde: Behavior in aqueous solution, reaction with proteins, and application to enzyme crosslinking. *Biotechniques* **2004**, *37*, 790–802. [CrossRef]
81. Wine, Y.; Cohenhadar, N.; Freeman, A.; Frolow, F. Elucidation of the mechanism and end products of glutaraldehyde crosslinking reaction by X-ray structure analysis. *Biotechnol. Bioeng.* **2010**, *98*, 711–718. [CrossRef] [PubMed]
82. Mateo, C.; Palomo, J.M.; Van Langen, L.M.; Van Rantwijk, F.; Sheldon, R.A. A new, mild cross-linking methodology to prepare cross-linked enzyme aggregates. *Biotechnol. Bioeng.* **2004**, *86*, 273–276. [CrossRef] [PubMed]
83. Zhen, Q.; Wang, M.; Qi, W.; Su, R.; He, Z. Preparation of β-mannanase CLEAs using macromolecular cross-linkers. *Catal. Sci. Technol.* **2013**, *3*, 1937–1941. [CrossRef]
84. Wang, A.; Zhang, F.; Chen, F.; Wang, M.; Li, H.; Zeng, Z.; Xie, T.; Chen, Z. A facile technique to prepare cross-linked enzyme aggregates using *p*-benzoquinone as cross-linking agent. *Korean J. Chem. Eng.* **2011**, *28*, 1090–1095. [CrossRef]
85. Velasco-Lozano, S.; López-Gallego, F.; Vázquez-Duhalt, R.; Mateos-Díaz, J.C.; Guisán, J.M.; Favela-Torres, E. Carrier-free immobilization of lipase from *Candida rugosa* with polyethyleneimines by carboxyl-activated cross-linking. *Biomacromolecules* **2014**, *15*, 1896–1903. [CrossRef] [PubMed]
86. Cabana, H.; Ahamed, A.; Leduc, R. Conjugation of laccase from the white rot fungus *Trametes versicolor* to chitosan and its utilization for the elimination of triclosan. *Bioresour. Technol.* **2011**, *102*, 1656–1662. [CrossRef] [PubMed]
87. Kunjukunju, S.; Roy, A.; Shekhar, S.; Kumta, P.N. Cross-linked enzyme aggregates of alginate lyase: A systematic engineered approach to controlled degradation of alginate hydrogel. *Int. J. Biol. Macromol.* **2018**, *115*, 176–184. [CrossRef] [PubMed]

88. Mehde, A.A.; Mehdi, W.A.; Özacar, M.; Özacar, Z.Z. Evaluation of different saccharides and chitin as eco-friendly additive to improve the magnetic cross-linked enzyme aggregates (CLEAs) activities. *Int. J. Biol. Macromol.* **2018**. [CrossRef] [PubMed]
89. Hu, X.; Liu, L.; Chen, D.; Wang, Y.; Zhang, J.; Shao, L. Co-expression of the recombined alcohol dehydrogenase and glucose dehydrogenase and cross-linked enzyme aggregates stabilization. *Bioresour. Technol.* **2016**, *224*, 531–535. [CrossRef] [PubMed]
90. Torabizadeh, H.; Tavakoli, M.; Safari, M. Immobilization of thermostable α-amylase from Bacillus licheniformis by cross-linked enzyme aggregates method using calcium and sodium ions as additives. *J. Mol. Catal. B Enzym.* **2014**, *108*, 13–20. [CrossRef]
91. Barbosa, O.; Ortiz, C.; Berenguer-Murcia, A.; Torres, R.; Rodrigues, R.C.; Fernandez-Lafuente, R. ChemInform Abstract: Glutaraldehyde in Bio-Catalysts Design: A Useful Crosslinker and a Versatile Tool in Enzyme Immobilization. *Cheminform* **2013**, *4*, 1583–1600. [CrossRef]
92. Reshmi, R.; Sugunan, S. Improved biochemical characteristics of crosslinked β-glucosidase on nanoporous silica foams. *J. Mol. Catal. B Enzym.* **2013**, *85*, 111–118. [CrossRef]
93. Šulek, F.; Fernández, D.P.; Knez, Ž; Habulin, M.; Sheldon, R.A. Immobilization of horseradish peroxidase as crosslinked enzyme aggregates (CLEAs). *Process Biochem.* **2011**, *46*, 765–769. [CrossRef]
94. Liao, Q.; Du, X.; Jiang, W.; Tong, Y.; Zhao, Z.; Fang, R.; Feng, J.; Tang, L. Cross-linked enzyme aggregates (CLEAs) of halohydrin dehalogenase from *Agrobacterium radiobacter* AD1: Preparation, characterization and application as a biocatalyst. *J. Biotechnol.* **2018**, *272–273*, 48–55. [CrossRef] [PubMed]
95. Kim, M.I.; Kim, J.; Lee, J.; Jia, H.; Na, H.B.; Youn, J.K.; Kwak, J.H.; Dohnalkova, A.; Grate, J.W.; Wang, P. Crosslinked enzyme aggregates in hierarchically-ordered mesoporous silica: A simple and effective method for enzyme stabilization. *Biotechnol. Bioeng.* **2007**, *96*, 210–218. [CrossRef] [PubMed]
96. Jung, D.; Paradiso, M.; Wallacher, D.; Brandt, A.; Hartmann, M. Formation of Cross-Linked Chloroperoxidase Aggregates in the Pores of Mesocellular Foams: Characterization by SANS and Catalytic Properties. *Chemsuschem* **2010**, *2*, 161–164. [CrossRef] [PubMed]
97. Rajendhran, J.; Gunasekaran, P. Application of cross-linked enzyme aggregates of Bacillus badius penicillin G acylase for the production of 6-aminopenicillanic acid. *Lett. Appl. Microbiol.* **2007**, *44*, 43–49. [CrossRef] [PubMed]
98. Arsenault, A.; Cabana, H.; Jones, J.P. Laccase-Based CLEAs: Chitosan as a Novel Cross-Linking Agent. *Enzyme Res.* **2011**, *2011*, 376015. [CrossRef] [PubMed]
99. Ji, Q.; Wang, B.; Tan, J.; Zhu, L.; Li, L. Immobilized multienzymatic systems for catalysis of cascade reactions. *Process Biochem.* **2016**, *51*, 1193–1203. [CrossRef]
100. Mateo, C.; Chmura, A.; Rustler, S.; Rantwijk, F.V.; Stolz, A.; Sheldon, R.A. Synthesis of enantiomerically pure (*S*)-mandelic acid using an oxynitrilase–nitrilase bienzymatic cascade: A nitrilase surprisingly shows nitrile hydratase activity. *Tetrahedron Asymmetry* **2006**, *17*, 320–323. [CrossRef]
101. Talekar, S.; Pandharbale, A.; Ladole, M.; Nadar, S.; Mulla, M.; Japhalekar, K.; Pattankude, K.; Arage, D. Carrier free co-immobilization of alpha amylase, glucoamylase and pullulanase as combined cross-linked enzyme aggregates (combi-CLEAs): A tri-enzyme biocatalyst with one pot starch hydrolytic activity. *Bioresour. Technol.* **2013**, *147*, 269–275. [CrossRef] [PubMed]
102. Ahumada, K.; Urrutia, P.; Illanes, A.; Wilson, L. Production of combi-CLEAs of glycosidases utilized for aroma enhancement in wine. *Food Bioprod. Process.* **2015**, *94*, 555–560. [CrossRef]
103. Stressler, T.; Ewert, J.; Eisele, T.; Fischer, L. Cross-linked enzyme aggregates (CLEAs) of PepX and PepN–production, partial characterization and application of combi-CLEAs for milk protein hydrolysis. *Biocatal. Agric. Biotechnol.* **2015**, *4*, 752–760. [CrossRef]
104. Mahmod, S.S.; Yusof, F.; Shah, H.; Jami, M.S.; Khanahmadi, S. Development of an immobilized biocatalyst with lipase and protease activities as a multipurpose cross-linked enzyme aggregate (multi-CLEA). *Process Biochem.* **2015**, *50*, 2144–2157. [CrossRef]
105. Li, H.; Xiao, W.; Xie, P.; Zheng, L. Co-immobilization of enoate reductase with a cofactor-recycling partner enzyme. *Enzyme Microb. Technol.* **2018**, *109*, 66–73. [CrossRef] [PubMed]
106. Nguyen, L.T.; Yang, K.L. Combined cross-linked enzyme aggregates of horseradish peroxidase and glucose oxidase for catalyzing cascade chemical reactions. *Enzyme Microb. Technol.* **2017**, *100*, 52–59. [CrossRef] [PubMed]

107. Ning, C.; Su, E.; Tian, Y.; Wei, D. Combined cross-linked enzyme aggregates (combi-CLEAs) for efficient integration of a ketoreductase and a cofactor regeneration system. *J. Biotechnol.* **2014**, *184*, 7–10. [CrossRef] [PubMed]
108. Sirimanne, S.R.; Patterson, D.G., Jr. A one-pot synthesis of (±)-(ring 13C6)-mandelic acid. *J. Label. Compd. Radiopharm.* **1993**, *33*, 725–731. [CrossRef]
109. Gao, J.; Wang, A.R.; Jiang, X.P.; Zhang, J.X.; Zhang, Y.W. Preparation of Expoxy-Functionalized Magnetic Nanoparticles for Immobilization of Glycerol Dehydrogenase. *J. Nanosci. Nanotechnol.* **2018**, 4852–4857. [CrossRef] [PubMed]
110. Ahumada, K.; Martínez-Gil, A.; Moreno-Simunovic, Y.; Illanes, A.; Wilson, L. Aroma Release in Wine Using Co-Immobilized Enzyme Aggregates. *Molecules* **2016**, *21*, 1485. [CrossRef] [PubMed]
111. Singh, R.K.; Zhang, Y.W.; Nguyen, N.P.; Jeya, M.; Lee, J.K. Covalent immobilization of β-1,4-glucosidase from Agaricus arvensis onto functionalized silicon oxide nanoparticles. *Appl. Microbiol. Biotechnol.* **2010**, *89*, 337–344. [CrossRef] [PubMed]

catalysts

MDPI

Article

Enhanced (−)-α-Bisabolol Productivity by Efficient Conversion of Mevalonate in *Escherichia coli*

Soo-Jung Kim [1], Seong Keun Kim [1], Wonjae Seong [1,2], Seung-Gyun Woo [1,2], Hyewon Lee [1], Soo-Jin Yeom [1], Haseong Kim [1,2], Dae-Hee Lee [1,2,*] and Seung-Goo Lee [1,2,*]

1 Synthetic Biology and Bioengineering Research Center, Korea Research Institute of Bioscience and Biotechnology (KRIBB), Daejeon 34141, Korea; bioksj@kribb.re.kr (S.-J.K.); draman97@kribb.re.kr (S.K.K.); winise@kribb.re.kr (W.S.); dntmdrbs12@kribb.re.kr (S.-G.W.); hlee@kribb.re.kr (H.L.); sujin258@kribb.re.kr (S.-J.Y.); haseong@kribb.re.kr (H.K.)

2 Department of Biosystems and Bioengineering, KRIBB School of Biotechnology, University of Science and Technology (UST), Daejeon 34113, Korea

* Correspondence: dhlee@kribb.re.kr (D.-H.L.); sglee@kribb.re.kr (S.-G.L.); Tel.: +82-42-879-8225 (D.-H.L.); +82-42-860-4373 (S.-G.L.)

Received: 30 March 2019; Accepted: 29 April 2019; Published: 9 May 2019

Abstract: (−)-α-Bisabolol, a naturally occurring sesquiterpene alcohol, has been used in pharmaceuticals and cosmetics owing to its beneficial effects on inflammation and skin healing. Previously, we reported the high production of (−)-α-bisabolol by fed-batch fermentation using engineered *Escherichia coli* (*E. coli*) expressing the exogenous mevalonate (MVA) pathway genes. The productivity of (−)-α-bisabolol must be improved before industrial application. Here, we report enhancement of initial (−)-α-bisabolol productivity to 3-fold higher than that observed in our previous study. We first harnessed a farnesyl pyrophosphate (FPP)-resistant mevalonate kinase 1 (MvaK1) from an archaeon *Methanosarcina mazei* (*M. mazei*) to create a more efficient heterologous MVA pathway that produces (−)-α-bisabolol in the engineered *E. coli*. The resulting strain produced 1.7-fold higher (−)-α-bisabolol relative to the strain expressing a feedback-inhibitory MvaK1 from *Staphylococcus aureus* (*S. aureus*). Next, to efficiently convert accumulated MVA to (−)-α-bisabolol, we additionally overexpressed genes involved in the lower MVA mevalonate pathway in *E. coli* containing the entire MVA pathway genes. (−)-α-Bisabolol production increased by 1.8-fold with reduction of MVA accumulation, relative to the control strain. Finally, we optimized the fermentation conditions including inducer concentration, aeration and enzymatic cofactor. The strain was able to produce 8.5 g/L of (−)-α-bisabolol with an initial productivity of 0.12 g/L h in the optimal fed-batch fermentation. Thus, the microbial production of (−)-α-bisabolol would be an economically viable bioprocess for its industrial application.

Keywords: (−)-α-bisabolol; mevalonate (MVA); mevalonate kinase 1; *Methanosarcina mazei*; fed-batch fermentation

---

1. Introduction

A monocyclic sesquiterpene alcohol, (−)-α-bisabolol, has been used in pharmaceuticals and cosmetics as it displays the beneficial effects of skin healing and anti-inflammation [1–5]. The global market of (−)-α-bisabolol is expected to reach $73 million by 2020, with an annual growth rate of 5.9% from 2016 [6]. Commercially available (−)-α-bisabolol is currently produced by the steam-distillation method using oils extracted from German chamomile or Brazilian candeia tree [7,8]. This process, however, has caused environmental issues, as well as economic concerns owing to a low extraction yield [8]. Natural (−)-α-bisabolol was obtained from the candeia tree with a yield of approximately 0.018 $g/g_{candeia\ power}$ through CO_2 supercritical extraction at 40 °C and 10 MPa [8]. Although a chemical process has been developed to produce (−)-α-bisabolol, it forms diastereomers of (−)-α-bisabolol ((+)-α-bisabolol and (±)-epi-α-bisabolol), and thus requires auxiliary purification steps [9]. In this context, the biological production of naturally occurring (−)-α-bisabolol using engineered microbes may be an attractive alternative to the current production processes of (−)-α-bisabolol.

(−)-α-Bisabolol can be synthesized from five-carbon building blocks of isopentenyl diphosphate (IPP) and its isomer dimethyl allylpyrophosphate (DMAPP) (Figure 1) [10]. Both universal precursors of terpenoids can be produced from the MVA or the 2-C-methyl-D-erythritol 4-phosphate (MEP) pathway. Exogenous MVA or endogenous MEP pathway has been employed in engineered *E. coli* for production of various terpenoids. Although the MEP pathway exhibits higher theoretical yield than the MVA pathway, the exogenous MVA pathway showed generally higher production than the endogenous MEP pathway [11]. In particular, the MVA pathway has been harnessed to efficiently convert acetyl-CoA to several terpenoids including (−)-α-bisabolol [12]. Both universal isoprene units, IPP and DMAPP are converted into farnesyl pyrophosphate (FPP), which is catalyzed by FPP synthase encoded by the endogenous *ispA* gene, which is then used for production of (−)-α-bisabolol by the (−)-α-bisabolol synthase (BBS, Figure 1).

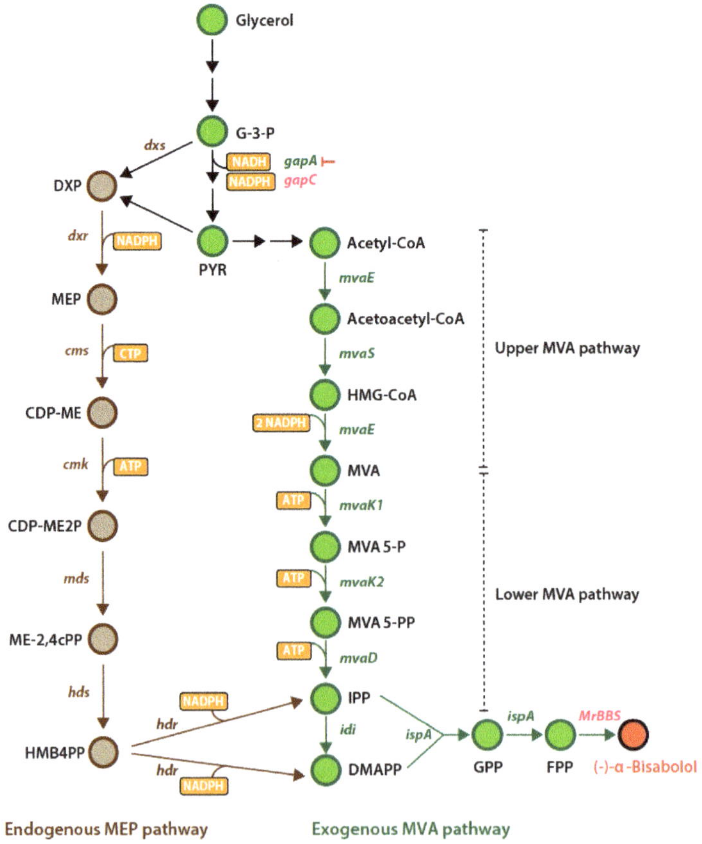

Figure 1. Biosynthetic pathway of (−)-α-bisabolol in engineered *E. coli*. The endogenous MEP) pathway consists of DXS (deoxyxylulose-5-phosphate synthase), DXR (deoxyxylulose 5-phosphate reductoisomerase), CMS (2-C-methylerythritol 4-phosphate cytidyl transferase), CMK (4-(cytidine 5′-diphospho)-2-C-methylerythritol kinase), MDS (2-C-methylerythritol 2,4-cyclodiphosphate synthase), HDS ((E)-4-hydroxy-3-methylbut-2-enyl diphosphate reductase), and HDR (hydroxymethylbutenyl

diphosphate reductase). The MEP pathway begins with the condensation of G-3-P (glyceraldehyde 3-phosphate) and pyruvate that is converted from G-3-P by endogenous nicotinamide adenine dinucleotide (NAD^{+})-dependent GAPDH (glyceraldehyde-3-phosphate dehydrogenase) coded by *gapA* and nicotinamide adenine dinucleotide phosphate ($NADP^{+}$)-dependent GAPDH coded by *gapC* from *Clostridium acetobutylicum* (*C. acetobutylicum*). The exogenous MVA pathway consists of MvaE (dual function of acetoacetyl-CoA thiolase and 3-hydroxy-3-methylglutaryl-CoA reductase), MvaS (3-hydroxy-3-methylglutaryl-CoA synthase), MvaK1 (MVA kinase), MvaK2 (phosphomevalonate kinase), MvaD (mevalonate 5-pyrophosphate decarboxylase), Idi (isopentenyl diphosphate isomerase), IspA (geranyl diphosphate synthase or FPP synthase), and *Mr*BBS ((−)-α-bisabolol synthase of *Matricaria recutita*). Endogenous and exogenous genes are depicted in brown and green, respectively.

The biological production of (−)-α-bisabolol has been explored using well-studied microbes such as *Saccharomyces cerevisiae* (*S. cerevisiae*) [9] and *E. coli* [12]. Because of the identification of BBS from German chamomile, *Matricaria recutita*, the microbes expressing the *Mr*BBS enzyme can synthesize an (−)-α-bisabolol as a major terpenoid product [9]. Previously, we engineered an *E. coli* strain to express the MrBBS enzyme and exogenous MVA pathway. The resulting *E. coli* produced 9.1 g/L of (−)-α-bisabolol with a productivity of 0.04 g/L h at early stage of fermentation (0–42 h) [12], whereas *S. cerevisiae* expressing the MrBBS enzyme alone produced 8 mg/L of (−)-α-bisabolol during four days of cultivation [9]. These studies showed the potential of (−)-α-bisabolol production by microbial fermentation. However, productivity remains to be improved for the industrial production of (−)-α-bisabolol using engineered microbes. In our empirical fermentation studies, the initial productivity (0–48 h) of (−)-α-bisabolol was critical to improving its overall productivity, because after 2 days of fermentation, the production rate of (−)-α-bisabolol showed no significant differences among various production strains and fermentation conditions.

In this study, we improve (−)-α-bisabolol productivity in engineered *E. coli*, which can serve as a promising platform strain for development of an economically feasible bioprocess of (−)-α-bisabolol production. To this end, we first introduced a heterologous MvaK1 from *M. mazei* that is resistant to FPP feedback inhibition. We then added a copy of the lower MVA pathway genes to the whole MVA pathway for the efficient conversion of MVA to (−)-α-bisabolol. Finally, we optimized the fermentation conditions of the engineered *E. coli* by tuning the inducer concentrations and aeration for MVA pathway expression and sufficient ATP supply, respectively. Overall, a fed-batch fermentation produced 8.5 g/L of (−)-α-bisabolol with 0.12 g/L h of initial productivity (0–46 h) in the engineered *E. coli*.

2. Results

2.1. Feedback-Resistant MvaK1

MvaK1 is responsible for the first step of the lower MVA pathway by converting MVA to mevalonate phosphate (MVA 5-P in Figure 1) [10] and is important for the regulation of the entire MVA pathway because it is inhibited by known feedback inhibitors: C5 (IPP and DMAPP), C15 (geranyl pyrophosphate (GPP) and FPP), and longer chain terpenoids [13,14]. FPP is a feedback inhibitor of the widely used *Staphylococcus aureus* MvaK1 (*Sa*MvaK1) for creating a heterologous MVA pathway [15]. Previously, we have also used *Sa*MvaK1 to produce (−)-α-bisabolol in engineered *E. coli* [16].

To avoid feedback inhibition of MvaK1 and subsequently improve (−)-α-bisabolol production, we replaced the *S. aureus mvak1* gene of the pTSN-Bisa-Sa plasmid with a feedback-resistant *mvaK1* gene (Figure S1) from the versatile methanogen *M. mazei*, which resulted in a pTSN-Bisa-Mm plasmid. The *E. coli* DH5α-pTSN-Bisa-Mm strain produced 555 mg/L of (−)-α-bisabolol, which is 1.7-fold higher than that of the *E. coli* DH5α-pTSN-Bisa-Sa strain (Figure 2B). This is consistent with MVA accumulation of *Mm*MvaK1 showing 1.7-fold less than that of *Sa*MvaK1 (Figure 2B, right panel), suggesting that the feedback-resistant *Mm*MvaK1 leads to an increase the MVA utilization efficiency.

Using the *E. coli* DH5α-pTSN-Bisa-Mm strain, we conducted a fed-batch fermentation by intermittently supplying glycerol (Figure 2C). Cells were grown exponentially for 24 h and produced

0.9 g/L of (−)-α-bisabolol and 4.9 g/L of MVA along with consuming initially supplied glycerol. A total of 8.2 g/L of (−)-α-bisabolol was yielded with a productivity of 0.06 g/L h, and 10.7 g/L of MVA was accumulated in 140 h. Overall, although feedback-resistant *Mm*MvaK1 was used for (−)-α-bisabolol production, a significant amount of MVA was still accumulated in the fed-batch fermentation.

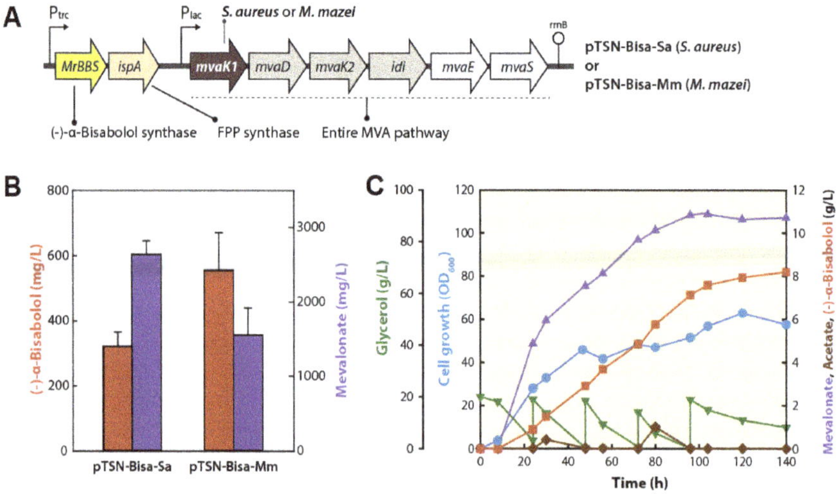

Figure 2. Introduction of MvaK1 from *M. mazei* for the improvement of (−)-α-bisabolol production. (**A**) Plasmid constructs for expressing the entire MVA pathway, FPP synthase and (−)-α-bisabolol synthase. The plasmid pTSN-Bisa-Sa and pTSN-Bisa-Mm have *mvak1* gene of *S. aureus* and *M. mazei*, respectively. (**B**) Improvement of (−)-α-bisabolol production in the engineered *E. coli* DH5α harboring pTSN-Bisa-Mm compared with the strain expressing pTSN-Bisa-Sa. Cells were grown in TB medium containing 10 g/L glycerol and 20% (v/v) of *n*-dodecane at 30 °C for 72 h without the addition of isopropyl β-D-1-thiogalactopyranoside (IPTG). The error bars represent the standard deviation of the concentrations of (−)-α-bisabolol and MVA from three biological replicates. (**C**) Fed-batch fermentation of *E. coli* DH5α harboring pTSN-Bisa-Mm. The fed-batch fermentation was performed in TB medium and 20% (v/v) of *n*-dodecane using two-phase culture in the absence of IPTG at 30 °C and pH 7.0. After depletion of glycerol initially added, glycerol was fed intermittently into the bioreactor during fermentation. An agitation speed of 280 rpm and an aeration rate of 1 vessel volume per minute (vvm) were maintained throughout the cultivation.

2.2. Overexpression of Entire MVA Pathway Genes

Enzymes responsible for (−)-α-bisabolol biosynthesis in *E. coli* DH5α-pTSN-Bisa-Mm strain are controlled by IPTG-inducible promoters; trc promoter for *MrBBS*, and *ispA* genes, and lac promoter for all MVA pathway genes (Figure 2A). To this end, we explored the effect of IPTG amount on (−)-α-bisabolol production and MVA accumulation in batch culture. When 0.025 mM IPTG was used for induction, (−)-α-bisabolol production increased by 1.8-fold along with a 1.4-fold decrease of MVA accumulation compared to those of the control that were not induced by IPTG (Figure 3A). To scrutinize the effect of IPTG on (−)-α-bisabolol production, a pSEVA231-Bisa-Mm was generated using a medium copy number plasmid, pSEVA231 (pBBR1 ori) (Figure 3B). Interestingly, the *E. coli* DH5α- pSEVA231-Bisa-Mm strain produced 926 mg/L of (−)-α-bisabolol without the accumulation of MVA under the induced condition (0.025 mM IPTG), which is 3.7-fold higher than the uninduced condition (Figure 3B). When the IPTG amount increased up to 0.1 mM, both pTSN-Bisa-Mm (high copy number), and pSEVA231-Bisa-Mm (medium copy number) showed a dramatic decrease (90%) in (−)-α-bisabolol production compared to those in the presence of 0.025 mM IPTG (Figure 3A,B).

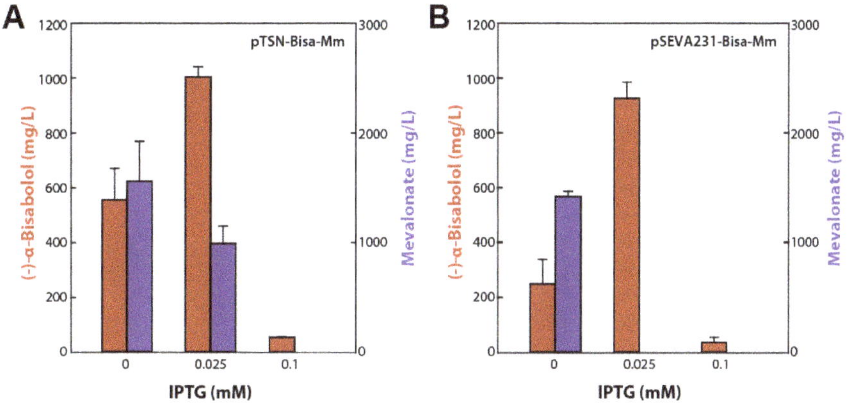

Figure 3. Effect of IPTG on (−)-α-bisabolol production. The concentrations of (−)-α-bisabolol and MVA produced by the *E. coli* DH5α harboring pTSN-Bisa-Mm, a high-copy plasmid (**A**) or pSEVA231-Bisa-Mm, a medium-copy plasmid (**B**). Cells were grown in TB medium containing 10 g/L glycerol and 20% (v/v) of *n*-dodecane in the presence of different IPTG concentrations (0, 0.025 and 0.1 mM) at 30 °C for 72 h. The error bars represent the standard deviation of the concentrations of (−)-α-bisabolol and MVA from three biological replicates.

2.3. *Reinforcement of the MVA Pathway*

Considering the high accumulation of MVA in the production of (−)-α-bisabolol, we reinforced the whole MVA pathway through the expression of an additional copy of lower MVA pathway genes. A newly generated plasmid, pSSN12Didi-MrBBS-IspA, contains the lower MVA pathway genes (*mvaK1*, *mvaK2*, *mvaD*, *idi*), *MrBBS*, and *ispA* (Figure 4A). The MvaK1 of pSSN12Didi-MrBBS-IspA plasmid was adopted from *Streptococcus pneumoniae* (*S. pneumoniae*), which has a 2.6-fold faster turnover number (k_{cat}) than that of *Mm*MvaK1 [17]. In the absence of IPTG, *E. coli* DH5α harboring both pTSN-Bisa-Mm and pSSN12Didi-MrBBS-IspA plasmids produced 1.2 g/L of (−)-α-bisabolol and 988 mg/L of MVA (Figure 4B,C), which are 2.2-fold higher and 1.4-fold lower than those of the *E. coli* DH5α containing the pTSN-Bisa-Mm plasmid alone, respectively. Because the metabolic flux was changed by the introduction of additional lower MVA pathway genes, we probed the effect of IPTG concentrations on the (−)-α-bisabolol production in the *E. coli* DH5α containing both pTSN-Bisa-Mm and pSSN12Didi-MrBBS-IspA plasmids. Unlike the results from the *E. coli* DH5α harboring the pTSN-Bisa-Mm alone (Figure 3A), production of both (−)-α-bisabolol and MVA decreased as the IPTG concentrations increased (Figure 4B,C), indicating that enzymes for (−)-α-bisabolol biosynthesis were sufficiently expressed in the absence of IPTG to increase the (−)-α-bisabolol production when the lower MVA pathway was reinforced.

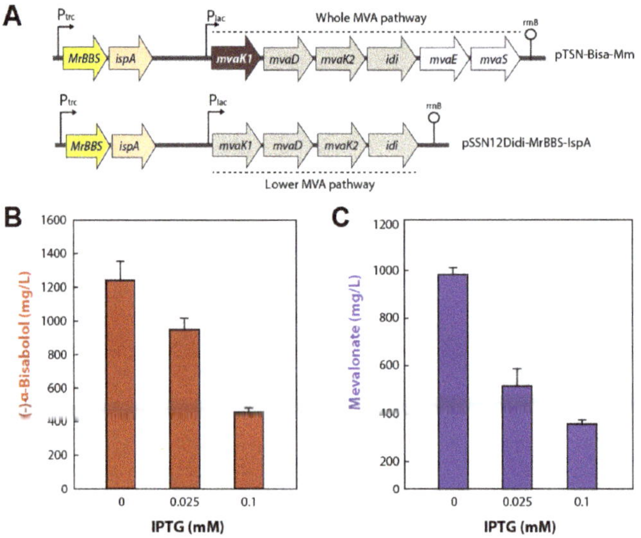

Figure 4. (−)-α-Bisabolol production in *E. coli* DH5α containing pTSN-Bisa-Mm and pSSN12Didi-MrBBS-IspA which has genes involved in the MVA lower pathway including *Sp*MvaK1, IspA and *Mr*BBS. (**A**) Plasmid constructs for expressing genes of entire (−)-α-bisabolol biosynthetic pathway or lower MVA pathway. The concentrations of (−)-α-bisabolol (**B**) and MVA (**C**) produced by the engineered *E. coli* DH5α harboring pTSN-Bisa-Mm and pSSN12Didi-MrBBS-IspA. Cells were cultivated in TB medium containing 10 g/L glycerol and 20% (v/v) of *n*-dodecane in the presence of different IPTG concentrations (0, 0.025 and 0.1 mM) at 30 °C for 72 h. The error bars represent the standard deviation of the concentrations of (−)-α-bisabolol and MVA from three biological replicates.

2.4. Sufficient Supply of reduced Nicotinamide Adenine Dinucleotide Phosphate (NADPH)

The engineered *E. coli* consumes 2 NADPH to convert 3-hydroxy-3-methyl-glutaryl-CoA (HMG)-CoA to MVA via the MvaE enzyme of the (−)-α-bisabolol biosynthetic pathway (Figure 1). Therefore, if the intracellular NADPH pool of engineered *E. coli* increases, it will improve (−)-α-bisabolol production. In *E. coli*, an endogenous GAPDH generates a reduced nicotinamide dinucleotide (NADH) to convert glyceraldehyde 3-phosphate into pyruvate. Therefore, we replaced the endogenous GAPDH gene (*gapA*) of *E. coli* with an $NADP^+$-dependent GAPDH gene (*gapC*) of *C. acetobutylicum* to increase the intracellular NADPH pool in *E. coli*. To do this, we inserted the *gapC* gene of *C. acetobutylicum* into the downstream of the *ispA* gene of the pTSN-Bisa-Mm plasmid, which resulted in a pTSN-Bisa-Mm-GapC plasmid (Figure 5A). The *E. coli*-pTSN-Bisa-Mm-GapC strain showed similar (−)-α-bisabolol production and cell growth to the *E. coli*-pTSN-Bisa-Mm strain (Figure 5B,C). This might be caused by competition between endogenous GapA and heterologous GapC in the *E. coli*-pTSN-Bisa-Mm-GapC strain. To investigate this, the *gapA* gene was repressed by clustered regularly interspaced short palindromic repeats (CRISPR) interference (CRISPRi). The CRISPRi system comprises L-rhamnose-inducible deactivated Cas9 (dCas9) and a constitutively expressed single guide RNA targeting *gapA* gene (sgRNA-GapA) by J23119 promoter, respectively (Figure 5A). The *E. coli* strain harboring both pTSN-Bisa-Mm-GapC and pdCas9-sgRNA-GapA plasmids produced 1.4-fold higher (−)-α-bisabolol compared to the *E. coli* strain containing the pTSN-Bisa-Mm-GapC plasmid alone. Interestingly, the cell growth of the *E. coli* strain repressing *gapA* by CRISPRi showed better cell growth than the control *E. coli* (Figure 5C). Given that the (−)-α-bisabolol/OD_{600} are similar between the two strains, it is likely that the increased production of (−)-α-bisabolol is due to increased cell mass.

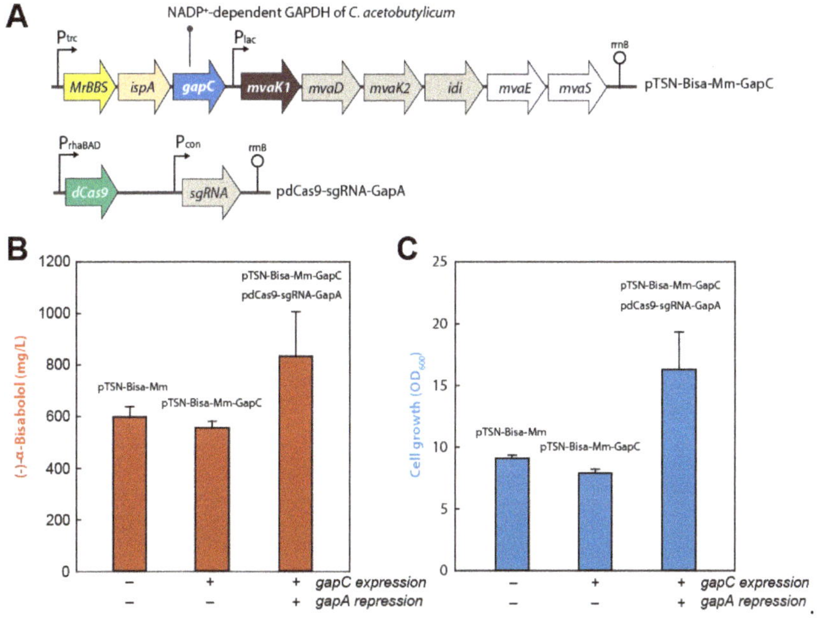

Figure 5. Overexpression of *gapC* gene encoding NADP$^+$-dependent GAPDH from *C. acetobutylicum* and repression of *gapA* gene coding for endogenous NAD$^+$-dependent GAPDH from *E. coli* using the CRISPRi system. (**A**) Plasmid constructs for expressing genes of the entire (−)-α-bisabolol biosynthetic pathway introducing the *gapC* gene downstream of the *ispA* gene (pTSN-Bisa-Mm-GapC) and for expressing inactivated Cas9 (dCas9) and sgRNA targeting the *gap*A gene (pdCas9-sgRNA-GapA). (**B**) Comparison of (−)-α-bisabolol concentrations produced by the *E. coli* DH5α harboring pTSN-Bisa-Mm, pTSN-Bisa-Mm-GapC, or both pTSN-Bisa-Mm-GapC and pCas9-sgRNA-GapA. (**C**) Comparison of cell growth of the strains. Cells were grown in TB medium containing 10 g/L glycerol and 20% (v/v) of *n*-dodecane in the presence of different IPTG concentrations (0, 0.025 and 0.1 mM) at 30 °C for 72 h. The error bars represent the standard deviation of the concentrations of (−)-α-bisabolol and OD_{600} from three biological replicates.

2.5. Effect of Aeration on (−)-α-Bisabolol Fermentation

We performed a fed-batch fermentation to produce the (−)-α-bisabolol in *E. coli* DH5α containing both pTSN-Bisa-Mm and pSSN12Didi-MrBBS-IspA plasmids. The yield of (−)-α-bisabolol was improved by 16% in 46 h compared to when the lower MVA pathway was not additionally overexpressed. However, the MVA still accumulated from the beginning of fermentation and reached 10.1 g/L in 68 h at 280 rpm despite reinforcing the lower MVA pathway (Figure 6A). It seems that there are other bottlenecks when MVA is converted to (−)-α-bisabolol through the lower MVA pathway. The synthetic MVA pathway requires 3 moles of ATP to convert MVA to (−)-α-bisabolol (Figure 1) and competes for the ATP with other essential cellular reactions involved in cell growth [18]. Because ATPs are efficiently generated under aerobic conditions using NADHs in oxidative phosphorylation [19], aeration effects were examined by controlling the agitation speed in fed-batch fermentation.

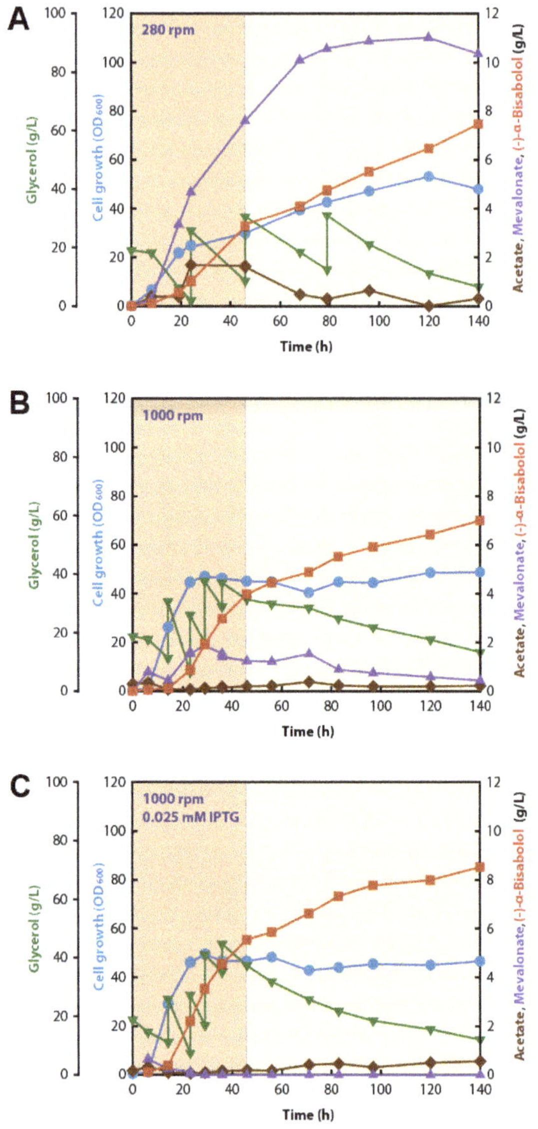

Figure 6. Fed-batch fermentation of the *E. coli* DH5α harboring both pTSN-Bisa-Mm and pSSN12Didi-MrBBS-IspA plasmids. The fed-batch fermentation was performed in TB medium and 20% (v/v) of *n*-dodecane using two-phase culture at 30 °C and pH 7.0. After depletion of glycerol initially added, glycerol was fed intermittently into the bioreactor during fermentation. An agitation speed of 280 rpm (**A**) or 1000 rpm (**B**,**C**) and an aeration rate of 1 vvm were maintained throughout the cultivation. After 6 h cultivation, 0.025 mM of IPTG was added to the bioreactor (**C**).

In fed-batch fermentation at an agitation speed of 1000 rpm, the *E. coli* strain harboring pTSN-Bisa-Mm and pSSN12Didi-MrBBS-IspA rapidly grew and reached the maximum cell growth within 23 h showing a 1.8-fold improved cell growth relative to those at 280 rpm. As expected, MVA accumulation was considerably reduced, but (−)-α-bisabolol production was not significantly improved (Figure 6B). It seems that acetyl-CoA was utilized for cell growth and other metabolism, or metabolites downstream of MVA, accumulated. To overcome this problem, we sought to enhance the overall (−)-α-bisabolol flux through overexpressing all genes in the (−)-α-bisabolol biosynthetic pathway by the addition of IPTG. We carried out the fed-batch fermentation at 1000 rpm and supplied IPTG at a concentration of 0.025 mM after 6 h of incubation. Similar to the culture in the absence of IPTG, cells rapidly grew for 22 h, but MVA did not accumulate during fermentation, supporting the lower pathway was intensified by IPTG addition. The yield and productivity of (−)-α-bisabolol were improved by 40% and 56%, respectively, as compared to the absence of IPTG, and the final titer of (−)-α-bisabolol reached 8.5 g/L (Figure 6C).

3. Discussion

In this study, *E. coli* was engineered for the efficient conversion of MVA to (−)-α-bisabolol using the feedback-resistant MvaK1 and reinforcement of the lower MVA pathway. The feedback-resistant MvaK1 was firstly identified in the archaeon *M. mazei*. However, only a handful of studies have been carried out on terpenoid production in microbes. Recently, feedback-resistant MvaK1 enzymes were identified and characterized from *Methanosaeta concilii* (McMvaK) and *Methanocella paludicola* (MpMvaK) [20]. The McMvaK and MpMvaK enzymes not only showed feedback resistance to DMAPP, IPP, GPP, and FPP, but also exhibited 4.9- and 5.5-fold higher affinity to MVA, respectively, than *Mm*MvaK1 [20]. Therefore, these MvaK1 enzymes may enable the enhancement of (−)-α-bisabolol production in engineered *E. coli*.

To find the optimal conditions to efficiently convert MVA to (−)-α-bisabolol, we examined the inducer concentrations, cofactor, ATP, and reinforcement of the lower MVA pathway. Adding the inducers for overexpression of MVA or MEP pathway genes has been a controversial issue in the microbial production of terpenoids [21–23]. Lycopene production was reduced under all IPTG-induced conditions in *E. coli* expressing the lower MVA pathway [23]. The leaky expression of all enzymes involved in the (−)-α-bisabolol production without IPTG addition exhibited the highest production among all tested IPTG concentrations [12]. The IPTG-induced overexpression of genes for (−)-α-bisabolol production can inhibit the essential cellular metabolism due to a deficiency of FPP or accumulation of toxic intermediates (IPP and HMG-CoA) of the heterologous MVA pathway [12]. In contrast, isoprene production increased as the IPTG concentration increased from 0.2 to 1.2 mM [24]. In this study, a small amount of IPTG was effective to increase (−)-α-bisabolol production in the engineered *E. coli*. Concerning the complex regulation of the MVA pathway, balancing the expression of multiple heterologous enzymes is crucial for the optimal production of (−)-α-bisabolol [10].

The availability of reducing cofactors such as NADH and NADPH strongly affects the yield and productivity of terpenoids in bacteria. The strengthening of the reducing power for the increased production of terpenoids has been attempted; modulation of glutamate dehydrogenase increased the production of β-carotene and lycopene through the increased supply of NADPH [25–27]. The overexpression of GAPDH of *C. acetobutylicum* also resulted in the improvement of isoprene production [24]. Moreover, the replacement of NAD^+-dependent GADPH of *E. coli* with the $NADP^+$-dependent GAPDH of *C. acetobutylicum* showed a 2.5-fold increase of lycopene productivity in the engineered *E. coli* [28]. We performed the fed-batch fermentation using the strain overexpressing *gapC* from *C. acetobutylicum* and repressing the *gapA* gene through the CRISPRi system under optimized conditions (1000 rpm and the addition of IPTG at a concentration of 0.025 mM) (Figure S2, Table 1). Contrary to the results in batch fermentations, both strains showed a negative effect on (−)-α-bisabolol production and had slightly reduced cell growth when compared with the strain that did not overexpress *gapC* and repress *gapA*. It appears that the overexpression of *gapC* and the repression of *gapA* were not effective when sufficient amounts of ATP and NADPH were supplied, owing to the activation of the citric acid cycle and respiration by increasing oxygen in the cells. This observation might be consistent with a previous study that lycopene production in *E. coli* was improved by decreasing the pentose phosphate pathway flux and increasing the tricarboxylic acid (TCA) cycle flux [26]. Additionally, NADPH has been shown to inhibit 3-hydroxy-3-methylglutaryl-CoA reductase (HMGR), which converts HMG-CoA to MVA [10]. Both the overexpression of *gapC* and aerobic condition might lead to an accumulation of excess NADPH, thereby repressing HMGR and causing a flux imbalance. As a result, overall (−)-α-bisabolol production decreased.

Results of fed-batch fermentations conducted in this study were summarized in Table 1. Compared with the previous report [12], the final titer of (−)-α-bisabolol (8.5 g/L) is similar in fed-batch fermentation using *E. coli* expressing pTSN-Bisa-Mm and pSSN12Didi-MrBBS-IspA at an agitation speed of 1000 rpm with IPTG added at a concentration of 0.025 mM. In particular, 5.5 g/L of (−)-α-bisabolol was obtained within two days, indicating that the productivity was improved by 3-fold compared to previous research. In result of fed-batch fermentations, (−)-α-bisabolol was continuously produced after cell growth ceased. Therefore, recycling resting cells is a promising strategy to further improve (−)-α-bisabolol productivity. Although promising results in productivity were obtained in this study, it is necessary to improve the final (−)-α-bisabolol titer for industrial applications. To achieve this, high-cell density culture experiments using living cells to continuously supply cofactors and enzymes should be conducted. Moreover, metabolic modeling of the system used in this study might provide further insight into bottlenecks for (−)-α-bisabolol production.

Table 1. Summary of fed-batch fermentations of (−)-α-bisabolol by engineered *E. coli*.

Plasmids	Agitation (rpm)	IPTG (mM)	Final Titer (g/L)	Initial Yield * (g/g)	Initial Productivity * (g/L h)
pTSN-Bisa-Mm	280	0	8.2	0.08	0.06
pTSN-Bisa-Mm pSSN12Didi-MrBBS-IspA	280 1000 1000	0 0 0.025	7.5 7.0 8.5	0.10 0.07 0.11	0.07 0.09 0.12
pTSN-Bisa-Mm-GapC pSSN12Didi-MrBBS-IspA	1000	0.025	5.3	0.07	0.07
pTSN-Bisa-Mm-GapC pSSN12Didi-MrBBS-IspA pdCas9-sgRNA-GapA	1000	0.025	3.6	0.05	0.05

* Initial yield and productivity are calculated based on values in the early cultivation period (0–46 h).

4. Materials and Methods

4.1. Strains and Culture Media

An *E. coli* DH5α strain (Enzynomics, Daejeon, Korea) was used for all experiments including gene cloning and (−)-α-bisabolol production. A lysogeny broth (LB) medium (10 g/L tryptone, 5 g/L yeast extract, and 5 g/L NaCl) (BD Bioscience, San Jose, CA, USA) was used for plasmid construction and pre-cultivation. Terrific broth (TB) medium containing glycerol (12 g/L enzymatic casein digest, 24 g/L yeast extract, 9.4 g/L K_2HPO_4, 2.2 g/L KH_2PO_4, and 1% (*w*/*v*) glycerol) was used for (−)-α-bisabolol production. All media were supplied with the appropriate antibiotics: ampicillin (100 μg/mL), chloramphenicol (34 μg/mL), and kanamycin (25 μg/mL). IPTG was used at concentrations of 0, 0.025, and 0.1 mM to induce gene expression involved in the (−)-α-bisabolol biosynthetic pathway.

4.2. Plasmid Construction

The plasmids and primers used in this study are listed in Table 2 and Table S1, respectively. Standard molecular biological techniques including genomic DNA preparation, restriction digestions of DNA, plasmid transformation were performed as previously described [29]. T4 DNA ligase, and all restriction enzymes were obtained from New England Biolabs (NEB, Ipswich, MA, USA). Polymerase chain reaction (PCR) was carried out following the manufacturer's protocols with a high fidelity KOD-Plus-Neo polymerase (Toyobo, Osaka, Japan). Kits for plasmid preparation and gel extraction were purchased from Promega (Madison, WI, USA) and oligonucleotide synthesis were conducted by Bioneer (Daejeon, Korea).

Table 2. Strains and plasmids used in this study.

Name	Description	References
Strains		
DH5α	F⁻, *Φ80lacZ·ΔM15·f(lacZYA−argF)U169 deoR recA1 endA1 hsdR17(rk−, mk+) phoA supE44 thi-1 gyrA96 relA1*	Enzynomics
Plasmids		
pTrc99A	P_{trc} promoter, Amp^R, $lacI^q$, pBR322 ori	GE Healthcare
pSTV28	P_{lac} promoter, Cm^R, p15A ori	Takara
pSECRi	P_{rhaBAD}::*cas9*(D10A, H840A) and constitutive sgRNA expression cassette in pSEVA221	[16]
pSNA-MrBBS-IspA	pTrc99A containing *mvaE* and *mvaS* of *Enterococcus faecalis*, *mvaK1* and *mvaK2* and *mvaD* of *S. pneumoniae*, *idi*, and *ispA* of *E. coli*, *MrBBS of M. recutita*	[12]
pTM-BBS	pTrc99A derivatives containing codon optimized *Matricaria recutita MrBBS*, *mvaK1* of *S. aureus*, *mvaD* and *mvaK2* of *S. pneumoniae*, *idi* of *E. coli*, *mvaE* and *mvaS* of *E. faecali*	[16]
pSSN12Didi	pSTV28 containing *mvaK1*, *mvaK2* and *mvaD* from *Streptococcus pneumoniae*, *idi* of *E. coli*	[12]
pTSN-Bisa-Sa	pTrc99A containing *mvaE* and *mvaS* of *Enterococcus faecalis*, *mvaK1* of *S. aureus*, *mvaK2*, and *mvaD* of *S. pneumoniae*, *idi*, and *ispA* of *E. coli*, *MrBBS of M. recutita*	This study
pTSN-Bisa-Mm	pTrc99A containing *mvaE* and *mvaS* of *Enterococcus faecalis*, *mvaK1* of *M. masei*, *mvaK2*, and *mvaD* of *S. pneumoniae*, *idi*, and *ispA* of *E. coli*, *MrBBS of M. recutita*	This study
pSEVA231-Bisa-Mm	pTSN-Bisa-Mm with pBBR1 ori instead of pBR322 ori	This study
pSSN12Didi-MrBBS-IspA	pSSN12Didi containing *ispA* of *E. coli* and *MrBBS of M. recutita*	This study
pTSN-Bisa-Mm-GapC	pTSN-Bisa-Mm containing *gapC* of *C. acetoburylicum*	This study
pdCas9-sgRNA-GapA	pSECRi containing gRNA targeting *gapA* gene	This study

The *E. coli* codon-optimized *mvaK1* gene of *M. mazei* (GenBank accession number: KKI06753.1) was synthesized by Bioneer (Figure S1). The synthesized *mvaK1* was PCR-amplified with MM-IF and MM-IR primers, and the plasmid backbone was amplified with the MM-VF and MM-VR primers from pSNA-MrBBS-IspA. The two PCR-amplicons were assembled via the Gibson Assembly Method [30] using Gibson Assembly Master Mix (NEB), resulting in the construction of the pTSN-Bisa-Mm plasmid.

The *mvaK1* from *S. aureus* was amplified with SA-IF and SA-IR primers from pTM-BBS, and the plasmid for the backbone was obtained from pSNA-MrBBS-IspA by PCR with a set of primers of SA-VF and SA-VR, followed by assembly with the Gibson assembly method. The resulting plasmid was named pTSN-Bisa-Sa.

For the construction of pSEVA231-Bisa-Mm, the first fragment containing the MVA pathway gene, *ispA* and *MrBBS* were amplified in pTSN-Bisa-Mm using primers of pBBR1-IF and pBBR1-IR. The second fragment harboring the kanamycin-resistant gene and pBBR1 origin was amplified from pSEVA231 as a template using pBBR1-VF and pBBR1-VR primers. The fragments were assembled via Gibson Assembly method.

To construct pSSN12Didi-MrBBS-IspA, the *MrBBS* and *ispA* including *trc* promoter gene were amplified using Didi-I-F and Didi-I-R primers from the pTSN-Bisa-Mm plasmid. The vector backbone containing genes encoding enzymes of the lower MVA pathway was amplified using Didi-V-F and Didi-V-R primers from the pSSN12Didi plasmid. Two amplified fragments were then assembled via the Gibson Assembly kit.

The *E. coli* codon-optimized *gapC* gene from *C. acetobutylicum* (GenBank accession number: NP_347346) including the ribosome binding site and SpeI/XbaI restriction enzyme sites was synthesized by Macrogen (Seoul, Korea). The synthesized DNA was then digested with SpeI/XbaI, and the fragment containing the *gapC* gene was gel-purified. The other fragment was prepared by digesting the pTSN-Bisa-Mm plasmid with XbaI. The two fragments were then ligated byg T4 DNA ligase, which created the plasmid pTSN-Bisa-Mm-GapC.

We used the primers of gapA-gRNA-F and gapA-gRNA-R for amplification of the whole pSECRi plasmid by PCR. The amplified DNA fragment was gel-purified and treated with T4 polynucleotide kinase to phosphorylate it. T4 DNA ligase was used ligate the PCR product. The sequences of all genes associated with the (−)-α-bisabolol biosynthetic pathway were verified by Sanger sequencing (Magcogen).

4.3. Batch and Fed-Batch Fermentation

To prepare the pre-culture, recombinant *E. coli* was cultured in 5 mL of LB medium supplied with appropriate antibiotics at 30 °C and 200 rpm overnight. The batch fermentation was carried out by inoculating 1% (*v*/*v*) of the pre-culture into 3 mL of the TB medium with 1% (*w*/*v*) of glycerol in a 50 mL mini-bioreactor (SPL Life Sciences, Gyeonggi-do, Korea). 20% (*v*/*v*) of *n*-dodecane was overlaid to extract (−)-α-bisabolol from all fermentation broths. The cultures were incubated at 30 °C and 200 rpm for 72 h. For fed-batch fermentation, the pre-culture was prepared in 5 mL of TB medium with 10 g/L of glycerol at 30 °C and 200 rpm overnight. 1% (*v*/*v*) of the cells were inoculated into 300 mL of TB medium supplied with 0.1% (*v*/*v*) of trace metal solution (27 g/L $FeCl_3·6H_2O$, 2 g/L $ZnCl_2·4H_2O$, 2 g/L $CoCl_2·6H_2O$, 2 g/L $Na_2MoO_4·2H_2O$, 1 g/L $CaCl_2·2H_2O$, 1.3 g/L $CuCl_2·6H_2O$, and 0.5 g/L H_3BO_3), 0.98 g/L of $MgSO_4$, 1% (*v*/*v*) vitamin, appropriate antibiotics, 20% (*v*/*v*) of *n*-dodecane in a 1 L fermenter (CNS Inc. Daejeon, Korea). 60% (*w*/*v*) of glycerol containing 9.8 g/L $MgSO_4$, 2% (*v*/*v*) of trace metal solution and 0.25% (*v*/*v*) of thiamine solution was fed intermittently during the fed-batch fermentation. The fed-batch fermentation was maintained at 30 °C, 1 vvm of air flow rate, and 280 or 1000 rpm of agitation. The pH was adjusted to pH 7.0 by adding 1 N HCl and 1 N NaOH solutions.

4.4. (−)-α-Bisabolol Quantification

Extraction of (−)-α-bisabolol proceeded in the *n*-dodecane phase which is initially added to the culture broth throughout the cultivation. The overlaid *n*-dodecane phase was collected after the pellet, supernatant, and a layer of *n*-dodecane were fractionated from the culture broth using centrifugation at 13,000 rpm for 3 min. Subsequently, the collected *n*-dodecane was analyzed for the determination of (−)-α-bisabolol concentration using a gas chromatograph (GC, 7890B, Agilent, SC, USA) which is supplied with a flame ionization detector (FID) with HP-5 column (30 m × 0.320 mm × 0.25 μm, Agilent, SC, USA). As the carrier gas, helium was used at a flow rate of 1 mL/min. Temperatures of an injector and an FID were maintained at 240 °C and 250 °C, respectively. The programmed temperature gradients controlled the column temperature: isotherm at 60 °C for 2 min; increase at a rate of 5 °C/min to 200 °C; isotherm at 200 °C for 2 min; increase at 50 °C/min to 300 °C; and isotherm at 300 °C for 5 min. For the generation of a standard curve, (−)-α-bisabolol was purchased from Sigma-Aldrich. In the GC analysis, there was a peak at 21.7 min in the *n*-dodecane phase sample of recombinant *E. coli* as a major peak (>95%) except for a peak of *n*-dodecane (11.5 min). The peak at 21.7 min corresponded to the standard (−)-α-bisabolol compound dissolved in *n*-dodecane. The (−)-α-bisabolol concentration produced was determined as follows:

$$(-)\text{–}\alpha\text{–Bisabolol}\ (\text{g/L}) = \frac{((-)\text{–}\alpha\text{–Bisabolol in } n\text{–dodecane}) \times (\text{Volume of } n\text{–dodecane})}{\text{Volume of medium}} \quad (1)$$

4.5. Determination of Cell Growth and Metabolites

Cell growth was monitored by measuring the absorbance at 600 nm (OD_{600}) using a spectrophotometer (Ultrospec 8000, GE Healthcare, Uppsala, Sweden). After the centrifugation of the culture broth at 13,000 rpm for 3 min, the overlaid *n*-dodecane phase was removed, and the remaining supernatant was used for analyzing metabolite concentrations. The concentrations of glycerol, acetate, and MVA were measured by high-performance liquid chromatography (HPLC, Agilent Technologies 1200 series) equipped with a refractive index detector (RID) with an Aminex HPX-87H column (1300 mm × 7.8 mm, Bio-Rad, Hercules, CA, USA). The column was eluted with 4 mM of sulfuric acid at a flow rate of 0.5 mL/min at 50 °C. All reagents for the standard solution were purchased from Sigma-Aldrich.

5. Conclusions

We improved (−)-α-bisabolol productivity from engineered *E. coli*, which can serve as a promising platform strain for the microbial production of (−)-α-bisabolol at an industrial scale. Metabolic engineering strategies used in this study, including feedback-resistance of MvaK1 enzyme, reinforcement of lower MVA pathway flux, balance of the NADPH and ATP pools, and optimization of fermentation, could be applied to enhance the terpenoid production from engineered microbes. Moreover, metabolic modeling based on genome-wide omics data might provide clues to identify unknown bottlenecks and interpret the results. This experiment will be conducted as a further study.

Supplementary Materials: The following are available online at http://www.mdpi.com/2073-4344/9/5/432/s1, Figure S1: Nucleotide sequence of the *E. coli* codon-optimized *mvaK1* gene derived from *M. mazei*, tableure S2: Fed-batch fermentation of the engineered *E. coli* DH5α, Table S1: List of primers used in this study.

Author Contributions: Conceptualization, S.-J.K., S.K.K., D.-H.L., and S.-G.L.; methodology, S.-J.K., S.K.K., W.S., S.-G.W., and H.K.; software, S.-J.K. and H.K.; validation, S.-J.K., S.K.K., and W.S.; formal analysis, S.-J.K., S.K.K., W.S., and S.-G.W.; investigation, S.-J.K., S.K.K., W.S., and S.-G.W.; resources, S.-J.K., S.K.K., and W.S.; data curation, S.-J.K. and S.K.K.; writing—original draft preparation, S.-J.K., S.K.K., D.-H.L., and S.-G.L.; writing—review and editing, S.-J.K., S.K.K., H.L., S.-J.Y., D.-H.L., and S.-G.L.; visualization, S.-J.K. and S.K.K.; supervision, D.-H.L. and S.-G.L.; project administration, D.-H.L. and S.-G.L.; funding acquisition, D.-H.L. and S.-G.L.

Acknowledgments: This research was funded by the Bio & Medical Technology Development Program, grant number 2018M3A9H3024746 and the Intelligent Synthetic Biology Center of Korea, grant number 2011-0031944 of the National Research Foundation (NRF) funded by the Ministry of Science and ICT of the Republic of Korea. The KRIBB Research Initiative Program also funded this research. The authors would like to thank Victor D. Lorenzo (Centro Nacional de Biotecnología—CSIC, Campus de Cantoblanco, Madrid, Spain) for the kind donation of the pSEVA plasmids and members of the Synthetic Biology Laboratory in the Synthetic Biology and Bioengineering Center at KRIBB for their valuable comments and helpful discussions.

Conflicts of Interest: The authors declare no conflict of interest.

References

1. Brehm-Stecher, B.F.; Johnson, E.A. Sensitization of *Staphylococcus aureus* and *Escherichia coli* to antibiotics by the sesquiterpenoids nerolidol, farnesol, bisabolol, and apritone. *Antimicrob. Agents Chemother.* **2003**, *47*, 3357–3360. [CrossRef] [PubMed]
2. Forrer, M.; Kulik, E.M.; Filippi, A.; Waltimo, T. The antimicrobial activity of α-bisabolol and tea tree oil against *Solobacterium moorei*, a gram-positive bacterium associated with halitosis. *Arch. Oral Biol.* **2013**, *58*, 10–16. [CrossRef] [PubMed]
3. Kamatou, G.P.; Viljoen, A.M. A review of the application and pharmacological properties of α-bisabolol and α-bisabolol-rich oils. *J. Am. Oil Chem. Soc.* **2010**, *87*, 1–7. [CrossRef]
4. Leite, G.d.O.; Leite, L.H.; Sampaio, R.d.S.; Araruna, M.K.A.; de Menezes, I.R.A.; da Costa, J.G.M.; Campos, A.R. (−)-α-Bisabolol attenuates visceral nociception and inflammation in mice. *Fitoterapia* **2011**, *82*, 208–211. [CrossRef] [PubMed]
5. Russell, K.; Jacob, S.E. Bisabolol. *Dermatitis* **2010**, *21*, 57–58.
6. Global One-Stop Reports Center. Available online: Http://www.Gosreports.Com/global-%ce%b1-bisabolol-market-worth-73-million-by-2020/ (accessed on 30 march 2019).
7. Albertti, L.A.G.; Delatte, T.L.; de Farias, K.S.; Boaretto, A.G.; Verstappen, F.; van Houwelingen, A.; Cankar, K.; Carollo, C.A.; Bouwmeester, H.J.; Beekwilder, J. Identification of the bisabolol synthase in the endangered candeia tree (*Eremanthus erythropappus (dc) mcleisch*). *Front. Plant Sci.* **2018**, *9*, 1340. [CrossRef]

8. de Souza, A.T.; Benazzi, T.L.; Grings, M.B.; Cabral, V.; da Silva, E.A.; Cardozo-Filho, L.; Antunes, O.A.C. Supercritical extraction process and phase equilibrium of candeia (*Eremanthus erythropappus*) oil using supercritical carbon dioxide. *J. Supercrit. Fluids* **2008**, *47*, 182–187. [CrossRef]
9. Son, Y.J.; Kwon, M.; Ro, D.K.; Kim, S.U. Enantioselective microbial synthesis of the indigenous natural product (−)-α-bisabolol by a sesquiterpene synthase from chamomile (*Matricaria recutita*). *Biochem. J.* **2014**, *463*, 239–248. [CrossRef]
10. Chatzivasileiou, A.O.; Stephanopoulos, G.; Ward, V.C.A. Metabolic engineering of *Escherichia coli* for the production of isoprenoids. *FEMS Microbiol. Lett.* **2018**, *365*, fny079.
11. Ajikumar, P.K.; Xiao, W.H.; Tyo, K.E.; Wang, Y.; Simeon, F.; Leonard, E.; Mucha, O.; Phon, T.H.; Pfeifer, B.; Stephanopoulos, G. Isoprenoid pathway optimization for taxol precursor overproduction in *Escherichia coli*. *Science* **2010**, *330*, 70–74. [CrossRef]
12. Han, G.H.; Kim, S.K.; Yoon, P.K.-S.; Kang, Y.; Kim, B.S.; Fu, Y.; Sung, B.H.; Jung, H.C.; Lee, D.H.; Kim, S.W. Fermentative production and direct extraction of (−)-α-bisabolol in metabolically engineered *Escherichia coli*. *Microb. Cell Fact.* **2016**, *15*, 185. [CrossRef]
13. Voynova, N.E.; Rios, S.E.; Miziorko, H.M. *Staphylococcus aureus* mevalonate kinase: Isolation and characterization of an enzyme of the isoprenoid biosynthetic pathway. *J. Bacteriol.* **2004**, *186*, 61–67. [CrossRef]
14. Andreassi, J.L.; Dabovic, K.; Leyh, T.S. *Streptococcus pneumoniae* isoprenoid biosynthesis is downregulated by diphosphomevalonate: An antimicrobial target. *Biochemistry* **2004**, *43*, 16461–16466. [CrossRef] [PubMed]
15. Kim, S.K.; Kim, S.H.; Subhadra, B.; Woo, S.G.; Rha, E.; Kim, S.W.; Kim, H.; Lee, D.H.; Lee, S.G. A genetically encoded biosensor for monitoring isoprene production in engineered *Escherichia coli*. *ACS Synth. Biol.* **2018**, *7*, 2379–2390. [CrossRef]
16. Kim, S.K.; Han, G.H.; Seong, W.; Kim, H.; Kim, S.W.; Lee, D.H.; Lee, S.G. CRISPR interference-guided balancing of a biosynthetic mevalonate pathway increases terpenoid production. *Metab. Eng.* **2016**, *38*, 228–240. [CrossRef] [PubMed]
17. Primak, Y.A.; Du, M.; Miller, M.C.; Wells, D.H.; Nielsen, A.T.; Weyler, W.; Beck, Z.Q. Characterization of a feedback-resistant mevalonate kinase from the archaeon *Methanosarcina mazei*. *Appl. Environ. Microbiol.* **2011**, *77*, 7772–7778. [CrossRef]
18. Kang, A.; George, K.W.; Wang, G.; Baidoo, E.; Keasling, J.D.; Lee, T.S. Isopentenyl diphosphate (IPP)-bypass mevalonate pathways for isopentenol production. *Metab. Eng.* **2016**, *34*, 25–35. [CrossRef] [PubMed]
19. Pontrelli, S.; Chiu, T.Y.; Lan, E.I.; Chen, F.Y.; Chang, P.C.; Liao, J.C. *Escherichia coli* as a host for metabolic engineering. *Metab. Eng.* **2018**, *50*, 16–46. [CrossRef]
20. Kazieva, E.; Yamamoto, Y.; Tajima, Y.; Yokoyama, K.; Katashkina, J.; Nishio, Y. Characterization of feedback-resistant mevalonate kinases from the methanogenic archaeons *Methanosaeta concilii* and *Methanocella paludicola*. *Microbiology* **2017**, *163*, 1283–1291. [CrossRef]
21. Martin, V.J.; Pitera, D.J.; Withers, S.T.; Newman, J.D.; Keasling, J.D. Engineering a mevalonate pathway in *Escherichia coli* for production of terpenoids. *Nat. Biotechnol.* **2003**, *21*, 796–802. [CrossRef] [PubMed]
22. Kizer, L.; Pitera, D.J.; Pfleger, B.F.; Keasling, J.D. Application of functional genomics to pathway optimization for increased isoprenoid production. *Appl. Environ. Microbiol.* **2008**, *74*, 3229–3241. [CrossRef] [PubMed]
23. Dahl, R.H.; Zhang, F.; Alonso-Gutierrez, J.; Baidoo, E.; Batth, T.S.; Redding-Johanson, A.M.; Petzold, C.J.; Mukhopadhyay, A.; Lee, T.S.; Adams, P.D. Engineering dynamic pathway regulation using stress-response promoters. *Nat. Biotechnol.* **2013**, *31*, 1039–1046. [CrossRef]
24. Liu, C.L.; Dong, H.G.; Zhan, J.; Liu, X.; Yang, Y. Multi-modular engineering for renewable production of isoprene via mevalonate pathway in *Escherichia coli*. *J. Appl. Microbiol.* **2019**, *126*, 1128–1139. [CrossRef]
25. Zhao, J.; Li, Q.; Sun, T.; Zhu, X.; Xu, H.; Tang, J.; Zhang, X.; Ma, Y. Engineering central metabolic modules of *Escherichia coli* for improving β-carotene production. *Metab. Eng.* **2013**, *17*, 42–50. [CrossRef]
26. Choi, H.S.; Lee, S.Y.; Kim, T.Y.; Woo, H.M. *In silico* identification of gene amplification targets for improvement of lycopene production. *Appl. Environ. Microbiol.* **2010**, *76*, 3097–3105. [CrossRef] [PubMed]
27. Alper, H.; Jin, Y.S.; Moxley, J.F.; Stephanopoulos, G. Identifying gene targets for the metabolic engineering of lycopene biosynthesis in *Escherichia coli*. *Metab. Eng.* **2005**, *7*, 155–164. [CrossRef] [PubMed]
28. Martinez, I.; Zhu, J.; Lin, H.; Bennett, G.N.; San, K.Y. Replacing *Escherichia coli* NAD-dependent glyceraldehyde 3-phosphate dehydrogenase (GAPDH) with a NADP-dependent enzyme from *Clostridium acetobutylicum* facilitates NADPH dependent pathways. *Metab. Eng.* **2008**, *10*, 352–359. [CrossRef] [PubMed]

29. Sambrook, J.; Russell, D.W. *Molecular Cloning: A Laboratory Manual*; Sambrook, J., Russell, D.W., Eds.; Cold Spring Harbor Laboratory Press: Cold Spring Harbor, NY, USA, 2001; Volume 3.
30. Gibson, D.G.; Young, L.; Chuang, R.Y.; Venter, J.C.; Hutchison III, C.A.; Smith, H.O. Enzymatic assembly of DNA molecules up to several hundred kilobases. *Nat. Methods* **2009**, *6*, 343. [CrossRef]

catalysts

MDPI

Article

Efficient Conversion of Acetate to 3-Hydroxypropionic Acid by Engineered *Escherichia coli*

Ji Hoon Lee [1,†], Sanghak Cha [2,†], Chae Won Kang [2], Geon Min Lee [2], Hyun Gyu Lim [2,*] and Gyoo Yeol Jung [1,2,*]

1 School of Interdisciplinary Bioscience and Bioengineering, Pohang University of Science and Technology, 77 Cheongam-Ro, Nam-Gu, Pohang, Gyeongbuk 37673, Korea; dlwlgns21c@postech.ac.kr

2 Department of Chemical Engineering and, Pohang University of Science and Technology, 77 Cheongam-Ro, Nam-Gu, Pohang, Gyeongbuk 37673, Korea; cktkdgkr@postech.ac.kr (S.C.); codnjs6897@postech.ac.kr (C.W.K.); lgm0417@postech.ac.kr (G.M.L.)

* Correspondence: hyungyu.lim@postech.ac.kr (H.G.L.); gyjung@postech.ac.kr (G.Y.J.); Tel.: +82-54-279-8335 (H.G.L.); +82-54-279-2391 (G.Y.J.)

† These authors contributed equally to this work.

Received: 17 October 2018; Accepted: 3 November 2018; Published: 7 November 2018

Abstract: Acetate, which is an abundant carbon source, is a potential feedstock for microbial processes that produce diverse value-added chemicals. In this study, we produced 3-hydroxypropionic acid (3-HP) from acetate with engineered *Escherichia coli*. For the efficient conversion of acetate to 3-HP, we initially introduced heterologous *mcr* (encoding malonyl-CoA reductase) from *Chloroflexus aurantiacus*. Then, the acetate assimilating pathway and glyoxylate shunt pathway were activated by overexpressing *acs* (encoding acetyl-CoA synthetase) and deleting *iclR* (encoding the glyoxylate shunt pathway repressor). Because a key precursor malonyl-CoA is also consumed for fatty acid synthesis, we decreased carbon flux to fatty acid synthesis by adding cerulenin. Subsequently, we found that inhibiting fatty acid synthesis dramatically improved 3-HP production (3.00 g/L of 3-HP from 8.98 g/L of acetate). The results indicated that acetate can be used as a promising carbon source for microbial processes and that 3-HP can be produced from acetate with a high yield (44.6% of the theoretical maximum yield).

Keywords: metabolic engineering; synthetic biology; 3-hydroxypropionic acid; microbial production; fatty acid synthesis; acetate

1. Introduction

Microbial conversion is a highly promising process for the production of diverse value-added chemicals and as an alternative to petroleum-based processes [1,2]. Specifically, it can utilize a variety of sugars such as glucose, galactose, xylose and glycerol, which are plentiful in nature and readily available as industrial waste, as a feedstock [3–5]. In addition to these sugars, acetate can be used as a carbon source. Acetate is cheap and greatly abundant, as it can be obtained from biomass hydrolysate or from the conversion of various single-carbon gases [6–8]. Therefore, the use of acetate may reduce the cost of feedstock and thereby facilitate the development of more economic processes. In this regard, several recent studies attempted to engineer microorganisms and demonstrated the successful conversion of acetate into value-added chemicals such as itaconic acid, succinic acid and fatty acid [9–11].

3-Hydroxypropionic acid (3-HP) is one of the important platform chemicals that can be produced by microbial fermentation [12,13]. As 3-HP consists of two functional groups (a hydroxyl and carboxylic group), it can be easily converted to other chemicals (e.g., acrylic acid, acrylamide, and propiolactone) for which there are huge markets [12]. Due to its production from sugars, several metabolic pathways have been suggested to date [12,13]. While the representative route is the coenzyme-B_{12}-dependent dehydration of glycerol [14,15], it is only applicable when the feedstock is glycerol. Alternatively, 3-HP can be produced via the reduction of malonyl-CoA using malonyl-CoA reductase (Figure 1) [16–18]. This pathway is suitable for most carbon sources, including acetate, because malonyl-CoA is a universal intermediate in cells [16,17]. Additionally, this pathway does not require an expensive cofactor, coenzyme B_{12}, which is a potential hurdle for the economic production of 3-HP [18,19].

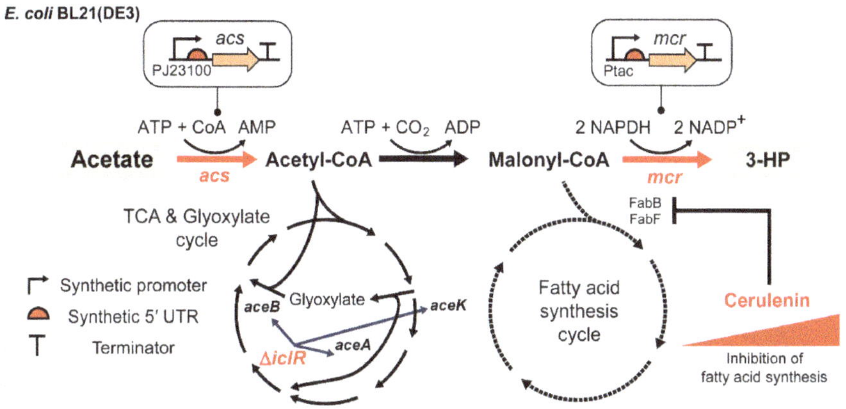

Figure 1. Schematic diagram of a metabolic pathway for the 3-HP production from acetate. 3-HP was synthesized by heterologous overexpression of *mcr* (encoding malonyl-CoA reductase from *C. aurantiacus*). For accelerated acetate assimilation, *acs* (encoding acetyl-CoA synthetase) was overexpressed and *iclR* (encoding glyoxylate shunt pathway repressor) was deleted. Deletion of *iclR* upregulates the expression of *aceA* (encoding isocitrate lyase), *aceB* (encoding malate synthase), and *aceK* (encoding isocitrate dehydrogenase kinase/phosphatase). Phosphorylation of isocitrate dehydrogenase (encoded by *icd*) results in its reduced activity. Different amounts of cerulenin were added to inhibit fatty acids biosynthesis.

To achieve the efficient conversion of acetate to 3-HP, acetate should be rapidly utilized. However, microorganisms slowly utilize acetate as a carbon source and exhibit reduced cell growth [20,21]. Therefore, acetate assimilation and biomass formation should be accelerated via genetic engineering [11,22–24]. Furthermore, once acetate is assimilated, malonyl-CoA has to be sufficiently converted to 3-HP. However, the primary use of malonyl-CoA in microorganisms is to synthesize fatty acids, which significantly reduces 3-HP production [25–27]. Thus, acetate consumption for fatty acid synthesis should be reduced to improve 3-HP production.

In this study, we demonstrated the efficient conversion of acetate to 3-HP by engineering a representative microorganism, *Escherichia coli*. Initially, we constructed a synthetic 3-HP production pathway with maximal expression of heterologous *mcr* (encoding malonyl-CoA reductase) from *Chloroflexus aurantiacus*. To accelerate acetate consumption, we activated both the acetate assimilating pathway and glyoxylate shunt pathway by amplifying *acs*, which encodes acetyl-CoA synthetase, and deleting *iclR*, which encodes the transcriptional repressor of the glyoxylate shunt pathway operon. Additionally, to enhance the conversion of malonyl-CoA to 3-HP, carbon flux into a competing pathway (i.e., fatty acid biosynthesis) was inhibited by adding cerulenin at different concentrations. Consequently, we demonstrated that 3-HP could be efficiently produced from acetate using the engineered microbial process.

2. Results

2.1. Heterologous Expression of mcr for 3-HP Production from Acetate

For 3-HP production, we introduced *mcr* from *C. aurantiacus* into the *E. coli* strain BL21(DE3) (Figure 1). To increase carbon flux toward 3-HP biosynthesis, we expressed *mcr* at a maximum level. Specifically, we developed a synthetic cassette with a strong inducible promoter (P_{tac}) and a synthetic 5′ UTR (Table 1) designed using UTR Designer [28] to ensure high transcription and translation levels. Furthermore, we used the pETDuet plasmid, which has a high copy number (~40 copies per cell), to ensure its overexpression. Additionally, we introduced 3 point mutations (N940V, K1106W, S1114R) known to enhance the activity of malonyl-CoA reductase [16].

Table 1. Synthetic 5′ UTR for gene expression.

Gene	5′ UTR sequence (5′−3′) [a]	Predicted Expression Level (a.u.)
mcr	AACAATTACTAGTAAGGAGAGGAGT	3,110,669.92
acs	AAAATCAGCGCCCAAGGAGTCACCG [b]	1,074,836.02

[a] 5′ UTR sequences were designed using UTR Designer [26]. [b] This sequence was originally designed in a previous study [9].

Following this, we cultivated the HJ1 strain (Table 2), which is BL21(DE3) with the constructed plasmid (pET-*mcr**), in modified minimal medium. After 48 h of fermentation, the HJ1 strain consumed 8.55 g/L of acetate and produced 95.7 mg/L of 3-HP (Figure 2). Although we successfully produced 3-HP from acetate with this engineered *E. coli* strain, the achieved titer was too low (1.49% of the theoretical maximum yield). Thus, the strain required further engineering to improve 3-HP production with efficient acetate utilization.

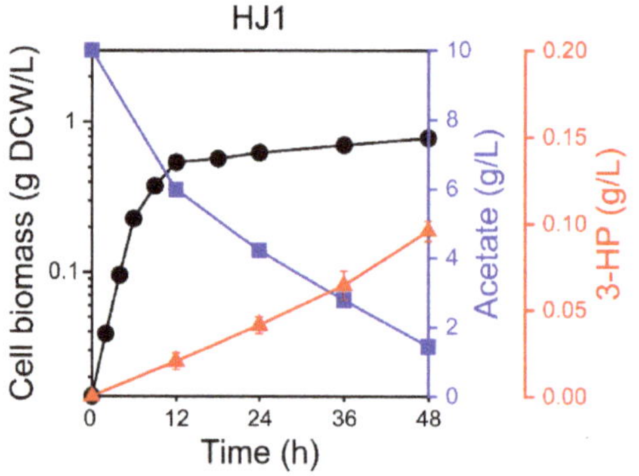

Figure 2. Time-course fermentation profile of the HJ1 strain. The left y-axis, right y-axis and right y-offset represent the cell biomass (g DCW/L), acetate (g/L) and 3-HP (g/L), respectively. The x-axis denotes time (h). Symbols: circles, cell biomass; squares, acetate; triangles, 3-HP. The error bars indicate standard deviations for measurements from three independent cultures. One OD_{600} unit corresponds to 0.31 g dry cell weight (g DCW/L).

2.2. Engineering the Acetate Assimilation and Glyoxylate Shunt Pathways

Compared to that of other sugars, the assimilation rate of acetate in microorganisms is relatively low [8]. Therefore, we investigated the effect of activating the acetate assimilation pathway on 3-HP production. To improve acetate uptake during the entire fermentation period, we expressed *acs* (encoding acetyl-CoA synthetase) with a synthetic expression cassette consisting of the strong constitutive promoter (P_{J23100}) and a synthetic 5′ UTR (Table 1). The synthetic expression cassette was inserted into the pACYCDuet plasmid, which has a moderate copy number (10–12 copies/cell), and the pACYC-Acs plasmid was introduced into the HJ1 strain, resulting in the HJ2 strain. Compared to the HJ1 strain, the HJ2 strain showed increased acetate consumption (a 1.11-fold increase) and cell biomass (a 1.90-fold increase), indicating that acetate assimilation was successfully expedited by *acs* overexpression (Figure 3A,B). Furthermore, the improved acetate assimilation in the HJ2 strain led to a 1.75-fold increase in 3-HP production (0.17 g/L, Figure 3C).

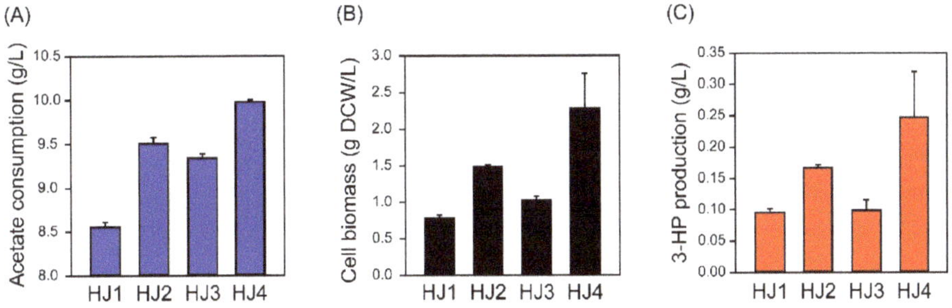

Figure 3. Genetic engineering to improve 3-HP production from acetate. Acetate consumption (g/L), cell biomass (g DCW/L) and 3-HP (g/L) production of the HJ1—4 strains after 48 h of fermentation. (**A**) Acetate consumption (g/L), (**B**) cell biomass (g DCW/L), and (**C**) 3-HP production (g/L) of engineered strains were compared. The error bars indicate standard deviations for measurements from three independent cultures.

We then evaluated the effect of activating the glyoxylate shunt pathway. Because the glyoxylate shunt pathway is responsible for a key anaplerotic reaction during acetate utilization, we expected that its activation would facilitate biomass formation and improve acetate consumption. To investigate this effect, we deleted chromosomal *iclR*, which is known to repress the expression of *aceBAK* in the glyoxylate shunt pathway [11,29,30], in the HJ1 strain. Similar to the results for the overexpression of *acs*, the acetate consumption and cell biomass of the resultant HJ3 strain were enhanced 1.09-fold and 1.31-fold, respectively (Figure 3A,B). These results indicated that *iclR* deletion successfully enhanced cell biomass synthesis from acetate, which resulted in increased overall acetate consumption.

When the overexpression of *acs* and deletion of *iclR* were combined, the acetate uptake rate (1.16-fold increase) and cell biomass (2.61-fold increase) were further enhanced (HJ4 strain, Figure 3A,B). Moreover, its 3-HP production was synergistically improved 2.54-fold (0.25 g/L) compared to that of the HJ1 strain (Figure 3C). Consequently, the combination of *acs* overexpression and *iclR* deletion resulted in the most significant improvement in acetate assimilation and 3-HP production. However, despite the improvement, only a small amount of acetate was converted to 3-HP (3.35% of the theoretical maximum yield), indicating that further flux control is required for the efficient conversion of acetate to 3-HP.

2.3. Improved 3-HP Production from Acetate by Inhibiting Fatty Acid Synthesis

Despite the elevated acetate consumption level, the 3-HP titer was still low, indicating the inefficient conversion of malonyl-CoA due to leakage toward fatty acid synthesis. Therefore, we decided to inhibit fatty acid synthesis to increase the intracellular malonyl-CoA pool for 3-HP production (Figure 1). Bacterial fatty acid synthesis can be inhibited by the addition of cerulenin, which binds to FabB and FabF and irreversibly inactivates them [31–33]. Thus, we cultivated the HJ4 strain with the addition of different levels of cerulenin (10, 25, 50 and 100 μM) to gradually reduce carbon flux from malonyl-CoA to fatty acids. As expected, the addition of cerulenin was unfavorable for cell growth (Figure 4A–C). On the other hand, notably, it was highly beneficial for 3-HP production, as the titers and yields were dramatically increased. Specifically, when 50 μM of cerulenin was added, the HJ4 strain produced 3.00 g/L of 3-HP while consuming 8.98 g/L of acetate (0.30 g/g, 44.6% of the maximum theoretical yield, Figure 4D). This titer was 12.0-fold higher compared to the 3-HP titer produced by the same strain without cerulenin addition. These results indicated that 3-HP production from acetate was successfully improved by inhibiting fatty acid biosynthesis and that flux control around malonyl-CoA was critical for the efficient conversion of acetate to 3-HP.

Table 2. Bacterial strains and plasmids used in this study.

Name	Description	Source
Strains		
Mach1-T1^R	*E. coli* F$^-$ φ80(*lacZ*)ΔM15 Δ*lac*X74 *hsd*R($r_K^-m_K^+$) Δ*rec*A1398 *end*A1 *ton*A	Invitrogen
BL21(DE3)	*E. coli* F$^-$ *ompT gal dcm lon hsd*SB (rB^- mB^-) λ(DE3)	Invitrogen
HJ1	BL21(DE3)/pET-*mcr**	This study
HJ2	BL21(DE3)/pET-*mcr**/pACYC-*acs*	This study
HJ3	BL21(DE3) Δ*iclR*/pET-*mcr**	This study
HJ4	BL21(DE3) Δ*iclR*/pET-*mcr**/pACYC-*acs*	This study
Plasmids		
PLB0110	Source of *mcr*	[15]
pETDuet	Expression vector, ColE1 ori, Amp^R	Novagen
pACYCDuet	Expression vector, p15A ori, Cm^R	Novagen
pKD46	Red recombinase expression vector, Amp^R	[32]
pFRT72$_{variant}$	Source of mutant FRT-kan^R-FRT	[33]
pCP20	FLP expression vector, Amp^R, Cm^R	[32]
pET-*mcr**	pETDuet/P$_{tac}$-SynUTR$_{mcr}$-*mcr* $^{N940V, K1106W, S1114R}$	This study
pACYC-*acs*	pACYCDuet/P$_{BBa_J23100}$-SynUTR$_{acs}$-*acs*	This study

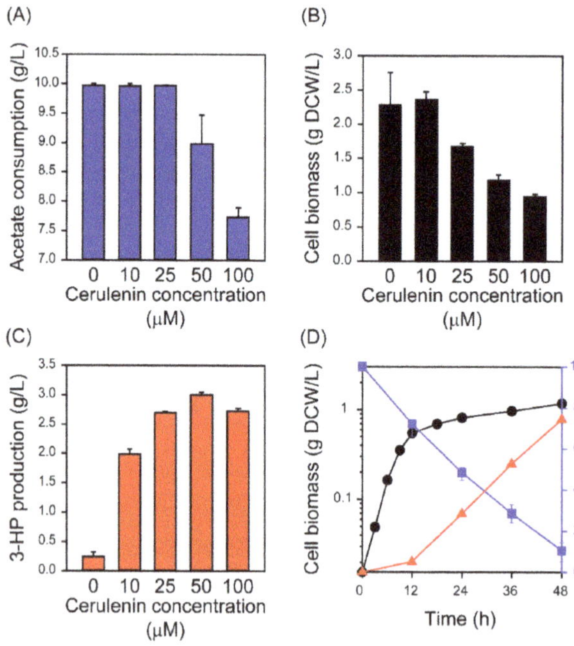

Figure 4. Improved 3-HP production with the addition of cerulenin. (**A**) Acetate consumption (g/L), (**B**) cell biomass (g DCW/L) and (**C**) 3-HP (g/L) production of the HJ4 strain with different cerulenin concentrations after 48 h fermentation. (**D**) Time-course fermentation profile (48 h) of the HJ4 strain with 50 μM of cerulenin in a modified acetate minimal medium. The left y-axis, right y-axis and right y-offset represent the cell biomass (g DCW/L), acetate (g/L) and 3-HP (g/L), respectively. The x-axis denotes time (h). Symbols: circles, cell biomass; squares, acetate; triangles, 3-HP. The error bars indicate standard deviations for measurements from three independent cultures.

3. Discussion

In this study, we developed a microbial process for the conversion of acetate to 3-HP. Initially, we introduced *mcr* at a maximum expression level to enable 3-HP production by an *E. coli* strain. In addition, acetate consumption was expedited by engineering the acetate assimilating pathway and glyoxylate shunt pathway. Furthermore, we increased carbon flux to the 3-HP production pathway by inhibiting fatty acid synthesis with the addition of cerulenin. These efforts allowed us to efficiently convert 8.98 g/L of acetate to 3.00 g/L of 3-HP. This is the first report on producing 3-HP from acetate, and the achieved yield (0.30 g/g, 44.6% of the maximum theoretical yield) was superior to the previously reported value (0.19 g/g) with glucose utilization [16].

To further improve the 3-HP production in *E. coli*, it could be more engineered. For example, the activity of heterologous malonyl-CoA reductase from *C. aurantiacus* could be improved by codon optimization of the *mcr* coding sequence. A number of previous studies have shown that codon optimization can elevate the activities of heterologous enzymes [34–36]. Thus, codon optimization of *mcr* would further enhance 3-HP production by expediting the conversion of malonyl-CoA to 3-HP. Furthermore, because production of 3-HP from acetate requires energy and reducing cofactor, their stable supplementation should be achieved by balancing between the TCA cycle and 3-HP synthesis. Therefore, the fine-tuning of flux toward the TCA cycle should be critical for 3-HP production. Moreover, based on the results obtained in this study, reduction of the flux to fatty acid synthesis may enhance the production of diverse malonyl-CoA-derived biochemicals in *E. coli*.

4. Materials and Methods

4.1. Reagents

Oligonucleotides, which are listed in Table 1, were synthesized by Cosmogenetech (Seoul, Korea). Plasmid DNA and genomic DNA were purified using Expin^{TM} Plasmid SV and Expin^{TM} Cell SV kits (GeneAll Biotechnology, Seoul, Korea). PCR products were purified using Expin^{TM} Gel SV kits (GeneAll). Restriction enzymes were purchased from New England Biolabs (Ipswich, MA, USA). Cerulenin was purchased from Cayman Chemical (Ann Arbor, MI, USA). 3-HP was obtained from Tokyo Chemical Industry (Tokyo, Japan). Other chemical reagents were purchased from Sigma-Aldrich (St. Louis, MO, USA).

4.2. Plasmid Cloning and Bacterial Strain Construction

All bacterial strains and plasmids used in this study are summarized in Table 2. Synthetic 5′ untranslated regions (5′ UTRs) were generated by UTR Designer [28] and are listed in Table 3. *E. coli* Mach1-T1^{R} (Invitrogen, Carlsbad, CA, USA) was used as a cloning host. To construct the pET-*mcr** plasmid, *mcr* was initially amplified from the PLB0110 plasmid [17] using the O-mcr-F1, O-mcr-F2, and O-mcr-B primers to attach a strong inducible promoter (P_{tac}) and a synthetic 5′ UTR. Then, the amplified fragment was digested with *BamHI* and *XhoI* endonucleases and inserted into the pETDuet plasmid. It should be noted that the coding sequence of *mcr* was modified from its original sequence via conventional site-directed mutagenesis using the O-N940V-F and O-N940V-B, O-K1106W-F and O-K1106W-B, and O-S1114R-F and O-S1114R-B primer pairs for the introduction of N940V(AAT to GTG), K1106W (AAG to TGG), and S1114R (AGT to CGT) mutations, respectively [16].

To construct the pACYC-*acs* plasmid, *acs* was amplified from the genomic DNA of *E. coli* BL21(DE3) and attached to the BBa_J23100 promoter obtained from the Registry of Standard Biological Parts (http://parts.igem.org) and a synthetic 5′ UTR using the O-Acs-F and O-Acs-B primers. The amplified fragment was digested using *EcoRI* and *SacI* restriction enzymes and introduced into the pACYCDuet plasmid.

Genome manipulation was conducted using the Lambda-Red recombination method with the pKD46 plasmid [37]. Chromosomal *iclR* was deleted by inserting a FRT-kan^{R}-FRT fragment that was amplified using the R-iclR-F and R-iclR-B primer pair. The integrated kanamycin resistance gene was removed by expression of the flippase from the pCP20 plasmid.

Table 3. Oligonucleotides used in this study.

Name	Sequence (5′−3′)
O-mcr-F1	GGAATTGTGAGCGGAT<u>AACAATTACTAGTAAGGAGAGGAGT</u> ATGAGCGGAACAGGACGACT
O-mcr-F2	GGATCCTTGACAATTAATCATCGGCTCGTATAATGTGTG GAATTGTGAGCGGATAACAATT
O-mcr-B	CTCGAGTGCGAAAAAACCCCGCCGAAGCGGGG TTTTTTGCGGCATGCTTACACGGTAATCGCCCGT
O-N940V-F	TATTACCTTGCCGACCGCAATGTCAGTGGTGAGACATTCC
O-N940V-B	GCGGTCGGCAAGGTAATAG
O-K1106W-F	ATTTCCGGGTAGCGCGCAAGATTGCCCTGAGTGATGGTG
O-K1106W-B	GCGCGCTACCCGGAAATG
O-S1114R-F	TGAGTGATGGTGCCAGTCTCGCGCTGGTCACTC
O-S1114R-B	AGACTGGCACCATCACTCAGGGC
O-acs-F	GAATTCTTGACGGCTAGCTCAGTCCTAGGTACAGTGCTAGC <u>AAAATCAGCGCCCAAGGAGTCACCG</u>ATGAGCCAAATTCACAAACACA
O-acs-B	GAGCTCAAAAAAAACCCCGCCCTGTCAGGGGCGGGG TTTTTTTTTTACGATGGCATCGCGATAG
R-iclR-F	TGCCACTCAGGTATGATGGGCAGAATATTGCC TCTGCCCGCCAGAAAAAGGCATGACCGGCGCGATGC
R-iclR-B	TAACAATAAAAATGAAAATGATTTCCACGAT ACAGAAAAAGGAGACTGTCGCTCAGCGGATCTCATGCGC
C-iclR-F	CAACATTAACTCATCGGATCAG
C-iclR-B	TCTATTGCCACTCAGGTATGATGGGC

4.3. Cultivation Methods

Cells were cultivated in a modified minimal medium consisting of 100 mM phosphate buffer (pH 7.0), 1.0 g/L NaCl, 1.0 g/L NH_4Cl, 0.5 g/L $MgSO_4 \cdot 7H_2O$, and 1.0 g/L yeast extract. Pre-neutralized 10 g/L acetate with NaOH was used as a carbon source. To initiate a culture, a single colony was inoculated into 3 mL of the medium in a test-tube. After an overnight incubation, the turbid culture was refreshed by re-inoculating into fresh medium. When the OD_{600} reached ~1.0, the refreshed seed was transferred to a 25 mL medium in a 300 mL flask until an OD_{600} of 0.05 was obtained. Genes under the P_{tac} promoter were expressed by the addition of 0.1 mM IPTG when the OD_{600} reached 1. At this time, different amounts of cerulenin were also added to inhibit fatty acid synthesis. The pH was adjusted by adding 5 M of HCl with a 12 h interval during 48 h of culture. The plasmids were maintained by adding 50 μg/mL ampicillin, 34 μg/mL chloramphenicol, and 50 μg/mL streptomycin. All cell cultures were conducted with three biological replicates.

4.4. Analytical Methods

To monitor the cell growth, OD_{600} was measured using a UV-1700 spectrophotometer (Shimadzu, Kyoto, Japan). Acetate and 3-HP were quantified using an Ultimate 3000 high-performance liquid chromatography (HPLC) system (Dionex, Sunnyvale, CA, USA). To separate the metabolites, Aminex HPX-87H (Bio-Rad Laboratories, Richmond, CA, USA) was used with 5 mM H_2SO_4 solution as a mobile phase (0.6 mL/min). The temperature of the column oven was set to 14 °C. Refractive index signals were measured using a Shodex RI-101 detector (Shodex, Klokkerfaldet, Denmark).

4.5. Calculation of the Theoretical Maximum Yield

Initially, 1 mol acetate could be converted into 1 mol acetyl-CoA with the generation of 1 mol AMP from 1 mol ATP via the Acs pathway. The generation of 1 mol AMP was equivalent to the consumption of 2 mol ATP (Equation (1)). Next, 1 mol malonyl-CoA was produced from 1 mol acetyl-CoA with the consumption of an additional 1 mol ATP. Finally, 1 mol malonyl-CoA was converted into 1 mol 3-HP with the consumption of 2 mol NADPH (Equation (3)). Therefore, 3 mol ATP and 2 mol NADPH were required to produce 3-HP from acetate (Equation (4)). They could be obtained from the TCA cycle with the oxidation of 1 mol acetate (1 mol NADH is equivalent to 2.5 mol ATP; 1 mol FADH is equivalent to 1.5 mol ATP; 1 mol GTP is equivalent to 1 mol ATP; 1 mol NADH is equivalent to 1 mol NADPH) (Equations (1), (5) and (6)). Consequently, the production of 1 mol 3-HP required 2 mol acetate (Equation (7), 50% mol/mol and 0.75 g/g).

$$\text{Acetate} + 2\ \text{ATP} + \text{CoA} \rightarrow \text{Acetyl-CoA} + 2\ \text{ADP} + 2\ \text{P}_\text{i} \tag{1}$$

$$\text{Acetyl-CoA} + \text{ATP} + \text{CO}_2 \rightarrow \text{Malonyl-CoA} + \text{ADP} + \text{P}_\text{i} \tag{2}$$

$$\text{Malonyl-CoA} + 2\ \text{NADPH} \rightarrow \text{3-HP} + 2\ \text{NADP}^+ + \text{CoA} \tag{3}$$

$$\text{Acetate} + \text{CO}_2 + 3\ \text{ATP} + 2\ \text{NADH} \rightarrow \text{3-HP} + 3\ \text{ADP} + 3\ \text{P}_\text{i} + 2\ \text{NAD}^+ \tag{4}$$

$$\text{Acetyl-CoA} + 3\ \text{NAD}^+ + \text{FAD}^+ + \text{GDP} + \text{P}_\text{i} \rightarrow 2\ \text{CO}_2 + 3\ \text{NADH} + \text{FADH} + \text{GTP} + \text{CoA} \tag{5}$$

$$\text{Acetate} + 2\ \text{NADP}^+ \rightarrow 2\ \text{CO}_2 + 2\ \text{NADPH} + 3\ \text{ATP} \tag{6}$$

$$2\ \text{Acetate} \rightarrow \text{3-HP} + \text{CO}_2 \tag{7}$$

Author Contributions: Conceptualization, J.H.L., S.C., H.G.L., and G.Y.J.; methodology, J.H.L., S.C., H.G.L., and G.Y.J.; software, J.H.L.; validation, J.H.L. and S.C.; formal analysis, J.H.L. and S.C.; investigation, J.H.L., S.C., C.W.K., and G.M.L.; resources, J.H.L., S.C., C.W.K., and G.M.L.; data curation, J.H.L., S.C., C.W.K., and G.M.L.; writing—original draft preparation, J.H.L., S.C., C.W.K., and G.M.L.; writing—review and editing, H.G.L. and G.Y.J.; visualization, J.H.L., S.C., H.G.L., and G.Y.J.; supervision, H.G.L. and G.Y.J.; project administration, G.Y.J.; funding acquisition, G.Y.J.

Funding: This research was supported by the C1 Gas Refinery Program (NRF-2018M3D3A1A01055754) and the Global Research Laboratory Program (NRF-2016K1A1A2912829) funded by the Ministry of Science and ICT. This research was also supported by the Korea Institute of Energy Technology Evaluation and Planning (KETEP) and the Ministry of Trade, Industry & Energy (MOTIE) of the Republic of Korea (no. 20174030201600).

Conflicts of Interest: The authors declare no conflicts of interest.

References

1. Kidanu, W.G.; Trang, P.T.; Yoon, H.H. Hydrogen and volatile fatty acids production from marine macroalgae by anaerobic fermentation. *Biotechnol. Bioprocess Eng.* **2017**, *22*, 612–619. [CrossRef]
2. Shrestha, B.; Dhakal, D.; Darsandhari, S.; Pandey, R.P.; Pokhrel, A.R.; Jnawali, H.N.; Sohng, J.K. Heterologous production of clavulanic acid intermediates in Streptomyces venezuelae. *Biotechnol. Bioprocess Eng.* **2017**, *22*, 359–365. [CrossRef]
3. Luo, X.; Ge, X.; Cui, S.; Li, Y. Value-added processing of crude glycerol into chemicals and polymers. *Bioresour. Technol.* **2016**, *215*, 144–154. [CrossRef] [PubMed]
4. Wei, N.; Quarterman, J.; Jin, Y.-S. Marine macroalgae: An untapped resource for producing fuels and chemicals. *Trends Biotechnol.* **2013**, *31*, 70–77. [CrossRef] [PubMed]
5. Ha, S.-J.; Galazka, J.M.; Kim, S.R.; Choi, J.-H.; Yang, X.; Seo, J.-H.; Glass, N.L.; Cate, J.H.D.; Jin, Y.-S. Engineered Saccharomyces cerevisiae capable of simultaneous cellobiose and xylose fermentation. *Proc. Natl. Acad. Sci. USA* **2011**, *108*, 504–509. [CrossRef] [PubMed]
6. Huang, B.; Yang, H.; Fang, G.; Zhang, X.; Wu, H.; Li, Z.; Ye, Q. Central pathway engineering for enhanced succinate biosynthesis from acetate in *Escherichia coli*. *Biotechnol. Bioeng.* **2018**, *115*, 943–954. [CrossRef] [PubMed]

7. Novak, K.; Pflügl, S. Towards biobased industry: Acetate as a promising feedstock to enhance the potential of microbial cell factories. *FEMS Microbiol. Lett.* **2018**, *365*, fny226. [CrossRef] [PubMed]
8. Lim, H.G.; Lee, J.H.; Noh, M.H.; Jung, G.Y. Rediscovering Acetate Metabolism: Its Potential Sources and Utilization for Biobased Transformation into Value-Added Chemicals. *J. Agric. Food Chem.* **2018**, *66*, 3998–4006. [CrossRef] [PubMed]
9. Leone, S.; Sannino, F.; Tutino, M.L.; Parrilli, E.; Picone, D. Acetate: Friend or foe? Efficient production of a sweet protein in *Escherichia coli* BL21 using acetate as a carbon source. *Microb. Cell Fact.* **2015**, *14*, 106. [CrossRef] [PubMed]
10. Yang, J.; Nie, Q. Engineering *Escherichia coli* to convert acetic acid to β-caryophyllene. *Microb. Cell Fact.* **2016**, *15*, 74. [CrossRef] [PubMed]
11. Noh, M.H.; Lim, H.G.; Woo, S.H.; Song, J.; Jung, G.Y. Production of itaconic acid from acetate by engineering acid-tolerant *Escherichia coli* W. *Biotechnol. Bioeng.* **2018**, *115*, 729–738. [CrossRef] [PubMed]
12. de Fouchécour, F.; Sánchez-Castañeda, A.-K.; Saulou-Bérion, C.; Spinnler, H.É. Process engineering for microbial production of 3-hydroxypropionic acid. *Biotechnol. Adv.* **2018**, *36*, 1207–1222. [CrossRef] [PubMed]
13. Kumar, V.; Ashok, S.; Park, S. Recent advances in biological production of 3-hydroxypropionic acid. *Biotechnol. Adv.* **2013**, *31*, 945–961. [CrossRef] [PubMed]
14. Seok, J.Y.; Yang, J.; Choi, S.J.; Lim, H.G.; Choi, U.J.; Kim, K.-J.; Park, S.; Yoo, T.H.; Jung, G.Y. Directed evolution of the 3-hydroxypropionic acid production pathway by engineering aldehyde dehydrogenase using a synthetic selection device. *Metab. Eng.* **2018**, *47*, 113–120. [CrossRef] [PubMed]
15. Lim, H.G.; Noh, M.H.; Jeong, J.H.; Park, S.; Jung, G.Y. Optimum Rebalancing of the 3-Hydroxypropionic Acid Production Pathway from Glycerol in *Escherichia coli*. *ACS Synth. Biol.* **2016**, *5*, 1247–1255. [CrossRef] [PubMed]
16. Liu, C.; Ding, Y.; Zhang, R.; Liu, H.; Xian, M.; Zhao, G. Functional balance between enzymes in malonyl-CoA pathway for 3-hydroxypropionate biosynthesis. *Metab. Eng.* **2016**, *34*, 104–111. [CrossRef] [PubMed]
17. Rathnasingh, C.; Raj, S.M.; Lee, Y.; Catherine, C.; Ashok, S.; Park, S. Production of 3-hydroxypropionic acid via malonyl-CoA pathway using recombinant *Escherichia coli* strains. *J. Biotechnol.* **2012**, *157*, 633–640. [CrossRef] [PubMed]
18. Cheng, Z.; Jiang, J.; Wu, H.; Li, Z.; Ye, Q. Enhanced production of 3-hydroxypropionic acid from glucose via malonyl-CoA pathway by engineered *Escherichia coli*. *Bioresour. Technol.* **2016**, *200*, 897–904. [CrossRef] [PubMed]
19. Liu, C.; Ding, Y.; Xian, M.; Liu, M.; Liu, H.; Ma, Q.; Zhao, G. Malonyl-CoA pathway: A promising route for 3-hydroxypropionate biosynthesis. *Crit. Rev. Biotechnol.* **2017**, *37*, 933–941. [CrossRef] [PubMed]
20. Enjalbert, B.; Millard, P.; Dinclaux, M.; Portais, J.-C.; Létisse, F. Acetate fluxes in Escherichia coli are determined by the thermodynamic control of the Pta-AckA pathway. *Sci. Rep.* **2017**, *7*, 42135. [CrossRef] [PubMed]
21. Rajaraman, E.; Agarwal, A.; Crigler, J.; Seipelt-Thiemann, R.; Altman, E.; Eiteman, M.A. Transcriptional analysis and adaptive evolution of *Escherichia coli* strains growing on acetate. *Appl. Microbiol. Biotechnol.* **2016**, *100*, 7777–7785. [CrossRef] [PubMed]
22. Chong, H.; Yeow, J.; Wang, I.; Song, H.; Jiang, R. Improving acetate tolerance of *Escherichia coli* by rewiring its global regulator cAMP receptor protein (CRP). *PLoS ONE* **2013**, *8*, e77422. [CrossRef] [PubMed]
23. Fernández-Sandoval, M.T.; Huerta-Beristain, G.; Trujillo-Martinez, B.; Bustos, P.; González, V.; Bolivar, F.; Gosset, G.; Martinez, A. Laboratory metabolic evolution improves acetate tolerance and growth on acetate of ethanologenic *Escherichia coli* under non-aerated conditions in glucose-mineral medium. *Appl. Microbiol. Biotechnol.* **2012**, *96*, 1291–1300. [CrossRef] [PubMed]
24. Kirkpatrick, C.; Maurer, L.M.; Oyelakin, N.E.; Yoncheva, Y.N.; Maurer, R.; Slonczewski, J.L. Acetate and formate stress: Opposite responses in the proteome of *Escherichia coli*. *J. Bacteriol.* **2001**, *183*, 6466–6477. [CrossRef] [PubMed]
25. Liu, T.; Vora, H.; Khosla, C. Quantitative analysis and engineering of fatty acid biosynthesis in *E. coli*. *Metab. Eng.* **2010**, *12*, 378–386. [CrossRef] [PubMed]
26. Yang, Y.; Lin, Y.; Li, L.; Linhardt, R.J.; Yan, Y. Regulating malonyl-CoA metabolism via synthetic antisense RNAs for enhanced biosynthesis of natural products. *Metab. Eng.* **2015**, *29*, 217–226. [CrossRef] [PubMed]
27. Zha, W.; Rubin-Pitel, S.B.; Shao, Z.; Zhao, H. Improving cellular malonyl-CoA level in *Escherichia coli* via metabolic engineering. *Metab. Eng.* **2009**, *11*, 192–198. [CrossRef] [PubMed]

28. Seo, S.W.; Yang, J.-S.; Kim, I.; Yang, J.; Min, B.E.; Kim, S.; Jung, G.Y. Predictive design of mRNA translation initiation region to control prokaryotic translation efficiency. *Metab. Eng.* **2013**, *15*, 67–74. [CrossRef] [PubMed]
29. Noh, M.H.; Lim, H.G.; Park, S.; Seo, S.W.; Jung, G.Y. Precise flux redistribution to glyoxylate cycle for 5-aminolevulinic acid production in *Escherichia coli*. *Metab. Eng.* **2017**, *43*, 1–8. [CrossRef] [PubMed]
30. Holms, W.H.; Bennett, P.M. Regulation of isocitrate dehydrogenase activity in *Escherichia coli* on adaptation to acetate. *J. Gen. Microbiol.* **1971**, *65*, 57–68. [CrossRef] [PubMed]
31. Moche, M.; Schneider, G.; Edwards, P.; Dehesh, K.; Lindqvist, Y. Structure of the complex between the antibiotic cerulenin and its target, beta-ketoacyl-acyl carrier protein synthase. *J. Biol. Chem.* **1999**, *274*, 6031–6034. [CrossRef] [PubMed]
32. Price, A.C.; Choi, K.H.; Heath, R.J.; Li, Z.; White, S.W.; Rock, C.O. Inhibition of beta-ketoacyl-acyl carrier protein synthases by thiolactomycin and cerulenin. Structure and mechanism. *J. Biol. Chem.* **2001**, *276*, 6551–6559. [CrossRef] [PubMed]
33. Rogers, J.K.; Church, G.M. Genetically encoded sensors enable real-time observation of metabolite production. *Proc. Natl. Acad. Sci. USA* **2016**, *113*, 2388–2393. [CrossRef] [PubMed]
34. Menzella, H.G. Comparison of two codon optimization strategies to enhance recombinant protein production in *Escherichia coli*. *Microb. Cell Fact.* **2011**, *10*, 15. [CrossRef] [PubMed]
35. Anthony, L.C.; Nowroozi, F.; Kwon, G.; Newman, J.D.; Keasling, J.D. Optimization of the mevalonate-based isoprenoid biosynthetic pathway in *Escherichia coli* for production of the anti-malarial drug precursor amorpha-4,11-diene. *Metab. Eng.* **2009**, *11*, 13–19. [CrossRef] [PubMed]
36. Yim, H.; Haselbeck, R.; Niu, W.; Pujol-Baxley, C.; Burgard, A.; Boldt, J.; Khandurina, J.; Trawick, J.D.; Osterhout, R.E.; Stephen, R.; et al. Metabolic engineering of *Escherichia coli* for direct production of 1,4-butanediol. *Nat. Chem. Biol.* **2011**, *7*, 445–452. [CrossRef] [PubMed]
37. Datsenko, K.A.; Wanner, B.L. One-step inactivation of chromosomal genes in *Escherichia coli* K-12 using PCR products. *Proc. Natl. Acad. Sci. USA* **2000**, *97*, 6640–6645. [CrossRef] [PubMed]

catalysts

Article

Construction of a *Vitreoscilla* Hemoglobin Promoter-Based Tunable Expression System for *Corynebacterium glutamicum*

Kei-Anne Baritugo [1,†], Hee Taek Kim [2,†], Mi Na Rhie [1,†], Seo Young Jo [1], Tae Uk Khang [2], Kyoung Hee Kang [2], Bong Keun Song [2], Binna Lee [3], Jae Jun Song [3], Jong Hyun Choi [3], Dae-Hee Lee [4], Jeong Chan Joo [2,*] and Si Jae Park [1,*]

1 Division of Chemical Engineering and Materials Science, Ewha Womans University, 52 Ewhayeodae-gil, Seodaemun-gu, Seoul 03760, Korea; 1204keianne@gmail.com (K.A.B.); mnrhie09@gmail.com (M.N.R.); kyjo3158@gmail.com (S.Y.J.)

2 Center for Bio-based Chemistry, Division of Convergence Chemistry, Korea Research Institute of Chemical Technology, P.O. Box 107, 141 Gajeong-ro, Yuseong-gu, Daejeon 34602, Korea; heetaek@krict.re.kr (H.T.K); tukhang@krict.re.kr (T.U.K.); coolcool@krict.re.kr (K.H.K.); bksong@krict.re.kr (B.K.S.)

3 Microbial Biotechnology Research Center, Jeonbuk Branch Institute, Korea Research Institute of Bioscience and Biotechnology (KRIBB), 181 Ipsin-gil, Jeongeup, Jeonbuk 56212, Korea; skanwodd@kribb.re.kr (B.L.); jjsong@kribb.re.kr (J.J.S.); jhchoi@kribb.re.kr (J.H.C.)

4 Synthetic Biology and Bioengineering Research Center, Korea Research Institute of Bioscience and Biotechnology (KRIBB), Daejeon 34141, Korea; dhlee@kribb.re.kr

* Correspondence: jcjoo@krict.re.kr (J.C.J.); parksj93@ewha.ac.kr (S.J.P.)

† These authors contributed equally to this work.

Received: 31 October 2018; Accepted: 15 November 2018; Published: 19 November 2018

Abstract: *Corynebacterium glutamicum* is an industrial strain used for the production of valuable chemicals such as L-lysine and L-glutamate. Although *C. glutamicum* has various industrial applications, a limited number of tunable systems are available to engineer it for efficient production of platform chemicals. Therefore, in this study, we developed a novel tunable promoter system based on repeats of the *Vitreoscilla* hemoglobin promoter (P_{vgb}). Tunable expression of green fluorescent protein (GFP) was investigated under one, four, and eight repeats of P_{vgb} (P_{vgb}, P_{vgb4}, and P_{vgb8}). The intensity of fluorescence in recombinant *C. glutamicum* strains increased as the number of P_{vgb} increased from single to eight (P_{vgb8}) repeats. Furthermore, we demonstrated the application of the new P_{vgb} promoter-based vector system as a platform for metabolic engineering of *C. glutamicum* by investigating 5-aminovaleric acid (5-AVA) and gamma-aminobutyric acid (GABA) production in several *C. glutamicum* strains. The profile of 5-AVA and GABA production by the recombinant strains were evaluated to investigate the tunable expression of key enzymes such as DavBA and $GadB_{mut}$. We observed that 5-AVA and GABA production by the recombinant strains increased as the number of P_{vgb} used for the expression of key proteins increased. The recombinant *C. glutamicum* strain expressing DavBA could produce higher amounts of 5-AVA under the control of P_{vgb8} (3.69 ± 0.07 g/L) than the one under the control of P_{vgb} (3.43 ± 0.10 g/L). The average gamma-aminobutyric acid production also increased in all the tested strains as the number of P_{vgb} used for $GadB_{mut}$ expression increased from single (4.81–5.31 g/L) to eight repeats (4.94–5.58 g/L).

Keywords: *Corynebacterium glutamicum*; P_{vgb}; tunable expression system; expression vectors; synthetic biology; *Vitreoscilla*; *vgb*

1. Introduction

Biorefinery processes have been developed to establish a sustainable alternative to produce chemicals, polymers, and fuels, and they currently rely on petroleum-based processes [1–17]. Development of recombinant microorganisms capable of converting a broader range of renewable biomass feedstock into bio-based chemicals, with properties comparable to those of conventional petrochemical products, have been extensively studied [18–27]. For example, processes for the production of bio-polymers, such as polylactic acid (PLA) and polybutylene succinate (PBS) in a biorefinery have been established [21–23]. Polymerization of microbial fermentation derived-monomers, such as diamines, dicarboxylic acids, and amino carboxylic acids, yields these bio-based polymers [4,5]. *Escherichia coli* is one of the most commonly used microorganisms for the production of monomers used to prepare bio-based polymers [1,4,5]. However, production of amino acid-derived products such as 5-aminovaleric acid (5-AVA) and gamma-aminobutyric acid (GABA) from glucose using recombinant *E. coli* is limited, due to its inability to accumulate important precursors such as L-lysine and L-glutamate [24,27]. In this regard, *Corynebacterium glutamicum*, which is a generally recognized as safe (GRAS) strain, is an excellent workhorse to develop because of its innate ability to produce large amounts of amino acids such as L-lysine and L-glutamate [1,4,5]. Several studies have demonstrated that *C. glutamicum* can produce high levels of bio-based nylon monomers such putrescine, succinic acid [4], GABA [27], cadaverine [28,29], 5-AVA and glutaric acid, [30]. Bio-based nylon-4, nylon-5 and nylon-6,5 can be produced from butyrolactam and Δ-valerolactam, which are derived from GABA and 5-AVA, respectively [4]. Biosynthetic production of both 5-AVA and GABA is more sustainable than chemical production because the fermentative production of 5-AVA and GABA is conducted under mild conditions with high catalytic efficiency [4,6,24–27,31]. In addition, 5-AVA and GABA have been successfully produced from biomass-derived sugars such as *Miscanthus* hydrolysate and empty fruit bunch biosugar solution [6,31].

To engineer *C. glutamicum* for biochemical production, its metabolic pathway has been manipulated using a plasmid-based expression system, which establishes the synthetic pathway for the production of biochemicals [5]. This is an important step for evaluating the success of the constructed pathway and for establishing the target metabolite production in recombinant strains. This also helps identify which key reactions can be improved in order to drive metabolic flux toward optimized chemical production [1,4,5]. For example, the production of GABA using *C. glutamicum* strains was established by using a synthetic promoter-based expression vector system ($P_{L26} < P_{I16} < P_{H36}$) capable of low, intermediate, and high-strength glutamate decarboxylase ($GadB_{mut}$) expression. It was demonstrated that the use of the high-strength promoter P_{H36} in the pHGmut strain (5.89 ± 0.35 g/L) enabled higher production of GABA than in pIGmut (5.32 ± 0.04 g/L) and pLGmut (4.87 ± 0.15 g/L) strains with expression of $GadB_{mut}$ under intermediate (P_{I16}) and low (P_{L26}) strength promoters, respectively [26]. Cadaverine production using a synthetic promoter-based expression system ($P_{L10} < P_{L26} < P_{I16} < P_{I64} < P_{H30} < P_{H36}$) was also evaluated, and it was observed that a high level of protein expression using the strongest promoter (P_{H36}) does not necessarily result in the highest level of metabolite production [28,29]. In case of batch fermentation for cadaverine production using recombinant *C. glutamicum* strains, the use of P_{H30} (23.8 g/L) for the expression of lysine decarboxylase produced a higher titer than that when the P_{H36} promoter was used (21.3 g/L) [28]. However, the repertoire of vector systems capable of different levels of protein expression in *C. glutamicum* is still limited [5]. Most promoters in plasmids derived from *E. coli* plasmids such as P_{tac}, P_{trc}, P_{lacUV5}, P_R, and P_L have been evaluated [4,5]. However, despite the adoption of the promoters from *E. coli* plasmids, the variety of genetic engineering tools applicable to *C. glutamicum* is still less compared to the tools available for *E. coli* [1,4,5]. Therefore, the discovery or creation of a new vector system and evaluation of its capability for tunable protein expression in a target host strain are important factors in developing recombinant strains for the production of biochemicals in biorefineries [4,5].

Recently, the use of five repeats of the P_{tac} promoter successfully enabled stable overexpression of *phaCAB* genes in recombinant *E. coli* and enhanced poly(R-3-hydroxybutyrate) (PHB) accumulation

5.6 times that of the control strain, which had a single copy of the P_{tac} promoter [32]. In another report, the use of the promoter from *Vitreoscilla* hemoglobin protein (P_{vgb}) was successfully demonstrated in *E. coli*. It was found that the expression vectors with increasing repeats of the P_{vgb} promoter enabled tunable expression of the PHB synthesis operon (*phaCAB*) and allowed enhanced accumulation of poly(hydroxybutyrate) (PHB) [33]. The highest accumulation of 90% PHB in 5.37 g/L CDW was achieved by the recombinant strain with eight repeats of the P_{vgb} promoter. The use of the P_{vgb} promoter system was also successfully demonstrated in recombinant *C. glutamicum*. It was used to investigate the effect of VHb protein expression on L-glutamate and L-glutamine production in the recombinant strains. However, it was observed that the expression of the VHb protein under the P_{tac} promoter was better than that with the use of a single repeat of P_{vgb} promoter (1.71 ± 0.08 nmol/g > 0.69 ± 0.10 nmol/g) [34]. Therefore, in this study, new tunable gene expression systems were developed based on repeats of P_{vgb}. The strength of the P_{vgb} promoter-based expression systems with increasing repeats of P_{vgb} were investigated by evaluating the expression level of GFP in recombinant strains. To demonstrate the use of the new tunable P_{vgb} promoter-based expression system as a novel platform for metabolic engineering of *C. glutamicum*, 5-AVA and GABA production were established in different strains of *C. glutamicum*. 5-AVA and GABA production were selected as model compounds, because they are derived from L-lysine and L-glutamate, respectively.

2. Results and Discussion

2.1. Construction of P_{vgb} Promoter-Based Tunable Expression Systems for C. glutamicum and Evaluation of Green Fluorescent Protein in the Recombinant Strains

To investigate the tunability of protein expression using P_{vgb} in recombinant *C. glutamicum*, several expression vector systems based on pCES208 plasmids were constructed. GFP was expressed in *C. glutamicum* under the control of increasing repeats of P_{vgb} (P_{vgb}, P_{vgb4}, P_{vgb8}). The strength of protein expression in the three constructs was investigated by measuring the intensity of fluorescence in the recombinant strains, by using the fluorescent activated cell sorting analysis (FACS) (Figure 1) [35]. The resulting recombinant *C. glutamicum* EGFPV1, EGFPV4, and EGPV8 strains harbored one (P_{vgb}), four (P_{vgb4}), and eight (P_{vgb8}) repeats of P_{vgb}, respectively. *C. glutamicum* KCTC 1857 was cultured and used as the negative control. Colonies were randomly picked and GFP expression in each cell was evaluated by using FACS. The measured fluorescence intensity obtained from colonies with P_{vgb}-promoter-based vectors ranged from 10^2–10^3, whereas the negative control cells did not show significant fluorescence. The fluorescence by clones with P_{vgb8} were more intense than the clones with P_{vgb4} and P_{vgb}, after 24 h of cultivation (Figure 2). However, after 48 h, no significant difference was observed in GFP expression between clones with P_{vgb4} and those with P_{vgb8} (data not shown). The increasing number of P_{vgb} in the constructed expression system was directly proportional to the intensity of the GFP fluorescence in the resulting recombinant strains (Figure 2). These observations were similar to the PHB production enhancement in recombinant *E. coli* strains harboring the PHB operon (*phaCAB*) under the control of P_{vgb8} compared to the use of P_{vgb}. Based on these results, we have demonstrated that the strength of protein expression using P_{vgb} was tunable by modulating the number of its copies in the constructed expression system. To further demonstrate tunability of the P_{vgb} promoter-based expression system in recombinant *C. glutamicum*, the developed constructs were used to produce 5-AVA and GABA from L-lysine and L-glutamate, respectively. 5-AVA and GABA were selected as model compounds to test the application of the P_{vgb} promoter-based expression system because their production in *C. glutamicum* has been extensively studied in recent times [6,25,26,29–31].

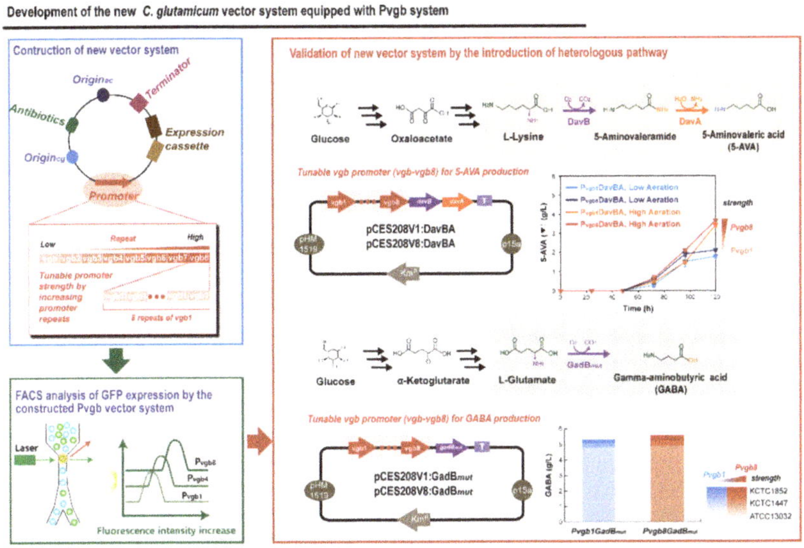

Figure 1. Schematic representation of the development of the *C. glutamicum* P_{vgb} promoter-based vector system equipped with P_{vgb} promoter.

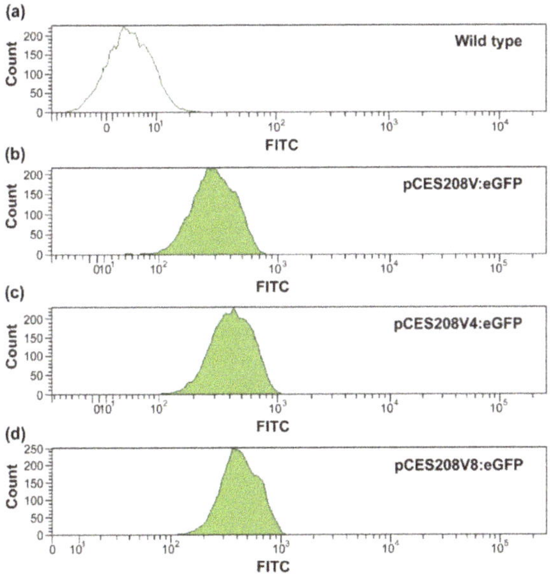

Figure 2. The fluorescence intensity of the recombinant EGFPV1 (**b**), EGFPV4 (**c**), and EGPV8 (**d**) strains after 24 h of cultivation. The wild type strain, *C. glutamicum* KCTC 1857 (**a**) was used as negative control.

2.2. 5-AVA Production Using Recombinant *C. glutamicum* with Tunable P_{vgb} Promoter-Based Expression Systems

To demonstrate that the P_{vgb} promoter-based expression system is a new platform for engineering better *C. glutamicum* strains, its feasibility to produce 5-AVA in *C. glutamicum* KCTC 1857 was evaluated [6,25,30]. Tunability of protein expression by P_{vgb} was verified by developing *C. glutamicium* 5AVA1 and 5AVA8 strains for the expression of lysine 2-monooxygenase (DavB) and delta-aminovaleramidase (DavA) under the control of single (P_{vgb}) and eight repeats (P_{vgb8}) of P_{vgb} [6,25,30]. 5-AVA production using the recombinant strains was evaluated in flask cultivation under low and high aeration conditions, as it was previously demonstrated that protein expression using P_{vgb} was enhanced under microaerobic conditions in recombinant *E. coli* strains [34]. The production of 5-AVA by the recombinant *C. glutamicum* 5AVA1 (3.43 ± 0.10 g/L) and 5AVA8 (3.68 ± 0.07 g/L) strains was better under high aeration conditions than at low aeration conditions (2.10–1.76 g/L) (Figure 3). Higher accumulation of L-lysine and glutarate was also detected after flask cultivation of 5AVA1 and 5AVA8 strains under high aeration conditions (4.41–4.46 g/L of L-lysine and 0.71–0.74 g/L of glutarate) than that at low aeration (3.99–4.58 g/L of L-lysine and 0.49–0.57 g/L of glutarate) (Figure 3c,d). 5-AVA production by all the recombinant strains were higher under high aeration conditions than that under flask cultivation with low aeration conditions. These results were similar to the trend of L-glutamate and L-glutamine production by recombinant *C. glutamicum* strains, which expressed the *Vitreoscilla* hemoglobin gene under P_{vgb} and P_{tac}, wherein chemical production was higher under high aeration than at low aeration [34]. This is because *C. glutamicum* is an obligate aerobic microorganism, and it requires high aeration for efficient amino acid production [1,4,5]. Additionally, because DavB requires oxygen as a co-substrate, higher aeration conditions are preferred for 5-AVA production [30].

Furthermore, the effect of eight repeats of P_{vgb} on 5-AVA production was evaluated to demonstrate the tunability of protein expression using the constructed P_{vgb} promoter-based vector system (Figure 3). Different levels of 5-AVA production were achieved when the number of P_{vgb} increased from single (P_{vgb}) to eight repeats (P_{vgb8}) (Figure 3). It was observed that 5-AVA production increased based on the number of repeats of P_{vgb}. 5-AVA production by the 5AVA8 strain (Figure 3d, 3.69 ± 0.07 g/L), which expressed DavBA under P_{vgb8}, was higher than that of the 5AVA1 strain (Figure 3c, 3.43 ± 0.10 g/L)), which harbored P_{vgb}. 5-AVA production by 5AVA1 and 5AVA8 strains (3.43–3.69 g/L) were comparably higher than that of the H30_AVA strain, which expressed DavBA under the strong synthetic promoter P_{H30} (1.4 ± 0.3 g/L g/L) (Table S1) [30]. This effect was similar to the observed trend of enhanced PHB accumulation in recombinant *E. coli* strains when the number of P_{tac} and P_{vgb} repeats increased to five and eight, respectively [32,33].

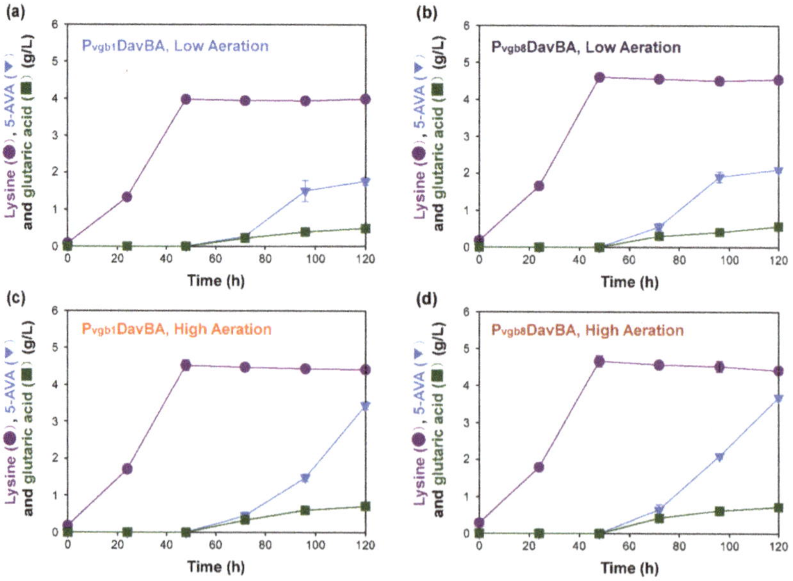

Figure 3. Time profile of 5-aminovaleric acid, L-lysine and glutaric acid production by recombinant *C. glutamicum* 5AVA1 (**a**, **c**) and 5AVA8 (**b**,**d**) strains under low (**a**,**b**) and high (**c**,**d**) aeration.

Finally, the effect of his-tagged DavB on 5-AVA production was also investigated, because we have previously reported that it enhanced 5-AVA production in *C. glutamcium* H30_AVA$_{His}$. In our previous study, we demonstrated enhanced 5AVA production by expressing his-tagged DavB along with DavA under the strong synthetic promoter P_{H30} in H30_AVA$_{His}$ strain, which produced 4.2 ± 0.9 g/L of 5-AVA [30]. This significantly increased 5-AVA production compared to that by the H30_AVA strain, which expressed DavB without his-tag (4.2 ± 0.9 g/L >1. 4 ± 0.3 g/L) (Figure 4) [30]. Therefore, to investigate the effect of this additional his-tag on 5-AVA production under the control of the P_{vgb} promoter system, recombinant strains 5AVA1$_{His}$ and 5AVA8$_{His}$ were constructed by inserting his-tagged DavB into the P_{vgb} promoter system with single (P_{vgb}) and eight (P_{vgb8}) repeats, respectively. 5-AVA production by the resulting recombinant strains were only evaluated under high aeration conditions, as the low aeration condition did not increase 5-AVA production in 5AVA1 and 5AVA8 strains (Figure 3). Under high aeration conditions, the 5AVA8$_{His}$ (3.31 ± 0.08 g/L) strain produced higher concentrations of 5-AVA than the 5AVA1$_{His}$ (2.82 ± 0.03 g/L) strain (Figure 4, Table S1). As shown in Figure 4, 5-AVA production increased as the number of P_{vgb} repeats increased from single to eight in 5AVA1$_{His}$ and 5AVA8$_{His}$ (3.31 ± 0.08 g/L > 2.82 ± 0.03 g/L). Glutaric acid production also increased in both 5AVA1$_{His}$ (0.64 ± 0.01 g/L) and 5AVA8$_{His}$ (0.67 ± 0.05g/L). Interestingly, both L-lysine accumulation and 5-AVA production increased as the number of repeats of P_{vgb} increased from a single repeat in 5AVA1$_{His}$ (4.00 ± 0.06 g/L) to eight repeats in the 5AVA8$_{His}$ (4.38 ± 0.03 g/L) strain. Based on these observations, the P_{vgb} promoter system was successfully used as a new platform for 5-AVA production in *C. glutamicum*.

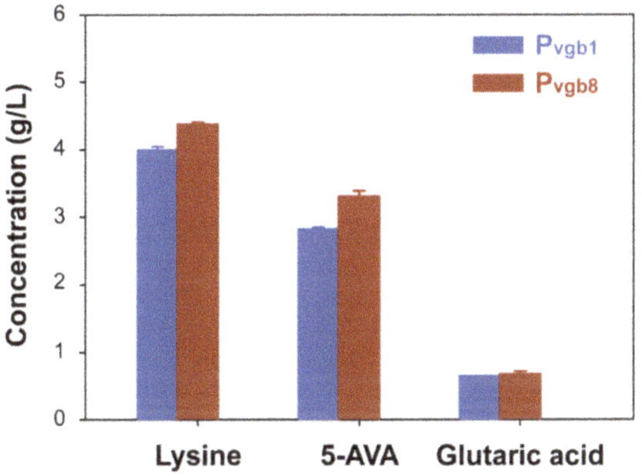

Figure 4. 5-Aminovaleric acid, L-lysine, and glutaric acid production by recombinant *C. glutamicum* 5AVA1$_{His}$ and 5AVA8$_{His}$ strains under high aeration conditions after 120 h flask cultivation.

2.3. *Gamma-Aminobutyric Acid Production Using Recombinant C. glutamicum with Tunable* P_{vgb} *Promoter-Based Expression Systems*

The constructed P_{vgb} promoter-based vector system was also used for the expression of mutated glutamate decarboxylase (GadB$_{mut}$) in order to investigate GABA production in the *C. glutamicum* strain ATCC 13032 and in high-L-glutamate producing strains *C. glutamicum* KCTC 1447 and *C. glutamicum* KCTC 1852 as host strains (Figure 5, Table S2) [26,31]. *C. glutamicum* KCTC 1447 and *C. glutamicum* KCTC 1852 were used because they are capable of high production of L-glutamate, an important precursor for GABA production [26,31]. *C. glutamicum* ATCC 13032 was also used as control host strain [30]. The effect of the eight repeats of P_{vgb} on GABA production was evaluated under high aeration conditions because we have previously demonstrated that low aeration did not increase chemical production by recombinant strains expressing DavBA under single and eight repeats of P_{vgb} during shake flask cultivation (Figure 3). It was observed that the average GABA production by *C. glutamicum* KCTC 1852-derived strains (5.31 g/L–5.58 g/L) was higher than that by recombinant

C. glutamicum ATCC 13032 and *C. glutamicum* KCTC 1447 strains (4.82–5.22 g/L) (Figure 5, Figure S3). This is because the host strain, *C. glutamicum* KCTC 1852, is naturally capable of higher L-glutamate production than *C. glutamicum* KCTC 1447 and *C. glutamicum* ATCC 13032 [30]. Higher GABA production was attributed to the efficient production of L-glutamate, an important pre-requisite for the development of strains for GABA production [26,31]. The level of GABA production in all the tested strains increased as the number of P_{vgb} repeats increased (Figure 5). For example, GABA production in V1GD1852 (5.31 ± 0.16 g/L) and V8GD1852 (5.58 ± 0.27 g/L) strains increased when $GadB_{mut}$ was expressed under P_{vgb} and P_{vgb8}, respectively. The same tendency was observed in V1GD13032 (4.82 ± 0.09 g/L) and V1GD1447 (5.09 ± 0.64 g/L) strains with respect to GABA production, under the control of the P_{vgb} promoter, and by V8GD13032 (4.94 ± 0.27 g/L) and V8GD1447 (5.22 ± 0.60 g/L) strains under the control of the P_{vgb8} promoter. Regarding GABA production using *C. glutamicum* ATCC 13032 strains (4.82–4.94 g/L), similar GABA production was achieved by pLGmut (4.87 ± 0.15 g/L), V1GD13032 (4.82 ± 0.09 g/L), and V8GD13032 (4.94 ± 0.27 g/L) strains, which expressed $GadB_{mut}$ under the synthetic promoters P_{L26}, P_{vgb}, and P_{vgb8}, respectively (Figure 4, Table S2) [26]. The level of GABA production using the constructed P_{vgb} promoter-based expression system (4.82–5.09 g/L) was comparable to the titers achieved by recombinant strains pHGmut (5.89 ± 0.35 g/L) and pIGmut (5.32 ± 0.04 g/L). The recombinant strains expressed $GadB_{mut}$ under synthetic promoters of high (P_{H36}) and intermediate (P_{I16}) strength [31]. The level of GABA production by the *C. glutamicum* KCTC 1852-derivative strains V1GD1852 (5.31 ± 0.16) and V8GD1852 (5.58 ± 0.27), expressing $GadB_{mut}$, were similar to the level of GABA production achieved when the strong synthetic promoter P_{H36} was used in the H36GM1852 strain (8.47 ± 0.06 g/L) (Table S2). This shows that the strength of $GadB_{mut}$ expression by the P_{vgb} promoter-based expression system is also tunable like previously used synthetic promoters (P_{L26}, P_{I16}, P_{H36}) (Table S2) [26,31]. Based on these results, we concur that the P_{vgb} -based expression system was capable of tunable expression of key proteins for 5-AVA (DavBA) and GABA ($GadB_{mut}$) production by increasing the number of P_{vgb} repeats. The addition of the newly constructed P_{vgb} promoter-based expression system into the repertoire of plasmids available for metabolic engineering of *C. glutamicum* provides an alternative method of fine-tuning protein expression levels for validating synthetic metabolic pathways and improving biochemical production in biorefineries.

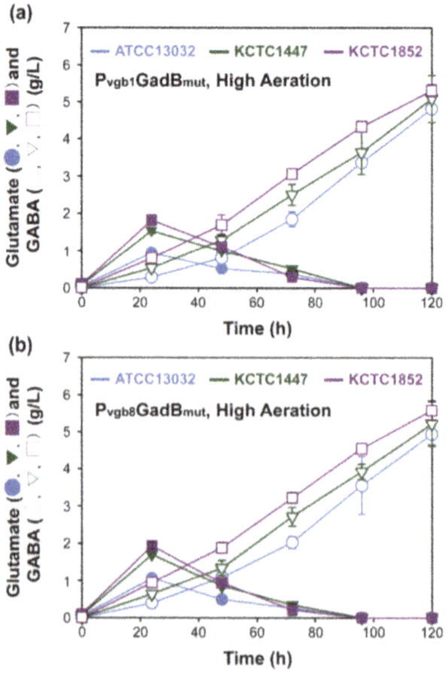

Figure 5. Time profile of Gamma-aminobutyric acid production by recombinant *C. glutamcium* strains expressing $GadB_{mut}$ under P_{vgb} (**a**) and P_{vgb8} (**b**) under high aeration condition.

3. Materials and Methods

3.1. Bacterial Strains and Plasmids

All the bacterial strains and plasmids used in this study are listed in Table 1. *E. coli* XL1-Blue (Stratagene, La Jolla, CA, USA) was used for the general gene cloning studies. *C. glutamicum* ATCC 13032, *C. glutamicum* KCTC 1447, *C. glutamicum* KCTC 1852, and *C. glutamicum* KCTC 1857 strains were purchased from the Korean Collection for Type Cultures (KCTC, Daejeon, Korea). The pCES208-based plasmids: pCES208H30:DavBA, pCES208H30:DavB$_{His}$A and pHG$_{mut}$ were constructed as previously described [26,30].

3.2. Plasmid Construction

All DNA manipulations were performed following standard procedures [36]. Polymerase chain reaction (PCR) was performed with the C1000 Thermal Cycler (Bio-Rad, Hercules, CA, USA). The primers used in this study were synthesized at Bioneer (Daejeon, Korea). The pHG$_{mut}$ plasmid was cut at KpnI and BamHI sites to replace the synthetic promoter P$_{H36}$ with P$_{vgb}$ promoter. The subsequent repeats of the P$_{vgb}$ promoter were inserted into the pCES208V:eGFP vector at BamHI/BglII sites to obtain pCES208V4:eGFP and pCES208V8:eGFP plasmids. The pCES208V:eGFP, pCES208V4:eGFP, and pCES208V8:eGFP plasmids for the expression of enhanced GFP were constructed by inserting GFP at the BamHI/NotI sites of modified pCES208-based plasmids with one, four, and eight repeats of P$_{vgb}$, respectively. The plasmids for the expression of DavBA, DavB$_{His}$A, and GadB$_{mut}$ were also inserted at the BamHI/NotI sites of pCES208V:eGFP and pCES208V8:eGFP (Table 1).

Table 1. Strains and plasmids used in the present study.

Strains and Plasmids	Relevant Characteristics	Reference
Strains		
E. coli XL1-Blue	*recA1 endA1 gyrA96 thi-1 hsdR17 supE44 relA1 lac* [FA1*proAB lacI^{q}ZΔM15* Tn*10* (Tet^R)]	Stratagene
C. glutamicum KCTC 1447	L-Glutamate producer	KCTC
C. glutamicum KCTC 1852	L-Glutamate producer	KCTC
C. glutamicum KCTC 1857	L-Lysine producer	KCTC
C. glutamicum ATCC 13032	Wild type	ATCC
EGFPV1	*C. glutamicum* KCTC 1857 with pCES208V:eGFP	This study
EGFPV4	*C. glutamicum* KCTC 1857 with pCES208V4:eGFP	This study
EGFPV8	*C. glutamicum* KCTC 1857 with pCES208V8:eGFP	This study
5AVA1	*C. glutamicum* KCTC 1857 with pCES208V:DavBA	This study
5AVA8	*C. glutamicum* KCTC 1857 with pCES208V8:DavBA	This study
5AVA1$_{His}$	*C. glutamicum* KCTC 1857 with pCES208V:DavB$_{His}$DavA	This study
5AVA8$_{His}$	*C. glutamicum* KCTC 1857 with pCES208V8:DavB$_{His}$DavA	This study
H30_AVA	*C. glutamicum* KCTC 1857 with pCES208H30:DavBA	30
H30_AVA$_{His}$	*C. glutamicum* KCTC 1857 with pCES208H30:DavB$_{His}$A	30
H36GM13032	*C. glutamicum* ATCC 13032 with pCES208H36:GadB$_{mut}$	This study
H36GM1447	*C. glutamicum* KCTC 1447 with pCES208H36:GadB$_{mut}$	This study
H36GM1852	*C. glutamicum* KTC 1852 with pCES208H36:GadB$_{mut}$	This study
H30GM13032	*C. glutamicum* ATCC 13032 with pCES208H30:GadB$_{mut}$	This study
H30GM1447	*C. glutamicum* KCTC 1447 with pCES208H30:GadB$_{mut}$	This study
H30GM1852	*C. glutamicum* KTC 1852 with pCES208H30:GadB$_{mut}$	This study
V1GD133032	*C. glutamicum* ATCC 13032 with pCES208V:GadB$_{mut}$	This study
V1GD1447	*C. glutamicum* KCTC 1447 with pCES208V:GadB$_{mut}$	This study
V1GD1852	*C. glutamicum* KTC 1852 with pCES208V:GadB$_{mut}$	This study
V8GD133032	*C. glutamicum* ATCC 13032 with pCES208V8:GadB$_{mut}$	This study
V8GD1447	*C. glutamicum* KCTC 1447 with pCES208V8:GadB$_{mut}$	This study
V8GD1852	*C. glutamicum* KTC 1852 with pCES208V8:GadB$_{mut}$	This study
Plasmids		
pCES208V:eGFP	pCES208 derivative; P$_{vgb}$ promoter, eGFP, Km^R	This study
pCES208V4:eGFP	pCES208 derivative; P$_{vgb4}$ promoter, eGFP, Km^R	This study
pCES208V8:eGFP	pCES208 derivative; P$_{vgb8}$ promoter, eGFP, Km^R	This study

Table 1. *Cont.*

Strains and Plasmids	Relevant Characteristics	Reference
pCES208V:DavBA	pCES208 derivative; P_{vgb} promoter, *Pseudomonas putida* KT2440 *davBA* genes, Km^R	This study
pCES208V8:DavBA	pCES208 derivative; P_{vgb8} promoter, *Pseudomonas putida* KT2440 *davBA* genes, Km^R	This study
pCES208V:DavB_{His}A	pCES208 derivative; P_{vgb} promoter, *Pseudomonas putida* KT2440 *davB*$_{His}$ and *davA* genes, Km^R	This study
pCES208V8:DavB_{His}A	pCES208 derivative; P_{vgb8} promoter, *Pseudomonas putida* KT2440 *davB*$_{His}$ and *davA* genes, Km^R	This study
pCES208H30:DavBA	pCES208 derivative; P_{H30} promoter, *Pseudomonas putida* KT2440 *davBA* genes, Km^R	30
pCES208H30:DavB_{His}A	pCES208 derivative; P_{H30} promoter, *Pseudomonas putida* KT2440 *davB*$_{His}$ and *davA* genes, Km^R	30
pCES208V:GadB_{mut}	pCES208 derivative; P_{vgb} promoter, *E. coli* mutated gadB (Glu89Gln/Δ452-466) gene, Km^R	This study
pCES208V8:GadB_{mut}	pCES208 derivative; P_{vgb8} promoter, *E. coli* mutated gadB (Glu89Gln/Δ452-466) gene, Km^R	This study
pHG_{mut} / pCES208H36:GadB_{mut}	pCES208 derivative; P_{H36} promoter, *E. coli* mutated gadB (Glu89Gln/Δ452-466) gene, Km^R	26
pCES208H30:GadB_{mut}	pCES208 derivative; P_{H30} promoter, *E. coli* mutated gadB (Glu89Gln/Δ452-466) gene, Km^R	This study

3.3. Culture Conditions

E. coli XL1-Blue, used for general gene cloning experiments, was cultured at 37 °C in Luria-Bertani (LB) medium (10 g/L tryptone, 5 g/L yeast extract, and 5 g/L NaCl). Flask cultures of the recombinant strains of *C. glutamicum* were obtained in triplicates by culturing at 30 °C and 250 rpm in a rotary shaker. High aeration conditions for flask cultivation were maintained by adding 20 mL of the appropriate culture medium into a 250-mL baffled flask. Low aeration conditions for flask cultivation were maintained by adding 100 mL of the culture medium into a 250-mL glass flask. Seed cultivation of *C. glutamicum* strains was carried out in 14-mL round-bottomed tubes containing 2 mL of Recovery Growth (RG) medium (10 g/L of glucose, 40 g/L of brain heart infusion, 10 g/L of beef extract, and 30 g/L of D-sorbitol) with incubation overnight at 30 °C and 250 rpm [29]. The main-flask cultures for 5-AVA production were grown in 250 mL baffled flasks containing 20 mL of CG50 medium for 120 h at 30 °C and 250 rpm. The CG50 medium for flask cultivation contained (per liter) 50 g glucose, 15 g yeast extract, 15 g $(NH4)_2SO_4·7H_2O$, 0.5 g KH_2PO4, 0.5 g $MgSO_4·7H_2O$, 0.01 g $MnSO_4·H_2O$, 0.01 g $FeSO_4·7H_2O$, and 20 μg/L of kanamycin (Km) for plasmid maintenance [30]. The GP1 medium optimized in our previous study [26] was used for GABA production. The GP1 medium for flask cultivation contained (per liter) 50 g $(NH_4)_2SO_4·7H_2O$, 1 g K_2HPO_4, 3 g urea, 0.4 g $MgSO_4·7H_2O$, 50 g peptone, 0.02 g $FeSO_4·7H_2O$, 0.007 g $MnSO_4·H_2O$, 200 μg thiamine, and 1 mM of pyridoxal 5′-phosphate hydrate (PLP) [26]. PLP was added to the culture medium as it is a cofactor of glutamate decarboxylase. Moreover, 0.1 mM of PLP was the optimum concentration for prolonging GABA production using recombinant *C. glutamicum* strains [26]. Kanamycin and biotin were added to the GP1 culture medium at 25 and 50 μg/L, respectively. Only 50 μg /L of biotin was used in flask cultivation for GABA production because biotin-limited condition promotes L-glutamate accumulation [26]. $CaCO_3$ was added to the culture medium at 10 g/L to minimize the pH change during cultivation.

3.4. Analysis

The concentrations of glucose and organic acids were determined by high performance liquid chromatography (HPLC). The standard and sample concentrations of 5-AVA, GABA, L-lysine, and L-glutamate were determined by HPLC using an Optimapak C18 column (RStech, DaeJeon, Korea) as previously reported [37].

3.5. Fluorescence-Activated cell Sorting Analysis (FACS) for Measuring GFP Expression by Recombinant Strains

FACS analysis was used to investigate the GFP expression by using the constructed P_{vgb} system. The recombinant EGFPV1, EGFPV4 and EGFPV8 strains were grown in Brain Heart Infusion (BHI) media for 24 h at 30 °C. The cells were then collected and diluted using PBS buffer. FACS analysis (BD Biosciences, San Jose, CA, USA) was performed for 100,000 clones of each samples using argon ion laser (blue, 488 nm) and band-pass filter (530 nm $\pm$ 15 nm) [35].

4. Conclusions

The P_{vgb} promoter-based expression system constructed in this study was capable of tunable expression of green fluorescent protein in recombinant *C. glutamcium* strains, when the repeats of P_{vgb} promoter increased from one (P_{vgb}) to four (P_{vgb4}) to eight (P_{vgb8}). Furthermore, GABA and 5-AVA production by recombinant *C. glutamicum* strains also increased when the expression of DavBA and $GadB_{mut}$ in the P_{vgb} promoter-based expression system increased from single to eight repeats of the P_{vgb} promoter. This shows that the strength of protein expression using the P_{vgb} promoter-based tunable system was comparable to that of previously established synthetic promoters (P_{L26}, P_{I16}, P_{H30} P_{H36}) [26,28–30]. Based on the different levels of 5-AVA and GABA production by all the tested strains, the P_{vgb} promoter-based expression system was capable of tunable expression of DavBA and $GadB_{mut}$ by mere manipulation in the number of P_{vgb} repeats. The newly constructed P_{vgb} promoter-based expression system developed in this study expands the repertoire of plasmids available for metabolic engineering of *C. glutamicum* and provides another method for fine-tuning levels of protein expression for convenient and rapid validation of synthetic metabolic pathways, ultimately improving biochemical production in biorefineries.

Supplementary Materials: The following are available online at http://www.mdpi.com/2073-4344/8/11/561/s1. Table S1: 5-AVA production using recombinant *C. glutamicum* strains expressing DavBA and $DavB_{His}$DavA under different synthetic promoters after 120 h of flask cultivation under high aeration. Table S2: GABA production by recombinant *C. glutamicum* strains expressing GadBmut under different synthetic promoters after 120 h of flask cultivation under high aeration.

Author Contributions: The study was conceptualized by P.S.J. The study was validated, and formal analysis was carried out by K.H.K., K.B., T.U.K., S.Y.J, B.L., J.J.S., D.H.L., B.K.S., and J.H.C. The original draft was prepared by K.B. The paper was edited and reviewed by H.T.K. and M.N.R. The resources for completion of this study were provided by P.S.J., H.T.K. and J.C.J.

Funding: Funding sources are declared in the acknowledgement section.

Acknowledgments: This work was supported by the Technology Development Program to Solve Climate Changes on Systems Metabolic Engineering for Biorefineries from the Ministry of Science and ICT (MSIT) through the National Research Foundation (NRF) of Korea (NRF-2015M1A2A2035810), the Bio & Medical Technology Development Program MSIT through the NRF of Korea (NRF-2018M3A9H3020459) and the Lignin Biorefinery from MSIT through the NRF of Korea (NRF-2017M1A2A2087634).

Conflicts of Interest: The authors declare that they have no competing interests.

References

1. Becker, J.; Wittmann, C. Advanced Biotechnology: Metabolically Engineered Cells for the Bio-Based Production of Chemicals and Fuels, Materials, and Health-Care Products. *Angew. Chem. Int. Ed.* **2015**, *54*, 3328–3350. [CrossRef] [PubMed]
2. Oh, Y.H.; Eom, I.Y.; Joo, J.C.; Yu, J.H.; Song, B.G.; Lee, S.H.; Hong, S.H.; Park, S.J. Recent advances in development of biomass pretreatment technologies used in biorefinery for the production of bio-based fuels, chemicals and polymers. *Korean J. Chem. Eng.* **2015**, *32*, 1945–1959. [CrossRef]
3. Lee, J.W.; Kim, H.U.; Choi, S.; Yi, J.H.; Lee, S.Y. Microbial production of building block chemicals and polymers. *Curr. Opin. Biotechnol.* **2011**, *22*, 758–767. [CrossRef] [PubMed]

4. Baritugo, K.; Kim, H.T.; David, Y.; Choi, J.H.; Choi, J.; Kim, T.W.; Park, C.; Hong, S.H.; Na, J.G.; Jeong, K.J.; et al. Recent advances in metabolic engineering of *Corynebacterium glutamicum* strains as potential platform microorganisms for biorefinery. *Biofuel Bioprod. Biorefin.* **2018**, *12*, 899–925. [CrossRef]
5. Baritugo, K.; Kim, H.T.; David, Y.; Choi, J.; Hong, S.H.; Jeong, K.J.; Joo, J.C.; Park, S.J. Metabolic engineering of *Corynebacterium glutamicum* for fermentative production of chemicals in biorefinery. *Appl. Microbiol. Biotechnol.* **2018**, *102*, 3915–3937. [CrossRef] [PubMed]
6. Joo, J.C.; Oh, Y.H.; Yu, J.H.; Hyun, S.M.; Khang, T.U.; Kang, K.H.; Song, B.K.; Park, K.; Oh, M.K.; Lee, S.Y.; et al. Production of 5-aminovaleric acid in recombinant *Corynebacterium glutamicum* strains from a *Miscanthus* hydrolysate solution prepared by a newly developed *Miscanthus* hydrolysis process. *Bioresour. Technol.* **2017**, *244*, 1692–1700. [CrossRef] [PubMed]
7. David, Y.; Joo, J.C.; Yang, J.E.; Oh, Y.H.; Lee, S.Y.; Park, S.J. Biosynthesis of 2-hydroxyacid-containing polyhydroxyalkanoates by employing butyryl-CoA transferases in metabolically engineered *Escherichia coli*. *Biotechnol. J.* **2017**, *12*. [CrossRef] [PubMed]
8. Choi, S.Y.; Parks, S.J.; Kim, W.J.; Yang, J.E.; Lee, H.; Shin, J.; Lee, S.Y. One-step fermentative production of poly(lactate-co-glycolate) from carbohydrates in *Escherichia coli*. *Nat. Biotechnol.* **2016**, *34*, 435–440. [CrossRef] [PubMed]
9. Kind, S.; Neubauer, S.; Becker, J.; Yamamoto, M.; Völkert, M.; Abendroth, G.; Zelder, O.; Wittmann, C. From zero to hero—Production of bio-based nylon from renewable re-sources using engineered *Corynebacterium glutamicum*. *Metab. Eng.* **2014**, *25*, 113–123. [CrossRef] [PubMed]
10. Chae, C.G.; Kim, Y.J.; Lee, S.J.; Oh, Y.H.; Yang, J.E.; Joo, J.C.; Kang, K.H.; Jang, Y.A.; Lee, H.; Park, A.R.; et al. Biosynthesis of poly(2-hydroxybutyrate-co-lactate) in metabolically engineered *Escherichia coli*. *Biotechnol. Bioprocess Eng.* **2016**, *21*, 169–174. [CrossRef]
11. David, Y.; Baylon, M.G.; Sudheer, P.D.V.N.; Baritugo, K.; Chae, C.G.; Kim, Y.J.; Kim, T.W.; Kim, M.; Na, J.G.; Park, S.J. Screening of microorganisms able to degrade low-rank coal in aerobic conditions: Potential coal bio-solubilization mediators from coal to biochemical. *Biotechnol. Bioprocess Eng.* **2017**, *22*, 178–185. [CrossRef]
12. Baylon, G.; David, Y.; Pamidimarri, S.; Baritugo, K.; Chae, C.G.; Kim, Y.J.; Wan, T.W.; Kim, M.S.; Na, J.G.; Park, S.J. Bio-solubilization of the untreated low rank coal by alkali-producing bacteria isolated from soil. *Korean J. Chem. Eng.* **2016**, *34*, 105–109. [CrossRef]
13. Sudheer, P.D.V.N.; David, Y.; Chae, C.; Kim, Y.J.; Baylon, M.G.; Baritugo, K.; Kim, T.W.; Kim, M.; Na, J.G.; Park, S.J. Advances in the biological treatment of coal for synthetic natural gas and chemicals. *Korean J. Chem. Eng.* **2016**, *10*, 2788–2801. [CrossRef]
14. Yang, J.E.; Park, S.J.; Kim, W.J.; Kim, H.J.; Kim, B.; Lee, H.; Shin, J.; Lee, S.Y. One-step fermentative production of aromatic polyesters from glucose by metabolically engineered *Escherichia coli* strains. *Nat. Commun.* **2018**, *9*, 79. [CrossRef] [PubMed]
15. Joo, J.C.; Khusnutdinova, A.N.; Flick, R.; Kim, T.H.; Bornscheuer, U.T.; Yakunin, A.F.; Mahadevan, R. Alkene hydrogenation activity of enoate reductases for an environmentally benign biosynthesis of adipic acid. *Chem. Sci.* **2017**, *8*, 1406–1413. [CrossRef] [PubMed]
16. Lee, J.H.; Lama, S.; Kim, J.R.; Park, S.H. Production of 1,3-Propanediol from Glucose by Recombinant *Escherichia coli* BL21(DE3). *Biotechnol. Bioprocess Eng.* **2017**, *23*, 250–258. [CrossRef]
17. Li, J.; Feng, R.; Wen, Z.; Zhang, A. Overexpression of *ARO10* in *pdc5*Δmutant resulted in higher isobutanol titers in *Saccharomyces cerevisiae*. *Biotechnol. Bioprocess Eng.* **2017**, *22*, 382–389. [CrossRef]
18. Zong, H.; Liu, X.; Chen, W.; Zhuge, B.; Sun, J. Construction of glycerol synthesis pathway in *Klebsiella pneumoniae* for bioconversion of glucose into 1,3-propanediol. *Biotechnol. Bioprocess Eng.* **2017**, *22*, 549–555. [CrossRef]
19. Kim, H.S.; Oh, Y.H.; Jang, Y.; Kang, K.H.; David, Y.; Yu, J.H.; Song, B.K.; Choi, J.; Chang, Y.K.; Joo, J.C.; et al. Recombinant *Ralstonia eutropha* engineered to utilize xylose and its use for the production of poly(3-hydroxybutyrate) from sunflower stalk hydrolysate solution. *Microb. Cell Factories* **2016**, *15*, 95. [CrossRef] [PubMed]
20. Park, S.J.; Jang, Y.A.; Noh, W.; Oh, Y.H.; Lee, H.; David, Y.; Baylon, M.G.; Shin, J.; Yang, J.E.; Choi, S.Y.; et al. Metabolic engineering of *Ralstonia eutropha* for the production of polyhydroxyalkanoates from sucrose. *Biotechnol. Bioeng.* **2015**, *112*, 638–643. [CrossRef] [PubMed]

21. Pang, X.; Zhuang, X.; Tang, Z.; Chen, X. Polylactic acid (PLA): Research, development and industrialization. *Biotechnol. J.* **2010**, *5*, 1125–1136. [CrossRef] [PubMed]
22. Xu, J.; Guo, B.H. Poly (butylene succinate) and its copolymers: Research, development and industrialization. *Biotechnol. J.* **2010**, *5*, 1149–1163. [CrossRef] [PubMed]
23. Kind, S.; Wittmann, C. Bio-based production of the platform chemical 1, 5-diaminopentane. *Appl. Microbiol. Biotechnol.* **2011**, *91*, 1287–1296. [CrossRef] [PubMed]
24. Park, S.J.; Kim, E.Y.; Noh, W.; Park, H.M.; Oh, Y.H.; Lee, S.H.; Song, B.K.; Jegal, J.; Lee, S.Y. Metabolic engineering of *Escherichia coli* for the production of 5-aminovalerate and glutarate as C5 platform chemicals. *Metab. Eng.* **2013**, *16*, 42–47. [CrossRef] [PubMed]
25. Shin, J.H.; Park, S.H.; Oh, Y.H.; Choi, J.W.; Lee, M.H.; Cho, J.S.; Jeong, K.J.; Joo, J.C.; Yu, J.; Park, S.J.; et al. Metabolic engineering of *Corynebacterium glutamicum* for enhanced production of 5-aminovaleric acid. *Microb. Cell Factories* **2016**, *15*, 174. [CrossRef] [PubMed]
26. Choi, J.W.; Yim, S.S.; Lee, S.H.; Kang, T.J.; Park, S.J.; Jeong, K.J. Enhanced production of gamma-aminobutyrate (GABA) in recombinant *Corynebacterium glutamicum* by expressing glutamate decarboxylase active in expanded pH range. *Microb. Cell Factories* **2015**, *14*, 21. [CrossRef] [PubMed]
27. Park, S.J.; Kim, E.Y.; Won, N.; Oh, Y.H.; Kim, H.Y.; Song, B.K.; Cho, K.M.; Hong, S.H.; Lee, S.H.; Jegal, J. Synthesis of nylon 4 from gamma-aminobutyrate (GABA) produced by recombinant *Escherichia coli*. *Bioprocess Biosyst. Eng.* **2013**, *36*, 885–892. [CrossRef] [PubMed]
28. Kim, H.T.; Baritugo, K.A.; Oh, Y.H.; Hyun, S.M.; Khang, T.U.; Kang, K.H.; Jung, S.H.; Song, B.K.; Park, K.; Kim, I.K.; et al. Metabolic engineering of *Corynebacterium glutamicum* for the high-level production of cadaverine that can be used for synthesis of biopolyamide 510. *ACS Sustain. Chem. Eng.* **2018**, *6*, 5296–5305. [CrossRef]
29. Oh, Y.H.; Choi, J.W.; Kim, E.Y.; Song, B.K.; Jeong, K.J.; Park, K.; Kim, I.; Woo, H.M.; Lee, S.H.; Park, S.J. Construction of Synthetic Promoter-Based Expression Cassettes for the production of Cadaverine in Recombinant *Corynebacterium glutamicum*. *Appl. Biochem. Biotech.* **2015**, *176*, 2065–2075. [CrossRef] [PubMed]
30. Kim, H.T.; Kang, T.U.; Baritugo, K.; Hyun, S.M.; Kang, K.H.; Jung, S.H.; Song, B.K.; Park, K.; Oh, M.; Kim, G.B.; et al. Metabolic engineering of *Corynebacterium glutamicum* for the production of glutaric acid, a C5 dicarboxylic acid platform chemical. *Metab. Eng.* **2018**. [CrossRef] [PubMed]
31. Baritugo, K.; Kim, H.T.; David, Y.; Khang, T.U.; Hyun, S.M.; Kang, K.H.; Yu, J.H.; Choi, J.H.; Song, J.J.; Joo, J.C.; et al. Enhanced production of gamma-aminobutyrate (GABA) in recombinant *Corynebacterium glutamicum* strains from empty fruit bunch biosugar solution. *Microb. Cell Factories* **2018**, *17*, 129. [CrossRef] [PubMed]
32. Li, M.; Wang, J.; Geng, Y.; Li, Y.; Wang, Q.; Liang, Q.; Qi, Q. A strategy of gene overexpression based on tandem repetitive promoters in *Escherichia coli*. *Microb. Cell Factories* **2012**, *11*, 19. [CrossRef] [PubMed]
33. Wu, H.; Wang, H.; Chen, J.; Chen, G.Q. Effects of cascaded vgb promoters on poly(hydroxybutyrate) (PHB) synthesis by recombinant *Escherichia coli* grown micro-aerobically. *Appl. Microbiol. Biotechnol.* **2014**, *98*, 10013–10021. [CrossRef] [PubMed]
34. Liu, Q.; Zhang, J.; Wei, X.X.; Ouyang, S.P.; Wu, Q.; Chen, G.Q. Microbial production of L-glutamate and L-glutamine by recombinant *Corynebacterium glutamicum* harboring *Vitreoscilla* hemoglobin gene *vgb*. *Appl. Microbiol. Biotechnol.* **2008**, *77*, 1297–1304. [CrossRef] [PubMed]
35. Ko, K.; Lee, B.; Cheong, D.; Han, Y.; Choi, J.H.; Song, J.J. Bacterial Cell Surface Display of a Multifunctional Cellulolytic Enzyme Screened from a Bovine Rumen Metagenomic Resource. *J. Microbiol. Biotechnol.* **2015**, *25*, 1835–1841. [CrossRef] [PubMed]
36. Sambrook, J.; Russell, D.W. *Molecular Cloning: A Laboratory Manual*; Cold Spring Harbor Laboratory Press: Cold Spring Harbor, NY, USA, 2001; Volume 3.
37. Cabezudo, M.D.; Hermosín, I.; Chicón, R.M. Free amino acid composition and botanical origin of honey. *Food Chem.* **2003**, *83*, 263–268. [CrossRef]

catalysts

Review

Advances in the Metabolic Engineering of *Escherichia coli* for the Manufacture of Monoterpenes

Si-si Xie †, Lingyun Zhu †, Xin-yuan Qiu, Chu-shu Zhu and Lv-yun Zhu *

Department of Biology and Chemistry, College of Liberal Arts and Sciences, National University of Defense Technology, Changsha 410073, Hunan, China; xiesisi_bio@126.com (S.-s.X.); lingyunzhu@nudt.edu.cn (L.Z.); qiuxinyuan12@nudt.edu.cn (X.-y.Q.); zhuchushu13@163.com (C.-s.Z.)

* Correspondence: zhulvyun@nudt.edu.cn; Tel.: +86-731-84574281

† These authors contributed equally to this work.

Received: 1 April 2019; Accepted: 6 May 2019; Published: 9 May 2019

Abstract: Monoterpenes are commonly applied as pharmaceuticals and valuable chemicals in various areas. The bioproduction of valuable monoterpenes in prokaryotic microbial hosts, such as *E. coli*, has progressed considerably thanks to the development of different outstanding approaches. However, the large-scale production of monoterpenes still presents considerable limitations. Thus, process development warrants further investigations. This review discusses the endogenous methylerythritol-4-phosphate-dependent pathway engineering and the exogenous mevalonate-dependent isoprenoid pathway introduction, as well as the accompanied optimization of rate-limiting enzymes, metabolic flux, and product toxicity tolerance. We suggest further studies to focus on the development of systematical, integrational, and synthetic biological strategies in light of the inter disciplines at the cutting edge. Our review provides insights into the current advances of monoterpene bioengineering and serves as a reference for future studies to promote the industrial production of valuable monoterpenes.

Keywords: monoterpene; prokaryotic microbial factory; metabolic engineering; MEP pathway; MEV pathway

1. Introduction

Terpenoids are widely distributed natural compounds that are extracted from plants, algae, mosses, and even insects and microbes. Terpenoids are composed of isoprene five-carbon units (C5) as the basic skeleton in accordance with the biogenetic isoprene rule and then further classified as monoterpenes (C10), sesquiterpenes (C15), diterpenes (C20), triterpenoids (C30), tetraterpenes (C40), and polyterpenes. These numerous compounds are normally used as medicines, insecticides, and fragrances [1]. For example, artemisinin is an endoperoxide sesquiterpene lactone isolated from *Artemisia annua Linn* and used as an anti-malarial drug [2]; paclitaxel is a cyclic diterpene hydrocarbon derived from the pacific yew and broadly applied in clinic as an anticancer drug [3,4]; squalene is the precursor of triterpenoids and is used as a pharmaceutical intermediate and bactericide [5,6]. All isoprenoids are synthesized from co-precursor isopentenyl pyrophosphate (IPP) and its isomer dimethylallyl pyrophosphate (DMAPP), and this process is catalyzed by a series of corresponding isoprenoid synthases. IPP and DMAPP are subsequently transformed to geranyl pyrophosphate (GPP), farnesyl pyrophosphate (FPP), or geranylgeranyl pyrophosphate (GGPP), which are the precursors of monoterpenes, sesquiterpenes, and diterpenes, respectively [7]. Two major pathways are involved in the natural synthesis of isoprenoid precursors IPP and DMAPP: the methylerythritol-4-phosphate-dependent pathway (MEP pathway), also termed as deoxyxylulose phosphate pathway (DXP pathway) and the mevalonate-dependent isoprenoid pathway (MEV pathway) [8,9]. Almost all eukaryotes and archaea use the MEV pathway, whereas most prokaryotes take advantage of the MEP pathway. Plants can utilize both biosynthetic pathways [10–12] (Figure 1).

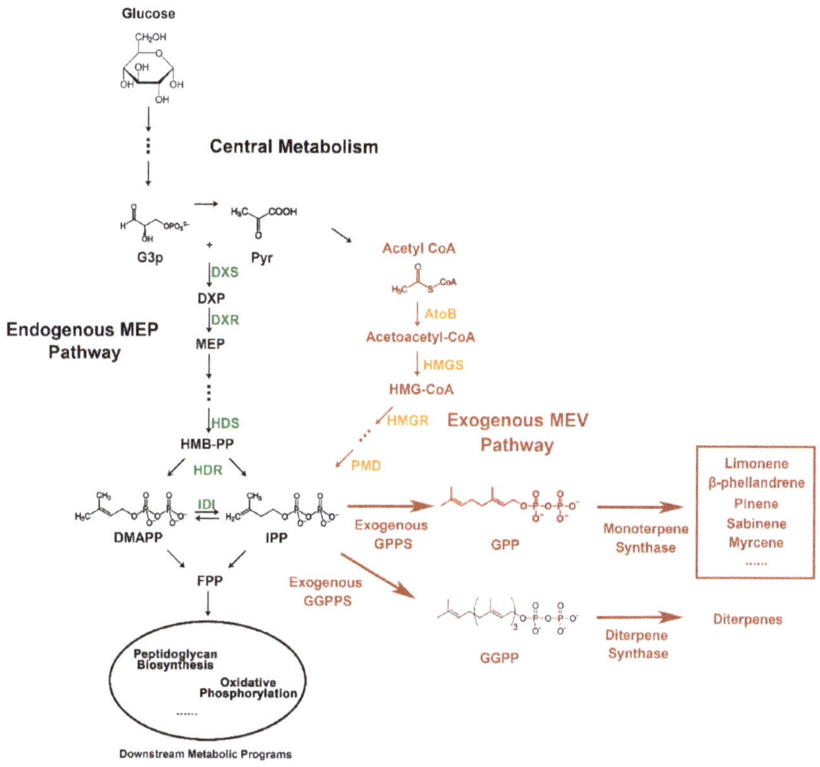

Figure 1. Monoterpene synthesis pathway in engineered prokaryotic host. GA3P, D-glyceraldehyde 3-phosphate; CoA, coenzyme A; MEP, methylerythritol 4-phosphate; DXS, 1-deoxyD-xylulose-5-phosphate synthase; DXR, 1-deoxy-D-xylulose-5phosphate reductoisomerase; AtoB, acetoacetyl-CoA synthase; HMGS, 3-hydroxy3-methylglutaryl-CoA synthase; HMGR, 3-hydroxy-3methylglutaryl-CoA reductase; IDI, isopentenyl diphosphate isomerase; DMAPP, dimethylallyl diphosphate; IPP, isopentenyl diphosphate.

Monoterpenes, the members with the smallest molecular weight in terpenoids, are gaining significant attention because of their various applications in medicines, biofuels, and agriculture in addition to the traditional use of essential oils and flavor production. Monoterpenes have been gradually recognized as essential medicines and prophylactic formulations because of their characteristics that can be easily absorbed by the body and transferred into blood and their ability to treat severe chronic diseases, including cancer [13]. Limonene, which is used to produce fragrance, flavor, and medicinal products, is a compound generally recognized as safe because of its earth-friendly cleaning performance [14,15]. Geraniol is often used in the production of perfumes and cosmetics, and can also be used as a clinical anticancer drug against pancreatic, colon, and other cancers [16–18]. Meanwhile, some monoterpenes and their derivatives, such as α-pinene, camphene, and limonene, have high calorific value of combustion and low freezing point; these advantages make them a favorable choice for next-generation clean jet-biofuels to replace the traditional jet fuels such as JP-10 and RJ-5 [19,20]. Moreover, some monoterpenes, such as carvacrol, p-cymene, and gamma-terpinene, are toxic to microbes and insects, and are thus often used as antibiotics and insecticides [21,22].

In the past, monoterpenes were mainly obtained from plant biomass, but this traditional production method restricts their wide applications because of its low yields, high costs, long reaction cycles and difficult purification [23]. Similar problems also exist in the subsequently developed chemical synthesis strategy with the complex reaction process and the environment pollution risks, even though the productivity could be increased. Moreover, such modes of production could hardly synthesize compounds with complex molecular structure and specific affinity and specificity [24]. To address these problems and achieve broad commercial applications, scholars have developed a series of biological manufacturing methods with the advantages of self-assembly, proliferation, mild reaction condition requirements, and environment-friendly features to produce valuable monoterpenes, especially in this rapid development era of synthetic biology and bioengineering [25,26]. Several microorganisms, such as *Escherichia coli* and *Saccharomyces cerevisiae*, are considered as perfect chassis and have been designed as microbial cell factories for the industrialized production of significant monoterpenoids [27,28]. Both MEV and MEP pathways and their downstream enzyme systems are available to be incorporated and engineered in these chassis with different strategies to produce various valuable monoprenoids with high yield. This review summarizes the advances of different strategies for the establishment and optimization of heterologous monoterpene synthesis in the prokaryotic cell factory (Table 1).

Table 1. Strategies used for monoterpene production.

Pathway	Strains of the *E. coli* Chassis	Origin of the Integrated Enzymes for the Monoterpene Production	Engineering Design	Monoterpene Product	Maximal Monoterpene Yield, Culture and Recovery Methods	Reference
MEP	BLR (DE3) Codon Plus-RIL cells	1. tGPPS from *Abies grandis* 2. tLS from *Mentha spicata*	1. Absence of enhanced MEP or MVA pathway 2. Adjusting promoter strength	Limonene	~5 mg/L, Steam distillation;	[29]
	BL21 (DE3)	1. tGPPS from *Abies grandis* 2. tLS from *Mentha spicata* 3. DXS and IDI from *E. coli* K12 MG1655	1. Codon optimization 2. Plasmid vector and enzyme arrangement selection 3. Integration of *gpps* and *ls* in one plasmid; integration of *dxs* and *idi* in another plasmid	Limonene	35.8 mg/L, two-phase culture of *n*-hexadecane organic layer	[30]
MEV	DH1 ΔacrAB	1. AACT and IDI from *E. coli* 2. HMGS and tHMGR from *Staphylococcus Aureus* 3. MVK, PMK, and PMD from *Saccharomyces cerevisiae* 4. tGPPS from *Abies grandis* 5. tLS from *Mentha spicata* 6. efflux pump from *Alcanivorax borkumensis*	1. Codon optimization 2. Exogenous pathway introduction 3. Replication origin and promoter strength selection 4. Integration of seven MEV pathway genes in one plasmid; integration of *gpps* and *ls* in another plasmid; efflux pump genes integrated in the last plasmid alone.	Limonene	~60 mg/L, two-phase culture of dodecane organic layer	[31]
	DH1	1. AtoB and IDI from *E. coli* 2. HMGS and tHMGR from *Staphylococcus Aureus* 3. MK, PMK, and PMD from *Saccharomyces cerevisiae* 4. tGPPS from *Abies grandis* 5. tLS from *Mentha spicata* 6. Cytochrome P450 from *Mycobacterium*	1. Codon optimization 2. Exogenous pathway introduction 3. Stronger promoter replacement 4. Integration of seven MEV pathway genes in one plasmid; integration of *gpps* and *ls* in one plasmid; integration of limonene-producing genes in one plasmid; integration of P450 system genes in one plasmid	Limonene, Perillyl alcohol	~435 mg/L, two-phase culture of dodecane organic layer; ~34 mg/L, *in situ* product recovery strategy based on Amberlite IRA 410 Cl (A)	[32]
	Rosetta	1. AtoB from *E. coli* 2.HMGS and HMGR from *Enterococcus faecalis* 3. FNI, MK, PMK and PMD from *Streptococcus pneumoniae R6* 4. tGPPS from *Picea abies* 5. PHLS from *Lavandula angustifolia*	1. Codon optimization 2. Exogenous pathway introduction 3. Integration of seven MEV pathway genes in one plasmid; integration of *gpps* and *phls* in one plasmid	β-phellandrene	25 mg/g_{dcw}, two-phase culture of hexane organic layer	[33]

Table 1. *Cont.*

Pathway	Strains of the *E. coli* Chassis	Origin of the Integrated Enzymes for the Monoterpene Production	Engineering Design	Monoterpene Product	Maximal Monoterpene Yield, Culture and Recovery Methods	Reference
	MG1655	1. AtoB and IDI from *E. coli* 2. HMGS and tHMGR from *Saccharomyces cerevisiae* 3. MK, PMK, and PMD from *Saccharomyces cerevisiae* 4. tGPPS and tPS from *Abies grandis*	1. Codon optimization 2. Exogenous pathway introduction 3. Fusion protein 4. Integration of seven MEV pathway genes in one plasmid; integration of *gpps* and *ps* in another plasmid	Pinene	32.4 mg/L, two-phase culture of dodecane organic layer	[34]
	BL21 (DE3)	1. AtoB and IDI from *E. coli* 2. HMGS and tHMGR from *Saccharomyces cerevisiae* 3. MK, PMK, and PMD from *Saccharomyces cerevisiae* 4. tGPPS from *Abies grandis* 5. tLS from *Mentha spicata*	1. Codon optimization 2. Exogenous pathway introduction 3. Integration of seven MEV pathway genes in one plasmid; integration of *gpps* and *ls* in another plasmid	Limonene	2.7 g/L, two-phase culture of diisonoylphtalate organic layer	[35]
	BL21(DE3)	1. MvaE and MvaS from *Enterococcus faecalis* 2. MK, PMK, PMD and IDI from *Saccharomyces cerevisiae* 3. GPPS from *Abies grandis* 4. SabS from *Salvia pomifera*	1. Codon optimization 2. Exogenous pathway introduction 3. Integration of three upper MEV pathway genes, *gpps* and *sabs* in one plasmid; integration of four lower MEV pathway genes in another plasmid	Sabinene	2.65 g/L	[36]
	MG1655	1. MvaE and MvaS from *Enterococcus faecalis* 2. MvaK1, MvaK2, MvaD from *Streptococcus pneumoniae* 3. IDI from *E. coli* 4. GPPS from site-directed mutation of FPPS 5. tGES from *Ocimum basilicum*	1. Codon optimization 2. Exogenous pathway introduction 3. Deletion of *E. coli* gene *yjgB* 4. Integration of seven MEV pathway genes in one plasmid; integration of *gpps* and *ges* in one plasmid	Geraniol	182.5 mg/L; two-phase culture of decane organic layer	[37]
	DH1	1. AtoB and IDI from *E. coli* 2. HMGS and tHMGR from *Saccharomyces cerevisiae* 3. MK, PMK, and PMD from *Saccharomyces cerevisiae* 4. tGPPS from *Abies grandis* 5. tMS from *Quercus ilex* L.	1. Codon optimization 2. Exogenous pathway introduction 3. Integration of seven MEV pathway genes and *gpps* in one plasmid; integration of *ms* alone in one plasmid	Myrcene	58.19 mg/L two-phase culture of dodecane organic layer	[38]
	BL21 (DE3)	1. MvaE and MvaS from *Enterococcus faecalis* 2. MK, PMK, PMD and IDI from *Saccharomyces cerevisiae* 3. tGPPS from *Abies grandis* 4. tGES from *Ocimum basilicum*	1. Codon optimization 2. Exogenous pathway introduction 3. Identification of the role of acetylesterase for converting geranyl acetate to geraniol 4. Integration of three upper MEV pathway genes, *gpps* and *ges* in one plasmid; integration of four lower MEV pathway genes in another plasmid	Geraniol	~ 2.0 g/L; two-phase culture of isopropyl myristate organic layer	[39]
	XL1-Blue	1. AtoB and IDI from *E. coli* 2. HMGS and tHMGR from *Staphylococcus Aureus* 3. MK, PMK, and PMD from *Saccharomyces cerevisiae* 4. tGPPS from *Abies grandis* 5. PS from *Pinus taeda*	1. Codon optimization 2. Exogenous pathway introduction 3. Directed evolution of PS 4. Integration of seven MEV pathway genes and *gpps* in one plasmid; integration of mutant *ps* alone in another plasmid	Pinene	140 mg/L; two-phase culture of dodecane organic layer	[40]

Abbreviations: AtoB, acetoacetyl-CoA synthase; HMGS, HMG-CoA synthase; HMGR, HMG-CoA reductase; MK, mevalonate kinase; PMK, phosphomevalonate kinase; PMD, mevalonate pyrophosphate decarboxylase; MvaE, bifunctional acetoacetyl-CoA thiolase and HMG-CoA reductase; MvaS, HMG-CoA synthase; MvaK1, mevalonate kinase; MvaK2, phosphomevalonate kinase; MvaD, mevalonate diphosphate decarboxylase; IDI, isopentenyl diphosphate isomerase; GPPS, geranyl diphosphate synthase; FPPS, farnesyl diphosphate synthase; LS, limonene synthase; PS, pinene synthase; SabS, sabinene synthase; GES, geraniol synthase; MS, myrcene synthase; AES, acetylesterase.

2. Engineering Endogenous MEP Pathway in Prokaryotic Chassis

As a pivotal microbial chassis for bioengineering, prokaryotic *E. coli* mainly use MEP pathway for isoprenoid biosynthesis, which usually unable to produce sufficient quantities of monoterpenes for industrial production. Although several groups indicated that the engineered MEP pathway could increase the level of isoprenoid production in *E. coli*, the common precursors IPP and DMAPP are primarily synthesized to FPP and higher polyprenyl diphosphates rather than to the intermediate of GPP and its downstream monoterpenoids [41,42]. Thus, introducing exogenous catalytic enzyme genes that encode GPP synthase (GPPS) and other monoterpene synthases to improve GPP production could be an appropriate solution. Carter et al. tested the function of a short exogenous metabolic pathway for the biosynthesis of the simple monoterpene carvone [29]. They incorporated four enzymes, namely, GPPS, limonene synthase (LS), cytochrome P450 limonene hydroxylase (L6H), and carveol dehydrogenase (CdH), into *E. coli*, which could theoretically catalyze IPP and DMAPP to carvone. However, the production of the intermediate limonene increased to nearly 5 mg/L, whereas the target product of carvone could hardly be detected. Intriguingly, feeding with exogenous limonene could push forward the carbon flux to the synthesis of carvone probably through increasing the supply of substrates. This strategy could improve the production of carvone into 0.25 mg/L. These results suggest that the endogenous supplies of the crucial precursors of IPP and DMAPP are at a relative low flux to the monoterpene synthesis. The requirement of the industrial production of monoterpenes may hardly be satisfied by simply incorporating several downstream enzymes to the MEP pathway.

Since the efficiency of the endogenous MEP pathway became a limiting factor, several groups attempted to engineer the MEP pathway to acquire higher titers of monoterpene. Enzymes 1-deoxy-d-xylulose-5-phosphate synthase (DXS), DXP reductoisomerase (DXR), and isopentenyl diphosphate isomerase (IDI) were demonstrated as the rate-limiting factors in the MEP pathway [43,44]. DXS catalyzes the formation of DXP, DXR reduces DXP to 2-C-methyl-D-erythritol-4-phosphate, and IDI catalyzes the conversion of the relatively unreactive IPP to the more-reactive electrophile DMAPP. The strategies of introducing exogenous rate-limiting enzymes with high expression by codon-optimization and/or increasing the expression levels of these endogenous enzymes through integrating with strong promoters could successfully control flux from the target precursors to the subsequent desired compounds. Du et al. embedded two exogenous genes encoding GPPS and LS in *E. coli* so that the production of limonene reaches 4.87 mg/L. Subsequently, they overexpressed DXS and IDI through plasmid transient transformation, which could enhance the production of limonene ultimately to 17.4 mg/L at 48 h. After a series of modifications to optimize the two-phase culture medium, the titer of limonene continuously elevated up to 35.8 mg/L, approximately 7-fold greater than the initial yield [30]. Despite the significant improvement in monoterpene biosynthesis, the wide applications of MEP pathway engineering are limited because of the presence of inherent regulation mechanisms and the unknown physiological control elements in the host cell, which cause the bottleneck of monoterpene production efficiency [45].

3. Introduction of Heterologous MEV Pathway

Bypassing the inherently metabolic synthesis pathway through replacing it with a heterologous mevalonate-dependent pathway provides a pioneering strategy for the production of valuable terpenoids [46]. For example, Martin's group introduced a heterologous MEV pathway into *E. coli* to increase amorphadiene titer greater than 100 mg/L [47]. This mechanism might be the overproduction of universal precursors IPP and DMAPP by the heterologous expression of MEV pathway enzymes with terpene synthases, which could enhance the conversion efficiency of IPP and DMAPP to relevant terpenoids. Similar strategies were inspired to increase the titers and yields of various monoterpenes. For instance, Gutierrez et al. engineered *E. coli* by introducing all seven enzyme-encoding genes involved in the MEV pathway and downstream limonene and perillyl alcohol (POH) synthesis-dependent genes encoding GPPS, LS, and cytochrome P450 [32]. Considering that the dispersion of the seven genes of the MEV pathway into multiple plasmids could aggravate the metabolic burden and hinder the target product synthesis in the host cell due to the raw materials and energy consumption [48,49], the authors integrated the seven genes into one plasmid and generated various versions of plasmid constructs for further study. The appropriately modified metabolic route increased the yield of limonene to 435 mg/L during 72 h of incubation with supplement of 1% glucose in the culture medium. Thus, the high titer of limonene could overcome its uptake and trafficking restrictions, which might promote the conversion efficiency of limonene to the target monoterpene such as carvone. Accordingly, the strain harboring the reasonably combinatory plasmids was induced to produce its derivative POH to about 34 mg/L. For α-pinene production, Yang et al. embedded the heterologous hybrid MEV pathway and the co-expressed GPPS and α-pinene synthase (PS).

The final biosynthesis production accumulated up to 5.44 mg/L and 0.97 g/L under the culture conditions of flask and fed-batch fermentation, respectively [50]. Another intriguing work indicated that the incorporation of GPPS and β-phellandrene synthase (PHLS) in *E. coli* could not result in the measurable yield of β-phellandrene. However, after introducing the MEV pathway in collaboration with the GPPS and the PHLS, the output of β-phellandrene reached 11 mg/g_{dcw} after over 20 h of incubation. This titer could be further improved to 25 mg/g_{dcw} by optimizing LB broth with 1% glucose supplement and then extending the incubation time to over 72 h [33]. Notably, when the endogenous MEP pathway and the exogenous MEV pathway were used individually, the production of sabinene generated from integrated GPPS and sabinene synthase in the MEV pathway was 20-fold higher than that in the MEP pathway. The final production via a series optimization for culture condition and inducer concentration achieved a maximum titer of 82.18 mg/L under shake-flask culture and 2.65 g/L under fed-batch culture [36]. However, the method of heterologous MEV pathway introduction is not always available for different monoterpene biosyntheses to reach an industrial grade, which is probably due to restrictions of the rate-determining step and the imbalance of the metabolic flux. These bottlenecks may result in the accumulation of some toxic intermediates and the limitation of the downstream essential products in the intracellular space.

4. Optimization of the Expression and Function of the Rate-Limiting Enzymes

A common method to break the restrictions of the rate-determining step is to increase the expression level of rate-limiting enzymes in prokaryotic hosts. Three strategies are available: (1) insertion of stronger promoters into the operons directing the enzyme gene expression; (2) codon optimization of enzyme-coding regions; and (3) screening of enzymes from different species for a higher compatible and efficient homolog [51,52]. For the integration of MEV pathway and monoterpene synthases, a series of strong promoters, including T7, lacUV5, and trc, was utilized [32,33]. In another way, the codon optimization scheme is also essential for the heterologous metabolic system due to relatively diminished expression efficiency by biased codon usage [53]. Moreover, for the enzyme homolog screening, GPPS is considered a crucial candidate, which could be classified as homomeric and heteromeric isoforms [54,55]. The heteromeric GPPS extracted from *Mentha piperita* is composed of two subunits with different sizes, which shows no single catalytic activity. This heterodimer could transform IPP and DMAPP to C_{10} GPP and C_{20} GGPP [56]. The homomeric GPPS from *Arabidopsis thaliana* is a polyprenyl pyrophosphate synthase that synthesizes multiple terpenes ranging from C_{25} to C_{45}, while another type from conifers, such as *Picea abies* and *Abies grandis*, yields only GPP [41,57,58]. Because the GPP specific-producing manner can avoid or suppress the byproduct generation and effectively control the metabolic flux into the monoterpenes, the fully characterized GPPSs from *P. abies* and *A. grandis* were confirmed the better synthetases for the monoterpene production in host microbes, which need further codon optimization and truncation of tail sequences encoding plastid signal peptide for prokaryotic cell integration.

Another striking method for the functional improvement of the enzymes is directed evolution, which relies on random mutagenesis and high-throughput screening [59]. It confers the enzymes with an unnatural powerful catalytic efficiency and effectively overcomes the metabolic flux nodes [60,61]. Notably, the monoterpene synthase-mediated reactions are widely considered as the rate-limiting step in production. The improved activity and stability of enzymes could benefit to compete for the essential intermediate GPP from FPPS, which will lead the carbon flow to the final monoterpene. For instance, Tashiro's group isolated a PS variant that outperforms the wild-type through the directed enzyme evolution. After co-expression of this variant with IDI, GPPS, and the MEV pathway, pinene productivity could reach 140 mg/L in flask culture [40]. Technically, the visualization of substrate consumption and the diversity of alternative variants could facilitate the establishment of the screening system for the directed evolution of enzymes [62,63]. Thus, this approach could rapidly select perfect mutants for advanced cellular performance, which is suggested to be suitable for many other enzymes in the metabolic route [59].

5. Controlling the Flux Distribution of Essential Intermediates

5.1. Fusion of Key Enzymes

Biofunctional fusion enzyme works as a multifunctional protein derived from a single nucleotide sequence that may contain different enzyme genes. This construction renders the active site of one enzyme face to another, which channels the intermediates directly though successive catalytic bioreactions. The spatial distance between enzymes lowers the substrate transmission loss and accelerates the reaction rate [64]. Meanwhile, the direction of internal enzyme components as well as the length and the amino acid composition of the linkers involved in the fusion enzymes are considered the optimizable parameters for the improvement of their catalytic efficiency [65,66]. Sarria et al. programed an engineered *E. coli* strain for pinene synthesis by introducing the exogenous MEV pathway and a series of GPPS-PS fusion enzymes with the components derived from three species. The highest-flux synthase combination elevated the concentration of pinene in host cells to about 28 mg/L, whereas the subsequent expression of assembled GPPS-PS fusion protein achieved a higher production of 32 mg/L, approximately 6-fold than that previously reported [34]. The high local accumulation and exchange of substances though the fusion enzyme system could theoretically overcome the metabolic burden of the essential intermediate leakage, such as GPP, and relieve or even eliminate its feedback inhibition effect and toxicity to the host. However, the critical drawback of this strategy is its high dependence to the accurate architecture of different enzyme components, which are restricted for expanding to higher-order enzyme fusion.

5.2. Spatial Organization of Heterologous Enzymes

A series of enzymes driving the engineered metabolic pathway could be spatially colocalized in a specific area of host cells by a programmable scaffold manner. This intriguing strategy has a significant scalability for enzyme population and can balance the overall pathway fluxes and alleviate metabolic burdens to optimize yields of target products [67]. The reason is that highly ordered enzymes reduce intermediate transmission time, protect them from diffusing or competing the bypass route, and circumvent undesirable equilibria and kinetics resulting from bulk-phase metabolite concentrations [68–70]. The synthetic scaffold strategy was initially inspired from the natural synthase complex exhibiting substrate channeling and the programmable nucleotide–protein interaction. Dueber et al. constructed synthetic protein scaffolds that spatially recruit three MEV biosynthetic enzymes to improve the mevalonate titers to 77-fold (~5 mM) with relatively low enzyme expression [71]. Another work arranged individual enzymes of the metabolic pathway at the proper stoichiometry via fusing enzymes to zinc-finger domains that specifically bind to corresponding DNA sequences. The titers of metabolites resveratrol, 1,2-propanediol, and mevalonate increased up to about 5-, 4.5-, and 2.5-fold compared with controls, respectively [72]. Our group demonstrated that fusing the acetoacetyl-CoA synthase (AtoB), 3-hydroxy3-methylglutaryl-CoA synthase (HMGS), and 3-hydroxy-3methylglutaryl-CoA reductase (HMGR) genes with rationally designed transcription activator-like effectors (TALEs) increases the mevalonate production by 3.7-fold [73]. RNA aptamers were also introduced to the scaffold systems for the designable colocalization of sequential metabolic enzymes [74,75]. The modularity and programmability of all these scaffold systems enable them to organize pathway enzymes in specific orientation and optimal stoichiometry. These assembled synthetic complexes could be conductive to the formation of a concentrated metabolic pool and benefit the manufacture of various multi-enzyme metabolic pathways.

5.3. Decrease the Flux of Essential Intermediates into Irrelevant Endogenous Pathways

In metabolic engineering, the irrelevant endogenous pathways utilizing the building blocks of introduced heterologous pathway in host could be the essential factor hindering the target monoterpene production. The loss could be avoided through blocking or rerouting irrelevant flux [76]. Zhou et al. introduced recombinant GPPS and the bottom portion of the MEV pathway into *E. coli* to yield geraniol up to 13.3 mg/L. With the combination of geraniol synthase (GES) heterologous expression, the geraniol production reached 105.2 mg/L. However, endogenous dehydrogenization and isomerization of geraniol into other geranoids restrained the production. After engineering the strain via deleting the microbial gene that is highly homologous to plant geraniol dehydrogenase, the conversion of geraniol significantly reduced and its productivity reached 129.7 mg/L. This titer could be further increased to 182.5 mg/L by the whole MEV pathway integration [37]. Liu's group explored an inverse conversion approach to acquire an approximate industrial productivity of geraniol. They identified that the acetylesterase (AES) from *E. coli* could transform geranyl acetate into geraniol; simultaneously, *E. coli* could reuse acetate as carbon source in the absence of glucose. By stopping glucose supply for the engineered strain after 48 h incubation, the geraniol production under fed-batch fermentation increased up to 2.0 g/L [39]. Thus, accompanied by the engineering of the MEV pathway, identification and deletion of the endogenous metabolic routes interfering the target product synthesis pathway will be a broadly applicable approach for controlling the metabolic flux and improving the production of various terpenoids.

6. Improvement of the Toxicity Tolerance for the Host Strain

Almost all monoterpenes exert critical toxicity to the host cell. Thus, continuously elevated production of monoterpenes may exacerbate the exhaustion of the cell factory. Their toxicity could inhibit the growth of producing hosts, lower the biotransformation activity of enzymes, and impede the availability of the metabolic route. In the end, the total production of monoterpenes by microbial strains is eventually reduced [21,77,78]. Commonly used approaches to improve toxicity tolerance include modification of membrane proteins, expression of efflux pumps, and activation of the stress response system, such as expression and regulation of heat shock proteins [79–81]. Dunlop et al. employed a cellular export system to alleviate biofuel toxicity and enhance the host's tolerance. They screened bacterial genomes and read out all the efflux pumps, a class of membrane transport proteins driven by the proton motive force [82–84]. Afterward, 43 of the pumps were selected and cloned into a plasmid library, which were subsequently transformed into *E. coli* for heterologous expression. This export mechanism shows significant protective effects of host survival for five chosen biofuels, including geranyl acetate, geraniol, α-pinene, limonene, and farnesyl hexanoate [31]. The elevated tolerance to the product toxicity and the significant improvement of product yields might be due to the efficient transportation of toxins out of the host and the maintenance of biomass accumulation by the efflux pumps. However, the combination of more than one type of efflux pumps could inhibit cell growth, implying that antagonistic mechanism is involved in the multi efflux pump effects [85]. Nevertheless, this approach has a profound effect on the toxicity tolerance of the prokaryotic hosts and has a great potential to be applied in monoterpene production. Alternatively, another intelligent strategy called in situ product recovery (ISPR) based on anion exchange resin was developed [86,87]. This method uses a column containing a fluidized bed of resin combined with a bioreactor to specifically trap toxic products and remove them from the culture media. This manner could relieve the effect of toxic inhibition on bioconversion and finally improve the production of the target monoterpene. Alonso-Gutierrez et al. tested four commercially available resins and then found the Amberlite IRA 410 Cl, a most suitable resin for POH recovery. Amberlite resin traps the POH, which can maintain a low final product concentration in the media and maintain a high rate of metabolic flux toward it. Their results showed 1.5-fold higher POH production than that in the resin-absent group. Moreover, combination of this method with overexpression of the cytochrome P450 system could increase the ultimate concentration of POH with 2.5-fold in total and 3.5-fold in specific production [32].

Meanwhile, a method using an aqueous-organic two-phase system for host cell culture produces similar protective effects. Rational utilization of in situ separation and extraction for two-phase culture medium not only alleviates the product toxicity to cells but also prevents the target monoterpene from volatilizing [88,89]. By taking advantage of the in situ two-phase extraction, the bacteria harboring the heterologous MEV pathway, GPPS and MS could produce myrcene to a maximum of 58.19 ± 12.13 mg/L [38]. However, these progresses for toxic tolerance improvement are mainly focused on the effective transfer of products. Other ways of optimizing the metabolic systems or engineering the host cells in a more robust condition should also be considered in future studies.

7. Conclusions

The metabolic engineering of monoterpene production has already achieved substantial progress in recent years. In early studies, the native MEP pathway as a primary regulatory objective acquired considerable attention. Most studies focused on the overexpression of key enzymes in this pathway, which limit productivity. Unfortunately, the implementations of such strategy could not achieve the expected effect for the industrial production of monoterpenes possibly because of the intrinsic barrier of the MEP pathway for supplying IPP and DMAPP, two essential universal monoterpene-building components. The introduction of a stronger heterologous MEV pathway partially compensated for this shortage. This strategy was subsequently designed to combine with a series of optimization approaches, such as functional improvement of the rate-limiting enzymes, control of the metabolic flux, and increase of the host's tolerance to the product toxicity as discussed above, for further productivity improvement. In general, most monoterpene biosyntheses remain far from the industrialization by now. To address this problem, systematic analysis of the complicated metabolic system should be considered in future studies for monoterpene manufacturing. One notable study designed a computational tool named principal component analysis of proteomics (PCAP), for the multi-dimensional engineering of global metabolic pathways. This rational mathematical tool interrogated the data of the targeted proteomics and products based on principal component analysis, which could help researchers to modify the expression of some specific enzymes, balance the metabolic pathways, and predict product yields. Thus, the overall strategies for engineering the higher-efficient monoterpene-producing chassis could be rationally designed [90]. Future studies should also consider the systematical integration of multiple approaches demonstrated to be effective in the prokaryotic chassis. An intriguing research took advantage of a multi-level analytical method to define and assess four engineering strategies for fermentative limonene production in *E. coli*: (1) Construction of a metabolic route by transformation of recombinant plasmids harboring heterologous MEV pathway enzymes and the subsequent synthases into the host strains; (2) Determination of the enzyme activity for choosing an appropriate GPPS-exploited cell-free system with the functional proteins extracted from the induced host and then adding extra substrates to the system; (3) Selection of the most suitable host by analyzing the physiological properties of different *E. coli* strains after introduction of the MEV pathway and the downstream steps; (4) Utilization of two liquid-phase carbon source fed-batch fermentation with glucose and glycerol, as well as adding a non-toxic organic solvent for the in situ extraction of monoterpenes [35]. This attempt of the integration scheme from genetic modification to process optimization surprisingly increased target productivity. Further studies are necessary to find other complementary advantages of various metabolic engineering approaches and systematically elongate the pipelines of engineering for monoterpene biomanufacturing.

Other interesting opportunities for future studies of monoterpene biomanufacturing will rely on the rapid development of the synthetic biology, which could interrogate the organizational principles of metabolic systems and easily rewire the prokaryotic cells into highly efficient cell factories. The growing strategies generated from this field, such as artificial life creation, gene circuit reconstruction, and metabolism redirection, would open new opportunities not only for the synthesis of monoterpenes but also for many other valuable important chemicals. Their applications in the construction and optimization of microbial cell factories will inevitably shorten the distance from laboratory trials into industrial production. Along with the continuous breakthrough of the bottlenecks and restrictive factors, we can expect monoterpenes to play valuable roles in health and environment applications over the coming years.

Author Contributions: S.-s.X. and L.-y.Z. wrote the manuscript, X.-y.Q. and L.-y.Z. design and visualize the figure, L.Z. and C.-s.Z. provided important advises.

Funding: This review was funded by the National Natural Science Foundation of China (grant numbers 31500686, 31870855), the Hunan Provincial Natural Science Foundation of China (grant number 2017JJ3358), and the National University of Defense Technology project (grant numbers ZK17-03-58).

Acknowledgments: The authors thank Shine Write for its linguistic assistance during the preparation of this manuscript.

Conflicts of Interest: The authors declare no conflict of interest.

References

1. Gershenzon, J.; Dudareva, N. The function of terpene natural products in the natural world. *Nat. Chem. Biol.* **2007**, *3*, 408–414. [CrossRef]
2. Fidock, D.A.; Rosenthal, P.J.; Croft, S.L.; Brun, R.; Nwaka, S. Antimalarial drug discovery: Efficacy models for compound screening. *Nat. Rev. Drug Discov.* **2004**, *3*, 509–520. [CrossRef] [PubMed]
3. Jennewein, S.; Croteau, R. Taxol: Biosynthesis, molecular genetics, and biotechnological applications. *Appl. Microbiol. Biotechnol.* **2001**, *57*, 13–19. [PubMed]
4. Croteau, R.; Ketchum, R.E.; Long, R.M.; Kaspera, R.; Wildung, M.R. Taxol biosynthesis and molecular genetics. *Phytochem. Rev.* **2006**, *5*, 75–97. [CrossRef]
5. Spanggord, R.J.; Wu, B.; Sun, M.; Lim, P.; Ellis, W.Y. Development and application of an analytical method for the determination of squalene in formulations of anthrax vaccine adsorbed. *J. Pharm. Biomed. Anal.* **2002**, *29*, 183–193. [CrossRef]
6. Arias, J.L.; Reddy, L.H.; Othman, M.; Gillet, B.; Desmaele, D.; Zouhiri, F.; Dosio, F.; Gref, R.; Couvreur, P. Squalene based nanocomposites: A new platform for the design of multifunctional pharmaceutical theragnostics. *ACS Nano* **2011**, *5*, 1513–1521. [CrossRef] [PubMed]
7. Tholl, D. Terpene synthases and the regulation, diversity and biological roles of terpene metabolism. *Curr. Opin. Plant Biol.* **2006**, *9*, 297–304. [CrossRef]
8. Lange, B.M.; Rujan, T.; Martin, W.; Croteau, R. Isoprenoid biosynthesis: The evolution of two ancient and distinct pathways across genomes. *Proc. Natl. Acad. Sci. USA* **2000**, *97*, 13172–13177. [CrossRef]
9. Kuzuyama, T. Mevalonate and nonmevalonate pathways for the biosynthesis of isoprene units. *Biosci. Biotechnol. Biochem.* **2002**, *66*, 1619–1627. [CrossRef]
10. Rohmer, M. The discovery of a mevalonate-independent pathway for isoprenoid biosynthesis in bacteria, algae and higher plants. *Nat. Prod. Rep.* **1999**, *16*, 565–574. [CrossRef] [PubMed]
11. Smit, A.; Mushegian, A. Biosynthesis of isoprenoids via mevalonate in Archaea: The lost pathway. *Genome Res.* **2000**, *10*, 1468–1484. [CrossRef]
12. Vranová, E.; Coman, D.; Gruissem, W. Network analysis of the MVA and MEP pathways for isoprenoid synthesis. *Annu. Rev. Plant Biol.* **2013**, *64*, 665–700. [CrossRef] [PubMed]
13. Murthy, K.N.C.; Jayaprakasha, G.K.; Mantur, S.M.; Patil, B.S. Citrus Monoterpenes: Potential Source of Phytochemicals for Cancer Prevention. In *Emerging Trends in Dietary Components for Preventing and Combating Disease*; American Chemical Society: Washington, DC, USA, 2012; pp. 545–558.
14. Sun, J. D-Limonene: Safety and clinical applications. *Altern. Med. Rev.* **2007**, *12*, 259–264.
15. Ciriminna, R.; Lomeli-Rodriguez, M.; Demma Cara, P.; Lopez-Sanchez, J.A.; Pagliaro, M. Limonene: A versatile chemical of the bioeconomy. *Chem. Commun.* **2014**, *50*, 15288–15296. [CrossRef]
16. Chen, W.; Viljoen, A. Geraniol—A review of a commercially important fragrance material. *S. Afr. J. Bot.* **2010**, *76*, 643–651. [CrossRef]
17. Kim, S.H.; Bae, H.C.; Park, E.J.; Lee, C.R.; Kim, B.J.; Lee, S.; Park, H.H.; Kim, S.J.; So, I.; Kim, T.W.; et al. Geraniol inhibits prostate cancer growth by targeting cell cycle and apoptosis pathways. *Biochem. Biophys. Res. Commun.* **2011**, *407*, 129–134. [CrossRef]

18. Sun, H.; Wang, Z.; Yakisich, J.S. Natural products targeting autophagy via the PI3K/Akt/mTOR pathway as anticancer agents. *Anticancer Agents Med. Chem.* **2013**, *13*, 1048–1056. [CrossRef]
19. George, K.W.; Alonso-Gutierrez, J.; Keasling, J.D.; Lee, T.S. Isoprenoid Drugs, Biofuels, and Chemicals-Artemisinin, Farnesene, and Beyond. *Adv. Biochem. Eng. Biotechnol.* **2015**, *148*, 355–389. [PubMed]
20. Gupta, P.; Phulara, S.C. Metabolic engineering for isoprenoid-based biofuel production. *J. Appl. Microbiol.* **2015**, *119*, 605–619. [CrossRef]
21. Trombetta, D.; Castelli, F.; Sarpietro, M.G.; Venuti, V.; Cristani, M.; Daniele, C.; Saija, A.; Mazzanti, G.; Bisignano, G. Mechanisms of antibacterial action of three monoterpenes. *Antimicrob. Agents Chemother.* **2005**, *49*, 2474–2478. [CrossRef] [PubMed]
22. Cristani, M.; D'Arrigo, M.; Mandalari, G.; Castelli, F.; Sarpietro, M.G.; Micieli, D.; Venuti, V.; Bisignano, G.; Saija, A.; Trombetta, D. Interaction of four monoterpenes contained in essential oils with model membranes: Implications for their antibacterial activity. *J. Agric. Food Chem.* **2007**, *55*, 6300–6308. [CrossRef] [PubMed]
23. Geron, C.; Rasmussen, R.; Arnts, R.R.; Guenther, A. A review and synthesis of monoterpene speciation from forests in the United States. *Atmos. Environ.* **2000**, *34*, 1761–1781. [CrossRef]
24. Koeller, K.M.; Wong, C.H. Enzymes for chemical synthesis. *Nature* **2001**, *409*, 232–240. [CrossRef]
25. Brenner, K.; You, L.; Arnold, F.H. Engineering microbial consortia: A new frontier in synthetic biology. *Trends Biotechnol.* **2008**, *26*, 483–489. [CrossRef] [PubMed]
26. Keasling, J.D. Synthetic biology for synthetic chemistry. *ACS Chem. Biol.* **2008**, *3*, 64–76. [CrossRef]
27. Keasling, J.D. Synthetic biology and the development of tools for metabolic engineering. *Metab. Eng.* **2012**, *14*, 189–195. [CrossRef]
28. Kim, J.; Salvador, M.; Saunders, E.; Gonzalez, J.; Avignone-Rossa, C.; Jimenez, J.I. Properties of alternative microbial hosts used in synthetic biology: Towards the design of a modular chassis. *Essays Biochem.* **2016**, *60*, 303–313. [CrossRef] [PubMed]
29. Carter, O.A.; Peters, R.J.; Croteau, R. Monoterpene biosynthesis pathway construction in Escherichia coli. *Phytochemistry* **2003**, *64*, 425–433. [CrossRef]
30. Du, F.-L.; Yu, H.-L.; Xu, J.-H.; Li, C.-X. Enhanced limonene production by optimizing the expression of limonene biosynthesis and MEP pathway genes in E. coli. *Bioresour. Bioprocess.* **2014**, *1*, 10. [CrossRef]
31. Dunlop, M.J.; Dossani, Z.Y.; Szmidt, H.L.; Chu, H.C.; Lee, T.S.; Keasling, J.D.; Hadi, M.Z.; Mukhopadhyay, A. Engineering microbial biofuel tolerance and export using efflux pumps. *Mol. Syst. Biol.* **2011**, *7*, 487. [CrossRef]
32. Alonso-Gutierrez, J.; Chan, R.; Batth, T.S.; Adams, P.D.; Keasling, J.D.; Petzold, C.J.; Lee, T.S. Metabolic engineering of Escherichia coli for limonene and perillyl alcohol production. *Metab. Eng.* **2013**, *19*, 33–41. [CrossRef] [PubMed]
33. Formighieri, C.; Melis, A. Carbon partitioning to the terpenoid biosynthetic pathway enables heterologous beta-phellandrene production in Escherichia coli cultures. *Arch. Microbiol.* **2014**, *196*, 853–861. [CrossRef] [PubMed]
34. Sarria, S.; Wong, B.; Garcia Martin, H.; Keasling, J.D.; Peralta-Yahya, P. Microbial synthesis of pinene. *ACS Synth. Biol.* **2014**, *3*, 466–475. [CrossRef]
35. Willrodt, C.; David, C.; Cornelissen, S.; Buhler, B.; Julsing, M.K.; Schmid, A. Engineering the productivity of recombinant Escherichia coli for limonene formation from glycerol in minimal media. *Biotechnol. J.* **2014**, *9*, 1000–1012. [CrossRef] [PubMed]
36. Zhang, H.; Liu, Q.; Cao, Y.; Feng, X.; Zheng, Y.; Zou, H.; Liu, H.; Yang, J.; Xian, M. Microbial production of sabinene—A new terpene-based precursor of advanced biofuel. *Microb. Cell Fact.* **2014**, *13*, 20. [CrossRef]
37. Zhou, J.; Wang, C.; Yoon, S.H.; Jang, H.J.; Choi, E.S.; Kim, S.W. Engineering Escherichia coli for selective geraniol production with minimized endogenous dehydrogenation. *J. Biotechnol.* **2014**, *169*, 42–50. [CrossRef] [PubMed]
38. Kim, E.M.; Eom, J.H.; Um, Y.; Kim, Y.; Woo, H.M. Microbial Synthesis of Myrcene by Metabolically Engineered Escherichia coli. *J. Agric. Food Chem.* **2015**, *63*, 4606–4612. [CrossRef]
39. Liu, W.; Xu, X.; Zhang, R.; Cheng, T.; Cao, Y.; Li, X.; Guo, J.; Liu, H.; Xian, M. Engineering Escherichia coli for high-yield geraniol production with biotransformation of geranyl acetate to geraniol under fed-batch culture. *Biotechnol. Biofuels* **2016**, *9*, 58. [CrossRef]

40. Tashiro, M.; Kiyota, H.; Kawai-Noma, S.; Saito, K.; Ikeuchi, M.; Iijima, Y.; Umeno, D. Bacterial Production of Pinene by a Laboratory-Evolved Pinene-Synthase. *ACS Synth. Biol.* **2016**, *5*, 1011–1020. [CrossRef] [PubMed]
41. Burke, C.; Croteau, R. Geranyl diphosphate synthase from Abies grandis: cDNA isolation, functional expression, and characterization. *Arch. Biochem. Biophys.* **2002**, *405*, 130–136. [CrossRef]
42. Burke, C.; Croteau, R. Interaction with the small subunit of geranyl diphosphate synthase modifies the chain length specificity of geranylgeranyl diphosphate synthase to produce geranyl diphosphate. *J. Biol. Chem.* **2002**, *277*, 3141–3149. [CrossRef]
43. Eisenreich, W.; Rohdich, F.; Bacher, A. Deoxyxylulose phosphate pathway to terpenoids. *Trends Plant Sci.* **2001**, *6*, 78–84. [CrossRef]
44. Eisenreich, W.; Bacher, A.; Arigoni, D.; Rohdich, F. Biosynthesis of isoprenoids via the non-mevalonate pathway. *Cell. Mol. Life Sci.* **2004**, *61*, 1401–1426. [CrossRef]
45. Banerjee, A.; Wu, Y.; Banerjee, R.; Li, Y.; Yan, H.; Sharkey, T.D. Feedback inhibition of deoxy-D-xylulose-5-phosphate synthase regulates the methylerythritol 4-phosphate pathway. *J. Biol. Chem.* **2013**, *288*, 16926–16936. [CrossRef] [PubMed]
46. Pitera, D.J.; Paddon, C.J.; Newman, J.D.; Keasling, J.D. Balancing a heterologous mevalonate pathway for improved isoprenoid production in Escherichia coli. *Metab. Eng.* **2007**, *9*, 193–207. [CrossRef] [PubMed]
47. Martin, V.J.; Pitera, D.J.; Withers, S.T.; Newman, J.D.; Keasling, J.D. Engineering a mevalonate pathway in Escherichia coli for production of terpenoids. *Nat. Biotechnol.* **2003**, *21*, 796–802. [CrossRef]
48. Bhattacharya, S.K.; Dubey, A.K. Metabolic burden as reflected by maintenance coefficient of recombinant Escherichia coli overexpressing target gene. *Biotechnol. Lett.* **1995**, *17*, 1155–1160. [CrossRef]
49. Rozkov, A.; Avignone-Rossa, C.A.; Ertl, P.F.; Jones, P.; O'Kennedy, R.D.; Smith, J.J.; Dale, J.W.; Bushell, M.E. Characterization of the metabolic burden on Escherichia coli DH1 cells imposed by the presence of a plasmid containing a gene therapy sequence. *Biotechnol. Bioeng.* **2004**, *88*, 909–915. [CrossRef]
50. Yang, J.; Nie, Q.; Ren, M.; Feng, H.; Jiang, X.; Zheng, Y.; Liu, M.; Zhang, H.; Xian, M. Metabolic engineering of Escherichia coli for the biosynthesis of alpha-pinene. *Biotechnol. Biofuels* **2013**, *6*, 60. [CrossRef]
51. Fell, D.A. Increasing the flux in metabolic pathways: A metabolic control analysis perspective. *Biotechnol. Bioeng.* **1998**, *58*, 121–124. [CrossRef]
52. Kholodenko, B.N.; Lyubarev, A.E.; Kurganov, B.I. Control of the metabolic flux in a system with high enzyme concentrations and moiety-conserved cycles. The sum of the flux control coefficients can drop significantly below unity. *Eur. J. Biochem.* **2010**, *210*, 147–153. [CrossRef]
53. Gustafsson, C.; Govindarajan, S.; Minshull, J. Codon bias and heterologous protein expression. *Trends Biotechnol.* **2004**, *22*, 346–353. [CrossRef]
54. Burke, C.C.; Wildung, M.R.; Croteau, R. Geranyl diphosphate synthase: Cloning, expression, and characterization of this prenyltransferase as a heterodimer. *Proc. Natl. Acad. Sci. USA* **1999**, *96*, 13062–13067. [CrossRef]
55. Schmidt, A.; Gershenzon, J. Cloning and characterization of two different types of geranyl diphosphate synthases from Norway spruce (Picea abies). *Phytochemistry* **2008**, *69*, 49–57. [CrossRef] [PubMed]
56. Chang, T.H.; Hsieh, F.L.; Ko, T.P.; Teng, K.H.; Liang, P.H.; Wang, A.H. Structure of a heterotetrameric geranyl pyrophosphate synthase from mint (Mentha piperita) reveals intersubunit regulation. *Plant Cell* **2010**, *22*, 454–467. [CrossRef]
57. Aubourg, S.; Lecharny, A.; Bohlmann, J. Genomic analysis of the terpenoid synthase (AtTPS) gene family of Arabidopsis thaliana. *Mol. Genet. Genom.* **2002**, *267*, 730–745. [CrossRef]
58. Schmidt, A.; Wächtler, B.; Temp, U.; Krekling, T.; Séguin, A.; Gershenzon, J. A bifunctional geranyl and geranylgeranyl diphosphate synthase is involved in terpene oleoresin formation in Picea abies. *Plant Physiol.* **2010**, *152*, 639–655. [CrossRef] [PubMed]
59. Furubayashi, M.; Ikezumi, M.; Kajiwara, J.; Iwasaki, M.; Fujii, A.; Li, L.; Saito, K.; Umeno, D. A high-throughput colorimetric screening assay for terpene synthase activity based on substrate consumption. *PLoS ONE* **2014**, *9*, e93317. [CrossRef]
60. Jürgens, C.; Strom, A.; Wegener, D.; Hettwer, S.; Wilmanns, M.; Sterner, R. Directed evolution of a (βα) 8-barrel enzyme to catalyze related reactions in two different metabolic pathways. *Proc. Natl. Acad. Sci. USA* **2000**, *97*, 9925–9930. [CrossRef] [PubMed]
61. Keasling, J.D. Manufacturing molecules through metabolic engineering. *Science* **2010**, *330*, 1355–1358. [CrossRef]

62. Lopez-Gallego, F.; Wawrzyn, G.T.; Schmidt-Dannert, C. Selectivity of fungal sesquiterpene synthases: Role of the active site's H-1 alpha loop in catalysis. *Appl. Environ. Microbiol.* **2010**, *76*, 7723–7733. [CrossRef] [PubMed]
63. Vardakou, M.; Salmon, M.; Faraldos, J.A.; O'Maille, P.E. Comparative analysis and validation of the malachite green assay for the high throughput biochemical characterization of terpene synthases. *MethodsX* **2014**, *1*, 187–196. [CrossRef] [PubMed]
64. Dale, G.E.; Oefner, C.; D'Arcy, A. The protein as a variable in protein crystallization. *J. Struct. Biol.* **2003**, *142*, 88–97. [CrossRef]
65. Chen, X.; Zaro, J.L.; Shen, W.C. Fusion protein linkers: Property, design and functionality. *Adv. Drug Deliv. Rev.* **2013**, *65*, 1357–1369. [CrossRef]
66. Reddy Chichili, V.P.; Kumar, V.; Sivaraman, J. Linkers in the structural biology of protein-protein interactions. *Protein Sci.* **2013**, *22*, 153–167. [CrossRef]
67. Lee, H.; DeLoache, W.C.; Dueber, J.E. Spatial organization of enzymes for metabolic engineering. *Metab. Eng.* **2012**, *14*, 242–251. [CrossRef] [PubMed]
68. Miles, E.W.; Rhee, S.; Davies, D.R. The molecular basis of substrate channeling. *J. Biol. Chem.* **1999**, *274*, 12193–12196. [CrossRef]
69. Zhang, Y.H. Substrate channeling and enzyme complexes for biotechnological applications. *Biotechnol. Adv.* **2011**, *29*, 715–725. [CrossRef]
70. Siu, K.H.; Chen, R.P.; Sun, Q.; Chen, L.; Tsai, S.L.; Chen, W. Synthetic scaffolds for pathway enhancement. *Curr. Opin. Biotechnol.* **2015**, *36*, 98–106. [CrossRef]
71. Dueber, J.E.; Wu, G.C.; Malmirchegini, G.R.; Moon, T.S.; Petzold, C.J.; Ullal, A.V.; Prather, K.L.; Keasling, J.D. Synthetic protein scaffolds provide modular control over metabolic flux. *Nat. Biotechnol.* **2009**, *27*, 753–759. [CrossRef]
72. Conrado, R.J.; Wu, G.C.; Boock, J.T.; Xu, H.; Chen, S.Y.; Lebar, T.; Turnsek, J.; Tomsic, N.; Avbelj, M.; Gaber, R.; et al. DNA-guided assembly of biosynthetic pathways promotes improved catalytic efficiency. *Nucleic Acids Res.* **2012**, *40*, 1879–1889. [CrossRef]
73. Xie, S.S.; Qiu, X.Y.; Zhu, L.Y.; Zhu, C.S.; Liu, C.Y.; Wu, X.M.; Zhu, L.; Zhang, D.Y. Assembly of TALE-based DNA scaffold for the enhancement of exogenous multi-enzymatic pathway. *J. Biotechnol.* **2019**, *296*, 69–74. [CrossRef] [PubMed]
74. Delebecque, C.J.; Lindner, A.B.; Silver, P.A.; Aldaye, F.A. Organization of intracellular reactions with rationally designed RNA assemblies. *Science* **2011**, *333*, 470–474. [CrossRef] [PubMed]
75. Sachdeva, G.; Garg, A.; Godding, D.; Way, J.C.; Silver, P.A. In vivo co-localization of enzymes on RNA scaffolds increases metabolic production in a geometrically dependent manner. *Nucleic Acids Res.* **2014**, *42*, 9493–9503. [CrossRef] [PubMed]
76. Zhao, J.; Li, C.; Zhang, Y.; Shen, Y.; Hou, J.; Bao, X. Dynamic control of ERG20 expression combined with minimized endogenous downstream metabolism contributes to the improvement of geraniol production in Saccharomyces cerevisiae. *Microb. Cell Fact.* **2017**, *16*, 17. [CrossRef] [PubMed]
77. Cowan, M.M. Plant products as antimicrobial agents. *Clin. Microbiol. Rev.* **1999**, *12*, 564–582. [CrossRef] [PubMed]
78. Cox, S.D.; Mann, C.M.; Markham, J.L.; Bell, H.C.; Gustafson, J.E.; Warmington, J.R.; Wyllie, S.G. The mode of antimicrobial action of the essential oil of Melaleuca alternifolia (tea tree oil). *J. Appl. Microbiol.* **2010**, *88*, 170–175. [CrossRef]
79. Shah, A.A.; Wang, C.; Chung, Y.R.; Kim, J.Y.; Choi, E.S.; Kim, S.W. Enhancement of geraniol resistance of Escherichia coli by MarA overexpression. *J. Biosci. Bioeng.* **2013**, *115*, 253–258. [CrossRef] [PubMed]
80. Wang, J.-F.; Xiong, Z.-Q.; Li, S.-Y.; Wang, Y. Enhancing isoprenoid production through systematically assembling and modulating efflux pumps in Escherichia coli. *Appl. Microbiol. Biotechnol.* **2013**, *97*, 8057–8067. [CrossRef]
81. Jongedijk, E.; Cankar, K.; Buchhaupt, M.; Schrader, J.; Bouwmeester, H.; Beekwilder, J. Biotechnological production of limonene in microorganisms. *Appl. Microbiol. Biotechnol.* **2016**, *100*, 2927–2938. [CrossRef]
82. Putman, M.; van Veen, H.W.; Konings, W.N. Molecular properties of bacterial multidrug transporters. *Microbiol. Mol. Biol. Rev.* **2000**, *64*, 672–693. [CrossRef]

83. Ramos, J.L.; Duque, E.; Gallegos, M.T.; Godoy, P.; Ramos-Gonzalez, M.I.; Rojas, A.; Teran, W.; Segura, A. Mechanisms of solvent tolerance in gram-negative bacteria. *Annu. Rev. Microbiol.* **2002**, *56*, 743–768. [CrossRef]
84. Nikaido, H.; Takatsuka, Y. Mechanisms of RND multidrug efflux pumps. *Biochim. Biophys. Acta* **2009**, *1794*, 769–781. [CrossRef] [PubMed]
85. Turner, W.J.; Dunlop, M.J. Trade-Offs in Improving Biofuel Tolerance Using Combinations of Efflux Pumps. *ACS Synth. Biol.* **2015**, *4*, 1056–1063. [CrossRef]
86. Newman, J.D.; Marshall, J.; Chang, M.; Nowroozi, F.; Paradise, E.; Pitera, D.; Newman, K.L.; Keasling, J.D. High-level production of amorpha-4, 11-diene in a two-phase partitioning bioreactor of metabolically engineered Escherichia coli. *Biotechnol. Bioeng.* **2006**, *95*, 684–691. [CrossRef] [PubMed]
87. Dafoe, J.T.; Daugulis, A.J. In situ product removal in fermentation systems: Improved process performance and rational extractant selection. *Biotechnol. Lett.* **2014**, *36*, 443–460. [CrossRef] [PubMed]
88. Schewe, H.; Holtmann, D.; Schrader, J. P450(BM-3)-catalyzed whole-cell biotransformation of alpha-pinene with recombinant Escherichia coli in an aqueous-organic two-phase system. *Appl. Microbiol. Biotechnol.* **2009**, *83*, 849–857. [CrossRef] [PubMed]
89. Brennan, T.C.; Turner, C.D.; Krömer, J.O.; Nielsen, L.K. Alleviating monoterpene toxicity using a two-phase extractive fermentation for the bioproduction of jet fuel mixtures in Saccharomyces cerevisiae. *Biotechnol. Bioeng.* **2012**, *109*, 2513–2522. [CrossRef] [PubMed]
90. Alonso-Gutierrez, J.; Kim, E.M.; Batth, T.S.; Cho, N.; Hu, Q.; Chan, L.J.G.; Petzold, C.J.; Hillson, N.J.; Adams, P.D.; Keasling, J.D.; et al. Principal component analysis of proteomics (PCAP) as a tool to direct metabolic engineering. *Metab. Eng.* **2015**, *28*, 123–133. [CrossRef]

catalysts

Review

Advancement of Metabolic Engineering Assisted by Synthetic Biology

Hyang-Mi Lee, Phuong N. L. Vo and Dokyun Na *

School of Integrative Engineering, Chung-Ang University, Seoul 06974, Korea; myhys84@cau.ac.kr (H.-M.L.); lamphuong2895@gmail.com (P.N.L.V.)

* Correspondence: blisszen@cau.ac.kr; Tel.: +82-2-820-5690

Received: 14 November 2018; Accepted: 27 November 2018; Published: 4 December 2018

Abstract: Synthetic biology has undergone dramatic advancements for over a decade, during which it has expanded our understanding on the systems of life and opened new avenues for microbial engineering. Many biotechnological and computational methods have been developed for the construction of synthetic systems. Achievements in synthetic biology have been widely adopted in metabolic engineering, a field aimed at engineering micro-organisms to produce substances of interest. However, the engineering of metabolic systems requires dynamic redistribution of cellular resources, the creation of novel metabolic pathways, and optimal regulation of the pathways to achieve higher production titers. Thus, the design principles and tools developed in synthetic biology have been employed to create novel and flexible metabolic pathways and to optimize metabolic fluxes to increase the cells' capability to act as production factories. In this review, we introduce synthetic biology tools and their applications to microbial cell factory constructions.

Keywords: synthetic biology; metabolic engineering; microbial cell factory; synthetic metabolic pathways

1. Introduction

Synthetic biology adopts the principles of electrical engineering to rationally design and engineer biological systems. In the early era of synthetic biology, various genetic circuits were constructed, including a genetic toggle switch and a repressilator, which began the rational design of man-made functional biological networks [1,2]. Recently, synthetic biologists have developed various biotechnological and computational methods and tools to manipulate the host genome and create various synthetic systems using various genetic components, such as DNA [3], RNA [4,5], and proteins [6], as well as complex synthetic circuits, such as bio-oscillators [7], toggle switches [1,8], and logic gates [9].

Synthetic systems have been widely applied to metabolic engineering for the biosynthesis of biofuels, commodity chemicals, and pharmaceutical molecules [10–13]. The integration of synthetic biology and metabolic engineering has provided a solution to the current increasing demand for sustainable and renewable resources in response to global concerns, such as fossil fuel depletion and rising healthcare expenses [14,15]. In particular, synthetic biology has provided advanced toolsets and novel systems, which significantly facilitate the development of metabolic engineering [10,16].

For example, recently, the metabolic production of artemisinin (widely used as an antimalarial drug) and its precursor, artemisinic acid, was successfully achieved. Since the supply of artemisinic acid from the plant *Artemisia annua* or via chemical synthesis was insufficient to meet the needs of the market, it was recently produced from the metabolically engineered microbial hosts *Escherichia coli* and *Saccharomyces cerevisiae* with higher productivity and an economically feasible titer and yield [13,17,18]. Micro-organisms have also been engineered to produce fuels for use as alternative and renewable sources of energy [19,20]. *E. coli* was metabolically engineered to produce isopropanol by introducing a synthetic metabolic pathway that converts acetyl-CoA to acetone and finally, to isopropanol. The pathway comprises enzymes originating from other bacterial species, namely, acetyl-coenzyme A [CoA] acetytransferase (*C. acetobutylicum thl*), acetoacetate decarboxylase (*C. acetobutylicum adc*), and secondary alcohol dehydrogenase (*C. beijerinckii adh*) [21].

Advancements in metabolic engineering assisted by synthetic biology have yielded unprecedented outputs in industrial and pharmaceutical biotechnology [22,23]. Herein, we review the recent advances in various synthetic biology tools and methods for constructing, manipulating, and optimizing metabolic synthetic pathways to achieve microbial cell factory constructions (Figure 1).

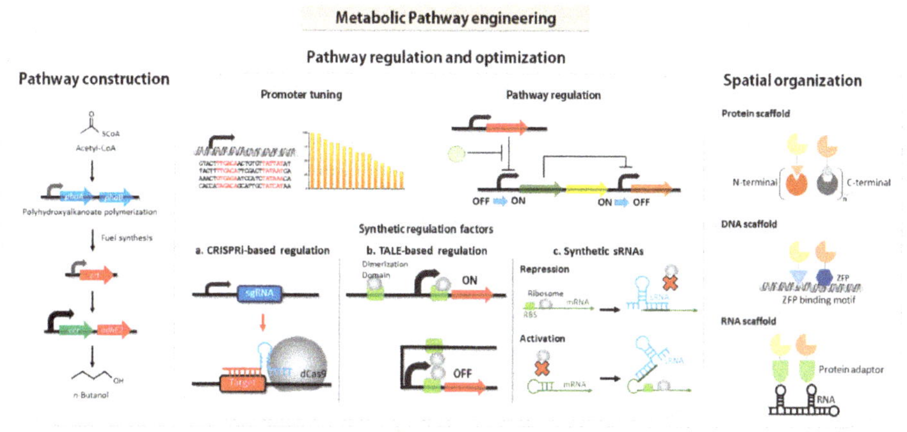

Figure 1. Synthetic biology tools adopted to metabolic engineering. *Pathway construction*: A synthetic metabolic pathway can be constructed by assembling heterologous genetic parts (e.g., promoters and coding sequences). An exemplary synthetic pathway is depicted, which consists of heterologous enzymes from three different organisms to convert a substrate (acetyl-CoA) into a desired product (*n*-butanol): blue, *R. eutrophus*; red, *C. acetobutylicum*; green, *S. collinus*. *Pathway regulation and optimization*: Synthetic metabolic pathways and their flux optimizations can be achieved by creating synthetic promoter libraries and by constructing a dynamic regulatory pathway. In addition, synthetic regulation factors, such as CRISPRi- and TALE-based transcription factors, and synthetic sRNAs for rational control over translation processes are also used. *Spatial organization*: Proteins, DNAs, and RNAs can be used as a spatial organization scaffold, where pathway enzymes are physically bound to the scaffold in a designable manner.

2. Synthetic Metabolic Pathway Construction and Flux Optimization

Desired chemical compounds produced from micro-organisms are often non-natural, and thus, novel metabolic pathways should be rationally constructed. Synthetic metabolic pathway construction is characterized by the assembly of multiple genes involved in a desired pathway from many different species [24–27]. Unlike natural metabolic pathways that have been optimized through evolution, synthetic pathways composed of enzymes from various species are not optimized in their fluxes. Therefore, toxic intermediates may form and accumulate, thus causing cell growth retardation and production titer reduction. Therefore, the designed synthetic pathways should be further optimized to maximize production of the target substance. Because of the complexity of gene expressions and enzyme kinetics, mathematical and computational methods have also been developed to predict the behavior of the designed pathways and circuits to ensure robustness [25,28–30]. The glossary explains the meaning of specialist terms, which are important for an understanding of the text (Box 1).

Box 1. Glossaries.

TAL effectors	Transcription activator-like effectors secreted by *Xanthomonas* bacteria, which recognize plant DNA sequences through a central repeat domain of ~34 amino acid repeats, and activate the expression of individual genes to aid bacterial infection
Bio-oscillators	Synthetic genetic circuits that mimic the natural genetic clocks of organisms, which induces the periodical behavior of a system
Toggle switch	Synthetic genetic circuits that display a stepwise function: ON or OFF
Logic gates	Synthetic genetic circuits consisting of multiple cellular sensors and actuators to perform precise logical operation (e.g., AND-, OR-, NOT-, and NAND-gates)
Phosphoketolase pathway	The pathway that contributes to carbon metabolism, in which a key enzyme, phosphoketolase, cleaves pentose phosphate into glyceraldehyde-3-phosphate and acetyl phosphate
Riboswitches	RNA sequences that resize in 5′ UTR and regulate transcription and/or translation by conformational changes upon ligand binding
Small regulatory RNAs (sRNAs)	Small non-coding RNAs that regulate translation by base-pairing with target mRNAs
Protospacer-adjacent motif (PAM)	5′-NGG-3′ sequence immediately following the target DNA sequence in the CRISPR-Cas system

2.1. Pathway Construction

There have been several reports on microbial cell factories that were metabolically engineered to produce bio-substances with enhanced metabolic pathway performance. For example, high-titer production of n-butanol from *E. coli* was achieved with a chimeric pathway assembled from three different organisms [31]. n-Butanol is obtained from acetyl-CoA by condensation of two acetyl-CoA units and subsequent conversions. A major challenge in metabolic pathways is reversibility. Bond-Watts et al. [31] constructed a synthetic pathway comprising enzymes from three different bacterial species: phaA (acetoacetyl-CoA thiolase/synthase) and phaB (3-hydroxybutyryl-CoA dehydrogenase) from *Ralstonia eutrophus*, crt (crotonase), adhE2 (bifunctional butyraldehyde and butanol dehydrogenase) from *C. acetobutylicum*, and ccr (crotonyl-CoA reductase) from *Streptomyces collinus*. The forward reactions of phaA and phaB are favored; thus, they are capable of providing sufficient metabolic flux toward n-butanol. To convert the product of phaB to n-butanol, *crt*, *ccr*, and *adhE2* genes were also introduced into the synthetic pathway. To shift the overall equilibrium further toward n-butanol, ccr and adhE2 were produced from a strong T7 promoter, while phaA, phaB, and crt were produced from a weak arabinose promoter.

There have been reports of the construction of CoA-dependent 1-butanol synthetic pathways in *E. coli* [32,33]. A novel 1-butanol synthetic pathway was designed with the modification of the (R)-1,3-butanediol pathway in order to supply more acetyl-CoA and NAD(P)H. The designed synthetic pathway consisted of heterologous enzymes from different species: butyraldehyde dehydrogenase (phaA, phaB, bld) from *Clostridium saccharoperbutylacetonicum*, inherent alcohol dehydrogenases (adhs) from the host *E. coli*, trans-enoyl-CoA reductase (ter) from *Treponema denticola*, and (R)-specific enoyl-CoA hydratase (phaJ) from *Aeromonas caviae*. Since inherent lactate dehydrogenase (ldhA) catalyzes the conversion of glucose to lactic acid under oxygen-depleted conditions, which disturbs 1-butanol production, the deletion of the *ldhA* gene eliminated the effect of the oxygen level on 1-butanol production and consequently, further increased the titer of 1-butanol.

In addition, biodiesel has gained great interest recently—specifically, fatty-acid-derived fuel molecules [34–36]—as they have lower toxicity to microbial hosts and higher energy density as compared with diesel fuel. Certain fatty acid-derived fuels, such as fatty acid methyl esters (FAMEs) and fatty acid ethyl esters (FAEEs), are usually formed via transesterification of fatty acyl-CoAs with alcohols. This reaction is primarily catalyzed by a wax ester synthase. Fatty acid derivatives have been well-demonstrated and produced in micro-organisms, such as *S. cerevisiae* [34] and *E. coli* [37].

De Jong et al. constructed synthetic pathways in *S. cerevisiae* for enhanced production of FAEEs (fatty acid ethyl esters) [34]. The production of acyl-CoA requires acetyl-CoA and NADPH. To increase the supply of acetyl-CoA and NADPH, the inherent ethanol degradation pathway was up-regulated, and a synthetic phosphoketolase pathway was additionally introduced into the host cell. The former pathway directed the carbon flux toward the biosynthesis of acetyl-CoA, resulting in 408 $\pm$ 270 μg of FAEE gCDW^{-1}. This was achieved by overexpressing ADH2 (alcohol dehydrogenase 2) and ALD6 (acetaldehyde dehydrogenase), which respectively catalyze the conversion of ethanol to acetaldehyde and acetaldehyde to acetate, along with a heterologous gene, $acs_{SE}{}^{L641P}$, from Salmonella enterica that encodes acetyl-CoA synthetase. The latter engineered pathway was the heterologous phosphoketolase pathway, which improved the production of both acetyl-CoA and NADPH. The engineered phosphoketolase pathway was constructed with two heterogeneous genes, *xpkA* (xylulose-5-phosphate phosphoketolase) and *ack* (acetate kinase), from *Aspergillus nidulans*. xpkA catalyzes the reaction where xylulose-5-phosphate, the precursor of the phosphoketolase pathway, is converted to acetyl-phosphate and glyceraldehyde-3-phosphate. Acetyl-phosphate is then converted into acetate as an intermediate by acetate kinase. Alternatively, acetyl-phosphate can be directly converted into acetyl-CoA by replacing ack with pta (phosphotransacetylase), which originates from Bacillus subtilis. The two phosphoketolase pathways were shown to result in increases in FAEE production to 5100 $\pm$ 509 and 4670 $\pm$ 379 μg FAEE gCDW^{-1}, respectively.

Recently, Sherkhanov et al. demonstrated a strategy to produce FAMEs in *E. coli* and reported a 35-fold improved FAME titer compared with that previously reported [37,38]. They constructed a synthetic metabolic pathway wherein FAMEs were produced through the direct methylation of fatty acids by a broader-range fatty acid methyl transferase [Drosophila melanogaster Juvenile Hormone Acid *O*-Methyltransferase]. This pathway was implemented in a β-oxidation-deficient and phospholipid synthesis-deficient *E. coli* strain that can tolerate high levels of medium-chain fatty acids expressed by acyl-ACP thioesterase (BTE) originating from *Umbellularia californica*. The resulting engineered strain produced the highest FAME titer ever observed in *E. coli*.

In addition to fueling biosynthesis, synthetic pathways have been built for the biosynthesis of various chemical compounds, such as succinic acid [39], 2,4-dihydroxybutyric acid [40], and maleate [41], and the engineered bio-factories have already been commercialized for the food industry.

However, in most cases, synthetic pathway construction requires laborious and tedious cloning work. Synthetic biology has developed several efficient multi-gene assembly strategies, such as Randomized BioBrick [30], which has been applied to construct an efficient lycopene biosynthesis pathway by random combinations of genetic components and has been shown to achieve a 30% increased titer in *E. coli*. In this method, various genetic parts, including promoters, RBSs, and terminators, are randomly assembled from PCR-amplified BioBricks simultaneously to identify the most efficient combination. The method successfully optimized a metabolic pathway that converts farnesyl diphosphate (FPP) to lycopene, which consists of three enzymes: geranylgeranyl diphosphate synthase (crtE), phytoene synthase (crtB), and phytoene desaturase (crtI). Other common methods are Golden Gate cloning [42], Golden Braid [43], MoClo [44], and MODAL [45]. However, such strategies have some disadvantages, such as the generation of scar sequences and the risk of PCR errors. Recently, Hochrein et al. introduced a user-friendly toolkit, AssemblX, to quickly create DNA structures with up to 25 functional units based on overlap-based cloning [46].

Furthermore, plasmid-based systems have drawbacks, such as the waste of cellular energy resources due to excessive metabolism [47,48] and plasmid maintenance. Recently, instead of using plasmids, chromosomal integration of genes has been favored for stable enzyme expression. Based on the clustered, regularly interspaced, short palindromic repeat (CRISPR) and CRISPR-associated systems (Cas), recently developed genomic editing methods have promoted the integration of metabolic pathways into the host genome. For example, a modified method of CRISPR-Cas, the delta integration CRISPR-Cas platform, can be used to increase the homologous recombination efficiency by designing guided RNA sequences that specifically cleaves the target sequence at multiple delta sites in the genome of *S. cerevisiae* [49]. In this delta integration CRISPR-Cas platform, the plasmid expressing a Cas9 protein and a delta-targeting guide RNA is co-transformed with the 8-to- 24-kb linear donor DNA fragments flanked between two homology arms of the delta sequences. The method has been used successfully to integrate a synthetic pathway composed of three enzymes in the (R,R)-2,3-butanediol production pathway: acetolactate synthase (alsS) and (R,R)-butanediol dehydrogenase (budA) from B. subtilis, and acetolactate decarboxylase (bdhA) from *Enterobacter aerogenes*. There have also been many other reports based on the use of CRISPR-Cas for chromosomal integration of heterologous pathways [50,51].

Novel synthetic pathways are widely used for metabolic engineering; thus, there are computational methods to design de novo metabolic pathways [52–54]. The in silico pathway design methods firstly enumerate possible metabolic pathways by connecting reactions obtained from various databases. As enzymes can often recognize molecules that are similar to their natural counterparts, potential and hypothetical reactions are also used during the pathway enumeration to extend the search space of metabolic pathways. Second, the enumerated pathways are pruned because the first step generates all combinations of pathways, which may include unnatural and unrealistic reactions and network structures. For example, generated pathways may include reactions that may not actually occur in nature, the lengths of generated pathways may be too long, or the pathways may form a cyclic structure wherein the target metabolite is used as an intermediate. Therefore, the enumerated pathways are evaluated to determine whether they comprise feasible enzymatic reactions and whether they can be tested and constructed in the laboratory. Third, the performance of the designed pathways is qualitatively evaluated in a given organism. In this step, the thermodynamics of enzyme reactions is considered, and several approaches have been developed, including the group contribution method and thermodynamics-based flux balance analysis. These methods finally suggest several feasible synthetic pathways that are predicted to perform the best in a given organism. The advances in synthetic metabolic pathway design methods facilitate the construction of metabolically engineered bio-factories that are capable of producing desired substances at high concentrations and can even produce non-natural chemical substances.

2.2. Enzyme Gene Expression Optimization

One of the challenges in synthetic metabolic pathway construction is to balance the kinetics of enzymes in the pathway. If a certain enzyme has lower expression or poor activity, it becomes a rate-limiting factor and induces abnormal accumulation of its substrates. Such intermediate metabolites are often toxic when accumulated, and thus, this accumulation may lead to growth retardation and eventually, to reduced production titer of the desired metabolite [48]. Furthermore, when a synthetic metabolic pathway is constructed, the metabolic flux within the pathway should be optimized by tuning the levels of enzyme gene expressions. There are an increasing number of studies working on understanding metabolic fluxes to address this obstacle.

Metabolic optimization through rational control over translation has been gathering great attention [55,56]. Enzyme genes should be expressed at a desired level to balance metabolic flux within a synthetic metabolic pathway to avoid the accumulation of intermediates and to maximize their overall performance. The gene expression process comprises two steps: transcription and translation. Since promoters are independent from downstream sequences, it is relatively easy to optimize the level of transcription. Conversely, translation is mediated by ribosome-binding sites (RBSs), and RBS sequences are structurally affected by downstream sequences, such as coding sequences. Since the structural effects on RBS cannot be predicted, so far, studies have focused on modeling translation processes and thereby, predicting accurate protein production levels.

Synthetic biologists have attempted to redesign RBS sequences for controlled enzyme expression [10,57]. Several synthetic tools have been developed that can predict the translation initiation rate and design RBS sequences to achieve balanced and robust gene expression. Existing tools include the RBS Calculator [58], RBSDesigner [59], and other innovative models for the automated design of synthetic RBSs [60–62]. RedLibs [63] was recently developed; it is a tool that generates a library of RBS sequences for pathway flux optimization to reduce experimental trials and errors. Such models have been used in many studies to facilitate the construction and optimization of protein expression levels in synthetic pathways [64–66].

Similarly, randomized RBS sequences have been used to generate various expression levels of enzymes (idi, crtE, crtB, and crtl) involved in lycopene biosynthetic pathway [67]. When 1080 colonies were examined out of a possible 3.8×10^6 combinations, the highest lycopene production found was an increase of up to 15.17 mg/g DCW.

2.3. Dynamic Metabolic Pathway Regulation

Another challenge in bio-factory construction is the optimization of the overall cellular metabolic fluxes toward the substance of interest. Certain metabolites often display toxicity when accumulated, and thereby, reduce cell growth or interfere with cellular functions. Introduced pathways may compete with cellular processes for cellular resources. These phenomena increase the burden on the host cell and eventually decrease the production titer [68]. Thus, cells use their own regulation and control of metabolic fluxes to balance the fitness of cellular machinery [69]. This suggests that the dynamic regulation of metabolic pathways becomes important for the efficient production of a desired substance by balancing metabolic fluxes without decreasing the cell growth rate.

Recently, dynamic regulatory systems have been developed for bio-factory construction [70,71]. The systems assist the engineered cells to adapt metabolic flux to fluctuations and changes within the host in real-time. The regulatory systems contain a sensor that detects the level of metabolite and an actuator that controls the metabolic flux (e.g., enzyme expressions or activities) [72]. For example, Zhang et al. developed a regulatory system that dynamically controls the expression of enzyme genes involved in bio-diesel production in response to acyl-CoA [73]. Overproduction of fatty acids would increase the production of bio-diesel but slow down the host cell growth. Bio-diesel is produced from fatty acyl-CoA that is produced from the natural fatty acid biosynthesis pathway. Thus, acyl-CoA is formed as an intermediate in the bio-diesel biosynthetic pathway. They used the FadR transcriptional regulator, which turns on a promoter in the absence of fatty acids/acyl-CoA and turns off in the presence of acyl-CoA. Thus, FadR was used to detect the intermediate concentration (fatty acids/acyl-CoA) and to dynamically control the expression of the enzyme genes involved in the diesel biosynthetic pathway. This regulatory system improved the stability of bio-diesel-producing strains and consequently, increased the bio-diesel titer. Liu et al. developed a similar strategy to over-produce fatty acids by regulating the expression of acetyl-CoA carboxylase in response to malonyl-CoA, since the overexpression of acetyl-CoA carboxylase induces cellular toxicity and slows down cell growth [74]. Several other sensors have also been used for bio-factory construction, such as aldehydes [75], butanol [76], and alkanes [77].

Although these models rely on a sensor for gene expression regulation, there are metabolites that do not have any appropriate sensor regulators. For instance, the heterologous mevalonate-based isoprenoid pathway produced two intermediates: 3-hydroxy-3-methylglutaryl-coenzyme A (HMG-CoA) and farnesyl pyrophosphate (FPP). These intermediates inhibit cell growth and affect fatty acid biosynthesis in *S. cerevisiae* and *E. coli* [78,79]. Thus, there have been reports recommending the use of synthetic promoters that are responsive to the two intermediates for the optimal regulation of the pathway. Dahl et al. developed stress response promoters to regulate the concentration of cellular HMG-CoA and FPP intermediates in the isoprenoid biosynthetic pathway [80].

Xu et al. constructed two synthetic promoters that are responsive to malonyl-CoA in opposite ways [81]. They introduced the FadR transcriptional regulator into *E. coli*, which binds to the fapO operator site when bound with malonyl-CoA. Interestingly, T7-fapO promoter was repressed by FapR, and malonyl-CoA relieved the repression. Conversely, pGAP-fapO promoter was activated by FapR, and this activation was removed by malonyl-CoA. Using these two promoters, they constructed a pathway switch that is toggled by malonyl-CoA. They constructed a pathway (*accADBC* genes under the control of the pGAP-fapO promoter) to supply malonyl-CoA and another pathway (*fabADGI* and *tesA′* genes under the control of the T7-fapO promoter) to convert the malonyl-CoA into fatty acids, and the expressions of the pathways were toggled by the malonyl-CoA-responsive synthetic promoters. This strategy to balance cell growth and product formation increased fatty acid synthesis in *E. coli*.

2.4. Spatial Organization

A recent strategy for maximizing the performance of synthetic pathways is the spatial organization of pathway enzymes. The spatial organization of enzymes in close proximity creates channels where molecules can move quickly through enzymes without diffusion and accumulation. This channeling effect greatly increases the overall pathway efficiency. Protein, DNA, and RNA molecules are used as a spatial organization scaffold, where pathway enzymes are physically bound to the scaffold in a designable manner [82].

Several protein scaffolds have been developed thus far and used to co-localize enzymes. For instance, a protein scaffold containing GBD, SH3, and PDZ domains was constructed, and three mevalonate biosynthesis enzymes (AtoB, HMGS, and HMGR) were engineered to contain ligands to bind to the domains. When the enzymes and the scaffold were co-expressed, the enzymes were co-localized to the scaffold protein by ligand–domain interactions. Owing to the artificial channeling, the scaffold system achieved a 77-fold increase in mevalonate production yield as compared with the non-scaffold pathway [6]. In addition, other attempts to efficiently improve the production titer using scaffold proteins have also been demonstrated to produce glucaric acid (5.0-fold increase) [83], butyrate (3.0-fold increase) [84], and resveratrol (5.0-fold increase) [85].

In addition to protein scaffolds, DNA- and RNA-based scaffolds have also been developed to enhance pathway performance. For example, the three enzymes (homoserine dehydrogenase, homoserine kinase, and threonine synthase) involved in the threonine biosynthetic pathway were engineered to fuse with DNA-binding domains (ZFPs); thus, the enzymes were able to bind to particular DNA sequences. Corresponding DNA sequences (ZFP-binding sites) were introduced into a plasmid, and these ZFP-binding sites were used to provide a DNA scaffold to arrange the three enzymes in a particular order. The improved pathway performance resulted in a 50% reduction in the threonine production time [84]. Similarly, transcription activator-like effectors (TALEs) have been used to bind DNA scaffolds. TALE domains and their binding DNA scaffold were used to colocalize tryptophan-2-mono-oxygenase (IAAM) and indole-3-acetimide hydrolase (IAAH), and the scaffold increased the production of indole-3-acetic acid (IAA) in *E. coli* [86]. In addition, Sachdeva et al. identified eight RNA binding domains and their corresponding RNA motifs [87]. RNA sequences can be designed using the RNA motifs to provide a geometric scaffold for enzyme arrangement [88,89]. When RNA scaffolds were used for synthetic pathways in *E. coli*, the RNA scaffolds improved the production titers of succinate and pentadecane [87,88].

Furthermore, in 2013, José et al. designed an enzyme compartmentalization technique for high production of three advanced biofuels, i.e., isobutanol, isopentanol, and 2-methyl-1-butanol, in engineered *S. cerevisiae* [90]. The natural metabolic pathway for isobutanol production consists of an upstream mitochondrial pathway and a downstream cytoplasmic pathway. The upstream pathway contains enzymes associated with the valine biosynthetic pathway, such as acetolactate synthase, ketolacid reductoisomerase, and dehydroxyacid dehydratase. The downstream pathway is the valine Ehrilich degradation pathway that contains α-ketoacid decarboxylase and alcohol dehydrogenase. The compartmentalization of the designed plasmid with both upstream and downstream pathways into mitochondria was achieved by tagging the N-terminal mitochondrial localization signal from subunit IV of the yeast cytochrome c oxidase to the enzymes in the pathways, which resulted in a substantial surge of isobutanol by 260% as compared with only a 10% yield increase when overexpressing the same pathway in the cytoplasm.

Recent successful applications of synthetic biology approaches to metabolic engineering are listed and summarized in Table 1.

Table 1. Successes of synthetic biology methods for metabolic engineering.

Methods		Description	Applications	References
Pathway construction	Synthetic engineered pathway *In silico* pathway design	Rapid construction and optimization of multi-gene pathways for higher-titer production	- n-Butanol - FAEEs, FAMEs - Succinic acid, 2,4-dihydroxybutyric acid, and maleate	[31–34,37,39–41,52–54]
	Randomized BioBrick AssemblX		- Lycopene production increased by 30%	[30]
	Chromosomal integration CRISPR-Cas-based method	Efficient and stable enzyme expression and higher production yield	- (*R*,*R*)-2,3-butanediol - Mevalonate and bisabolene titers increased by 41-fold and 5-fold, respectively	[49–51]
Enzyme expression optimization	RBS Calculator RBSDesigner RedLibs	Control over protein expression levels through ribosome-binding sites	- Optimization of isoprenoid production - L-tyrosine production - N-acetylneuraminate biosynthesis - Lycopene synthesis	[64–67]
Pathway regulation	Dynamic regulation system	Dynamic regulation and balancing of metabolic systems as well as redirection of fluxes to achieve high production of desired protein	- A fatty acyl-CoA biosensor FadR - Fatty acids, aldehydes, butanol, and alkanes - Isoprenoid biosynthetic	[73–77,80,81]
Spatial organization	Protein scaffolding	Concentrating intermediates and rapidly directing them through the metabolic pathway	- 77-fold increase in mevalonate production - 5-fold increase in glucaric acid synthesis - 3-fold increase in butyrate production - 5-fold increase in resveratrol synthesis	[6,83–85]
	DNA and RNA scaffolds		- 50% reduction in production time of I-threonine - Production of IAA - Succinate - Pentadecane	[84,86–88]

Table 1. *Cont.*

Methods			Description	Applications	References
Synthetic components for expression regulation	Promoter tuning		Optimal regulation of the gene expression strength in a dynamic range up to hundreds-of-fold activation	- Production of PepN protein at about 10–15% of the total cellular protein	[91]
	Promoter regulation				
	Synthetic transcription factors	CRISPR-based regulation	Regulation of gene expression using a nuclease-deficient Cas9 (dCas9)	- Production of 2.0 g/L β-carotene (fed-batch fermentation) - Production control of P (3HB-co-4HB) in *E. coli*	[92,93]
		Engineering transcription factors	Engineering AraC protein through altering effector specificity using a method of saturation mutagenesis	- 20-fold increased production of triacetic acid lactone in *E. coli*	[94]
		Chimeric transcription factor	Fusion of the target metabolite recognition domain and the promoter regulatory domain	- Tyrosine - Isoprenoid	[95]
	Synthetic RNAs		Regulation of gene expression based on physicochemical models of RNA	- Development of high-yield VB12 production strains - Tight transcriptional regulation of the toxic protein SacB	[96,97]

3. Synthetic Components for Gene Expression

From DNA to protein: The regulation of the transcription and translation processes is a fundamental theme for the development of practical tools for metabolic engineering. The engineering of microbial cell factories requires the introduction of heterologous enzyme genes and/or the modification of endogenous enzyme genes. For optimal performance, the expression of enzyme genes should be finely regulated [98]. Gene expression can be regulated at transcription and/or translation. Thus, the various techniques introduced in this section represent the most recent and influential synthetic tools for controlling the transcription and translation processes that can be applied to the metabolic engineering of various micro-organisms. There have been attempts to design synthetic promoters and their regulators using the TAL effector and CRISPR for transcriptional regulation. For translational regulation, riboswitches and small regulatory RNAs (sRNAs) have been developed and used for metabolic engineering.

3.1. Transcription Regulation

3.1.1. Promoters for Tuning

Various synthetic promoters have been developed to optimally tune the enzyme expression levels for the efficient production of a desired substance. Conventional inducible promoters have been frequently used to acquire different levels of gene expression by controlling the concentrations of inducers; however, there are only few promoters available, and their strengths are not diverse. Thus, to fine-tune enzyme expression for metabolic flux optimization, synthetic promoter libraries have been developed to identify the best combination of promoters for enzyme genes in a synthetic metabolic pathway. As mentioned in Section 2.2 of this paper, randomly generated libraries of synthetic promoters have also been used to fine-tune the metabolic flux within a synthetic pathway.

A wide range of synthetic promoter activities was created by the randomization of spacer sequences between −35 to −10 consensus sequences [99,100]. Such randomized libraries have been used for the fine-tuning of enzyme expression and accordingly, to achieve high-level production of compounds. For example, Rud et al. developed a synthetic promoter for the gram-positive bacterium *Lactobacillus plantarum* by randomizing the non-consensus sequence of the rRNA promoter. This randomization generated 33 different constitutive promoter sequences, and this promoter library showed various relative expression levels ranging from 1 to 160. Using the library, they identified an optimal promoter that showed stable expression of PepN and GusA [91]. In another study, a library of synthetic promoters was randomly generated in actinomycetes. The non-consensus nucleotides of the ermEp1 promoter were randomized, and accordingly, 56 synthetic promoters were generated. The library of synthetic promoters showed 2% to 319% activity compared with the native ermEp1 promoter, and their strengths were confirmed by RNA-seq analysis. As a proof of concept, the strongest promoter was used to express the *rppA* gene encoding for a type III polyketide synthase, which converts five malonyl-CoA compounds to 1,3,6,8-tetrahydroxynaphthalene, which is spontaneously oxidized to flaviolin. Consequently, a 3.3-fold increase in the production of flaviolin was achieved with the synthetic promoter [101]. Synthetic promoter libraries have also been commonly used to fine-tune enzyme expressions and thus obtain higher titers of substances, including 2,3-butanediol in *E. coli*, actinorhodin in *Streptomyces coelicolor* A3(2) [102], and endoxylanase in *Corynebacterium glutamicum* [103].

Furthermore, there have also been attempts to introduce synthetic promoters into the genome. Braatsch et al. used Red/ET recombination to replace a genomic native promoter with synthetic promoters. Synthetic promoter sequences were PCR-amplified with a kanamycin resistance gene flanked with Flp recognition targets. These amplified sequences were then replaced with the genomic sequence by homologous recombination. After appropriate selection, the kanamycin resistance gene was excised by Flp recombinase [104]. They replaced the native pgi promoter with synthetic promoters exhibiting a wide spectrum of strengths ranging from 25% to 570% of the native pgi promoter.

Thus far, various computational methods have been developed to decipher the effects of DNA sequences on promoter strength and to design de novo design synthetic promoter sequences. Position weight matrix models were used to predict the strength of *E. coli* core promoter sequences recognized by the sigma factor σ^E [105]. The models were then improved by incorporating the effect of upstream elements into promoter strength [106]. However, since position weight matrix models are highly reliant on experimental data obtained from promoter-strength studies, they are only applicable to well-studied micro-organisms, such as *E. coli*. Other models have also been developed to predict promoter strengths based on the partial least squares [107] and artificial neural network [108] methods among other methods. Computation models have been used for the de novo design of synthetic promoters to fine-tune the enzyme genes involved in the deoxyxylulose phosphate pathway in *E. coli* [108].

3.1.2. Insulated Promoters

The expression of enzymes downstream of regulatory promoter could be interfered with by cellular factors; thus, it is often required for synthetic systems to be insulated from cellular systems for reliable operation. The recently developed SELEX-based screening of bacterial genomes identified the complex regulatory targets of the characterized 116 transcription factors via 156 transcription profiles [109]. Because the interactions have not yet been completely elucidated, when constructing a synthetic system, DNA sequences that may interact with cellular factors could be included. Because of this unexpected inclusion of interacting DNA sequences, synthetic systems may behave unexpectedly and lose their robustness. To resolve this concern, Davis et al. devised a simple strategy to insulate a promoter by flanking the promoter sequence with insulating sequences [110] because upstream or downstream sequences of a promoter may contain regulatory sequences that can affect the promoter. Briefly, they used the *E. coli* rrnB P1 promoter as the core and added insulating sequences covering −105 to +55. Randomization of the space nucleotides of the rrnB P1 promoter generated a library of promoters showing different levels of strength. When evaluated with the *GFP* gene under the control of the insulated promoter, the promoter's strength was not affected by upstream or downstream sequences due to the long flanking sequences. Zong et al. also improved the insulation-based engineering strategy by identifying insulated promoter cores and operators and predicting the consequence of their combinations via a biophysical model of synthetic transcription [111]. They demonstrated a modular system that correctly programs synthesis circuits by designing 83 combined promoters that randomly combine 53 different promoter cores and 36 synthesis operators to encode the NOT gate function in *E. coli* DH10B.

3.1.3. Synthetic Transcription Factors

Endogenous enzyme genes often need to be regulated for optimized metabolic flux regulation. However, there are only a limited number of regulator proteins available and they require specific recognition sequences. Thus, conventional regulators cannot be used for the regulation of endogenous genes that have very diverse promoter sequences. There have been many attempts to develop customizable synthetic transcription factors [92,93,112–115]. For example, TALE proteins can be reprogrammed to activate or inhibit transcription by binding at a particular sequence. More recently, clustered, regularly interspaced, short palindromic repeat (CRISPR)-Cas9 systems have been used to regulate gene expression more efficiently than conventional methods. Therefore, these synthetic transcription factors could be the key to synthetic biology and metabolic engineering.

CRISPR-Based Regulation

Recently, there has been growing evidence that the CRISPR/Cas9-based genome editing approach enables efficient, intricately controlled gene expression and versatile cross-species genome editing in various micro-organisms [5,113,114]. Briefly, the action mechanism of the CRISPR/Cas9 system is that the complex of small crRNAs produced by the coordinated action of a small, trans-activating crRNA (tracrRNA), the Cas9 nuclease, and the host RNaseIII directs Cas9 to cleave the target DNA, which is followed by an adjacent protospacer-adjacent motif (PAM). The target DNA is recognized by its alignment with the 12–15 nucleotides at the 3′ end of the crRNA guide sequence as well as the NGG sequence of PAM. Unlike the effort to manipulate other transcription factors, the CRISPR/Cas9 system can be simply applied to genome editing and metabolic pathway engineering by only altering the target sequence without any protein engineering. For example, Jiang et al. reported the introduction of point mutations and codon changes in the *E. coli* genome using the CRISPR/Cas9 method [5]. For instance, the β-carotene biosynthetic pathway, the methylerythritol-phosphate pathway, and the central metabolic pathways in *E. coli* were systematically optimized for β-carotene production by iterative editing of the 2-day editing cycle, and the best production was 2.0 g/L β-carotene (fed-batch fermentation) [92].

With respect to the regulation of gene expression, CRISPRi-based interference (CRISPRi) has also been applied to metabolic engineering using a nuclease-deficient Cas9 (dCas9) that contains two point mutations, one each in the RuvC (D10A) and HNH (H840A) domains [4,113]. Due to the gRNA binding ability of dCas9, it has been used as a programmable transcription repressor by circumventing the binding of RNA polymerase to the promoter sequence or as a transcription terminator by binding to the target gene. Recently, CRISPRi was applied to target the $P_{T7/LacO1}$ promoter for preventing the leaky expression of toxic proteins [115]. Conversely, the fusion of RNA polymerase omega subunit (ω) and dCas9 achieved programmable transcriptional activation by enhancing the binding stability of RNA polymerase to the upstream promoter region. Therefore, these RNA-guided transcriptional reprogramming approaches have a high potential for fine-tuning the production of target genes in metabolic engineering. According to Lv et al., the repression of multiple genes in polyhydroxyalkanoate (PHA) biosynthesis using CRISPRi enables the production control of P (3HB-co-4HB) in *E. coli* [93].

TALE-Based Regulation

As a new genetic switch tool, TALE proteins, which can bind to specific DNA target sequences, have been engineered to produce TALE Dimers (TALEDs) by replacing the nuclease domain of a TALE with a well-studied FKBP dimerization domain [112]. The engineered TALEDs contain a DNA-binding domain and dimerization domain; therefore, they can bend the DNA structure by binding to two TALED-binding sites and by subsequent dimerization. Based on the DNA looping paradigm, engineered TALEDs have been implemented to switch the transcription initiation of the lac operon system of *E. coli*.

Engineering/Chimeric Transcription Factors

In addition to CRISPR- and TALE-based regulation, there have been other attempts to develop customizable synthetic transcription factors by engineering native transcription factors. For example, by altering effector specificity using a method of inducing saturation mutagenesis, Tang et al. engineered AraC to sense triacetic acid lactone (TAL) and used the engineered AraC in the directed evolution of *Gerbera hybrida* 2-pyrone synthase to catalyze the synthesis of TAL [94]. As a result, they were able to identify an improved *G. hybrida* 2-pyrone synthase variant and by using the synthase, they achieved 20-fold increased production of TAL. Frei et al. also developed engineered AraC variants that respond to new inducer compounds, vanillin and salicylic acid, for molecular sensing and reporting [116]. These designed transcription factors may expand the sensing spectrum of native transcription factors to regulate and optimize synthetic pathways. In addition, Chou et al. introduced a new adaptive control system called feedback-regulated evolution of phenotype, in which two modules dynamically control the mutation rate (an actuator module) based on the concentration of a target molecule (a sensor module) [95]. For this system, a chimeric transcription factor has been developed that not only can bind to the target metabolite but also regulate the promoter for tyrosine and isoprenoid production in *E. coli*.

3.2. Expression Regulation

The diversity, specificity, and kinetics of regulatory RNAs contribute to the regulation of diverse physiological responses, including transcription, translation, and mRNA stability in living cells [96,97,117–122]. The structures, functions, and mechanisms of regulatory RNAs have been elucidated in many bacterial species for decades. The well-known bacterial regulatory RNAs called riboswitches, which are located at the 5′ untranslated regions of mRNAs, comprise a ligand-binding aptamer region and the expression platform. Upon binding of a metabolite ligand, the riboswitches regulate the expression of a downstream coding sequence through the formation of terminator/anti-terminator structures or masking/unmasking Shine–Dalgarno (SD) sequences in the mRNA [120,121]. Riboswitches repress gene expression as cis-acting RNA regulators. Engineered riboswitches have been used as a genetic toolkit to obtain desired gene production or toxic product regulation in bacterial species. For example, the application of engineered VB12 riboswitches to develop high-yield VB12 production strains using a flow cytometry high-throughput screening system in Salmonella typhimurium has been reported [96]. Moreover, six different theophylline-responsive riboswitches have been used for intricate transcriptional regulation of the toxic protein SacB in several cyanobacterial species; these riboswitches have shown more efficient performances than isopropyl-D-thiogalactopyranoside (IPTG) induction of the lacI^{q}-Ptrc promoter system [97].

Synthetic regulatory RNAs have also been specifically developed to regulate gene expression during the post-transcription process [117–119]. Isaacs et al. developed two types of artificial riboregulator: cis-repressed mRNA (crRNA), which interferes with ribosome binding through intramolecular interaction with SD, and trans-activating RNA (taRNA), which activates gene expression via an intermolecular interaction that releases the SD sequence [118]. In addition, there are trans-acting sRNAs that regulate the translation and stability of target mRNA through base pair complementation. Like natural sRNAs, synthetic sRNAs mainly comprise two parts: a scaffold sequence and a target-binding sequence [119]. The scaffold sequence is able to recruit the Hfq protein, which not only promotes effective hybridization of the synthetic sRNA and its target mRNA but also facilitates the degradation of the target mRNA. The target binding sequence that is complementary to the translation initiation region of the target mRNA competes with ribosomes to achieve efficient translation inhibition. Using this mechanism, synthetic sRNAs have been applied to metabolic engineering for efficient gene knockdown for tyrosine and phenol production in *E. coli* [119,123] and butanol production in *C. acetobutylicum* [122]. Recently, orchestrated regulation during both transcription and translation has been reported to achieve a high muconic acid titer of up to 1.8 g/L [124].

4. Summary

Metabolic engineering for the microbial production of chemicals and pharmaceuticals has been accelerated with the help of various genetic tools in synthetic biology. In this review, we introduced key technologies that are used in synthetic biology for metabolic pathway engineering, such as the design, construction, and optimization of metabolic processes as well as the engineering of synthetic components, which could make cellular processes predictable and robust to achieve desirable outcomes. Various strategies and tools that are used in synthetic biology to develop promising pathways and modules enable the establishment of intellectual foundations for cell factory construction in metabolic engineering.

One of the future goals of synthetic biology that is applicable to metabolic engineering is the artificial design of a novel micro-organism that possesses the capability to produce desired metabolites most efficiently and in large quantities like a factory. However, the prediction of whole cell physiologies from DNA sequences and thereby, the design of novel organisms from scratch is a great challenge. Even simple protein expression processes are very complex because of interruptions by other molecules. In addition, to our knowledge there are still no methods to accurately predict the kinetics of proteins (e.g., enzymes, interactions among proteins, DNAs, metabolites, etc.) solely from DNA sequences. Thus, synthetic biology is still at the early stages of development. To fulfil the goal, more knowledge on cellular machineries should be accumulated to allow the prediction of cell behaviors from DNA nucleotides, and new technologies should also be developed to produce designed micro-organisms.

Author Contributions: Conceptualization, H.-M.L. and P.N.L.V.; Writing-Original Draft Preparation, H.-M.L. and P.N.L.V.; Writing-Review & Editing, D.N.; Supervision, D.N.

Funding: This work was supported by C1 Gas Refinery Program through the National Research Foundation of Korea (NRF) funded by the Ministry of Science, ICT and Future Planning (NRF-2016M3D3A1A01913244) and was also supported by the Chung-Ang University Research Grants in 2017.

Conflicts of Interest: The authors declare no conflict of interest.

References

1. Gardner, T.S.; Cantor, C.R.; Collins, J.J. Construction of a genetic toggle switch in *Escherichia coli*. *Nature* **2000**, *403*, 339–342. [CrossRef] [PubMed]
2. Elowitz, M.B.; Leibler, S. A synthetic oscillatory network of transcriptional regulators. *Nature* **2000**, *403*, 335–338. [CrossRef] [PubMed]
3. Lee, J.H.; Jung, S.C.; Bui le, M.; Kang, K.H.; Song, J.J.; Kim, S.C. Improved production of L-threonine in *Escherichia coli* by use of a DNA scaffold system. *Appl. Environ. Microbiol.* **2013**, *79*, 774–782. [CrossRef] [PubMed]
4. Bikard, D.; Jiang, W.; Samai, P.; Hochschild, A.; Zhang, F.; Marraffini, L.A. Programmable repression and activation of bacterial gene expression using an engineered CRISPR-Cas system. *Nucleic Acids Res.* **2013**, *41*, 7429–7437. [CrossRef] [PubMed]
5. Jiang, W.; Bikard, D.; Cox, D.; Zhang, F.; Marraffini, L.A. RNA-guided editing of bacterial genomes using CRISPR-Cas systems. *Nat. Biotechnol.* **2013**, *31*, 233–239. [CrossRef] [PubMed]
6. Dueber, J.E.; Wu, G.C.; Malmirchegini, G.R.; Moon, T.S.; Petzold, C.J.; Ullal, A.V.; Prather, K.L.; Keasling, J.D. Synthetic protein scaffolds provide modular control over metabolic flux. *Nat. Biotechnol.* **2009**, *27*, 753–759. [CrossRef] [PubMed]
7. Tomazou, M.; Barahona, M.; Polizzi, K.M.; Stan, G.B. Computational re-design of synthetic genetic oscillators for independent amplitude and frequency modulation. *Cell Syst.* **2018**, *6*, 508–520. [CrossRef]
8. Lugagne, J.B.; Sosa Carrillo, S.; Kirch, M.; Kohler, A.; Batt, G.; Hersen, P. Balancing a genetic toggle switch by real-time feedback control and periodic forcing. *Nat. Commun.* **2017**, *8*, 1671. [CrossRef]
9. Cameron, D.E.; Bashor, C.J.; Collins, J.J. A brief history of synthetic biology. *Nat. Rev. Microbiol.* **2014**, *12*, 381–390. [CrossRef]
10. Keasling, J.D. Synthetic biology and the development of tools for metabolic engineering. *Metab. Eng.* **2012**, *14*, 189–195. [CrossRef]
11. Chubukov, V.; Mukhopadhyay, A.; Petzold, C.J.; Keasling, J.D.; Martin, H.G. Synthetic and systems biology for microbial production of commodity chemicals. *NPJ Syst. Biol. Appl.* **2016**, *2*, 16009. [CrossRef] [PubMed]
12. Ajikumar, P.K.; Xiao, W.H.; Tyo, K.E.; Wang, Y.; Simeon, F.; Leonard, E.; Mucha, O.; Phon, T.H.; Pfeifer, B.; Stephanopoulos, G. Isoprenoid pathway optimization for taxol precursor overproduction in *Escherichia coli*. *Science* **2010**, *330*, 70–74. [CrossRef] [PubMed]
13. Paddon, C.J.; Keasling, J.D. Semi-synthetic artemisinin: A model for the use of synthetic biology in pharmaceutical development. *Nat. Rev. Microbiol.* **2014**, *12*, 355–367. [CrossRef] [PubMed]
14. Lee, J.W.; Na, D.; Park, J.M.; Lee, J.; Choi, S.; Lee, S.Y. Systems metabolic engineering of microorganisms for natural and non-natural chemicals. *Nat. Chem. Biol.* **2012**, *8*, 536–546. [CrossRef] [PubMed]
15. Yadav, V.G.; De Mey, M.; Lim, C.G.; Ajikumar, P.K.; Stephanopoulos, G. The future of metabolic engineering and synthetic biology: Towards a systematic practice. *Metab. Eng.* **2012**, *14*, 233–241. [CrossRef]
16. Stephanopoulos, G. Synthetic biology and metabolic engineering. *ACS Synth. Biol.* **2012**, *1*, 514–525. [CrossRef] [PubMed]
17. Westfall, P.J.; Pitera, D.J.; Lenihan, J.R.; Eng, D.; Woolard, F.X.; Regentin, R.; Horning, T.; Tsuruta, H.; Melis, D.J.; Owens, A.; et al. Production of amorphadiene in yeast, and its conversion to dihydroartemisinic acid, precursor to the antimalarial agent artemisinin. *Proc. Natl. Acad. Sci. USA* **2012**, *109*, E111–E118. [CrossRef] [PubMed]

18. Paddon, C.J.; Westfall, P.J.; Pitera, D.J.; Benjamin, K.; Fisher, K.; McPhee, D.; Leavell, M.D.; Tai, A.; Main, A.; Eng, D.; et al. High-level semi-synthetic production of the potent antimalarial artemisinin. *Nature* **2013**, *496*, 528–532. [CrossRef]
19. Hirokawa, Y.; Suzuki, I.; Hanai, T. Optimization of isopropanol production by engineered cyanobacteria with a synthetic metabolic pathway. *J. Biosci. Bioeng.* **2015**, *119*, 585–590. [CrossRef] [PubMed]
20. Lee, S.K.; Chou, H.; Ham, T.S.; Lee, T.S.; Keasling, J.D. Metabolic engineering of microorganisms for biofuels production: From bugs to synthetic biology to fuels. *Curr. Opin. Biotechnol.* **2008**, *19*, 556–563. [CrossRef]
21. Hanai, T.; Atsumi, S.; Liao, J.C. Engineered synthetic pathway for isopropanol production in *Escherichia coli*. *Appl. Environ. Microbiol.* **2007**, *73*, 7814–7818. [CrossRef] [PubMed]
22. Keasling, J.D. Manufacturing molecules through metabolic engineering. *Science* **2010**, *330*, 1355–1358. [CrossRef] [PubMed]
23. Quin, M.B.; Schmidt-Dannert, C. Designer microbes for biosynthesis. *Curr. Opin. Biotechnol.* **2014**, *29*, 55–61. [CrossRef] [PubMed]
24. Ellis, T.; Adie, T.; Baldwin, G.S. DNA assembly for synthetic biology: From parts to pathways and beyond. *Integr. Biol. (Camb.)* **2011**, *3*, 109–118. [CrossRef]
25. Matsumoto, T.; Tanaka, T.; Kondo, A. Engineering metabolic pathways in *Escherichia coli* for constructing a "microbial chassis" for biochemical production. *Bioresour. Technol.* **2017**, *245*, 1362–1368. [CrossRef] [PubMed]
26. Pandey, R.P.; Parajuli, P.; Koffas, M.A.G.; Sohng, J.K. Microbial production of natural and non-natural flavonoids: Pathway engineering, directed evolution and systems/synthetic biology. *Biotechnol. Adv.* **2016**, *34*, 634–662. [CrossRef] [PubMed]
27. Yan, H.; Sun, L.; Huang, J.; Qiu, Y.; Xu, F.; Yan, R.; Zhu, D.; Wang, W.; Zhan, J. Identification and heterologous reconstitution of a 5-alk(en)ylresorcinol synthase from endophytic fungus *Shiraia* sp. Slf14. *J. Microbiol.* **2018**, *56*, 805–812. [CrossRef] [PubMed]
28. Carbonell-Ballestero, M.; Garcia-Ramallo, E.; Montanez, R.; Rodriguez-Caso, C.; Macia, J. Dealing with the genetic load in bacterial synthetic biology circuits: Convergences with the Ohm's law. *Nucleic Acids Res.* **2016**, *44*, 496–507. [CrossRef]
29. Colloms, S.D.; Merrick, C.A.; Olorunniji, F.J.; Stark, W.M.; Smith, M.C.; Osbourn, A.; Keasling, J.D.; Rosser, S.J. Rapid metabolic pathway assembly and modification using serine integrase site-specific recombination. *Nucleic Acids Res.* **2014**, *42*, e23. [CrossRef]
30. Sleight, S.C.; Sauro, H.M. Randomized BioBrick assembly: A novel DNA assembly method for randomizing and optimizing genetic circuits and metabolic pathways. *ACS Synth. Biol.* **2013**, *2*, 506–518. [CrossRef]
31. Bond-Watts, B.B.; Bellerose, R.J.; Chang, M.C. Enzyme mechanism as a kinetic control element for designing synthetic biofuel pathways. *Nat. Chem. Biol.* **2011**, *7*, 222–227. [CrossRef] [PubMed]
32. Kataoka, N.; Vangnai, A.S.; Pongtharangkul, T.; Tajima, T.; Yakushi, T.; Matsushita, K.; Kato, J. Construction of CoA-dependent 1-butanol synthetic pathway functions under aerobic conditions in *Escherichia coli*. *J. Biotechnol.* **2015**, *204*, 25–32. [CrossRef] [PubMed]
33. Pásztor, A.; Kallio, P.; Malatinszky, D.; Akhtar, M.K.; Jones, P.R. A synthetic O_2-tolerant butanol pathway exploiting native fatty acid biosynthesis in *Escherichia coli*. *Biotechnol. Bioeng.* **2015**, *112*, 120–128. [CrossRef] [PubMed]
34. De Jong, B.W.; Shi, S.; Siewers, V.; Nielsen, J. Improved production of fatty acid ethyl esters in *Saccharomyces cerevisiae* through up-regulation of the ethanol degradation pathway and expression of the heterologous phosphoketolase pathway. *Microb. Cell Fact.* **2014**, *13*, 39. [CrossRef] [PubMed]
35. Goh, E.B.; Baidoo, E.E.K.; Burd, H.; Lee, T.S.; Keasling, J.D.; Beller, H.R. Substantial improvements in methyl ketone production in *E. coli* and insights on the pathway from in vitro studies. *Metab. Eng.* **2014**, *26*, 67–76. [CrossRef] [PubMed]
36. Yan, J.; Yan, Y.; Madzak, C.; Han, B. Harnessing biodiesel-producing microbes: From genetic engineering of lipase to metabolic engineering of fatty acid biosynthetic pathway. *Crit. Rev. Biotechnol.* **2017**, *37*, 26–36. [CrossRef]
37. Sherkhanov, S.; Korman, T.P.; Clarke, S.G.; Bowie, J.U. Production of FAME biodiesel in *E. coli* by direct methylation with an insect enzyme. *Sci. Rep.* **2016**, *6*, 24239. [CrossRef]
38. Nawabi, P.; Bauer, S.; Kyrpides, N.; Lykidis, A. Engineering *Escherichia coli* for biodiesel production utilizing a bacterial fatty acid methyltransferase. *Appl. Environ. Microbiol.* **2011**, *77*, 8052–8061. [CrossRef]

39. Zhu, X.; Tan, Z.; Xu, H.; Chen, J.; Tang, J.; Zhang, X. Metabolic evolution of two reducing equivalent-conserving pathways for high-yield succinate production in *Escherichia coli*. *Metab. Eng.* **2014**, *24*, 87–96. [CrossRef]
40. Walther, T.; Topham, C.M.; Irague, R.; Auriol, C.; Baylac, A.; Cordier, H.; Dressaire, C.; Lozano-Huguet, L.; Tarrat, N.; Martineau, N.; et al. Construction of a synthetic metabolic pathway for biosynthesis of the non-natural methionine precursor 2,4-dihydroxybutyric acid. *Nat. Commun.* **2017**, *8*, 15828. [CrossRef]
41. Noda, S.; Shirai, T.; Mori, Y.; Oyama, S.; Kondo, A. Engineering a synthetic pathway for maleate in *Escherichia coli*. *Nat. Commun.* **2017**, *8*, 1153. [CrossRef] [PubMed]
42. Engler, C.; Kandzia, R.; Marillonnet, S. A one pot, one step, precision cloning method with high throughput capability. *PLoS ONE* **2008**, *3*, e3647. [CrossRef] [PubMed]
43. Sarrion-Perdigones, A.; Falconi, E.E.; Zandalinas, S.I.; Juarez, P.; Fernandez-del-Carmen, A.; Granell, A.; Orzaez, D. GoldenBraid: An iterative cloning system for standardized assembly of reusable genetic modules. *PLoS ONE* **2011**, *6*, e21622. [CrossRef] [PubMed]
44. Weber, E.; Engler, C.; Gruetzner, R.; Werner, S.; Marillonnet, S. A modular cloning system for standardized assembly of multigene constructs. *PLoS ONE* **2011**, *6*, e16765. [CrossRef] [PubMed]
45. Casini, A.; MacDonald, J.T.; De Jonghe, J.; Christodoulou, G.; Freemont, P.S.; Baldwin, G.S.; Ellis, T. One-pot DNA construction for synthetic biology: The Modular Overlap-Directed Assembly with Linkers (MODAL) strategy. *Nucleic Acids Res.* **2014**, *42*, e7. [CrossRef] [PubMed]
46. Hochrein, L.; Machens, F.; Gremmels, J.; Schulz, K.; Messerschmidt, K.; Mueller-Roeber, B. AssemblX: A user-friendly toolkit for rapid and reliable multi-gene assemblies. *Nucleic Acids Res.* **2017**, *45*, e80. [CrossRef]
47. Keasling, J.D. Synthetic biology for synthetic chemistry. *ACS Chem. Biol.* **2008**, *3*, 64–76. [CrossRef]
48. Wu, G.; Yan, Q.; Jones, J.A.; Tang, Y.J.; Fong, S.S.; Koffas, M.A.G. Metabolic burden: Cornerstones in synthetic biology and metabolic engineering applications. *Trends Biotechnol.* **2016**, *34*, 652–664. [CrossRef]
49. Shi, S.; Liang, Y.; Zhang, M.M.; Ang, E.L.; Zhao, H. A highly efficient single-step, markerless strategy for multi-copy chromosomal integration of large biochemical pathways in *Saccharomyces cerevisiae*. *Metab. Eng.* **2016**, *33*, 19–27. [CrossRef]
50. Jakociunas, T.; Bonde, I.; Herrgard, M.; Harrison, S.J.; Kristensen, M.; Pedersen, L.E.; Jensen, M.K.; Keasling, J.D. Multiplex metabolic pathway engineering using CRISPR/Cas9 in *Saccharomyces cerevisiae*. *Metab. Eng.* **2015**, *28*, 213–222. [CrossRef]
51. Alonso-Gutierrez, J.; Koma, D.; Hu, Q.; Yang, Y.; Chan, L.J.G.; Petzold, C.J.; Adams, P.D.; Vickers, C.E.; Nielsen, L.K.; Keasling, J.D.; et al. Toward industrial production of isoprenoids in *Escherichia coli*: Lessons learned from CRISPR-Cas9 based optimization of a chromosomally integrated mevalonate pathway. *Biotechnol. Bioeng.* **2018**, *115*, 1000–1013. [CrossRef] [PubMed]
52. Hadadi, N.; Hatzimanikatis, V. Design of computational retrobiosynthesis tools for the design of *de novo* synthetic pathways. *Curr. Opin. Chem. Biol.* **2015**, *28*, 99–104. [CrossRef] [PubMed]
53. Feng, F.; Lai, L.; Pei, J. Computational chemical synthesis analysis and pathway design. *Front. Chem.* **2018**, *6*, 199. [CrossRef] [PubMed]
54. Medema, M.H.; van Raaphorst, R.; Takano, E.; Breitling, R. Computational tools for the synthetic design of biochemical pathways. *Nat. Rev. Microbiol.* **2012**, *10*, 191–202. [CrossRef]
55. Na, D.; Kim, T.Y.; Lee, S.Y. Construction and optimization of synthetic pathways in metabolic engineering. *Curr. Opin. Microbiol.* **2010**, *13*, 363–370. [CrossRef]
56. Erb, T.J.; Jones, P.R.; Bar-Even, A. Synthetic metabolism: Metabolic engineering meets enzyme design. *Curr. Opin. Chem. Biol.* **2017**, *37*, 56–62. [CrossRef]
57. Jones, J.A.; Toparlak, O.D.; Koffas, M.A. Metabolic pathway balancing and its role in the production of biofuels and chemicals. *Curr. Opin. Biotechnol.* **2015**, *33*, 52–59. [CrossRef]
58. Salis, H.M. The ribosome binding site calculator. In *Synthetic Biology, Part B—Computer Aided Design and DNA Assembly*; Voigt, C., Ed.; Academic Press: Cambridge, MA, USA, 2011; Volume 498, pp. 19–42.
59. Na, D.; Lee, D. RBSDesigner: Software for designing synthetic ribosome binding sites that yields a desired level of protein expression. *Bioinformatics* **2010**, *26*, 2633–2634. [CrossRef]
60. Salis, H.M.; Mirsky, E.A.; Voigt, C.A. Automated design of synthetic ribosome binding sites to control protein expression. *Nat. Biotechnol.* **2009**, *27*, 946–950. [CrossRef]

61. Seo, S.W.; Yang, J.S.; Cho, H.S.; Yang, J.; Kim, S.C.; Park, J.M.; Kim, S.; Jung, G.Y. Predictive combinatorial design of mRNA translation initiation regions for systematic optimization of gene expression levels. *Sci. Rep.* **2014**, *4*, 4515. [CrossRef]
62. Seo, S.W.; Yang, J.S.; Kim, I.; Yang, J.; Min, B.E.; Kim, S.; Jung, G.Y. Predictive design of mRNA translation initiation region to control prokaryotic translation efficiency. *Metab. Eng.* **2013**, *15*, 67–74. [CrossRef] [PubMed]
63. Jeschek, M.; Gerngross, D.; Panke, S. Rationally reduced libraries for combinatorial pathway optimization minimizing experimental effort. *Nat. Commun.* **2016**, *7*, 11163. [CrossRef] [PubMed]
64. Nowroozi, F.F.; Baidoo, E.E.; Ermakov, S.; Redding-Johanson, A.M.; Batth, T.S.; Petzold, C.J.; Keasling, J.D. Metabolic pathway optimization using ribosome binding site variants and combinatorial gene assembly. *Appl. Microbiol. Biotechnol.* **2014**, *98*, 1567–1581. [CrossRef] [PubMed]
65. Kim, S.C.; Min, B.E.; Hwang, H.G.; Seo, S.W.; Jung, G.Y. Pathway optimization by re-design of untranslated regions for L-tyrosine production in *Escherichia coli*. *Sci. Rep.* **2015**, *5*, 13853. [CrossRef] [PubMed]
66. Yang, P.; Wang, J.; Pang, Q.; Zhang, F.; Wang, J.; Wang, Q.; Qi, Q. Pathway optimization and key enzyme evolution of *N*-acetylneuraminate biosynthesis using an in vivo aptazyme-based biosensor. *Metab. Eng.* **2017**, *43*, 21–28. [CrossRef] [PubMed]
67. Zhang, S.; Zhao, X.; Tao, Y.; Lou, C. A novel approach for metabolic pathway optimization: Oligo-linker mediated assembly (OLMA) method. *J. Biol. Eng.* **2015**, *9*, 23. [CrossRef] [PubMed]
68. Ceroni, F.; Algar, R.; Stan, G.B.; Ellis, T. Quantifying cellular capacity identifies gene expression designs with reduced burden. *Nat. Methods* **2015**, *12*, 415–418. [CrossRef]
69. Gerosa, L.; Sauer, U. Regulation and control of metabolic fluxes in microbes. *Curr. Opin. Biotechnol.* **2011**, *22*, 566–575. [CrossRef]
70. Venayak, N.; Anesiadis, N.; Cluett, W.R.; Mahadevan, R. Engineering metabolism through dynamic control. *Curr. Opin. Biotechnol.* **2015**, *34*, 142–152. [CrossRef]
71. Brockman, I.M.; Prather, K.L. Dynamic metabolic engineering: New strategies for developing responsive cell factories. *Biotechnol. J.* **2015**, *10*, 1360–1369. [CrossRef]
72. Liu, D.; Evans, T.; Zhang, F. Applications and advances of metabolite biosensors for metabolic engineering. *Metab. Eng.* **2015**, *31*, 35–43. [CrossRef] [PubMed]
73. Zhang, F.; Carothers, J.M.; Keasling, J.D. Design of a dynamic sensor-regulator system for production of chemicals and fuels derived from fatty acids. *Nat. Biotechnol.* **2012**, *30*, 354–359. [CrossRef] [PubMed]
74. Liu, D.; Xiao, Y.; Evans, B.S.; Zhang, F. Negative feedback regulation of fatty acid production based on a malonyl-CoA sensor-actuator. *ACS Synth. Biol.* **2015**, *4*, 132–140. [CrossRef] [PubMed]
75. Frazao, C.R.; Maton, V.; Francois, J.M.; Walther, T. Development of a metabolite sensor for high-throughput detection of aldehydes in *Escherichia coli*. *Front. Bioeng. Biotechnol.* **2018**, *6*, 118. [CrossRef] [PubMed]
76. Dietrich, J.A.; Shis, D.L.; Alikhani, A.; Keasling, J.D. Transcription factor-based screens and synthetic selections for microbial small-molecule biosynthesis. *ACS Synth. Biol.* **2013**, *2*, 47–58. [CrossRef] [PubMed]
77. Reed, B.; Blazeck, J.; Alper, H. Evolution of an alkane-inducible biosensor for increased responsiveness to short-chain alkanes. *J. Biotechnol.* **2012**, *158*, 75–79. [CrossRef] [PubMed]
78. Martin, V.J.; Pitera, D.J.; Withers, S.T.; Newman, J.D.; Keasling, J.D. Engineering a mevalonate pathway in *Escherichia coli* for production of terpenoids. *Nat. Biotechnol.* **2003**, *21*, 796–802. [CrossRef]
79. Kizer, L.; Pitera, D.J.; Pfleger, B.F.; Keasling, J.D. Application of functional genomics to pathway optimization for increased isoprenoid production. *Appl. Environ. Microbiol.* **2008**, *74*, 3229–3241. [CrossRef]
80. Dahl, R.H.; Zhang, F.; Alonso-Gutierrez, J.; Baidoo, E.; Batth, T.S.; Redding-Johanson, A.M.; Petzold, C.J.; Mukhopadhyay, A.; Lee, T.S.; Adams, P.D.; et al. Engineering dynamic pathway regulation using stress-response promoters. *Nat. Biotechnol.* **2013**, *31*, 1039–1046. [CrossRef]
81. Xu, P.; Li, L.; Zhang, F.; Stephanopoulos, G.; Koffas, M. Improving fatty acids production by engineering dynamic pathway regulation and metabolic control. *Proc. Natl. Acad. Sci. USA* **2014**, *111*, 11299–11304. [CrossRef]
82. Agapakis, C.M.; Boyle, P.M.; Silver, P.A. Natural strategies for the spatial optimization of metabolism in synthetic biology. *Nat. Chem. Biol.* **2012**, *8*, 527–535. [CrossRef] [PubMed]
83. Moon, T.S.; Dueber, J.E.; Shiue, E.; Prather, K.L. Use of modular, synthetic scaffolds for improved production of glucaric acid in engineered *E. coli*. *Metab. Eng.* **2010**, *12*, 298–305. [CrossRef]

84. Baek, J.M.; Mazumdar, S.; Lee, S.W.; Jung, M.Y.; Lim, J.H.; Seo, S.W.; Jung, G.Y.; Oh, M.K. Butyrate production in engineered *Escherichia coli* with synthetic scaffolds. *Biotechnol. Bioeng.* **2013**, *110*, 2790–2794. [CrossRef] [PubMed]
85. Wang, Y.; Yu, O. Synthetic scaffolds increased resveratrol biosynthesis in engineered yeast cells. *J. Biotechnol.* **2012**, *157*, 258–260. [CrossRef] [PubMed]
86. Zhu, L.Y.; Qiu, X.Y.; Zhu, L.Y.; Wu, X.M.; Zhang, Y.; Zhu, Q.H.; Fan, D.Y.; Zhu, C.S.; Zhang, D.Y. Spatial organization of heterologous metabolic system in vivo based on TALE. *Sci. Rep.* **2016**, *6*, 26065. [CrossRef] [PubMed]
87. Sachdeva, G.; Garg, A.; Godding, D.; Way, J.C.; Silver, P.A. In vivo co-localization of enzymes on RNA scaffolds increases metabolic production in a geometrically dependent manner. *Nucleic Acids Res.* **2014**, *42*, 9493–9503. [CrossRef] [PubMed]
88. Qiu, X.Y.; Xie, S.S.; Min, L.; Wu, X.M.; Zhu, L.Y.; Zhu, L. Spatial organization of enzymes to enhance synthetic pathways in microbial chassis: A systematic review. *Microb. Cell Fact.* **2018**, *17*, 120. [CrossRef]
89. Lee, H.; DeLoache, W.C.; Dueber, J.E. Spatial organization of enzymes for metabolic engineering. *Metab. Eng.* **2012**, *14*, 242–251. [CrossRef]
90. Avalos, J.L.; Fink, G.R.; Stephanopoulos, G. Compartmentalization of metabolic pathways in yeast mitochondria improves the production of branched-chain alcohols. *Nat. Biotechnol.* **2013**, *31*, 335–341. [CrossRef]
91. Rud, I.; Jensen, P.R.; Naterstad, K.; Axelsson, L. A synthetic promoter library for constitutive gene expression in *Lactobacillus plantarum*. *Microbiology* **2006**, *152*, 1011–1019. [CrossRef]
92. Li, Y.; Lin, Z.; Huang, C.; Zhang, Y.; Wang, Z.; Tang, Y.J.; Chen, T.; Zhao, X. Metabolic engineering of *Escherichia coli* using CRISPR-Cas9 meditated genome editing. *Metab. Eng.* **2015**, *31*, 13–21. [CrossRef] [PubMed]
93. Lv, L.; Ren, Y.L.; Chen, J.C.; Wu, Q.; Chen, G.Q. Application of CRISPRi for prokaryotic metabolic engineering involving multiple genes, a case study: Controllable P(3HB-co-4HB) biosynthesis. *Metab. Eng.* **2015**, *29*, 160–168. [CrossRef] [PubMed]
94. Tang, S.Y.; Qian, S.; Akinterinwa, O.; Frei, C.S.; Gredell, J.A.; Cirino, P.C. Screening for enhanced triacetic acid lactone production by recombinant *Escherichia coli* expressing a designed triacetic acid lactone reporter. *J. Am. Chem. Soc.* **2013**, *135*, 10099–10103. [CrossRef] [PubMed]
95. Chou, H.H.; Keasling, J.D. Programming adaptive control to evolve increased metabolite production. *Nat. Commun.* **2013**, *4*, 2595. [CrossRef] [PubMed]
96. Cai, Y.; Xia, M.; Dong, H.; Qian, Y.; Zhang, T.; Zhu, B.; Wu, J.; Zhang, D. Engineering a vitamin B12 high-throughput screening system by riboswitch sensor in *Sinorhizobium meliloti*. *BMC Biotechnol.* **2018**, *18*, 27. [CrossRef] [PubMed]
97. Ma, A.T.; Schmidt, C.M.; Golden, J.W. Regulation of gene expression in diverse cyanobacterial species by using theophylline-responsive riboswitches. *Appl. Environ. Microbiol.* **2014**, *80*, 6704–6713. [CrossRef]
98. Bradley, R.W.; Buck, M.; Wang, B. Tools and principles for microbial gene circuit engineering. *J. Mol. Biol.* **2016**, *428*, 862–888. [CrossRef] [PubMed]
99. Hammer, K.; Mijakovic, I.; Jensen, P.R. Synthetic promoter libraries–tuning of gene expression. *Trends Biotechnol.* **2006**, *24*, 53–55. [CrossRef]
100. Rytter, J.V.; Helmark, S.; Chen, J.; Lezyk, M.J.; Solem, C.; Jensen, P.R. Synthetic promoter libraries for *Corynebacterium glutamicum*. *Appl. Microbiol. Biotechnol.* **2014**, *98*, 2617–2623. [CrossRef]
101. Siegl, T.; Tokovenko, B.; Myronovskyi, M.; Luzhetskyy, A. Design, construction and characterisation of a synthetic promoter library for fine-tuned gene expression in actinomycetes. *Metab. Eng.* **2013**, *19*, 98–106. [CrossRef]
102. Sohoni, S.V.; Fazio, A.; Workman, C.T.; Mijakovic, I.; Lantz, A.E. Synthetic promoter library for modulation of actinorhodin production in *Streptomyces coelicolor* A3(2). *PLoS ONE* **2014**, *9*, e99701. [CrossRef] [PubMed]
103. Yim, S.S.; An, S.J.; Kang, M.; Lee, J.; Jeong, K.J. Isolation of fully synthetic promoters for high-level gene expression in *Corynebacterium glutamicum*. *Biotechnol. Bioeng.* **2013**, *110*, 2959–2969. [CrossRef] [PubMed]
104. Braatsch, S.; Helmark, S.; Kranz, H.; Koebmann, B.; Jensen, P.R. *Escherichia coli* strains with promoter libraries constructed by Red/ET recombination pave the way for transcriptional fine-tuning. *Biotechniques* **2008**, *45*, 335–337. [CrossRef] [PubMed]

105. Rhodius, V.A.; Mutalik, V.K. Predicting strength and function for promoters of the *Escherichia coli* alternative sigma factor, sigmaE. *Proc. Natl. Acad. Sci. USA* **2010**, *107*, 2854–2859. [CrossRef] [PubMed]
106. Rhodius, V.A.; Mutalik, V.K.; Gross, C.A. Predicting the strength of UP-elements and full-length *E. coli* sigmaE promoters. *Nucleic Acids Res.* **2012**, *40*, 2907–2924. [CrossRef]
107. Jonsson, J.; Norberg, T.; Carlsson, L.; Gustafsson, C.; Wold, S. Quantitative sequence-activity models (QSAM)—Tools for sequence design. *Nucleic Acids Res.* **1993**, *21*, 733–739. [CrossRef] [PubMed]
108. Meng, H.; Wang, J.; Xiong, Z.; Xu, F.; Zhao, G.; Wang, Y. Quantitative design of regulatory elements based on high-precision strength prediction using artificial neural network. *PLoS ONE* **2013**, *8*, e60288. [CrossRef] [PubMed]
109. Ishihama, A.; Shimada, T.; Yamazaki, Y. Transcription profile of *Escherichia coli*: Genomic SELEX search for regulatory targets of transcription factors. *Nucleic Acids Res.* **2016**, *44*, 2058–2074. [CrossRef]
110. Davis, J.H.; Rubin, A.J.; Sauer, R.T. Design, construction and characterization of a set of insulated bacterial promoters. *Nucleic Acids Res.* **2011**, *39*, 1131–1141. [CrossRef]
111. Zong, Y.; Zhang, H.M.; Lyu, C.; Ji, X.; Hou, J.; Guo, X.; Ouyang, Q.; Lou, C. Insulated transcriptional elements enable precise design of genetic circuits. *Nat. Commun.* **2017**, *8*, 52. [CrossRef]
112. Becker, N.A.; Schwab, T.L.; Clark, K.J.; Maher, L.J., 3rd. Bacterial gene control by DNA looping using engineered dimeric transcription activator like effector (TALE) proteins. *Nucleic Acids Res.* **2018**, *46*, 2690–2696. [CrossRef] [PubMed]
113. Jakociunas, T.; Jensen, M.K.; Keasling, J.D. CRISPR/Cas9 advances engineering of microbial cell factories. *Metab. Eng.* **2016**, *34*, 44–59. [CrossRef] [PubMed]
114. Sander, J.D.; Joung, J.K. CRISPR-Cas systems for editing, regulating and targeting genomes. *Nat. Biotechnol.* **2014**, *32*, 347–355. [CrossRef] [PubMed]
115. McCutcheon, S.R.; Chiu, K.L.; Lewis, D.D.; Tan, C. CRISPR-Cas expands dynamic range of gene expression from T7RNAP promoters. *Biotechnol. J.* **2018**, *13*, e1700167. [CrossRef] [PubMed]
116. Frei, C.S.; Qian, S.; Cirino, P.C. New engineered phenolic biosensors based on the AraC regulatory protein. *Protein Eng. Des. Sel.* **2018**, *31*, 213–220. [CrossRef] [PubMed]
117. Rodrigo, G.; Landrain, T.E.; Jaramillo, A. *De novo* automated design of small RNA circuits for engineering synthetic riboregulation in living cells. *Proc. Natl. Acad. Sci. USA* **2012**, *109*, 15271–15276. [CrossRef] [PubMed]
118. Isaacs, F.J.; Dwyer, D.J.; Ding, C.; Pervouchine, D.D.; Cantor, C.R.; Collins, J.J. Engineered riboregulators enable post-transcriptional control of gene expression. *Nat. Biotechnol.* **2004**, *22*, 841–847. [CrossRef] [PubMed]
119. Na, D.; Yoo, S.M.; Chung, H.; Park, H.; Park, J.H.; Lee, S.Y. Metabolic engineering of *Escherichia coli* using synthetic small regulatory RNAs. *Nat. Biotechnol.* **2013**, *31*, 170–174. [CrossRef]
120. Breaker, R.R. Prospects for riboswitch discovery and analysis. *Mol. Cell* **2011**, *43*, 867–879. [CrossRef]
121. Waters, L.S.; Storz, G. Regulatory RNAs in bacteria. *Cell* **2009**, *136*, 615–628. [CrossRef]
122. Cho, C.; Lee, S.Y. Efficient gene knockdown in *Clostridium acetobutylicum* by synthetic small regulatory RNAs. *Biotechnol. Bioeng.* **2017**, *114*, 374–383. [CrossRef] [PubMed]
123. Kim, B.; Park, H.; Na, D.; Lee, S.Y. Metabolic engineering of *Escherichia coli* for the production of phenol from glucose. *Biotechnol. J.* **2014**, *9*, 621–629. [CrossRef] [PubMed]
124. Yang, Y.; Lin, Y.; Wang, J.; Wu, Y.; Zhang, R.; Cheng, M.; Shen, X.; Wang, J.; Chen, Z.; Li, C.; et al. Sensor-regulator and RNAi based bifunctional dynamic control network for engineered microbial synthesis. *Nat. Commun.* **2018**, *9*, 3043. [CrossRef] [PubMed]

catalysts

MDPI

Review

A Review of the Enhancement of Bio-Hydrogen Generation by Chemicals Addition

Yong Sun [1,2,*], Jun He [1], Gang Yang [3,*], Guangzhi Sun [2] and Valérie Sage [4,*]

1 Department of Chemical and Environmental Engineering, University of Nottingham Ningbo, Ningbo 315100, China; Jun.He@nottingham.edu.cn
2 School of Engineering, Edith Cowan University, 270 Joondalup Drive, Joondalup, WA 6027, Australia; g.sun@ecu.edu.au
3 State key Laboratory of Biochemical Engineering, Institute of Process Engineering, Chinese Academy of Sciences, Beijing 100190, China
4 The Commonwealth Scientific and Industrial Research Organization (CSIRO), Energy Business Unit, Canberra, WA 6151, Australia
* Correspondence: yong.sun@nottingham.edu.cn or y.sun@ecu.edu.au (Y.S.); yanggang@home.ipe.ac.cn (G.Y.); valerie.sage@csiro.au (V.S.)

Received: 10 March 2019; Accepted: 7 April 2019; Published: 11 April 2019

Abstract: Bio-hydrogen production (BHP) produced from renewable bio-resources is an attractive route for green energy production, due to its compelling advantages of relative high efficiency, cost-effectiveness, and lower ecological impact. This study reviewed different BHP pathways, and the most important enzymes involved in these pathways, to identify technological gaps and effective approaches for process intensification in industrial applications. Among the various approaches reviewed in this study, a particular focus was set on the latest methods of chemicals/metal addition for improving hydrogen generation during dark fermentation (DF) processes; the up-to-date findings of different chemicals/metal addition methods have been quantitatively evaluated and thoroughly compared in this paper. A new efficiency evaluation criterion is also proposed, allowing different BHP processes to be compared with greater simplicity and validity.

Keywords: hydrogenase; bio-hydrogen; chemicals addition; review

1. Introduction

To effectively curb the world emissions from fossil-based energy by 2030 [1,2], attempts of exploring alternative renewable energy have been made worldwide in both scientific and industrial communities in the past decades [3,4]. Due to its great features, such as having the highest energy density among other fuels and complete cleanness after combustion, hydrogen has attracted a lot of attention as an energy carrier [5]. However, the current existing hydrogen generation processes have been dominated by the conventional routes of natural gas steam reforming (SR), natural gas thermal cracking, coal gasification, and partial oxidation of the heavier-than-naphtha hydrocarbons, which use fossil fuel as feedstock, are energy-intensive, and less environmentally friendly [6,7]. Although direct water-splitting via a semi-conductive photocatalyst to produce renewable hydrogen has recently attracted much interest [8], the significant bottom neck of very low efficiency still remains a big technical hurdle to be overcome for its short- and middle-term industrial application. On the other hand, biological processes for hydrogen generation possess many intrinsic appealing advantages, such as simplicity in operation, wide availability of renewable feed stocks (such as agricultural waste and food waste), carbon neutrality, and cost-effectiveness in operation [9–14]. Bio-hydrogen production can be achieved by two kinds of biological processes: (1) light-dependent, and (2) light-independent. For photo-dependent processes, it could be further divided into the photolysis and photo-fermentation subcategories. For the photo-independent processes, hydrogen generation is achieved by dark fermentation (DF). For all of these processes, the bio-catalyst hydrogenase ([FeFe], [NiFe], [Fe]) might be the most significant catalyst for the evolution of hydrogen. Among all of these above-mentioned processes, DF is one of the most promising due to its appealing features of simplicity of operation, relatively high hydrogen conversion, flexibility in cultivation, and simultaneous realization of hydrogen production and organic waste consumptions [15]. Therefore, the effective enhancement of BHP during DF has become a research focus among scholars in the last decades. Many approaches have been found to effectively enhance hydrogen generation during the DF, which include pretreatment (e.g., ultrasonic, acid/base, enzyme hydrolysis), optimized operation (e.g., hydraulic retention time), co-fermentation, genetic engineering, and chemical addition [16–18]. Some of those approaches will directly or indirectly affect the hydrogenase biocatalyst, while some others might affect the other metabolic pathways or the growth of microbes, which ultimately accelerate or inhibit hydrogen generation [19,20]. From a practical perspective, the chemical addition is more feasible compared with other approaches mentioned above. This is the reason why the numbers of reports in regard to process intensification by chemical addition have been growing very rapidly in recent years [21]. Therefore, this motivated us to review the recent progress of chemical addition, such as metal monomers, metal oxides, nanoparticles (NPs), and synergistic factors that potentially affected the activities of a hydrogenase biocatalyst and consequently led to increased hydrogen generation. The hydrogen production rates will be quantitatively compared among these different works. In this review paper, to avoid repetitive summary and discussion that had been addressed by other scholars, we only focus on chemical addition that could potentially affect the activity of hydrogenase during DF for BHP.

2. Enzyme System in Bio-Hydrogen Generation

Hydrogen generation via biological processes can be achieved by a series of biological electrochemical reactions. These reactions are facilitated by a series of biocatalyst enzymes that are found to play critical roles during the BHP. There are three main bio-hydrogen production and consumption enzymes, which are responsible for the net bio-hydrogen evolution. These three different enzymes are reversible hydrogenase, membrane-bounded uptake hydrogenase, and nitrogenase enzymes. Among them, nitrogenase and hydrogenase are the two pivotal biocatalysts [22].

2.1. Functions of Nitrogenase

Hydrogen generation can be catalyzed by nitrogenase under an anaerobic environment at photofermentation conditions from photosynthetic bacteria. Nitrogenase is well-known for fixing the nitrogen molecule, and is commonly found in archaea and bacteria. While the nitrogen molecule is catalyzed into ammonia by the nitrogenase, hydrogen gas is generated as a by-product, and the entire chemical redox balance is maintained during this biological catalytic nitrogen fixation process, which is summarized in Equation (1) below:

$$\mathrm{N_2} + 8H^+ + 8e^- \xrightarrow{Nitrogenase} 2\mathrm{NH_3} + H_2 \uparrow \tag{1}$$

Hydrogen generation catalyzed by nitrogenase is thermodynamically regarded as an energy-intensive and irreversible reaction, which consumes four moles of adenosine triphosphate (ATP) per mole of bio-hydrogen produced. Ammonia (product) removal and an anaerobic condition is critical to hydrogen generation. A schematic diagram of the structure of nitrogenase is shown in Figure 1.

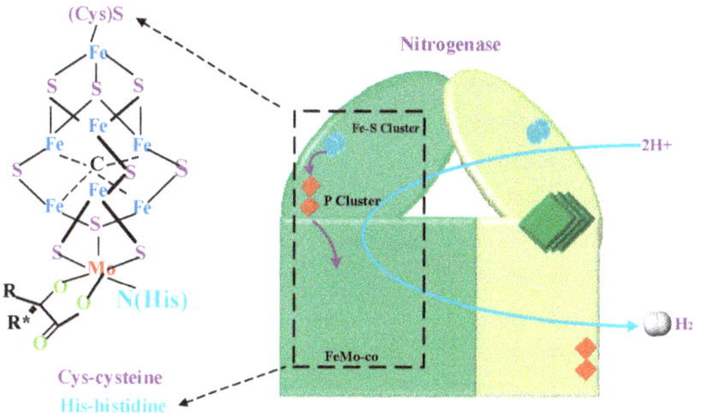

Figure 1. Schematic structure of nitrogenase, where R and R* are the ligands. The figure was rearranged from Seefeldt et al. [23].

The typical structure of nitrogenase consists of three metal-containing cofactors, which are the iron-sulfur cluster, P cluster, and FeMo cluster. The iron–sulfur cluster serves a critical role in delivering electrons to the FeMo cluster via the P cluster. The FeMo serves as an active site for dinitrogen reduction to ammonia. The nitrogenase enzyme widely exists in the photofermentation in archaea and bacteria. Factors such as chemical additions that either enhance or suppress the activity of nitrogenase will result in a variation of hydrogen evolution. Taking the purple non-sulfur bacteria (PNSB), for example, under nitrogen-deficient conditions, the turnover from the nitrogenase is continuous, reducing the protons into H_2. During each circle, at the Fe–S cluster associated with FeMo-co, 2 ATP are hydrolyzed with the transfer of one electron to the MoFe protein and the complex dissociates. The entire turnover is extremely slow at 6.4 s^{-1}, and added to its additional great deal of enzymatic machinery, energy, and time used for the biosynthesis of these complex metal centers, it consequently results in low efficiency [24].

2.2. Functions and Classification of Hydrogenase

The uncovering of the molecular structure of hydrogenase began from the first report of the atomic structure of the D.gigas enzyme [25]. The metal centers, which are the active sites of the biocatalyst, can be broadly classified into three different types, namely the [NiFe], [FeFe], and [Fe] types [26,27]. The [FeFe] hydrogenase catalyzes the oxidation of H_2, as well as the reduction of H^+, but the enzyme is mainly found in the H_2 generating process, whereas the [NiFe] hydrogenase catalyzes the consumption of hydrogen [28]. The detailed schematic diagram of the molecular structure of hydrogenase is shown in Figure 2. The active site of [NiFe] hydrogenase is a dinuclear thiolate-bridged Ni-Fe complex. The [FeFe]-hydrogenases' active sites are organized into modular domains with accessory clusters functioning as inter- and intra-molecular electron-transfer centers electronically linked to the catalytic H-cluster. Hydrogenases, especially the [FeFe] hydrogenase, are sensitive to the presence of oxygen (which is only active under strictly anaerobic conditions). However, studies have shown that the [NiFe] hydrogenases present better O_2 tolerance than the hydrogenase with [FeFe] metal centers [29]. In the metal center of a [NiFe] type hydrogenase, the active site usually contains two cis nickel (Ni) coordination sites available for substrate binding, a bridging site and a terminal Ni site, with the Ni site terminally bound to the thiolate of Cysteine 530 [30]. [NiFe] widely exists in bacteria during hydrogen fermentation, while the [FeFe]-type hydrogenase can only be found in a few microbial species, such as green algea Chlamydomonas reinhardtii [31,32].

Although the two biocatalysts—namely, the nitrogenase and the hydrogenase—show completely different structures and catalyze hydrogen generation via completely different reaction pathways, these two types of biocatalyst sometimes coexist within the cell of one microbe. Therefore, the addition of metal elements such as nickel or iron will affect the activity of these metal-based biocatalysts, which, in turn, will enhance or inhibit the hydrogen generation.

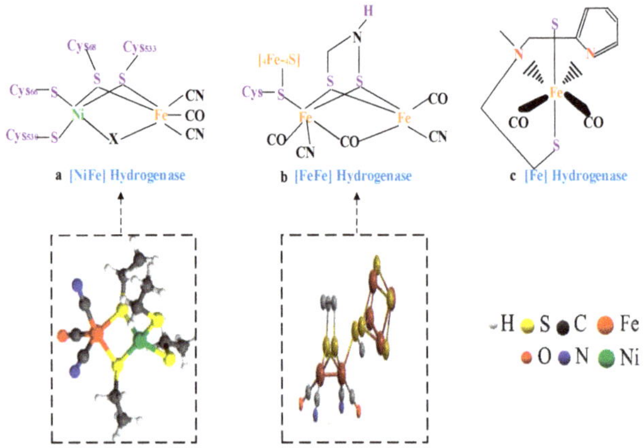

Figure 2. Schematic structure of different types of hydrogenase: (**a**) [NiFe] hydrogenase; (**b**) [FeFe] hydrogenase; and (**c**) [Fe] hydrogenase. The figure was adapted and rearranged from Fontecilla-Camps et al. [29].

3. Bio-Hydrogen Production Pathways

Bio-hydrogen production can be achieved by two major processes: (1) light-dependent processes, and (2) light-independent processes. The light-dependent processes can be accomplished by biophotolysis and photofermentation processes, while the light-independent processes can be realized by dark anaerobic fermentation. A conceptual illustration of the bio-hydrogen production pathway is shown in Figure 3.

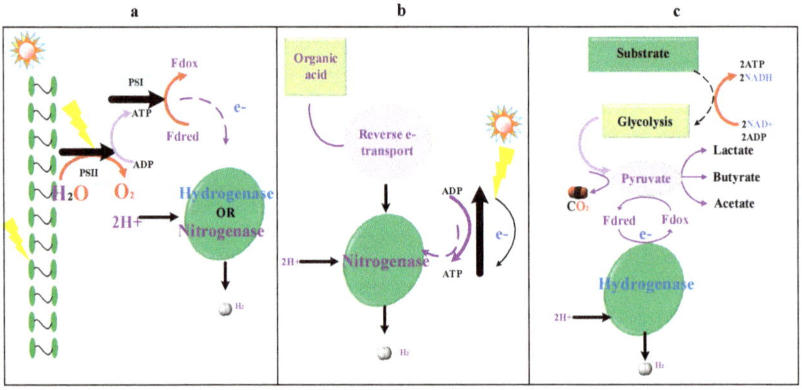

Figure 3. Conceptual illustration of the bio-hydrogen generation pathways: (**a**) biophotolysis, (**b**) photofermentation. (**c**) DF (dark fermentation), PSI represents photosynthesis system 1, PSII is photosynthesis system 2, Fdox is the oxidized Ferredoxin, and Fdred is the reduced Ferredoxin.

3.1. Biophotolysis Process

In biophotolysis (BP), the hydrogen ion is catalyzed either by nitrogenase or hydrogenase ([FeFe] [Fe]) to produce hydrogen gas in the presence of light within the cells of microbes (Figure 3a). Species such as algae and cyanobacteria ([NiFe]-type hydrogenase) are able to adopt this pathway to produce this zero-emission hydrogen gas from sunlight radiations [33]. BP can be further classified into two subcategories—direct biophotolysis, and indirect photolysis [34]. Many microbes, such as green algae or cyanobacterium, which are able to harvest solar energy to drive the water-splitting process to produce O_2 and reduce the ferredoxin-an electron carrier in the chloroplasts, are able to perform biophotolysis via this direct BP pathway [35]. The water-splitting reaction is catalyzed by the photosynthesis system 2 (PSII) under anaerobic conditions, leading to the formation of hydrogen. The amount of electrons is linearly transferred from water to the ferredoxin, driven by the light energy harvested by PSI and PSII in the absence of oxygen. The reduced ferredoxin then donates the obtained electrons from PSI to the enzyme (hydrogenase or nitrogenase) to form hydrogen gas from protons. This entire pathway is shown in Figure 3a.

The microbes that are able to produce hydrogen via the BP process includes the following: *Chlamydomonas reinhardtii*, *Chlorella fusca*, *Scenedesmus obliquus*, *Chlorococcum littorale*, and *Platymonas subcordiformis* [36]. Due to the powerful suppressive effect of the oxygen as a by-product of PSII, the entire hydrogen generation process, including gene expression, mRNA stability, and enzymatic catalysis, will be strongly negatively influenced. Therefore, effective approaches in enhancing direct BP should focus on how to effectively remove or purge the oxygen produced from the system [37,38].

Another route of BP is indirect biophotolysis, of which oxygenic photosynthesis and hypoxic nitrogen fixation reactions are spatially separated from each other. Indirect biophotolysis is widely adopted by cyanobacteria. These are mostly filamentous, and nitrogen fixing in the specialized cell is known as heterocysts. Genera, such as Nostoc, Anabaena, Calothrix, Oscillatoria, are able to produce hydrogen via this indirect photolysis. Some non-nitrogen fixing genera, such as Synechocystis, Synechococcus, and Gloebacter, are also reported to possess this indirect BP pathway [33,36,39].

3.2. Photofermentation Process

In photofermentation (PF), the reduction of ferredoxins and generation of ATP is achieved via the reverse electron flow, driven by harvested solar energy, with the purple non-sulfur photosynthetic bacterium (PNS) under anaerobic conditions [40,41]. Instead of obtaining electrons from water-splitting reactions, as in the direct photolysis that exists in microalgae or cyanobacteria, the organic compounds, such as organic acid, acts as an electron donor under anaerobic conditions in the PNS bacterium. The schematic diagram illustrating this indirect photofermentation is shown in Figure 3b. The hydrogen generated via this pathway has appealing advantages: (1) complete substrate conversion to H_2 and CO_2; (2) removal of the adverse effect of oxygen that inhibits the activity of [FeFe] hydrogenase, hoxEFUYH [NiFe] hydrogenase, and nitrogenase enzymes [42–45]; (3) effective utilization of sunlight in both visible (400–700 nm) and near-infrared (700–950 nm) regions of the solar light spectrum; (4) wide availability of organic compounds used as an electron donor; (5) a relatively lower energy barrier to overcome, compared with water-splitting in direct photolysis [46]. Species that are able to produce the hydrogen via this photofermentation process include the following: *Rhodospirillum rubrum*, *Rhodopseudomonas palustris*, *Rhodobacter sphaeroides*, and *Rhodobacter capsulatus* [47,48].

3.3. Dark Fermentation Process

The essence of dark fermentation (DF) is the catalytic reaction of converting organic substrates into hydrogen under anaerobic conditions. Instead of harvesting the energy from solar light, the energy used to drive the neutralization reaction between the proton (H^+) and electrons (e^-) to form hydrogen comes from the microbial metabolic oxidation of organic substrates, such as glycolysis of glucose to the pyruvate intermediate. Complicated metabolic products are produced during DF. The product distribution of metabolic products varies significantly with the varieties of microbes, the oxidation of the substrate, and the environmental conditions, such as pH, hydrogen partial pressure, and level of nutrition [49,50]. Taking the glycolysis pathways as an example (Figure 3c), the ATP is generated through the substrate phosphorylation and energy-yielding reactions, including the formation of redox equivalents, such as the reduction of oxidized nicotinamide adenine dinucleotide (NAD^+) to nicotinamide adenine dinucleotide (NADH). The produced pyruvate intermediate is then reduced by the produced redox equivalents to form intermediary metabolites, and eventually leads to lactate, CO_2, and ethanol formation [51]. Another fermentation pathway includes the transformation of pyruvate to acetyl-coenzyme A (Acetyl-CoA), accompanied by the formation of an additional redox equivalent,

CO_2 and formate, and eventually leading to the splitting of Acetyl-CoA, and generation of ATPs and acetate [52,53]. The routes for forming molecular hydrogen can be expressed by Equations (2)–(4):

$$C_6H_{12}O_6 + 2NAD^+ \rightarrow 2CH_3COCOOH + 2NADH + 2H^+ \quad (2)$$

$$2NADH + H^+ + 2Fd^{2+} \rightarrow 2Fd^+ + NAD^+ + 2H^+ \quad (3)$$

$$2Fd^+ + 2H^+ \rightarrow 2Fd^{2+} + 2H_2 \quad (4)$$

where Fd represents ferredoxins.

These hydrogen-generation reactions are catalyzed by hydrogenase under an anaerobic condition. There are many appealing advantages of generating hydrogen via these DF pathways: (1) relative simplicity in hydrogen generation, with higher conversion, production efficiency, and lower energy input; (2) versatile feedstock, such as organic food waste or inorganic waste, used for the fermentation; (3) the anaerobic conditions will create a favorable state for maintaining better activity of the biocatalyst for both [NiFe] and [FeFe] hydrogenase, and result in a relatively larger yield of hydrogen; (4) the bio-hydrogen fermentation process is flexible to create either a pure or mixed cultivation of the microbes. A brief comparison of different bio-hydrogen pathways, their corresponding technical challenges, and their effective approaches for hydrogen generation enhancement are summarized in Table 1.

Table 1. Summary of bio-hydrogen pathways catalyzed by hydrogenase.

Pathways	Challenges	Microbes Strains	Hydrogen Enhancement
BP	- Low light conversion efficiency [54] - Incompatibility to simultaneously produce hydrogen and oxygen [55] - High cost for product removal (impermeable hydrogen bioreactor) [56]	- *Scenedesmus obliquus Chlamydomonas reinhardii* (green algae) [57] - *Anabaena variabilis* (cyanobacteria) [58]	- Simultaneous separation or removal of aversive effect to the hydrogenase produced from oxygen [59,60] - Co-culture optimization [61]
PF	- Low photo chemical efficiency [62] - Relative lower hydrogen productivity [63] - Higher energy demand required from nitrogenase	PNS bacteria, such as *Rhodopseudomonas genus* [64]	Process optimization, such as: - Batch cycled arrangement [65] - Recombined DNA techniques [66] - Immobilization of microbes [67] - Chemical additions, such as Ni, EDTA, DMSO [68]
DF	- Relatively poor yield [69] - Metabolic products inhabitation [70] - Lack of research using continuous fermentation	- Thermococcus onnurineus - Enterobacter asburiae - Bacillus coagulans - Thermotoga neapolitana - *Clostridium* sp - *Escherichia coli* - *Bacteroides splanchincus* [51,71,72]	- Hybrid cultivation [73,74] - Chemical additions, such as metal monomers, metal ions and metal oxides [21,75] - Nanoparticles [76] - Membrane reactor [77]

Although technical hurdles and challenges still remain for these three hydrogen generation routes, the DF is still among the most promising technical route for BHP, which attracts great research interests and has even been successfully established at a pilot scale [26,78]. Therefore, DF will be the focus of our subsequent discussion for the enhancement of BHP by chemical additions.

4. Metal Additives

Although many attempts at process intensification, such as pretreatment, process optimization, and co-fermentation have been found to be effective in enhancing hydrogen production, the supplementation of additives have attracted much attention in DF due to its simplicity and cost-effectiveness compared with other approaches of process intensification [21]. Among different kinds of supplementation of additives, metal additives are among one of the most widely employed. It has been found that trace metals play a significant role during the anaerobic fermentation process, especially for the activities of the hydrogenase [79]. The addition of metal into fermentation media has been identified to have the following beneficial effects: (1) facilitation of intracellular electron transportation, and (2) provision of essential nutrition for microbial growth. In this paper, attention will be focused on the effects of metal addition on bio-hydrogen generation. For convenience of discussion, the chemical additions are further divided into subcategories, including monomer, metal ion, metal oxide, and others, such as chemical addition, together with the combination of different types of operations, such as immobilizations.

4.1. Metal Monomers

The addition of metal monomers, such as Fe^0 and Ni^0 during DF, were found to be able to enhance hydrogen generation. The effects of these added metal monomers could be broadly divided into the two categories: (1) directly affects the activity of the biocatalyst; (2) affects the complicated metabolic pathways during DF that leads to enhanced hydrogen generation. Results for the addition of various metal monomers are shown in Table 2. With the addition of different metal monomers, the hydrogen yield was enhanced by different factors, from 10% to 110%, compared with that of the control test without metal addition, depending on the specific conditions such as different inoculum, substrates, or fermentative conditions.

Table 2. Summary and comparisons of bio-hydrogen production with addition of metal monomers and nanoparticles.

Metal	Conc/mg L^{-1}	Feed	Organism	Process	Temp/°C	Yielda/	Reference
Au (NPs)	10 nM	Sucrose	MC	Batch	35	4.47 (+61.7%)	[80]
Ag (NPs)	20 nM	Glucose	MC	Batch	37	2.48 (+67.6%)	[81]
Ni^0	2.5	Glucose	AS	Batch	37	57 [c] (+79.8%)	[82]
Ni (NPs)	5.7	Glucose	AS	Batch	33	2.54 (+22.7%)	[76]
Ni (NPs)	60	Wastewater	AS	Batch	55	24.7 [b] (+22%)	[32]
Fe^0	2000	OWM	AS	Batch	30	102 [b] (+46%)	[83]
Fe^0	400	Sucrose	AS	Batch	30	1.2 (+37%)	[84]
Fe^0	550	Sludge	AS	CSTR	37	650 [d] (+10%)	[85]
Fe^0	100	DS	AS	Batch	37	26 [c] (+16%)	[86]
Fe (NPs)	400	Grass	CB	Batch	37	65 [c] (+44%)	[87]
Cu (NPs)	2.5	Hexose	CA	Batch	30	1.7 (−3.5%)	[88]
Fe (NPs)	200	SJ	AS	Batch	30	1.15 (+62%)	[89]
Ni + Fe (NPs)	37.5 + 37.5	Starch	AS	Batch	37	250 [b] (110%)	[82]
Ni (NPs)	35	Glucose	CB	Batch	35	212 [b] (+32%)	[70]
Ni (NPs) + BC	35	Glucose	CB	Batch	35	238 [b] (+49%)	[70]

OMW: organic market waste: DS: dewatered sludge; SJ: sugarcane juice. MC: mixed consortia; AS: Anaerobic sludge; CB: *Clostridium. Butyricum*; CA: *Clostridium acetobutylicum*; [a] mol/mol of hexose; [b] L/kg TSS or COD or VSS (TSS: total suspended solids, COD chemical oxygen demand, VSS volatile suspended solids); [c] mL/g-dry grass; [d] ml/L.d.

Among the different types of metal monomers, the iron metal monomers are the most promising due to their appealing advantages of relative low cost, and effectiveness in affecting the activity of hydrogenase [21]. In addition, the oxidative-reductive potential (ORP) of fermentation solution could be reduced by the addition of zero-valent iron, which in turn creates a thermodynamically favorable environment for the growth of bacteria. Besides zero-valent metal monomers, the addition of nanosize zero-valent metal monomers, such as iron, nickel, or gold nanoparticles (NPs) began to attract attention due to the unique surface size and quantum size effect. The addition of iron or nickel NPs will facilitate the acceleration of electron transfer between the ferredoxin and hydrogenase to drive hydrogen generation. In addition, the added zero-valent Fe or Ni NPs could be oxidized into metal ions, such as Fe^{2+} or Ni^{2+}, via the anaerobic corrosion process, which will potentially produce very similar beneficial effects upon BHP as those metal ions of Fe^{2+} or Ni^{2+} addition do during the fermentation.

Based upon current reports, one of the highest improvements in hydrogen generation (+110%) could be achieved by adding Ni (37.5 mg/L) and Fe (37.5 mg/L) NPs together during the DF [82]. This surely indicates that the improved electron transfer enhances the overall activity of hydrogenase. However, instead of continuously enhancing the hydrogen generation, an overdose of the metal monomer starts to produce adverse effects upon hydrogen production, as reported in previous research [70]. This indicates that too high a concentration of metal monomers can be harmful for both the activities of hydrogenase, and for other metabolic pathways that indirectly affect hydrogen generation. Therefore, the optimal condition that meets both maximum performance and cost-effectiveness of operation exists. From Table 2, it is not difficult to identify the existing challenges and limitations: (1) there is no consistent quantitative evaluation standard for the assessment of BHP, leading to difficulties in comparing the performance of the different metal monomers; (2) most of the DF focuses on using the sugars, such as glucose or sucrose, and there are very limited efforts at investigating the effect of an addition of metal monomers to the biocatalyst hydrogenase using other types of organic substrates, such as food wastes; (3) the operation of DF was mostly conducted in batch operation, which ends up with continuous inhibitory intermediates accumulation during the DF [90].

4.2. Metal Ions

Metal ion is one of the most common additives that could be used to enhance the catalysts' performances during the DF. The iron ion is widely employed, not only because of its relative cost-effectiveness compared with other metal ions, but also because of its essential role in the constitutions of hydrogenase and ferredoxin. Like the functionality of iron ions, the role of nickel ion in enhancing the activities of hydrogenase also appears to be obvious. According to the works reported by Grafe and Friedrich, the nickel ion has been found in several hydrogenase and nickel-dependent uptake hydrogenase [91]. According to Zhang et al. [92], the addition of nickel ion directly stimulated the activity of hydrogenase. According to the different hypothesis available, the availability of nickel to a cell may affect the activity of the biocatalyst itself or affect the synthesis of other protein, which, in the end, will contribute to the enhancement of hydrogen evolution [91,92].

A summary of the usage of different metal ions, mainly Fe^{2+}/Fe^{3+} and Ni^{2+}, as additives during DF using different types of substrates, such as sugars, wastewater, and food waste is shown in Table 3. From this, it can be seen that the addition of metal irons, especially Fe^{2+} and Ni^{2+}, are effective in enhancing bio-hydrogen fermentation. The role of the metal irons, such as Fe^{2+}/Fe^{3+} and Ni^{2+}, is found to facilitate both the increase of biomass (cell growth) and hydrogen production during the DF. From the work reported by Hisham et al. [93], it was found that the addition of metal elements, such as Ca^{2+} or Mg^{2+} metal ions, led to a significant decrease in hydrogen generation (−30%, −70%), while the biomass experienced a steady increase up to 40%. According to [94], by adding ferrous chloride during DF, the hydrogen generation was enhanced by 650% (increased to 130 $ml.g^{-1}$). Although the improvement compared to the baseline was significant in that work, the absolute value of hydrogen produced (650 mL of cumulative hydrogen production) was marginal compared with other literature reports (which is often over 1000 mL cumulative hydrogen production within the similar duration of cultivation) [90]. Apart from the addition of singular ion, the hybrid mixtures, such as Fe-Ni or Ni-Mg-Al (hydrotalcite), were also found to be effective in enhancing the hydrogen production. Their addition was found to increase the hydrogen generation to about 70–80%.

From the above discussion, it can be suggested that the roles of the different types of metal ions during DF are completely different. The addition of Fe^{2+}/Fe^{3+}, Ni^{2+} or the mixture Fe–Ni seems to directly affect the activity of hydrogenase, and therefore the bio-hydrogen generation process could be directly manipulated by adding these types of metal ions. However, in regard to other metal ions, such as Ca^{2+}, Mg^{2+}, Cu^{2+}, Na^{+}, it seems that these metal ions tend to affect the growth of cell mass or indirectly influence other relevant metabolic pathways during the DF, of which no obvious hydrogen production improvement was observed by these types of metal ion additions.

Table 3. Comparisons of bio-hydrogen production with addition of nickel and other metal nanoparticles.

Metal Ion	Opt/mg L^{-1}	Organism	Feed	Process	Temp/°C	Yielda	Reference
$FeCl_3$	60 uM	Cyanobacteria	BG	Batch	30	0.06 ^{b} (+25%)	[95]
$FeSO_4$	300	AS	Glucose	Batch	35	302 (+56%)	[96]
$FeSO_4$	300	CB	Glucose	Batch	37	2.4 ^{c} (+30%)	[97]
$FeSO_4$	25	CB	Glucose	Batch	30	408 (+4.3%)	[93]
$FeSO_4$	63	AS	PS	Batch	30	226 (52%)	[98]
$FeSO_4$	100	HTS	Glucose	Batch	35	2.6 ^{c} (+13%)	[99]
$FeCl_2$	353	AS	Sucrose	Batch	37	132 (+650%)	[94]
$FeCl_2$	50	AS	Glucose	Batch	37	216 (+23.4%)	[100]
$FeCl_3$	213	EA	Glucose	Batch	30	1.7 ^{c} (+55%)	[101]
Ni^{2+}	0.6	CD	Sucrose	Batch	35	2.1 ^{c} (+107%)	[102]
Ni^{2+}	0.2	MC	Glucose	Batch	35	2.4 ^{c} (+75%)	[103]
$NiCl_2$	0.1	AS	Glucose	Batch	35	289 (+55%)	[103]
$NiCl_2$	16	AS	SW	Batch	34	1120 (+500%)	[104]
$MgCl_2$	200	MC	Glucose	FB	35	1.75 (+600%)	[105]
Na_2CO_3	2000	AS	Sucrose	UASB	37	40 (+300%)	[106]
$MgCl_2$	500	CB	Glucose	Batch	30	209 (−30%)	[93]
$CaCl_2$	500	CB	Glucose	Batch	30	82 (−72%)	[93]
NaCl	5000	CA	Glucose	Batch	30	2.7 ^{c} (−29%)	[107]
Hydrotalcite	250	HTS	Sucrose	UASB	37	3.4 ^{c} (+80%)	[108]
Ni-Fe	50 + 25	CB	Glucose	ACSTR	30	300 (+70%)	[109]

Hydrotalcite: Ni-Mg-Al; AS: anaerobic sludge: CB: Clostridium. Butyricum; HTS: heat treated sludge; EA: *Enterobacter aerogenes*; MC: mixed consortia; CA: Clostridium acetobutylicum; CD: Cow dung; BG: $BG11_0$ media; PS: potato starch; SW: synthetic wastewater. FB: fed-batch; UASB: upflow anaerobic blanket; ACSTR: anaerobic continuous stirred tank reactor. ^{a} ml; ^{b} μmol.mg^{-1} h^{-1}; ^{c} mol.mol^{-1}; ^{d} imL L^{-1} h^{-1}.

4.3. Metal Oxide

Metal oxides play a very similar role as to metal ions. In recent studies, it has been found that the reduced size of metal oxides (nanoparticle size) will be favorable to the electron transfer between ferredoxin and [NiFe] or [FeFe]-based hydrogenase, which in turn accelerates the catalytic reactions of hydrogen generation [110]. The summary of adding metal oxides and their corresponding BHP performance is shown in Table 4. Various kinds of metal oxides, such as TiO_2, CoO, Fe_2O_3, NiO, and their mixtures (Fe_2O_3/NiO), and substrates such as glucose, organic wastewater, glucose, and starch, were used in previous studies, conducted mostly in batch operation under mesophilic conditions. The addition of NiO_2 NPs, together with a co-addition of other NPs, were also found to enhance hydrogen generation due to facilitations of the electron transfer between ferredoxin and hydrogenase [111]. Therefore, the addition of metal oxides, especially with nanoparticle size, is another effective approach in directly enhancing the activities and performances of hydrogenase, which, in turn, will boost BHP.

Table 4. Impact of metal oxide upon the activity of nickel-contained hydrogenase for hydrogen production.

Metal Ion	Opt/mg L^{-1}	Organism	Feed	Process	Temp/°C	Yielda	Reference
TiO_2	100	BA	Glucose	Batch	30	160 (+46%)	[112]
CoO (NPs)	1	AS	POME	Batch	37	0.5 ^{b} (+10%)	[113]
TiO_2	300	RS	SM	Batch	32	1900 (+54%)	[114]
γ-Fe_2O_3	25	AS	SB	Batch	30	0.9 ^{c} (+62%)	[115]
α-Fe_2O_3	63	MC	Inorganic salt	Batch	30	3.6 ^{c} (33%)	[116]
Fe_2O_3	175	CA	CL	Batch	37	2.3 ^{c} (+18%)	[117]
NiO (NPs)	10	MC	CDW	Batch	37	13 ^{b} (+33%)	[118]
NiO (NPs)	1.5	BA	POME	Batch	37	25 ^{b} (+15%)	[113]
NiO (NPs)	100	AS	GS	Batch	35	2.1 ^{c} (+107%)	[119]
Fe_2O_3/NiO	50/10	MC	CDW	Batch	37	17 ^{b} (+45%)	[118]
Fe_2O_3/NiO (NPs)	200/5	AS	DW	Batch	37	19 ^{b} (+25%)	[120]

BA: Bacillus anthracis; AS: anaerobic sludge: RS: Rhodobacter sphaeroides; MC: mixed consortia; CA: Clostridium acetobutylicum. POME: Palm oil mill effluent; SM: Sistrom's medium; SB: Sugarcane bagasse; CL: curry leaf; CDW: complex dairy wastewater; GS: glucose and starch, DW: distillery water. ^{a} mL; ^{b} μmol mg^{-1} h^{-1}; ^{c} mol mol^{-1}; ^{d} mL L^{-1} h^{-1}.

4.4. Others

A summary of BHP with synergistic effects are compared in Table 5. Microbial immobilization is one of the most widely employed approaches used to prevent biomass wash-out when hydrogen evolution rate (HRE) is low during continuous operation. The appealing advantages of employing microbial or cell biocatalyst immobilization include: (a) tolerance toward the perturbation of environmental factors, such as temperature, pH, and accumulation of inhibitory intermediates; (b) higher bio-catalytic activity; and (c) higher process stability [121]. Table 6 summarizes the performances of different microbial immobilizer additions during DF [56]. Clearly, with the implementation of immobilization, the BHP is enhanced at different levels. Many researchers have found that immobilization supports, such as activated carbon (AC) or biochar (BC), tend to form a favorable thermodynamic chemical redox potential, which makes the hydrogen generation catalyzed by the hydrogenase run more effectively [70].

Table 5. Comparisons of bio-hydrogen production with synergistic effects of adding biomass immobilization and metals.

Additives	Opt/mg L^{-1}	Organism	Feed	Process	Temp/°C	Yield[a]	Reference
AC	200	AS	Glucose	Batch	60	1.77 [c] (+106%)	[122]
BC	10	AS	Food waste	Batch	35	1475 [d] (+41%)	[123]
Fe^0 + AC	100	HTS	CBR	Batch	30	83 [b] (+48%)	[124]
Fe (NPs) + CAB	1000	AS	PW	Batch	38	298 [b] (+400%)	[125]
Gel	0.2% (*w*/*v*)	EA	Glucose	Batch	30	1.77 [c] (+80%)	[126]
PF	2500	AS	Glucose	Continuous	37	0.6 [c] (+21%)	[127]
Sponge	-	EA	Starch	Continuous	40	3.03 [c] (+37%)	[128]
Foam + Fe	0.7 (g)	EA	Molasses	Continuous	37	3.5 [c] (+77%)	[129]
Fe^2 ++ BC	300	HTS	Glucose	Batch	35	234 [b] (+48%)	[130]
Ni (NPs) + BC	35	CB	Glucose	Batch	35	238 [b] (+49%)	[70]

AC: activate carbon; BC: biochar; CBR: corn-bran residue, PF: polyurethane foam. AS: anaerobic sludge; HTS: heat treated sludge; EA: Enterobacter aerogenes; CB: Clostridium. Butyricum. CBR: corn-bran residue; PW: potato waste. [a] mL; [b] mL g^{-1}; [c] mol mol^{-1}; [d] mL L^{-1} h^{-1}.

Apart from the immobilization approach, the conditions of operation are also found to be effective in enhancing BHP. For example, the hydrogen production yield was higher in continuous BHP, compared with the batch operation [129,131]. This is possibly due to the effective removal of inhibitory metabolic intermediates, which creates a favorable chemical environment for hydrogenase to catalyze the hydrogen formation reaction [132,133]. Nonetheless, although continuous operation could be more appealing compared to batch operation for large-scale production, cells washing-out is one of the critical problems that need to be carefully handled during continuous operation. Furthermore, it is also interesting to find that synergistic effects, such as the addition of immobilized support, together with nanoparticles such as nickel NPs, have a positive effect on BHP. This suggests that these additions directly affect the activities of biocatalyst hydrogenase, the electron transport, and the endurance to environmental perturbation, which in turn boosts hydrogen generation during the DF.

4.5. Results Comparison

In this work, we tried to summarize all the relevant reported works on BHP that we cited in order to find out some quantitative trends on the basis of substrate conversion efficiency ($Y_{H2/S}$) expressed in the mole of hydrogen produced per mole of substrate in mol mol^{-1}, hydrogen evolution rate (HER) expressed in mmol L^{-1} h^{-1}, and specific hydrogen production rate (qH_2) expressed in mmol g^{-1} h^{-1}. It has been addressed by many scholars that the C-molar-based mass balance is necessary when hydrogen yield and rate are expressed during DF [78]. The failure to present mass balance and kinetic data can lead to poor quality assurance and difficulties in quantitative comparison for dark fermentative BHP. It is necessary to set up presentation standards for hydrogen yield and rate, for the convenience of communication and cross-referencing throughout the scientific community.

Due to the limited number of literature and some inconsistences in the presentation of yields and rates due to the omission of mass balances during DF in some reported works, we only used data from literature reports with complete mass balance and rate expressions, and made a limited number of comparisons in regard to $Y_{H2/S}$ versus HER, and $Y_{H2/S}$ versus qH_2 during DF. The results are presented in Figure 4. In this work, for the convenience of comparisons, we divided the chemical additions into the five different categories, which are metal NPs, metal monomers, metal irons, metal oxide, and others (metals other than iron or nickel), respectively. In addition, among these five different categories for BHP, the different kinds of synergistic effects, such as operations (continuous stirred tank reactor CSTR) or microbial immobilization, were all considered and counted. Within the same category, the $Y_{H2/S}$, HER, and qH_2 were averaged, then based upon the calculated mean values ($Y_{H2/S}$, HER and qH_2), the deviations were calculated using the following:

$$\bar{s} = \sqrt{\frac{1}{N-1}\sum_{i=1}^{N}(x_i - \bar{x})^2} \quad (5)$$

where N represents the numbers of sample size, $\bar{x}$ is the sample mean, and x_i is one sample value. The detailed calculations of those values of $Y_{H2/S}$, HER, and qH_2, together with their corresponding deviations, could be found from previous reports [51,90]. In regard to the variation of the deviations calculated from those reported, there are many different factors that could be attributed to the large deviations observed with the different metal additions for BHP: (a) different microbial strains, (b) different ways of operation, such as batch or CSTR, (c) different substrates, such as glucose and sucrose employed during DF.

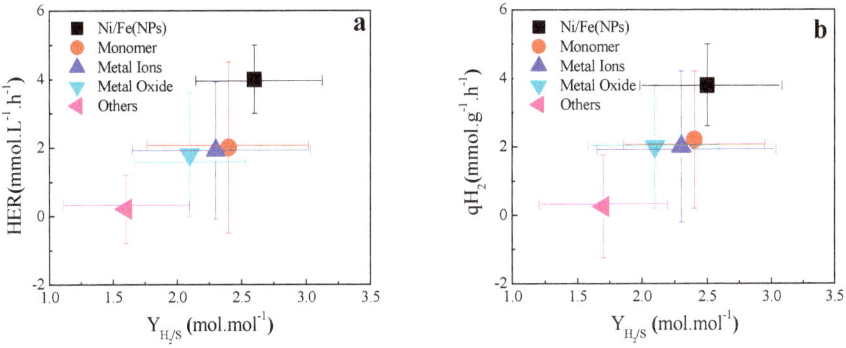

Figure 4. Data illustrating the mean and standard deviation of different metal additions, such as NPs, mental monomers, metal ions, metal oxides, and other metals on BHP during DF: (**a**) HER as a function of $Y_{H2/S}$; (**b**) qH_2 as a function of $Y_{H2/S}$. Note: Others exclude reports using immobilized supporters.

From the comparison, the enhancement of the activities of nickel contained hydrogenase can be broadly divided into three different regions. In the first region, the addition of nano-size particles (together with other synergistic factors, such as adding immobilized supports like AC or BC) is found to be relatively more effective in facilitating the bio-catalytic hydrogen reactions, on the basis of both substrate conversion efficiency and hydrogen evolution rate. The added NPs (such as Ni, Fe, or Ni/Fe) not only facilitate the electrons transfer, but also are engaged in other synergistic factors, such as immobilization and improved operations, that ultimately improve the BHP under anaerobic conditions [70]. In the second region, the metal monomer, metal ions, and metal oxides show very similar performances in enhancing the activities of nickel- or iron-containing biocatalysts during BHP. However, their effects tend to be complex: not only do they possibly affect the activities of the hydrogenase, but these additions might also affect other metabolisms or pathways and cell growth, which will eventually contribute to the increase of BHP. The third region is for those metals other than Ni, Fe, which are relatively less effective in enhancing the BHP, and it was found that some of them are even toxic to either the activities of the hydrogenase or the growth of microbes. Therefore, the addition of these materials is not recommended for the enhancement of the activities of hydrogenase during BHP.

5. Economic Perspective of Different Hydrogen Generation Routes

At the present time, hydrogen is predominantly produced by thermal technologies on the commercial scale, via SR, partial oxidation (POX), and autothermal reforming (ATR). The most widely used and economical approach of hydrogen production is via the steam reforming of methane (natural gas) (SMR), which nearly accounts for 90% of the world's hydrogen generation, at a cost of U.S.$ 7/GJ [7,134]. One of the thorny challenges for these thermochemical processes lies in the simultaneous generation of greenhouse gases that needed to be captured and stored [135,136] or indirectly converting the produced CO_2 back into hydrocarbons via catalytic processes, such as Fischer-Tropsch (FT) synthesis [137,138], which will inevitably increase the cost of the entire hydrogen generation process by about 20 to 40% [139]. The alternative route of replacing fossil-based hydrocarbons with carbon-neutral biomass leads to a doubling of the cost of hydrogen production (about US $14–15/GJ, depending on the types of feedstock and conversion routes), which makes the biomass thermochemical process much less competitive and alluring. Another promising technical route of hydrogen production on a large scale is by the electrolysis of water [140]. However, converting higher-grade electrical energy into relatively lower-grade chemical energy, such as hydrogen and oxygen, is found to be contradictory to the practice of energy cascade utilization, let alone it being more mature and cost-effective to store and transport electricity compared to the hydrogen.

Apart from conventional centralized hydrogen production, on-site and decentralized small-scale hydrogen generation, which possesses the advantages of lowering the prices of transport and onsite utilization of non-usable biomass with high water content, has begun to attract more and more interest. The biological hydrogen generation process is found to be perfectly suitable for small-scale and decentralized hydrogen production using those non-usable biomasses with high water content, with the cost of hydrogen production varying from 10 to 20 U.S.$/GJ, which could be further improved via the R&D impetus in the foreseeable future. Holladay et al. recently made a comparison among different technologies for hydrogen generation synoptically [56] and the results are summarized in Table 6.

Table 6. The comparisons of different technical routes for hydrogen production, and their effectiveness.

Route	Feedstock	Energy Efficiency/%
SR	Hydrocarbons	70–85 [a]
POX	Hydrocarbons	60–75 [a]
ATR	Hydrocarbons	60–75 [a]
Plasma reforming	Hydrocarbons	9–80 [b]
Pyrolysis	Coal	50 [a]
Co-Pyrolysis	Coal + Waste material	80 [a]
Photolysis	Solar + water	0.5 [c]
DF	Biomass	60–80 [d]
Photofermentation	Biomass + Solar	0.1 [e]
Microbial electrolysis	Biomass + Electric	78 [f]
PWS	Water + Solar	12.4 [g]

PWS photo-electrochemical water-splitting. [a] Thermal efficiency based on the higher heating values; [b] Does not include hydrogen purification; [c] Conversion of solar energy to hydrogen by water-splitting excluding hydrogen purification; [d] Theoretical maximum of 4 mol H_2 for 1 mol of glucose; [e] Conversion of solar energy to hydrogen by organic materials excluding hydrogen purification; [f] Total energy efficiency including applied voltage and energy in the substrate, excluding hydrogen purification; [g] Conversion of solar energy to hydrogen by water-splitting excluding hydrogen purification.

Clearly, BHP using biomass as feedstock shows very appealing effectiveness, let alone if further considering competitive and beneficial characteristics, such as the reduced environmental impact, and relative simplicity in operation compared with the thermochemical and electrochemical processes.

In addition, compared with photo-fermentation, DF presents very high efficiency, has a lower footprint, and is independent of solar energy. Therefore, it is envisioned that the effective approach of enhancing the activities of biocatalysts for BHP via DF is pivotal for highly efficient hydrogen production.

6. Future Perspectives

It is apparent that the enhancement of BHP during DF by process intensification and optimization has begun to approach its technical bottleneck at the current stage. From an energy cascade utilization and material recycling and reused perspective, the future for hydrogen production needs to implement multistage processes to further maximize the harvesting of solar energy [141]. The schematic diagram of a multistage procedure, comprised of four or five different steps or approaches, is proposed in Figure 5. In this system, the feeding flows of this multistage process are solar lights, renewable biomass, and water, and outflows are produced hydrogen gas, oxygen gas, and processed biomasses that could be further converted into value-added organic fertilizer [142]. Within this multistage conversion process, the hydrogen production is initiated by photo-fermentation and photocatalysis (solar water-splitting) with feeding-water and organic substrates. Within the system, the cascade utilization of organic substrates could further maximize the hydrogen production in each individual processing step. Theoretically, it is possible to acquire a maximum hydrogen production rate of 12 moles of hydrogen from 1 mole of substrate (glucose) through this combined approach, using purple non-sulfur photosynthetic bacteria and anaerobic bacteria by integrating DF with the photo-fermentations [143]. In addition, the proposed process also integrates the photocatalysis and micro-electrolysis processes for the sake of maximizing hydrogen productions of the entire process. The challenges of this proposed integrated process lies in: (a) the pH swing between the steps of the photofermentation stage, where ammonia will be generated continuously during the photofermentation catalyzed by nitrogenase, and the nearly neutral pH value of DF; (b) how to best optimize the feeding concentration of organic substrates (C/N/O ratio) [144] and control different metabolic pathways on the level of genetic expressions, as this will significantly affect the level of genetic diversity expressions during the fermentations [145–147]; (c) eco-friendly access to the water available.

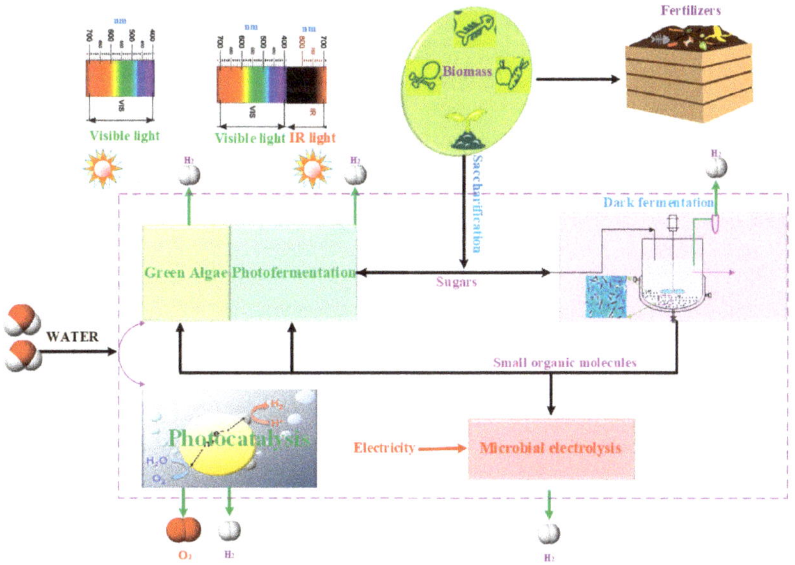

Figure 5. Simplified conceptual illustration diagram of an integrated and multi-coupling process of BHP (biohydrogen production) with photocatalysis using biomass as the substrate.

7. Conclusions

In this paper, biological hydrogen generation produced from renewable bio-resources was found to be a practical route for hydrogen production. Among different BHP routes, the DF has been found to be a practical approach in BHP, especially when it is enhanced by chemical addition. Among the different approaches of chemical addition to improve the activity of hydrogenase, the addition of NPs (Ni, Fe) was found to be relatively more efficient due to its direct effects of facilitating the electron transport between the ferredoxin and the hydrogenase. The order of effectiveness in enhancing the activities of hydrogenase on the basis of substrate conversion efficiency ($Y_{H2/S}$) and hydrogen evolution rate (HER) follows the order of metal NPs > metal monomers/metal ions/metal oxides > other metals (other than Ni, Fe). In order to make the BHP process more feasible and economical enough for industrial applications, future endeavors should focus on the optimized integration of different hydrogen production processes with the energy cascade utilization and material recycling and recovering. By appropriately integrating different approaches, it is potentially possible to approach the theoretical maximum hydrogen yield (12 mol H_2 per 1 mol Glucose consumption). These novel approaches of process intensifications, and integration and appropriate combination of several hydrogen generation processes, such as photocatalysis, photofermentation, and DF processes, will eventually facilitate large-scale curbing of the emission footprint and the cost of BHP in the foreseeable future.

Funding: The authors would like to appreciate the funding support from funding support from National Key R&D Program of China (2018YFC1903500), Edith Cowan University for staff support grant and staff excellent awards grant. The Faculty Inspiration Grant of University of Nottingham and Qianjiang Talent Scheme-Grant/Award Number: QJD1803014 are also highly appreciated.

Acknowledgments: The Acid-Based coupled production group at Institute of Process Engineering Chinese Academy of Science is also highly appreciated. The critical and insightful comments raised from three anonymous reviewers in significantly improving the quality of this work were also highly appreciated.

Conflicts of Interest: All authors have no conflict of interest in this work.

References

1. Haszeldine, R.S. Carbon Capture and Storage: How Green Can Black Be? *Science* **2009**, *325*, 1647–1652. [CrossRef] [PubMed]
2. Sakimoto, K.K.; Wong, A.B.; Yang, P.D. Self-photosensitization of nonphotosynthetic bacteria for solar-to-chemical production. *Science* **2016**, *351*, 74–77. [CrossRef]
3. Wargacki, A.J.; Leonard, E.; Win, M.N.; Regitsky, D.D.; Santos, C.N.S.; Kim, P.B.; Cooper, S.R.; Raisner, R.M.; Herman, A.; Sivitz, A.B.; et al. An Engineered Microbial Platform for Direct Biofuel Production from Brown Macroalgae. *Science* **2012**, *335*, 308–313. [CrossRef] [PubMed]
4. Al Sadat, W.I.; Archer, L.A. The O-2-assisted Al/CO_2 electrochemical cell: A system for CO_2 capture/conversion and electric power generation. *Sci. Adv.* **2016**, 2. [CrossRef]
5. Navlani-Garcia, M.; Mori, K.; Kuwahara, Y.; Yamashita, H. Recent strategies targeting efficient hydrogen production from chemical hydrogen storage materials over carbon-supported catalysts. *NPG Asia Mater.* **2018**, *10*, 277–292. [CrossRef]
6. Benemann, J. Hydrogen biotechnology: Progress and prospects. *Nat. Biotechnol.* **1996**, *14*, 1101–1103. [CrossRef]
7. Shown, I.; Hsu, H.C.; Chang, Y.C.; Lin, C.H.; Roy, P.K.; Ganguly, A.; Wang, C.H.; Chang, J.K.; Wu, C.I.; Chen, L.C.; et al. Highly Efficient Visible Light Photocatalytic Reduction of CO_2 to Hydrocarbon Fuels by Cu-Nanoparticle Decorated Graphene Oxide. *Nano Lett.* **2014**, *14*, 6097–6103.
8. Liu, J.; Liu, Y.; Liu, N.Y.; Han, Y.Z.; Zhang, X.; Huang, H.; Lifshitz, Y.; Lee, S.T.; Zhong, J.; Kang, Z.H. Metal-free efficient photocatalyst for stable visible water-splitting via a two-electron pathway. *Science* **2015**, *347*, 970–974. [CrossRef]
9. Sun, Y.; Yang, G.; Jia, Z.H.; Wen, C.; Zhang, L. Acid Hydrolysis of Corn Stover Using Hydrochloric Acid: Kinetic Modeling and Statistical Optimization. *Chem. Ind. Chem. Eng. Q.* **2014**, *20*, 531–539. [CrossRef]
10. Sun, Y.; Zhang, J.P.; Yang, G.; Li, Z.H. Production of activated carbon by H3PO4 activation treatment of corncob and its performance in removing nitrobenzene from water. *Environ. Prog.* **2007**, *26*, 78–85. [CrossRef]

11. Sun, Y.; Wei, J.; Yao, M.S.; Yang, G. Preparation of activated carbon from furfural production waste and its application for water pollutants removal and gas separation. *Asia-Pac. J. Chem. Eng.* **2012**, *7*, 547–554. [CrossRef]
12. Bandyopadhyay, A.; Stöckel, J.; Min, H.; Sherman, L.A.; Pakrasi, H.B. High rates of photobiological H2 production by a cyanobacterium under aerobic conditions. *Nat. Commun.* **2010**. [CrossRef]
13. Han, W.; Yan, Y.T.; Shi, Y.W.; Gu, J.J.; Tang, J.H.; Zhao, H.T. Biohydrogen production from enzymatic hydrolysis of food waste in batch and continuous systems. *Sci. Rep.* **2016**, *6*, 38395. [CrossRef]
14. Hwang, J.H.; Kim, H.C.; Choi, J.A.; Abou-Shanab, R.A.I.; Dempsey, B.A.; Regan, J.M.; Kim, J.R.; Song, H.; Nam, I.H.; Kim, S.N.; et al. Photoautotrophic hydrogen production by eukaryotic microalgae under aerobic conditions. *Nat. Commun.* **2014**, *5*, 3234. [CrossRef]
15. Mishra, P.; Krishnan, S.; Rana, S.; Singh, L.; Sakinah, M.; Wahid, Z. Outlook of fermentative hydrogen production techniques: An overview of dark, photo and integrated dark-photo fermentative approach to biomass. *Energy Strategy Rev.* **2019**, *24*, 27–37. [CrossRef]
16. Trchounian, K.; Sawers, R.G.; Trchounian, A. Improving biohydrogen productivity by microbial dark- and photo-fermentations: Novel data and future approaches. *Renew. Sust. Energy Rev.* **2017**, *80*, 1201–1216. [CrossRef]
17. Gadhe, A.; Sonawane, S.S.; Varma, M.N. Enhanced biohydrogen production from dark fermentation of complex dairy wastewater by sonolysis. *Int. J. Hydrog. Energy* **2015**, *40*, 9942–9951. [CrossRef]
18. Gadhe, A.; Sonawane, S.S.; Varma, M.N. Evaluation of ultrasonication as a treatment strategy for enhancement of biohydrogen production from complex distillery wastewater and process optimization. *Int. J. Hydrog. Energy* **2014**, *39*, 10041–10050. [CrossRef]
19. Hassan, A.H.S.; Mietzel, T.; Brunstermann, R.; Schmuck, S.; Schoth, J.; Kuppers, M.; Widmann, R. Fermentative hydrogen production from low-value substrates. *World J. Microb. Biot.* **2018**, *34*, 176. [CrossRef]
20. Yao, Z.T.; Su, W.P.; Wu, D.D.; Tang, J.H.; Wu, W.H.; Liu, J.; Han, W. A state-of-the-art review of biohydrogen producing from sewage sludge. *Int. J. Energy Res.* **2018**, *42*, 4301–4312. [CrossRef]
21. Yang, J.W.G. Various additives for improving dark fermentative hydrogen production: A review. *Renew. Sustain. Energy Rev.* **2018**, *95*, 130–146.
22. Das, D.; Veziroglu, T.N. Hydrogen production by biological processes: A survey of literature. *Int. J. Hydrog. Energy* **2001**, *26*, 13–28. [CrossRef]
23. Seefeldt, L.C.; Peters, J.W.; Beratan, D.N.; Bothner, B.; Minteer, S.D.; Raugei, S.; Hoffman, B.M. Control of electron transfer in nitrogenase. *Curr. Opin. Chem. Biol.* **2018**, *47*, 54–59. [CrossRef]
24. Hallenbeck, P.C.; Benemann, J.R. Biological hydrogen production; fundamentals and limiting processes. *Int. J. Hydrog. Energy* **2002**, *27*, 1185–1193. [CrossRef]
25. Trofanchuk, O.; Stein, M.; Gessner, C.; Lendzian, F.; Higuchi, Y.; Lubitz, W. Single crystal EPR studies of the oxidized active site of [NiFe] hydrogenase from Desulfovibrio vulgaris Miyazaki F. *J. Biol. Inorg. Chem.* **2000**, *5*, 36–44. [CrossRef]
26. Morra, S.; Arizzi, M.; Allegra, P.; la Licata, B.; Sagnelli, F.; Zitella, P.; Gilardi, G.; Valetti, F. Expression of different types of [FeFe]-hydrogenase genes in bacteria isolated from a population of a bio-hydrogen pilot-scale plant. *Int. J. Hydrog. Energy* **2014**, *39*, 9018–9027. [CrossRef]
27. Peters, J.W.; Schut, G.J.; Boyd, E.S.; Mulder, D.W.; Shepard, E.M.; Broderick, J.B.; King, P.W.; Adams, M.W.W. [FeFe]- and [NiFe]-hydrogenase diversity, mechanism, and maturation. *Bba-Mol. Cell Res.* **2015**, *1853*, 1350–1369. [CrossRef]
28. Grunwald, P. *Biocatalysis: Biochemical Fundamentals and Applications*; Wold Scientific Publishing Co. Pte. Ltd.: London, UK, 2018.
29. Fontecilla-Camps, J.C.; Volbeda, A.; Cavazza, C.; Nicolet, Y. Structure/function relationships of [NiFe]- and [FeFe]-hydrogenases. *Chem. Rev.* **2007**, *107*, 4273–4303. [CrossRef]
30. Volbeda, A.; Fontecilla-Camps, J.C. The active site and catalytic mechanism of NiFe hydrogenases. *Dalton Trans.* **2003**, 4030–4038. [CrossRef]
31. Kothari, R.; Singh, D.P.; Tyagi, V.V.; Tyagi, S.K. Fermentative hydrogen production—An alternative clean energy source. *Renew. Sust. Energy Rev.* **2012**, *16*, 2337–2346. [CrossRef]

32. Elreedy, A.; Ibrahim, E.; Hassan, N.; El-Dissouky, A.; Fujii, M.; Yoshimura, C.; Tawfik, A. Nickel-graphene nanocomposite as a novel supplement for enhancement of biohydrogen production from industrial wastewater containing mono-ethylene glycol. *Energy Convers. Manag.* **2017**, *140*, 133–144. [CrossRef]
33. Nagarajan, D.; Lee, D.J.; Kondo, A.; Chang, J.S. Recent insights into biohydrogen production by microalgae - From biophotolysis to dark fermentation. *Bioresour. Technol.* **2017**, *227*, 373–387. [CrossRef]
34. Azwar, M.Y.; Hussain, M.A.; Abdul-Wahab, A.K. Development of biohydrogen production by photobiological, fermentation and electrochemical processes: A review. *Renew. Sust. Energy Rev.* **2014**, *31*, 158–173. [CrossRef]
35. Happe, T.; Hemschemeier, A.; Winkler, M.; Kaminski, A. Hydrogenases in green algae: Do they save the algae's life and solve our energy problems? *Trends Plant Sci.* **2002**, *7*, 246–250. [CrossRef]
36. Eroglu, E.; Melis, A. Photobiological hydrogen production: Recent advances and state of the art. *Bioresour. Technol.* **2011**, *102*, 8403–8413. [CrossRef]
37. Melis, A.; Neidhardt, J.; Benemann, J. Dunaliella salina (Chlorophyta) with small chlorophyll antenna sizes exhibit higher photosynthetic productivities and photon use efficiencies than normally pigmented cells. *J. Appl. Phycol.* **1998**, *10*, 515–525. [CrossRef]
38. Williams, C.R.; Bees, M.A. Mechanistic modeling of sulfur-deprived photosynthesis and hydrogen production in suspensions of Chlamydomonas reinhardtii. *Biotechnol. Bioeng.* **2014**, *111*, 1–16. [CrossRef]
39. Das, D.; Veziroglu, T.N. Advances in biological hydrogen production processes. *Int. J. Hydrog. Energy* **2008**, *33*, 6046–6057. [CrossRef]
40. Liu, X.M.; Ren, N.Q.; Song, F.N.; Yang, C.P.; Wang, A.J. Recent advances in fermentative biohydrogen production. *Prog. Nat. Sci.* **2008**, *18*, 253–258. [CrossRef]
41. Hallenbeck, P.C.; Ghosh, D. Advances in fermentative biohydrogen production: The way forward? *Trends Biotechnol.* **2009**, *27*, 287–297. [CrossRef]
42. Cheng, H.L.J.; Zhang, J.; Ding, L.; Lin, R. Enhanced dark hydrogen fermentation of Enterobacter aerogenes/HoxEFUYH with carbon cloth. *Int. J. Hydrog. Energy* **2019**, in press. [CrossRef]
43. Kovacs, K.L.; Maroti, G.; Rakhely, G. A novel approach for biohydrogen production. *Int. J. Hydrog. Energy* **2006**, *31*, 1460–1468. [CrossRef]
44. Sarma, S.J.; Brar, S.K.; le Bihan, Y.; Buelna, G.; Soccol, C.R. Mitigation of the inhibitory effect of soap by magnesium salt treatment of crude glycerol—A novel approach for enhanced biohydrogen production from the biodiesel industry waste. *Bioresour. Technol.* **2014**, *151*, 49–53. [CrossRef]
45. Boboescu, I.Z.; Gherman, V.D.; Lakatos, G.; Pap, B.; Biro, T.; Maroti, G. Surpassing the current limitations of biohydrogen production systems: The case for a novel hybrid approach. *Bioresour. Technol.* **2016**, *204*, 192–201. [CrossRef]
46. Basak, N.; Das, D. The prospect of purple non-sulfur (PNS) photosynthetic bacteria for hydrogen production: The present state of the art. *World J. Microb. Biot.* **2007**, *23*, 31–42. [CrossRef]
47. Willey, J.M.; Sherwood, L.M.; Woolverton, C.J. *Prescott's Microbiology*; McGraw Hill Education: New York, NY, USA, 2017.
48. Ozturk, Y.; Yucel, M.; Daldal, F.; Mandaci, S.; Gunduz, U.; Turker, L.; Eroglu, I. Hydrogen production by using Rhodobacter capsulatus mutants with genetically modified electron transfer chains. *Int. J. Hydrog. Energy* **2006**, *31*, 1545–1552. [CrossRef]
49. de la Cueva, S.C.; Guzman, C.L.A.; Hernandez, V.E.B.; Rodriguez, A.D. Optimization of biohydrogen production by the novel psychrophilic strain N92 collected from the Antarctica. *Int. J. Hydrog. Energy* **2018**, *43*, 13798–13809. [CrossRef]
50. Kumari, S.; Das, D. Improvement of biohydrogen production using acidogenic culture. *Int. J. Hydrog. Energy* **2017**, *42*, 4083–4094. [CrossRef]
51. Ergal, I.; Fuchs, W.; Hasibar, B.; Thallinger, B.; Bochmann, G.; Rittmann, S.K.M.R. The physiology and biotechnology of dark fermentative biohydrogen production. *Biotechnol. Adv.* **2018**, *36*, 2165–2186. [CrossRef]
52. Buckel, W.; Thauer, R.K. Energy conservation via electron bifurcating ferredoxin reduction and proton/Na+ translocating ferredoxin oxidation. *Bba-Bioenerg.* **2013**, *1827*, 94–113. [CrossRef]
53. Schafer, T.; Selig, M.; Schonheit, P. Acetyl-Coa Synthetase (Adp Forming) in Archaea, a Novel Enzyme Involved in Acetate Formation and Atp Synthesis. *Arch. Microbiol.* **1993**, *159*, 72–83. [CrossRef]
54. Miura, Y. Hydrogen-Production by Biophotolysis Based on Microalgal Photosynthesis. *Process. Biochem.* **1995**, *30*, 1–7. [CrossRef]

55. Show, K.Y.; Yan, Y.G.; Ling, M.; Ye, G.X.; Li, T.; Lee, D.J. Hydrogen production from algal biomass—Advances, challenges and prospects. *Bioresour. Technol.* **2018**, *257*, 290–300. [CrossRef]
56. Holladay, J.D.; Hu, J.; King, D.L.; Wang, Y. An overview of hydrogen production technologies. *Catal Today* **2009**, *139*, 244–260. [CrossRef]
57. Khetkorn, W.; Rastogi, R.P.; Incharoensakdi, A.; Lindblad, P.; Madamwar, D.; Pandey, A.; Larroche, C. Microalgal hydrogen production—A review. *Bioresour. Technol.* **2017**, *243*, 1194–1206. [CrossRef]
58. Esper, B.; Badura, A.; Rogner, M. Photosynthesis as a power supply for (bio-)hydrogen production. *Trends Plant Sci.* **2006**, *11*, 543–549. [CrossRef]
59. Aslam, M.; Ahmad, R.; Yasin, M.; Khan, A.L.; Shahid, M.K.; Hossain, S.; Khan, Z.; Jamil, F.; Rafiq, S.; Bilad, M.R.; et al. Anaerobic membrane bioreactors for biohydrogen production: Recent developments, challenges and perspectives. *Bioresour. Technol.* **2018**, *269*, 452–464. [CrossRef]
60. Kapdan, I.K.; Kargi, F. Bio-hydrogen production from waste materials. *Enzym. Microb. Technol.* **2006**, *38*, 569–582. [CrossRef]
61. Laurinavichene, T.V.; Kosourov, S.N.; Ghirardi, M.L.; Seibert, M.; Tsygankov, A.A. Prolongation of H(2) photoproduction by immobilized, sulfur-limited Chlamydomonas reinhardtii cultures. *J. Biotechnol.* **2008**, *134*, 275–277. [CrossRef]
62. Abo-Hashesh, M.; Desaunay, N.; Hallenbeck, P.C. High yield single stage conversion of glucose to hydrogen by photofermentation with continuous cultures of Rhodobacter capsulatus JP91. *Bioresour. Technol.* **2013**, *128*, 513–517. [CrossRef]
63. Xia, A.; Cheng, J.; Ding, L.K.; Lin, R.C.; Huang, R.; Zhou, J.H.; Cen, K.F. Improvement of the energy conversion efficiency of Chlorella pyrenoidosa biomass by a three-stage process comprising dark fermentation, photofermentation, and methanogenesis. *Bioresour. Technol.* **2013**, *146*, 436–443. [CrossRef]
64. Larimer, F.W.; Chain, P.; Hauser, L.; Lamerdin, J.; Malfatti, S.; Do, L.; Land, M.L.; Pelletier, D.A.; Beatty, J.T.; Lang, A.S.; et al. Complete genome sequence of the metabolically versatile photosynthetic bacterium Rhodopseudomonas palustris. *Nat. Biotechnol.* **2004**, *22*, 55–61. [CrossRef]
65. Machado, R.G.; Moreira, F.S.; Batista, F.R.X.; Ferreira, J.S.; Cardoso, V.L. Repeated batch cycles as an alternative for hydrogen production by co-culture photofermentation. *Energy* **2018**, *153*, 861–869. [CrossRef]
66. Kars, G.; Gunduz, U.; Yucel, M.; Rakhely, G.; Kovacs, K.L.; Eroglu, I. Evaluation of hydrogen production by Rhodobacter sphaeroides OU001 and its hupSL deficient mutant using acetate and malate as carbon sources. *Int. J. Hydrog. Energy* **2009**, *34*, 2184–2190. [CrossRef]
67. Ren, N.Q.; Liu, B.F.; Ding, J.; Xie, G.J. Hydrogen production with R. faecalis RLD-53 isolated from freshwater pond sludge. *Bioresour. Technol.* **2009**, *100*, 484–487. [CrossRef]
68. Budiman, P.M.; Wu, T.Y. Role of chemicals addition in affecting biohydrogen production through photofermentation. *Energy Convers. Manag.* **2018**, *165*, 509–527. [CrossRef]
69. Keskin, T.; Hallenbeck, P.C. Hydrogen production from sugar industry wastes using single-stage photofermentation. *Bioresour. Technol.* **2012**, *112*, 131–136. [CrossRef]
70. Sun, Y.; Yang, G.; Zhang, J.P.; Wen, C.; Sun, Z. Optimization and kinetic modeling of an enhanced bio-hydrogen fermentation with the addition of synergistic biochar and nickel nanoparticle. *Int. J. Energy Res.* **2019**, *43*, 983–999. [CrossRef]
71. Kim, M.S.; Fitriana, H.N.; Kim, T.W.; Kang, S.G.; Jeon, S.G.; Chung, S.H.; Park, G.W.; Na, J.G. Enhancement of the hydrogen productivity in microbial water gas shift reaction by Thermococcus onnurineus NA1 using a pressurized bioreactor. *Int. J. Hydrog. Energy* **2017**, *42*, 27593–27599. [CrossRef]
72. Lee, J.; Jung, N.; Shin, J.H.; Park, L.H.; Sung, Y.E.; Park, T.H. Enhancement of hydrogen production and power density in a bio-reformed formic acid fuel cell (BrFAFC) using genetically modified Enterobacter asburiae SNU-1. *Int. J. Hydrog. Energy* **2014**, *39*, 11731–11737. [CrossRef]
73. Zagrodnik, R.; Laniecki, M. Hydrogen production from starch by co-culture of Clostridium acetobutylicum and Rhodobacter sphaeroides in one step hybrid dark-and photofermentation in repeated fed-batch reactor. *Bioresour. Technol.* **2017**, *224*, 298–306. [CrossRef]
74. Shen, N.; Dai, K.; Xia, X.Y.; Zeng, R.J.; Zhang, F. Conversion of syngas (CO and H-2) to biochemicals by mixed culture fermentation in mesophilic and thermophilic hollow-fiber membrane biofilm reactors. *J. Clean. Prod.* **2018**, *202*, 536–542. [CrossRef]
75. Hwang, P.S.Y.; Lee, M.-K.; Yun, Y.-M.; Kim, D.-H. Enhanced hydrogen fermentation by zero valent iron addition. *Int. J. Hydrog. Energy* **2019**, *44*, 3387–3394. [CrossRef]

76. Pugazhendhi, S.S.A.; Nguyen, D.D.; Banu, J.R.; Kumar, G. Application of nanotechnology (nanoparticles) in dark fermentative hydrogen production. *Int. J. Hydrog. Energy* **2019**, *44*, 1431–1440. [CrossRef]
77. Ren, H.Y.; Kong, F.Y.; Ma, J.; Zhao, L.; Xie, G.J.; Xing, D.F.; Guo, W.Q.; Liu, B.F.; Ren, N.Q. Continuous energy recovery and nutrients removal from molasses wastewater by synergistic system of dark fermentation and algal culture under various fermentation types. *Bioresour. Technol.* **2018**, *252*, 110–117. [CrossRef]
78. Khan, S.Y.Z.; Ahmad, M.M.; Chok, V.S.; Uemura, Y.; Sabil, K.M. Review on Hydrogen Production Technologies in Malaysia. *Int. J. Eng. Technol.* **2010**, *10*, 1–8.
79. Sun, Y.; Zhang, J.P.; Yang, G.; Li, Z.H. Analysis of trace elements in corncob by microwave Digestion-ICP-AES. *Spectrosc. Spect. Anal.* **2007**, *27*, 1424–1427.
80. Taherdanak, M.; Zilouei, H.; Karimi, K. The effects of Fe-0 and Ni-0 nanoparticles versus Fe2+ and Ni2+ ions on dark hydrogen fermentation. *Int. J. Hydrog. Energy* **2016**, *41*, 167–173. [CrossRef]
81. Rittmann, S.; Herwig, C. A comprehensive and quantitative review of dark fermentative biohydrogen production. *Microb. Cell Fact.* **2012**, *11*, 115. [CrossRef]
82. Zhang, Y.F.; Shen, J.Q. Enhancement effect of gold nanoparticles on biohydrogen production from artificial wastewater. *Int. J. Hydrog. Energy* **2007**, *32*, 17–23. [CrossRef]
83. Patel, S.K.S.; Kalia, V.C.; Choi, J.H.; Haw, J.R.; Kim, I.W.; Lee, J.K. Immobilization of Laccase on SiO2 Nanocarriers Improves Its Stability and Reusability. *J. Microbiol. Biotechnol.* **2014**, *24*, 639–647. [CrossRef]
84. Camacho, C.E.; Romano, F.I.; Ruggeri, B. Macro approach analysis of dark biohydrogen production in the presence of zero valent powered Fe degrees. *Energy* **2018**, *159*, 525–533. [CrossRef]
85. Zhang, L.; Zhang, L.X.; Li, D.P. Enhanced dark fermentative hydrogen production by zero-valent iron activated carbon micro-electrolysis. *Int. J. Hydrog. Energy* **2015**, *40*, 12201–12208. [CrossRef]
86. Yu, L.; Yu, Y.; Jiang, W.T.; Wei, H.Z.; Sun, C.L. Integrated treatment of municipal sewage sludge by deep dewatering and anaerobic fermentation for biohydrogen production. *Environ. Sci. Pollut. Res.* **2015**, *22*, 2599–2609. [CrossRef]
87. Yu, L.; Jiang, W.T.; Yu, Y.; Sun, C.L. Effects of dilution ratio and Fe-0 dosing on biohydrogen production from dewatered sludge by hydrothermal pretreatment. *Environ. Technol.* **2014**, *35*, 3092–3104. [CrossRef]
88. Yang, G.; Wang, J.L. Improving mechanisms of biohydrogen production from grass using zero-valent iron nanoparticles. *Bioresour. Technol.* **2018**, *266*, 413–420. [CrossRef]
89. Khan, M.M.; Lee, J.; Cho, M.H. Electrochemically active biofilm mediated bio-hydrogen production catalyzed by positively charged gold nanoparticles. *Int. J. Hydrog. Energy* **2013**, *38*, 5243–5250. [CrossRef]
90. Reddy, K.; Nasr, M.; Kumari, S.; Kumar, S.; Gupta, S.K.; Enitan, A.M.; Bux, F. Biohydrogen production from sugarcane bagasse hydrolysate: Effects of pH, S/X, Fe2+, and magnetite nanoparticles. *Environ. Sci. Pollut. Res.* **2017**, *24*, 8790–8804. [CrossRef]
91. Zhang, X.K.; Tabita, F.R.; Vanbaalen, C. Nickel Control of Hydrogen-Production and Uptake in Anabaena Spp Strain-Ca and Strain-1f. *J. Gen. Microbiol.* **1984**, *130*, 1815–1818.
92. Zhang, X.K.; Haskell, J.B.; Tabita, F.R.; Vanbaalen, C. Aerobic Hydrogen-Production by the Heterocystous Cyanobacteria Anabaena Spp Strain-Ca and Strain-1f. *J. Bacteriol.* **1983**, *156*, 1118–1122.
93. Alshiyab, H.; Kalil, M.S.; Hamid, A.A.; Yusoff, W.M.W. Trace Metal Effect on Hydrogen Production Using C.acetobutylicum. *J. Biol. Sci.* **2008**, *8*, 1–9. [CrossRef]
94. Lee, Y.J.; Miyahara, T.; Noike, T. Effect of iron concentration on hydrogen fermentation. *Bioresour. Technol.* **2001**, *80*, 227–231. [CrossRef]
95. Taikhao, S.; Phunpruch, S. Effect of Metal Cofactors of Key Enzymes on Biohydrogen Production by Nitrogen Fixing Cyanobacterium Anabaena siamensis TISIR 8012. In Proceedings of the 2017 International Conference on Alternative Energy in Developing Countries and Emerging Economies, Thailand, Bangkok, 25–26 May 2017; pp. 360–365.
96. Wang, J.L.; Wan, W. Effect of Fe2+ concentration on fermentative hydrogen production by mixed cultures. *Int. J. Hydrog. Energy* **2008**, *33*, 1215–1220. [CrossRef]
97. Chong, M.L.; Rahman, N.A.; Yee, P.L.; Aziz, S.A.; Rahim, R.A.; Shirai, Y.; Hassan, M.A. Effects of pH, glucose and iron sulfate concentration on the yield of biohydrogen by Clostridium butyricum EB6. *Int. J. Hydrog. Energy* **2009**, *34*, 8859–8865. [CrossRef]
98. Vi, L.V.T.; Salakkam, A.; Reungsang, A. Optimization of key factors affecting bio-hydrogen production from sweet potato starch. *Energy Procedia* **2017**, *138*, 973–978.

99. Mullai, P.; Rene, E.R.; Sridevil, K. Biohydrogen Production and Kinetic Modeling Using Sediment Microorganisms of Pichavaram Mangroves, India. *BioMed Res. Int.* **2013**, *1*, 1–9. [CrossRef]
100. Dhar, B.R.; Elbeshbishy, E.; Nakhla, G. Influence of iron on sulfide inhibition in dark biohydrogen fermentation. *Bioresour. Technol.* **2012**, *126*, 123–130. [CrossRef]
101. Karthic, P.; Joseph, S.; Arun, N. Optimization of Process Variables for Biohydrogen Production from Glucose by Enterobacter aerogenes. *Open Access Sci. Rep.* **2012**, *1*, 1–6.
102. Gou, C.Y.; Guo, J.B.; Lian, J.; Guo, Y.K.; Jiang, Z.S.; Yue, L.; Yang, J.L. Characteristics and kinetics of biohydrogen production with Ni2+ using hydrogen-producing bacteria. *Int. J. Hydrog. Energy* **2015**, *40*, 161–167. [CrossRef]
103. Wang, J.; Wan, W. Influence of Ni2+ concentration on biohydrogen production. *Bioresour. Technol.* **2008**, *99*, 8864–8868. [CrossRef]
104. Srikanth, S.; Mohan, S.V. Regulatory function of divalent cations in controlling the acidogenic biohydrogen production process. *RSC Adv.* **2012**, *2*, 6576–6589. [CrossRef]
105. Calli, B.; Boënne, W.; Vanbroekhoven, K. *Bio-Hydrogen Potential of Easily Biodegradable Substrate Through Dark Fermentation*; WHEC 16: Lyon, France, 2006.
106. Xiaolong, H.; Minghua, Z.; Hanqing, Y.; Qinqin, S.; Lecheng, L. Effect of Sodium Ion Concentration on Hydrogen Production from Sucrose by Anaerobic Hydrogen-producing Granular Sludge. *Chin. J. Chem. Eng.* **2006**, *14*, 511–517.
107. Alshiyab, H.; Kalil, M.S.; Hamid, A.A.; Yusoff, W.M.W. Effects of salt addition on hydrogen production by C.acetobutylicum. *Pak. J. Biol. Sci.* **2008**, *11*, 2193–2200. [CrossRef]
108. HongLe, D.T.; Nitisoravut, R. Ni-Mg-Al Hydrotalcite for Improvement of Dark Fermentative Hydrogen Production. *Energy Procedia* **2015**, *79*, 301–306.
109. Karadag, D.; Puhakka, J.A. Enhancement of anaerobic hydrogen production by iron and nickel. *Int. J. Hydrog. Energy* **2010**, *35*, 8554–8560. [CrossRef]
110. Sekoai, P.T.; Ouma, C.N.M.; Preez, S.P.D.; Modisha, P.; Engelbrecht, N.; Bessarabov, D.G.; Ghimire, A. Application of nanoparticles in biofuels: An overview. *Fuel* **2019**, *237*, 380–397. [CrossRef]
111. Patel, S.K.S.; Lee, J.K.; Kalia, V.C. Nanoparticles in Biological Hydrogen Production: An Overview. *Indian J. Microbiol.* **2018**, *58*, 8–18. [CrossRef]
112. Zhao, Y.X.; Chen, Y.G. Nano-TiO2 Enhanced Photofermentative Hydrogen Produced from the Dark Fermentation Liquid of Waste Activated Sludge. *Environ. Sci. Technol.* **2011**, *45*, 8589–8595. [CrossRef]
113. Mishra, P.; Thakur, S.; Mahapatra, D.M.; Wahid, Z.A.; Liu, H.; Singh, L. Impacts of nano-metal oxides on hydrogen production in anaerobic digestion of palm oil mill effluent—A novel approach. *Int. J. Hydrog. Energy* **2018**, *43*, 2666–2676. [CrossRef]
114. Pandey, A.; Gupta, K.; Pandey, A. Effect of nanosized TiO2 on photofermentation by Rhodobacter sphaeroides NMBL-02. *Biomass Bioenerg.* **2015**, *72*, 273–279. [CrossRef]
115. Nasr, M.; Tawfik, A.; Ookawara, S.; Suzuki, M.; Kumari, S.; Bux, F. Continuous biohydrogen production from starch wastewater via sequential dark-photo fermentation with emphasize on maghemite nanoparticles. *J. Ind. Eng. Chem.* **2015**, *21*, 500–506. [CrossRef]
116. Han, H.L.; Cui, M.J.; Wei, L.L.; Yang, H.J.; Shen, J.Q. Enhancement effect of hematite nanoparticles on fermentative hydrogen production. *Bioresour. Technol.* **2011**, *102*, 7903–7909. [CrossRef]
117. Mohanraj, S.; Kodhaiyolii, S.; Rengasamy, M.; Pugalenthi, V. Green Synthesized Iron Oxide Nanoparticles Effect on Fermentative Hydrogen Production by Clostridium acetobutylicum. *Appl. Biochem. Biotechnol.* **2014**, *173*, 318–331. [CrossRef]
118. Gadhe, A.; Sonawane, S.S.; Varma, M.N. Enhancement effect of hematite and nickel nanoparticles on biohydrogen production from dairy wastewater. *Int J. Hydrogen Energ* **2015**, *40*, 4502–4511. [CrossRef]
119. Engliman, N.S.; Abdul, P.M.; Wu, S.Y.; Jahim, J.M. Influence of iron (II) oxide nanoparticle on biohydrogen production in thermophilic mixed fermentation. *Int. J. Hydrog. Energy* **2017**, *42*, 27482–27493. [CrossRef]
120. Gadhe, A.; Sonawane, S.S.; Varma, M.N. Influence of nickel and hematite nanoparticle powder on the production of biohydrogen from complex distillery wastewater in batch fermentation. *Int. J. Hydrog. Energy* **2015**, *40*, 10734–10743. [CrossRef]
121. Sivagurunathan, P.; Pugazhendhi, A.; Kumar, G.; Park, J.H.; Kim, S.H. Biohydrogen fermentation of galactose at various substrate concentrations in an immobilized system and its microbial correspondence. *J. Biosci. Bioeng.* **2018**, *125*, 559–564. [CrossRef]

122. Tanisho, S.; Ishiwata, Y. Continuous hydrogen production from molasses by fermentation using urethane foam as a support of flocks. *Int. J. Hydrog. Energy* **1995**, *20*, 541–545. [CrossRef]
123. Muri, P.; Marinsek-Logar, R.; Djinovic, P.; Pintar, A. Influence of support materials on continuous hydrogen production in anaerobic packed-bed reactor with immobilized hydrogen producing bacteria at acidic conditions. *Enzym. Microb. Technol.* **2018**, *111*, 87–96. [CrossRef]
124. Chookaew, T.; O-Thong, S.; Prasertsan, P. Biohydrogen production from crude glycerol by immobilized Klebsiella sp TR17 in a UASB reactor and bacterial quantification under non-sterile conditions. *Int. J. Hydrog. Energy* **2014**, *39*, 9580–9587. [CrossRef]
125. Yokoi, H.; Maki, R.; Hirose, J.; Hayashi, S. Microbial production of hydrogen from starch-manufacturing wastes. *Biomass Bioenerg* **2002**, *22*, 389–395. [CrossRef]
126. Jamali, N.S.; Jamaliah, M.J.; Isahak, W.N.R.W. Biofilm formation on granular activated carbon in xylose and glucose mixture for thermophilic biohydrogen production. *Int. J. Hydrog. Energy* **2016**, *41*, 21617–21627. [CrossRef]
127. Sunyoto, N.; Zhu, M.; Zhang, Z.; Zhang, D. Effect of biochar addition and initial pH on hydrogen production from the first phase of two-phase anaerobic digestion of carbohydrates food waste. *Energy Procedia* **2017**, *105*, 379–384. [CrossRef]
128. Zhang, J.; Zhang, J.; Zang, L. Thermophilic bio-hydrogen production from corn-bran residue pretreated by calcined-lime mud from papermaking process. *Bioresour. Technol.* **2015**, *198*, 564–570. [CrossRef]
129. Sekoai, P.T.; Daramola, M.O. Effect of metal ions on dark fermentative biohydrogen production using suspended and immobilized cells of mixed bacteria. *Chem. Eng. Commun.* **2018**, *205*, 1011–1022. [CrossRef]
130. Satar, I.; Ghasemi, M.; Aljlil, S.A.; Isahak, W.N.R.W.; Abdalla, M.; Alam, J.; Daud, W.R.W.; Yarmo, M.A.; Akbarzadeh, O. Production of hydrogen by Enterobacter aerogenes in an immobilized cell reactor. *Int. J. Hydrog. Energy* **2017**, *42*, 9024–9030. [CrossRef]
131. Nakatani, H.; Ding, N.; Ohara, Y.; Hori, K. Immobilization of Enterobacter aerogenes by a Trimeric Autotransporter Adhesin, AtaA, and Its Application to Biohydrogen Production. *Catalysts* **2018**, *8*, 159. [CrossRef]
132. Palazzi, E.; Fabiano, B.; Perego, P. Process development of continuous hydrogen production by Enterobacter aerogenes in a packed column reactor. *Bioprocess. Eng.* **2000**, *22*, 205–213. [CrossRef]
133. Zhang, J.S.; Fan, C.F.; Zang, L.H. Improvement of hydrogen production from glucose by ferrous iron and biochar. *Bioresour. Technol.* **2017**, *245*, 98–105. [CrossRef]
134. Rami, H.Ö.; El-Emam, S. Comprehensive Review on the Techno-Economics of Sustainable Large-Scale Clean Hydrogen Production. *J. Clean. Prod.* **2019**, in press.
135. Sun, Y.; Yao, M.S.; Zhang, J.P.; Yang, G. Indirect CO_2 mineral sequestration by steelmaking slag with NH4Cl as leaching solution. *Chem. Eng. J.* **2011**, *173*, 437–445. [CrossRef]
136. Sun, Y.; Zhang, J.P.; Zhang, L. NH4Cl selective leaching of basic oxygen furnace slag: Optimization study using response surface methodology. *Environ. Prog. Sustain.* **2016**, *35*, 1387–1394. [CrossRef]
137. Sun, Y.; Yang, G.; Wen, C.; Zhang, L.; Sun, Z. Artificial neural networks with response surface methodology for optimization of selective CO_2 hydrogenation using K-promoted iron catalyst in a microchannel reactor. *J. CO_2 Util.* **2018**, *23*, 10–21. [CrossRef]
138. Sun, Y.; Yang, G.; Zhang, L.; Sun, Z. Fischer-Tropsch synthesis in a microchannel reactor using mesoporous silica supported bimetallic Co-Ni catalyst: Process optimization and kinetic modeling. *Chem. Eng. Process.* **2017**, *119*, 44–61. [CrossRef]
139. Sun, Y.; Lin, Z.; Peng, S.H.; Sage, V.; Sun, Z. A Critical Perspective on CO_2 Conversions into Chemicals and Fuels. *J. Nanosci. Nanotechnol.* **2019**, *19*, 1–13. [CrossRef]
140. Schmidt, O.; Gambhir, A.; Staffell, I.; Hawkes, A.; Nelson, J.; Few, S. Future cost and performance of water electrolysis: An expert elicitation study. *Int. J. Hydrog. Energy* **2017**, *42*, 30470–30492. [CrossRef]
141. Trchounian, A. Mechanisms for hydrogen production by different bacteria during mixed-acid and photo-fermentation and perspectives of hydrogen production biotechnology. *Crit. Rev. Biotechnol.* **2015**, *35*, 103–113. [CrossRef]
142. Arizzi, M.; Morra, S.; Pugliese, M.; Guilin, M.L.; Gilardi, G.; Valetti, F. Biohydrogen and biomethane production sustained by untreated matrices and alternative application of compost waste. *Waste Manag.* **2016**, *56*, 151–157. [CrossRef]

143. Hawkes, F.R.; Hussy, I.; Kyazze, G.; Dinsdale, R.; Hawkes, D.L. Continuous dark fermentative hydrogen production by mesophilic microflora: Principles and progress. *Int. J. Hydrog. Energy* **2007**, *32*, 172–184. [CrossRef]
144. Hamilton, C.; Hiligsmann, S.; Calusinska, M.; Beckers, L.; Masset, J.; Wilmotte, A.; Thonart, P. Investigation of hydrogenase molecular marker to optimize hydrogen production from organic wastes and effluents of agro-food industries. *Biotechnol. Agron. Soc. Environ.* **2010**, *14*, 574–575.
145. Quéméneur, M.; Hamelin, J.; Marie-Thérèse, S.B.; EricLatrille, G.-O.; EricTrably, J.-P. Changes in hydrogenase genetic diversity and proteomic patterns in mixed-culture dark fermentation of mono-, di- and tri-saccharides. *Int. J. Hydrog. Energy* **2011**, *36*, 11654–11665. [CrossRef]
146. Hamilton, C.; Calusinska, M.; Baptiste, S.; Masset, J.; Beckers, L.; Thonart, P.; Hiligsmann, S. Effect of the nitrogen source on the hydrogen production metabolism and hydrogenases of Clostridium butyricum CWBI1009. *Int. J. Hydrog. Energy* **2018**, *43*, 5451–5462. [CrossRef]
147. Masset, J.; Calusinska, M.; Hamilton, C.; Hiligsmann, S.; Joris, B.; Wilmotte, A.; Thonart, P. Fermentative hydrogen production from glucose and starch using pure strains and artificial co-cultures of *Clostridium* spp. *Biotechnol. Biofuels* **2012**, *5*, 35. [CrossRef]

catalysts

MDPI

Article

Mass Transfer Performance of a String Film Reactor: A Bioreactor Design for Aerobic Methane Bioconversion

Rina Mariyana [1], Min-Sik Kim [2], Chae Il Lim [3], Tae Wan Kim [4], Si Jae Park [5], Byung-Keun Oh [3], Jinwon Lee [3] and Jeong-Geol Na [3,*]

1 PT Rekayasa Industri, Kalibata Timur 1 Street No.36, Jakarta 12740, Indonesia; na.mariyana@gmail.com
2 Biomass and Waste Energy Laboratory, Korea Institute of Energy Research, 152 Gajeong-ro, Yuseong-Gu, Daejeon 34129, Korea; kms0540@kier.re.kr
3 Department of Chemical and Biomolecular Engineering, Sogang University, Seoul 04107, Korea; codlf0620@naver.com (C.I.L.); bkoh@sogang.ac.kr (B.-K.O.); jinwonlee@sogang.ac.kr (J.L.)
4 Department of Biotechnology and Bioengineering, Chonnam National University, Gwangju 61186, Korea; chekimtw@chonnam.ac.kr
5 Division of Chemical Engineering and Materials Science, Ewha Womans University, Seoul 03760, Korea; parksjdr@gmail.com
* Correspondence: narosu@sogang.ac.kr; Tel.: +82-2-705-7955

Received: 25 September 2018; Accepted: 22 October 2018; Published: 24 October 2018

Abstract: The mass transfer performance of a string film reactor (SFR)—a bioreactor design for the aerobic bioconversion of methane—was investigated. The results showed that the SFR could achieve high mass transfer performance of gases, and the highest values of the mass transfer coefficients for oxygen and methane were 877.1 h^{-1} and 408.0 h^{-1}, respectively. There were similar mass transfer coefficients for oxygen and methane in absorption experiments using air, methane, and air–methane mixed gas under the same liquid flow rate conditions, implying that each gas is delivered into the liquid without mutual interaction. The mass transfer performance of the SFR was significantly influenced by the liquid flow rate and the hydrophilicity of the string material, whereas the magnitude of the gas flow rate effect on the mass transfer performance depended on both the tested liquid flow rate and the gas flow rate. Furthermore, the mass transfer performance of the SFR was compared with those of other types of bioreactors.

Keywords: aerobic methane bioconversion; bioreactor; string film reactor; mass transfer performance

1. Introduction

Natural gas conversion into liquid chemicals has become attractive [1] due to the rapid rise in natural gas production [2,3], the tremendous demand for liquid transportation fuel [2], and its compatibility with current vehicle engines and infrastructure [1,4]. However, the current conversion approach, which employs the Fischer–Tropsch process, still faces several obstacles preventing its commercialization owing to its high capital requirements and energy-intensive nature [1,4]. The aerobic bioconversion of methane—the main component of natural gas—by using methanotrophic microorganisms as biocatalysts has shown good potential as an alternative to the Fischer–Tropsch process due to its high selectivity, ambient operating temperature, one-step direct process, and reduced technological complexity [1,4–7]. Therefore, it can be applied to monetize small sources of natural gas, such as stranded and flared natural gas. Recently, microorganisms, as host strains for bio-based production, have been extensively engineered to make them efficient microbial cell factories that are compatible with currently available production and product purification processes by using alternative feedstocks to petroleum [8]. For example, based on these metabolic engineering strategies, *Methylomicrobium alcaliphilum* 20Z was successfully engineered to produce 2,3-butanediol from methane [9]. Even though host strains can be successfully developed to produce target products by performing metabolic engineering strategies, appropriate fermentation processes of these host strains should also be developed to realize the efficient production of target products by considering several factors to increase the efficiency of product formation. Thus, insufficient supply of methane and oxygen to the microbial catalysts because of low mass transfer rate and the solubility of the gas in the aqueous phase should be addressed in order to realize the commercialization of the aerobic bioconversion of methane, which has been limited by the low titer of the product, low productivity, and low production yields. These have resulted from slow process kinetics and low metabolic energy efficiency, among other factors [1].

A reactor design and process development deliberately reflecting microbial characteristics can help overcome the current limit on microbial catalyst performance, especially in fermentation using C1 gas as a carbon substrate. The hydrogen productivity of *Thermococcus onnurineus* NA1, an archaeal strain which has good potential to endure high pressure, was dramatically enhanced by developing a pressurizing bioreactor [10]. Thus, a well-developed reactor design is very significant for maximizing the activity potential of engineered microbial catalysts. To deal with poor characteristics of gas transfer, either the volumetric liquid-side mass transfer coefficient k_La or the concentration driving force [10], which are the two main factors affecting the mass transfer rate as in the following equation, should be improved.

$$\frac{dC}{dt} = k_La(C^* - C) - Q_Cx \quad (1)$$

The common approaches that are employed to improve the volumetric liquid-side mass transfer coefficient k_La involve the optimization of operating parameters and the modification of the bioreactor configuration. A number of approaches have been employed in various types of bioreactors, for example, by increasing the mechanical agitation rate [11,12], investigating several types of impeller configuration [12], and applying a spinning disk microbubble generator [13] in a stirred tank reactor; by testing various types of membrane material [11,14] and adding an external diffuser [15] to a hollow fiber membrane reactor; by modifying the geometry of the reactor [16], testing various gas distributor designs [16], and by applying vibration excitement [17] in a bubble column reactor [18]; by modifying the reactor configuration with the addition of a semipermeable membrane [19] and by optimizing the geometry of the reactor [20] in an air-lift reactor.

The improvement in the value of k_La was confirmed for various bioreactor types, but the bioreactors that were utilized have major drawbacks that can reduce their feasibility for aerobic methane bioconversion application in terms of their operability. These include their high mechanical burden and membrane fouling problems [21–23]. Most strategies have focused on the transfer of oxygen in aerobic cultures, where the main objective of gas transfer is to maintain the dissolved oxygen concentration at a desired level. Unlike aerobic fermentation with sugar, the goal of performing the bioconversion of methane is to produce bulk chemicals from low-priced raw materials efficiently and economically, and it is therefore important to achieve a low energy consumption and high conversion of feed gas. In this respect, the direct application of conventional cases requires careful consideration.

To develop a mass transfer system for methane bioconversion while addressing the obstacles of existing systems, a string film reactor (SFR) was designed and implemented in this study. An SFR is a column reactor in which the contact between liquid and gas takes place on the strings (Figure 1). The SFR was designed to tackle the drawbacks of existing bioreactors through the use of hydrophilic strings that are able to direct and to guide the flow of liquid in a desired manner, thereby preventing nonuniform liquid flow; the formation of a thin film on the surface of the strings, which can decrease the mass transfer resistance; and the use of cell immobilization on the strings as a biofilm to enable high-cell-density culture inside the reactor without causing fouling problems. Furthermore, the SFR can be easily scaled up or scaled down to the desired production volume by specifying the column size and the number of strings.

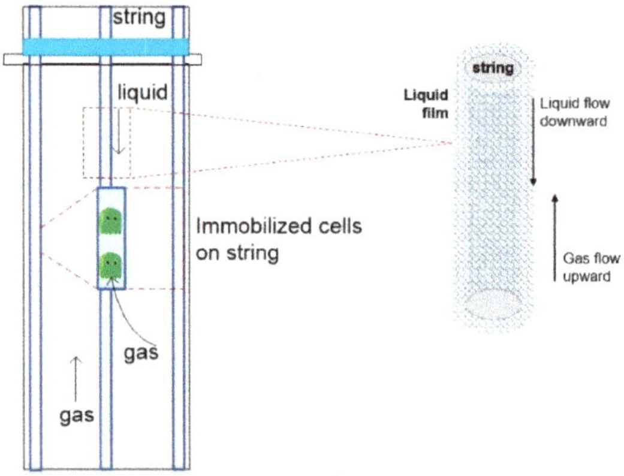

Figure 1. The concept of the string film reactor (SFR).

To assess the mass transfer performance of the SFR, this study explored the effects of the liquid flow rate, gas flow rate, and string materials on the k_La value of oxygen. In addition, because both methane and oxygen participate in the aerobic bioconversion of methane as limiting substrates, the mass transfer characteristics of each gas must be considered independently. In this study, considering the real environment of the system, we investigated the mass transfer behavior of the gases when methane, air, or a mixed gas composed of both methane and air were applied as feed.

2. Results

2.1. *Effect of Liquid Flow Guidance by Strings on Mass Transfer Performance*

To investigate the effect of varying the liquid flow direction on the mass transfer efficiency, the mass transfer coefficients for oxygen were measured at various liquid flow rates with and without the string in the system. As shown in Figure 2, the mass transfer performance of the SFR was considerably influenced by the liquid flow rate. It was observed that k_La values increased by 3.39 times and 2.83 times with and without the string, respectively, when the liquid flow rate increased from 100 mL/min to 500 mL/min.

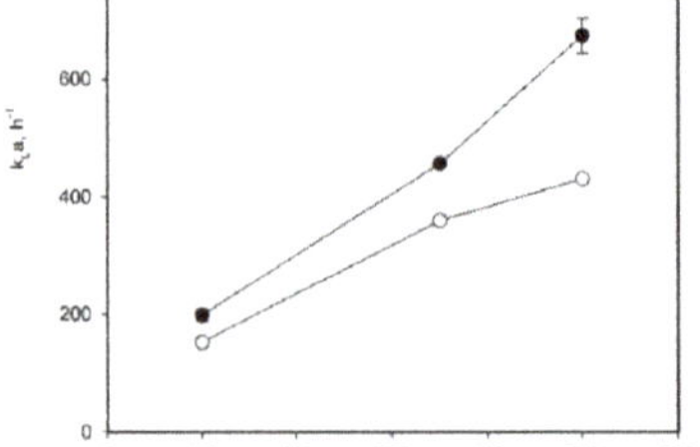

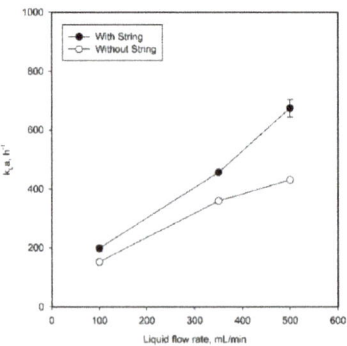

Figure 2. Volumetric mass transfer coefficients for oxygen for the SFR with the string (closed circles) and without the string (open circles).

The presence of strings also influenced the mass transfer performance, and the mass transfer coefficients with strings increased significantly compared to the corresponding values without the string. At a gas flow rate of 350 mL/min, the DO (dissolved oxygen) value in the steady state without the string was only 60.5%, while it was 72.5% with strings, exhibiting enhanced oxygen transfer. As a result, the k_La value was 455.72 h^{-1}, which is 1.26 times larger than that obtained without the string.

2.2. Investigation of Mass Transfer Characteristics in SFR Using Methane–Air Mixed Gas

The mass transfer coefficients obtained for methane were measured at various liquid flow rates. As in the experiments with air, the mass transfer coefficients increased with the liquid flow rates (Figure 3a), but the extent of this increase was reduced with the liquid flow rates, while the variances of the k_La values under the same conditions increased. The mass transfer coefficient of methane at the liquid flow rate of 100 mL/min was 79.0 h^{-1} and increased to 408.0 h^{-1} at a liquid flow rate of 500 mL/min.

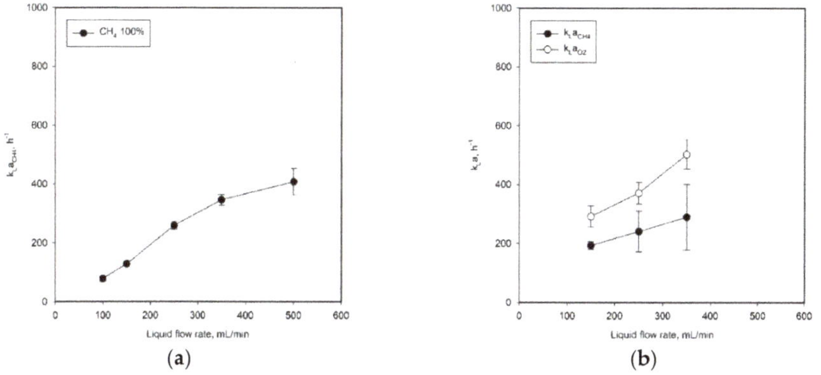

Figure 3. Volumetric mass transfer coefficients for (**a**) methane in the system using 100% methane gas and (**b**) methane (closed circles) and oxygen (open circles) using the gas containing 30% methane and 70% air.

To evaluate the feasibility of the SFR system for methane conversion and to investigate the mass transfer characteristics of the system using the gas containing air and methane, the mass transfer coefficients for each gas were determined at various liquid flow rates. A mixed gas composed of 30% methane and 70% air, which is commonly used for the bioconversion of methane [5], was supplied to the system as feed gas. Although there were some fluctuations—possibly caused by a disturbance in the measurement of dissolved methane by the gas chromatography, owing to the presence of impurities including air in the sample tube (see Section 4.2 for the detailed method of measurement)—the k_La values were similar to those obtained in the cases using single-component gas (Figure 3b). When the liquid flow rate was 350 mL/min, the mass transfer coefficients of methane and oxygen were 289.7 h^{-1} and 503.2 h^{-1}, respectively.

2.3. Improvement of Mass Transfer Performance by Using Hydrophilic Porous Strings

Absorption experiments for oxygen were conducted to investigate the effect of the hydrophilicity of the string on the mass transfer performance. Two types of string material were tested. One was of the fabric that was used in the previous sections and the other was of felt. Because hydrophilic behavior is correlated with the critical surface tension of the material [24], the felt material, which consists purely of cotton, should be more hydrophilic than the fabric material, which consists of a mixture of nylon and cotton, owing to the higher critical surface tension value of cotton (60–70 mN/m) [25] relative to that of nylon (30–44 mN/m) [26]. Also, the felt material used in this study was porous and was able to absorb larger amounts of water than the fabric material.

The hydrophilic property of the string material is in accordance with the hydrophilic behavior of the material that was observed in this experiment. The felt material was able to be perfectly wetted by the water and could maintain sufficient contact with water without forming droplets on its surfaces. However, for the case involving fabric strings, the liquid flow was not stable and the droplets were splashed on the wall surface of the system; this tendency became more severe with the increase of the liquid flow rate. The mass transfer performance of the SFR was affected by the hydrophilicity of the string material, as shown in Figure 4a. The felt material, which is more hydrophilic, achieved k_La values that were about 1.3 times higher than those of the fabric material.

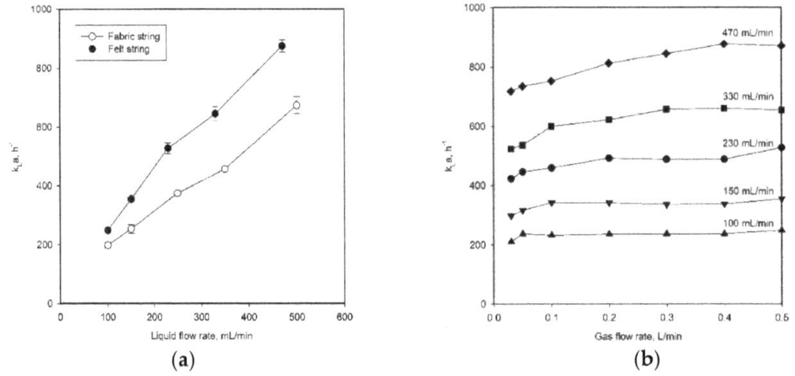

Figure 4. Mass transfer performance of the SFR using felt strings: (**a**) comparison of the volumetric mass transfer coefficients for oxygen with fabric strings (open circles) and felt strings (closed circles) and (**b**) effect of the gas flow rate on the mass transfer coefficients.

The effects of the gas flow rates on k_La values were observed in the SFR using the felt string material. The experiments were conducted by gradually decreasing the gas flow rate while the liquid flow rate was kept constant. As shown in Figure 4b, the mass transfer performance of the SFR was not affected by the gas flow rate above 0.4 L/min for each tested liquid flow rate. However, at a gas flow rate below 0.4 L/min, the magnitude of the effects of the gas flow rate on the mass transfer performance depends on both of the liquid flow rate and the gas flow rate. It was also observed that the k_La value was independent of the gas flow rate after it reached a particular gas flow rate, which is defined as the critical gas flow rate. Therefore, it should be noted that it is important to operate the SFR at the critical gas flow rate in order to obtain the optimum value of k_La for low gas throughputs. For each tested liquid flow rate, the critical gas flow rates obtained in this study are presented in Table 1.

Table 1. Critical gas flow rate at each tested liquid flow rate.

Liquid Flow Rate (mL/min)	Critical Gas Flow Rate (L/min)
100	0.05
150	0.10
250	0.20
350	0.30
500	0.40

3. Discussion

In this study, the SFR system aimed to increase the mass transfer rate by efficiently manipulating the contact between the gas and liquid. The experiment results obtained indicate that the liquid flow rate and the hydrophilicity of the string material are parameters that need to be considered for the design and operation of the SFR.

The positive effects of the liquid flow rates on k_La could be well explained by the physical concept behind Higbie's model. According to Higbie's model, there is a continual attachment wherein the liquid element is in contact with the gas for particular exposure time θ and absorbs the gas by molecular diffusion which increases rapidly initially and decreases with time [27]. At a high liquid flow rate, the periodic replacement of the liquid element at the gas–liquid interface is more frequent, and the rate of gas absorption therefore becomes higher.

The flow pattern of liquid on the surface of the strings is also very important with regards to realizing an improved mass transfer performance, as can be observed clearly from the comparison of k_La values for the cases with and without string, as well as for the cases with the strings having different hydrophilicities. When the direction of flow of the liquid was varied by the strings present in the SFR, the contact of gas and liquid at the interface was stabilized, thus allowing improved mass transfer performance. These effects were further enhanced when the hydrophilic porous string was used. This may be responsible for the increase in the interfacial area of the contact, as was also reported by Onda et al. [28] and Han et al. [29]. According to their results, a positive correlation between the hydrophilicity of the packing material and the contact area was observed in their mass transfer equation for a packed bed reactor.

The biological conversion of methane was carried out primarily in the aerobic condition, and both methane and oxygen should therefore be sufficiently supplied to the system. Most studies on the gas transfer systems that have been reported in the literature have been evaluated by studying the performance with oxygen, and very little empirical work has been done to investigate methane or methane–oxygen transfer.

According to the results presented in this study, the k_La values for oxygen and methane in the system obtained from 100% air and 100% methane were similar to those of the mixed gas containing 30% methane (Figure 5a), implying that the k_La values obtained for oxygen and methane in the system are independent of the composition of the gas. However, the mass transfer coefficient values obtained for methane were lower than those for oxygen under the same conditions. A strategy to complement the low efficiency of methane transfer should be established. It is generally known that the k_La value of a gas is closely related to its diffusivity [30]. Based on the diffusivity of methane and oxygen [31], the k_La value for methane was estimated and compared with the measured values (Figure 5b). The mass transfer coefficients for methane at the liquid flow rates of 250 mL/min and 350 mL/min were about 70% of the corresponding values for oxygen, but values that were lower than 70% of the k_La value for oxygen were obtained at lower liquid flow rates, while higher values were obtained at higher flow rates. Further study is required to better understand the differences in the mass transfer coefficients for different gases.

To assess the mass transfer capability of the SFR, the highest k_La value obtained in this study for oxygen was compared with those obtained from other studies for various types of bioreactors. To enable an equivalent comparison among reactors, operating conditions, including the gas flow, liquid flow, and liquid volume, are presented. As can be seen in Table 2, the k_La value of the SFR was higher than that of the stirred tank reactor reported by Orgill et al. [11] and Karimi et al. [12], whereas the stirred tank reactor was operated at a high agitation speed of 900–1000 rpm. This confirmed that the SFR was able to efficiently transfer the gas to the liquid with lower energy consumption than the stirred tank reactor, owing to its capability. This could minimize mass transfer resistance by the formation of liquid film without relying on a high agitation speed as in the stirred tank reactor. In addition, the k_La value of the SFR was higher than that of the bubble column reactor reported by Lau et al. [15], Khrisna and Ellenberger [17], Bekassy et al. [20], and Budzynski et al. [32], despite the lower superficial gas velocity of the SFR. The packed bed–trickle flow reactor should be compared with the SFR at a comparable value of gas flow rate in order to have an equivalent comparison because the mass transfer of the packed bed–trickle flow reactor is significantly affected by the gas flow rate [11,27]. For a gas flow rate of around 0.1 L/min, the SFR outperformed the packed bed–trickle flow reactor, with k_La values of 752.0 h^{-1} (not shown in Table 2) and 421 h^{-1}, respectively. The SFR exhibited lower performance compared with the hollow fiber membrane reactor. However, there is an advantage in terms of simplicity and operability.

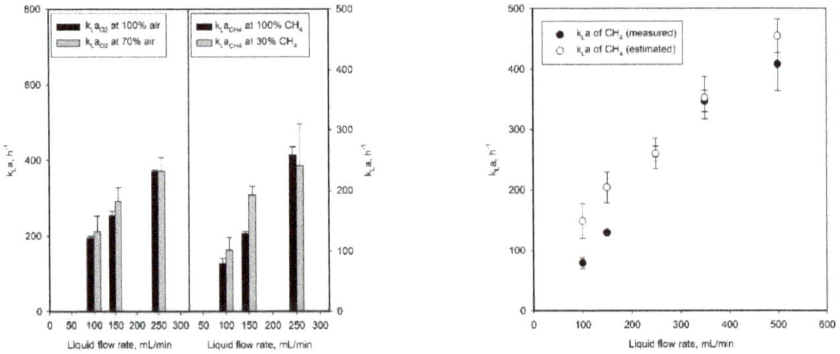

(a) (b)

Figure 5. Comparison of the mass transfer characteristics in the SFR using different feeding gases: (**a**) the volumetric mass transfer coefficients for oxygen (left side) and methane (right side) and (**b**) the measured (closed circle) and estimated (open circle) volumetric mass transfer coefficients obtained for methane.

Table 2. Mass transfer coefficients for oxygen for the SFR and other reactor types.

Reactor Type	Operating Condition					Highest k_La (h^{-1}) Obtained for Oxygen	Reference
	Gas Flow Rate (mL/min)	Liquid Flow Rate (L/min)	Superficial Gas Velocity (cm/s)	Reactor Volume (L)	Liquid Volume (L)		
String film (SFR)	500	0.4	0.36	0.486	0.06	874.67 [L]	This study
Stirred tank (900 rpm)	400	-	-	3.5	2.5	114 [R]	[11]
Stirred tank (1000 rpm)	5000	-	-	2.44	1.77	216 [R]	[12]
Bubble column	-	-	1.2	8.64	6.90	180 [R]	[17]
Bubble column	-	-	10.8	22	4.0	360 [R]	[16]
Bubble column	10,000	-	0.93	32	11	126 [R]	[32]
Air-lift	-	-	1.77–7.07	1.75	0.85	360 [R]	[20]
Packed bed–trickle flow	131	0.05	-	1.2	0.0081	421 [L]	[11]
Hollow fiber membrane	1000	0.4	-	-	0.018	1062 [L]	[11]

[R] Reactor volume based; [L] Liquid volume based.

To evaluate the SFR, it should be considered that the effective volume of liquids for gas absorption in the entire system is small. The reactor-volume-based k_La is reduced when the system volume is used as the basis of the calculation, while the liquid-volume-based k_La is very high. It is expected that many of the drawbacks resulting from a low effective volume can be addressed easily by increasing the number of strings in the system. Nevertheless, by comparing the performance of the SFR with a stirred tank, bubble column, or air-lift reactor that injects gas directly into the system, the required costs for system installation and energy consumption, as well as the mass transfer performance, should be assessed thoroughly.

4. Materials and Methods

4.1. Schematic of the String Film Reactor (SFR)

A schematic of the SFR is shown in Figure 6. The SFR was made of acrylic plastic with an inner diameter of 5.08 cm and a total height of 44 cm. It was equipped with a water jacket to maintain the temperature.

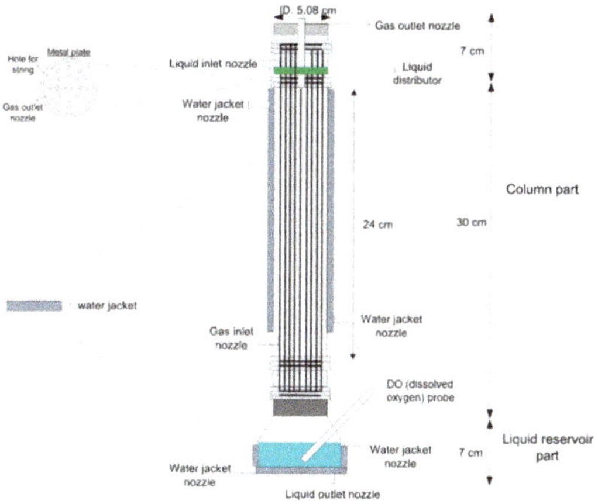

Figure 6. Schematic diagram of the string film reactor.

The SFR had two main parts, i.e., a column part and a liquid reservoir part. The column part, where the contact between liquid and gas took place, had a total volume of 486 mL with a void fraction of 0.92, and consisted of 20 strings with a diameter of 0.3175 cm and an effective length of 24 cm. The strings were held using a metal plate that was placed on the top and lower parts of the column part; the plate consisted of 20 holes with a diameter of 0.4 cm, as shown in Figure 6. The bottom of the SFR acted as a liquid reservoir for dissolved oxygen and dissolved methane measurements. The liquid distributor, which was made of hydrophilic natural pulp material, was placed on the metal plate.

4.2. Experimental Procedure for Gas–Liquid Contact Using the SFR System

The mass transfer performance of the SFR was determined in the absence of microorganism cells. The experimental setup employed for measuring the mass transfer performance of the SFR is shown in Figure 7. The experiments were performed at 30 °C and atmospheric pressure, which is the process condition typically used in the aerobic bioconversion of methane. Double-distilled water from a water holding tank, which was initially purged using nitrogen to obtain a DO value close to 0%, was fed to the SFR using a peristaltic pump (BT301L, Lead Fluid, Baoding, China) at a desired flow rate. The air or air–methane from the gas tank was fed directly to the SFR by adjusting the flow rate using a mass flow controller. The SFR was operated in counter current mode. Various gas (0.03–0.5 L/min) and liquid (100–500 mL/min) flow rates were tested.

In the case of air, a DO probe (Oxyprobe D500, Broadly James, Irvine, CA, USA) was placed in the liquid reservoir and connected to a fermentor control unit (CNS, South Korea) for data acquisition. The DO values were observed and recorded every 30 s until the system reached steady state, which was determined when the DO value did not change within a 3 min period. In the case of methane, its dissolved concentration was measured using a collection tube (BD vacutainer®, Becton Dickinson, Franklin Lakes, NJ, USA) and a gas chromatograph (6890N, Agilent, Santa Clara, CA, USA). When the system reached a steady state, a specific amount of the liquid in the reservoir was drawn by a BD vacutainer, which is a plastic test tube with a rubber stopper creating a vacuum seal inside of the tube. The amount of dissolved methane concentration was calculated by comparing using Gas Chromatography (GC) the gas composition before and after sampling. Runs were performed in triplicate for each test condition.

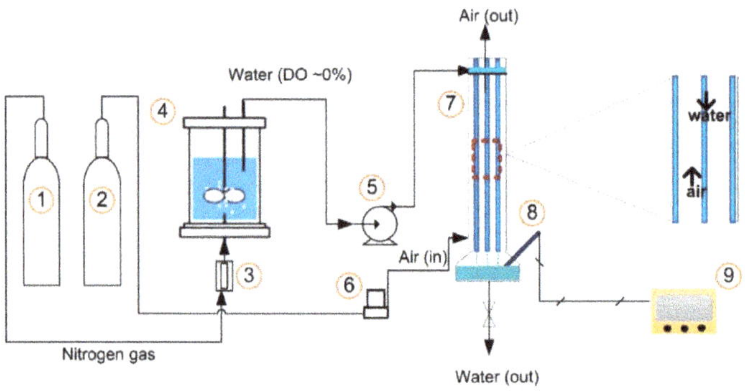

Figure 7. Mass transfer experiment setup: (1) Nitrogen gas tank; (2) Air (or air–methane) gas tank; (3) Nitrogen gas rotameter; (4) Water holding tank; (5) Peristaltic pump; (6) Mass flow controller; (7) SFR; (8) DO electrode; and (9) Data acquisition unit.

4.3. Determination of Volumetric Mass Transfer Performance

The mass transfer performance was determined by calculating the value of the volumetric liquid-side mass transfer coefficient k_La. The formula used to calculate the volumetric liquid-side mass transfer coefficient k_La was obtained by deriving the oxygen mass balance in the liquid, as described below.

$$\frac{dC}{dt} = k_La(C^* - C) \quad (2)$$

Integrating each side of Equation (1), Equation (2) could be obtained as follows:

$$\ln\frac{C^* - C_0}{C^* - C_f} = k_La\left(t_f - t_0\right) \quad (3)$$

where C_0 and C_f are the dissolved gas concentrations at the starting point and end point in the system, respectively. In the SFR, C_f does not change with time when it is in steady state, and t_f denotes the residence time of liquid in the system; k_La can be derived using the liquid volume, the liquid flow rate, and the variation in the dissolved gas concentrations:

$$\therefore k_La = \frac{Q_L}{V}\ln\frac{C^* - C_0}{C^* - C_s} \quad (4)$$

where Q_L is the liquid flow rate and C_s is the dissolved concentration of the gases in the steady state.

5. Conclusions

The SFR was able to perform gas transfer to the liquid with a simple configuration and achieved a large volumetric liquid-side mass transfer coefficient k_La. When comparing the mass transfer performance for the cases with air, methane, and air–methane mixed gas, the mass transfer coefficient did not change significantly, showing the independent transfer behavior of oxygen and methane.

The mass transfer of the SFR was determined using three main parameters:

1. The liquid flow rate, which had a significant effect on the mass transfer performance;
2. The gas flow rate, where the mass transfer performance increased with gas flow rate values below the critical gas flow rate, while it was almost constant above the critical rate. The value of the critical gas flow rate depends on the liquid flow rate;
3. The hydrophilicity of the string material, where a more hydrophilic material could significantly improve the mass transfer performance.

Therefore, for either the operation or further development of the SFR, the above three parameters should be considered. Furthermore, the SFR exhibited good potential as an alternative to the existing types of bioreactor owing to its high mass transfer performance and simplicity.

Author Contributions: Conceptualization, J.-G.N.; Data curation, R.M. and J.-G.N.; Formal analysis, T.W.K. and B.-K.O.; Investigation, R.M., M.-S.K., C.I.L. and T.W.K.; Methodology, M.-S.K. and C.I.L.; Supervision, J.L. and J.-G.N.; Validation, T.W.K., S.J.P., B.-K.O., J.L. and J.-G.N.; Writing—original draft, R.M.; Writing—review & editing, S.J.P. and J.-G.N.

Funding: This research was supported by the C1 Gas Refinery Program through NRF funded by the Ministry of Science and ICT (NRF-2015M3D3A1A01064926) and "Human Resources Program in Energy Technology" of the Korea Institute of Energy Technology Evaluation and Planning (KETEP), granted financial resources from the Ministry of Trade, Industry & Energy, Republic of Korea (No. 20174010201150).

Conflicts of Interest: The authors declare no conflict of interest.

References

1. Fei, Q.; Guarnieri, M.T.; Tao, L.; Laurens, L.M.L.; Dowe, N.; Pienkos, P.T. Bioconversion of natural gas to liquid fuel: Opportunities and challenges. *Biotechnol. Adv.* **2014**, *32*, 596–614. [CrossRef] [PubMed]
2. U.S. Energy Information Administration, Annual Energy Outlook 2012 with Projections to 2035 (EIA Publication 0383, 2012). Available online: https://www.eia.gov/outlooks/archive/aeo12/ (accessed on 23 October 2018).
3. Kidanu, W.G.; Trang, P.T.; Yoon, H.H. Hydrogen and volatile fatty acids production from marine macroalgae by anaerobic fermentation. *Biotechnol. Bioproc. Eng.* **2017**, *22*, 612–619. [CrossRef]
4. Conrado, R.J.; Gonzalez, R. Envisioning the bioconversion of methane to liquid fuels. *Science* **2014**, *343*, 621–623. [CrossRef] [PubMed]
5. Hur, D.H.; Na, J.-G.; Lee, E.Y. Highly efficient bioconversion of methane to methanol using a novel type I Methylomonas sp. DH-1 newly isolated from brewery waste sludge. *J. Chem. Technol. Biotechnol.* **2017**, *92*, 311–318. [CrossRef]
6. Canul-Chan, M.; Chable-Naal, J.; Rojas-Herrera, R.; Zepeda, A. Hydrocarbon degradation capacity and population dynamics of a microbial consortium obtained using a sequencing batch reactor in the presence of molasses. *Biotechnol. Bioproc. Eng.* **2017**, *22*, 170–177. [CrossRef]
7. David, Y.; Baylon, M.G.; Pamidimarri, S.D.V.N.; Baritugo, K.-A.; Chae, C.G.; Kim, Y.J.; Kim, T.W.; Kim, M.S.; Na, J.-G.; Park, S.J. Screening of microorganisms able to degrade low-rank coal in aerobic conditions: Potential coal biosolubilization mediators from coal to biochemical. *Biotechnol. Bioproc. Eng.* **2017**, *22*, 178–185. [CrossRef]
8. Lee, J.W.; Kim, H.U.; Choi, S.; Yi, J.; Lee, S.Y. Microbial production of building block chemicals and polymers. *Curr. Opin. Biotechnol.* **2011**, *22*, 758–767. [CrossRef] [PubMed]
9. Nguyen, A.D.; Hwang, I.Y.; Lee, O.K.; Kim, D.; Kalyuzhnaya, M.G.; Mariyana, R.; Hadiyati, S.; Kim, M.S.; Lee, E.Y. Systematic metabolic engineering of Methylomicrobium alcaliphilum 20Z for 2,3-butanediol production from methane. *Metab. Eng.* **2018**, *47*, 323–333. [CrossRef] [PubMed]
10. Kim, M.S.; Fitriana, H.N.; Kim, T.W.; Kang, S.G.; Jeon, S.G.; Chung, S.H.; Park, G.W.; Na, J.-G. Enhancement of the hydrogen productivity in microbial water gas shift reaction by Thermococcus onnurineus NA1 using a pressurized bioreactor. *Int. J. Hydrogen Energy.* **2017**, *42*, 27593–27599. [CrossRef]
11. Orgill, J.J.; Atiyeh, H.K.; Devarapalli, M.; Phillips, J.R.; Lewis, R.S.; Huhnke, R.L. A comparison of mass transfer coefficients between trickle-bed, hollow fiber membrane and stirred tank reactors. *Bioresour. Technol.* **2013**, *133*, 340–346. [CrossRef] [PubMed]
12. Karimi, A.; Golbabaei, F.; Mehrnia, M.R.; Neghab, M.; Mohammad, K. Oxygen mass transfer in a stirred tank bioreactor using different impeller configurations for environmental purposes. *Iranian J. Environ. Health Sci. Eng.* **2013**, *10*. [CrossRef] [PubMed]
13. Bredwell, M.D.; Worden, R.M. Mass-transfer properties of microbubbles. 1. Experimental Studies. *Biotechnol. Prog.* **1998**, *14*, 31–38. [CrossRef] [PubMed]
14. Munasinghe, P.C.; Khanal, S.K. Syngas fermentation to biofuel: evaluation of carbon monoxide mass transfer and analytical modeling using a composite hollow fiber (chf) membrane bioreactor. *Bioresour. Technol.* **2012**, *122*, 130–136. [CrossRef] [PubMed]

15. Lee, P.-H.; Ni, S.-Q.; Chang, S.-Y.; Sung, S.; Kim, S.-H. Enhancement of carbon monoxide mass transfer using an innovative external hollow fiber membrane (hfm) diffuser for syngas fermentation: experimental studies and model development. *Chem. Eng. J.* **2012**, *184*, 268–277. [CrossRef]
16. Lau, R.; Lee, P.H.V.; Chen, T. Mass transfer studies in shallow bubble column reactors. *Chem. Eng. Process.* **2012**, *62*, 18–25. [CrossRef]
17. Krishna, R.; Ellenberger, J. Improving gas–liquid contacting in bubble columns by vibration excitement. *Int. J. Multiph. Flow* **2002**, *28*, 1223–1234. [CrossRef]
18. Kim, Z.H.; Park, Y.S.; Ryu, Y.J.; Lee, C.G. Enhancing biomass and fatty acid productivity of Tetraselmis sp. in bubble column photobioreactors by modifying light quality using light filters. *Biotechnol. Bioproc. Eng.* **2017**, *22*, 397–404. [CrossRef]
19. Jajuee, B.; Margaritis, A.; Karamanev, D.; Bergougnou, M.A. Mass transfer characteristics of a novel three-phase airlift contactor with a semipermeable membrane. *Chem. Eng. J.* **2006**, *125*, 119–126. [CrossRef]
20. Bekassy-Molnar, E.; Majeed, J.G.; Vatai, G. Overall volumetric oxygen transfer coefficient and optimal geometry of airlift tube reactor. *Chem. Eng. J.* **1997**, *68*, 29–33. [CrossRef]
21. Bredwell, M.D.; Srivastava, P.; Worden, R.M. Reactor design issues for synthesis-gas fermentations. *Biotechnol. Prog.* **1999**, *15*, 834–844. [CrossRef] [PubMed]
22. Satterfield, C.N.; Pelossof, A.A.; Sherwood, T.K. Mass transfer limitations in a trickle-bed reactor. *AIChE J.* **1969**, *15*, 226–234. [CrossRef]
23. Qureshi, N.; Annous, B.A.; Ezeji, T.C.; Karcher, P.; Maddox, I.S. Biofilm reactors for industrial bioconversion processes: Employing potential of enhanced reaction rates. *Microb. Cell. Fact.* **2005**, *4*. [CrossRef] [PubMed]
24. Arkles, B. Hydrophobicity, Hydrophilicity and Silanes. Paint & Coat. Ind. 2006. Available online: https://www.pcimag.com/ext/resources/PCI/Home/Files/PDFs/Virtual_Supplier_Brochures/Gelest_Additives.pdf (accessed on 23 October 2018).
25. Vasiljevic, J.; Tomsic, B.; Jerman, I.; Simoncic, B. Organofunctional Trialkoxysilane Sol-Gel Precursors for Chemical Modification of Textile Fibres. *Tekstilec* **2017**, *60*, 198–213. [CrossRef]
26. Critical Surface Tension and Contact Angle with Water for Various Polymers. Available online: https://www.accudynetest.com/polytable_03.html?sortby_contact_angle (accessed on 13 October 2018).
27. Charpentier, J.C. Mass transfer in gas-liquid absorbers and reactors. *Adv. Chem. Eng.* **1981**, *11*, 1–123. [CrossRef]
28. Onda, K.; Takeuchi, H.; Okumoto, Y. Gas absorption with chemical reaction in packed column. *J. Chem. Eng. Jpn.* **1968**, *1*, 62–66. [CrossRef]
29. Han, M.W.; Lee, W.K.; Choi, D.K. Effect of shape and wettability of packing materials on the efficiency of packed column. *Korean J. Chem. Eng.* **1985**, *2*, 25–31. [CrossRef]
30. Garcia-Ochoa, F.; Gomez, E. Bioreactor scale-up and oxygen transfer rate in microbial processes: An overview. *Biotechnol. Adv.* **2009**, *27*, 153–176. [CrossRef] [PubMed]
31. Mass Diffusivity. Available online: https://en.wikipedia.org/wiki/Mass_diffusivity (accessed on 24 September 2018).
32. Budzynski, P.; Gwiazda, A.; Dziubinski, M. Intensification of mass transfer in a pulsed bubble column. *Chem. Eng. Process.* **2017**, *112*, 18–30. [CrossRef]

MDPI
St. Alban-Anlage 66
4052 Basel
Switzerland
Tel. +41 61 683 77 34
Fax +41 61 302 89 18
www.mdpi.com

Catalysts Editorial Office
E-mail: catalysts@mdpi.com
www.mdpi.com/journal/catalysts

www.ingramcontent.com/pod-product-compliance
Lightning Source LLC
LaVergne TN
LVHW071939080526
838202LV00064B/6640

* 9 7 8 3 0 3 9 2 1 5 7 4 4 *